Jochen Mau, Sergey I. Mukhin, Guanyu Wang, Shuhua Xu, Željko Čupić (Eds.)
BIOKYBERNETIKA II

De Gruyter Series in Mathematics and Life Sciences

Volume 14

BIOKYBERNETIKA II

Thermodynamics, Functional Physiology, Process Control, and Society Impact

Edited by
Jochen Mau, Sergey I. Mukhin, Guanyu Wang, Shuhua Xu,
Željko Čupić

DE GRUYTER

d|u|p düsseldorf university press

Mathematics Subject Classification 2020
Primary: 90-10, 91-10, 92-10, 93-10; Secondary: 91B76, 92B(05,25), 92C(05,30,37,48,55,60), 92D(25,40), 92E10, 93A13

Editors
Jochen Mau is professor (em) of Biomathematics and Statistics in Medicine, Heinrich Heine University Düsseldorf, Germany. He holds a PhD in Mathematics from University of Mainz, Germany.

Sergey Mukhin is professor of Lomonosov Moscow State University. He holds a PhD of Physical and Mathematical Sciences.

Guanyu Wang is associate professor at Chinese University of Hong Kong, China. He holds PhDs of both Engineering (Zhejiang University) and Medical Sciences (University of Cologne).

Shuhua Xu is Distinguished Professor of Human Population Genetics, Fudan University, China. He holds a PhD from Fudan University.

Željko Čupić, PhD, is Principal Research Fellow and President of the Scientific Council at the University of Belgrade - Institute of Chemistry, Technology and Metallurgy - National Institute of the Republic of Serbia.

ISSN 2195-5530
ISSN 2195-5549 (DIGITAL)
ISBN 978-3-11-914480-3
ISBN 978-3-11-221662-0 (PDF)
ISBN 978-3-11-221681-1 (E-PUB)
DOI https://doi.org/10.1515/9783112216620

Library of Congress Control Number: 2024943388

Bibliographic information published by the Deutsche Nationalbibliothek
The Deutsche Nationalbibliothek lists this publication in the Deutsche Nationalbibliografie; detailed bibliographic data are available on the Internet at https://dnb.dnb.de.

De Gruyter, d|u|p düsseldorf university press and Walter de Gruyter GmbH are part of De Gruyter Brill.
www.degruyterbrill.com
dup.degruyter.com

Questions about General Product Safety Regulation: productsafety@degruyterbrill.com

Printing: CPI books GmbH, Leck

FSC www.fsc.org
MIX
Papier | Fördert gute Waldnutzung
FSC® C083411

To the Young Science who join their talents and ambitions across the sciences for collaboration in health-intent research and evaluation to benefit their people.

Contents

Preface —— IX

Overview —— XI

Part I: **Theories**

Eva Déli
The role of thermodynamics in the evolution of biological complexity —— 5

Jochen Mau
Structure-function-energy, I: Structure —— 29

Idan Sorin, Michael Zaks, and Alexander Nepomnyashchy
Seasonal forcing in rock–paper–scissors population dynamics —— 43

Part II: **Person's life-sphere**

Natalia Levashova, Evgenii Generalov, Alexey Goltsov, and Alla Sidorova
Three-dimensional spatiotemporal dynamics of tumor growth —— 75

Aleksey Lutsenko and Alla Sidorova
Quantitative assessment of chirality of supersecondary protein structures —— 93

Denis Shpigun, Alla Sidorova, and Pavel Levashov
The relationship between chirality of secondary helical structures and the stability of lysozyme and myoglobin —— 111

Pavel Levashov
Mathematics of supercooperativity in biochemistry —— 123

Alexander Khrulenko and Fedor Popko
Modeling circulatory substance dynamics with integrated hepatic clearance —— 139

Alexander N. Churilov and John G. Milton
Modeling pulsatile and diurnal rhythmicity of free and total cortisol —— 153

Anastasiia Mozokhina and Vitaly Volpert
Investigation of virus infection progression in tissue with reaction–diffusion models —— 177

Masoud Saade, Samiran Ghosh, Malay Banerjee, and Vitaly Volpert
Recent developments in epidemiological delay models — 205

Jochen Mau and Wenhua Tian
Shanghai 2022 outbreak of SARS-CoV-2 omicron revisited — 243

Alla Sidorova and Natalia Levashova
A model of territorial development of cities as autowave self-organization of conjugated natural and anthropogenic subsystems — 291

Emmanuelle Augeraud-Véron, Arnaud Ducrot, and Ousmane Seydi
Game-theoretic optimization in spatial control of invasive plants — 313

Pierre Auger, Ali Moussaoui, Tri Nguyen-Huu, and Doanh Nguyen-Ngoc
Enhancing maximum sustainable yield in multisite single-species and predator–prey fisheries — 345

Wenhua Tian
The effects of social support to frailty of elderly people in a longevity township of China — 367

Part III: **Technologies**

Arthur I. Sabirov and Tatyana V. Zakharova
Solving the magnetoencephalography inverse problem for continuously distributed source models — 387

Alexei Shishkin and Stepan Denisov
Classification of intracytoplasmic sperm injection videos — 403

Pavel V. Panov and Tatyana V. Zakharova
Bioequivalence testing based on the Kullback–Leibler divergence — 421

Part IV: **Clinical medicine**

Evgenii N. Khailov, Nikolay L. Grigorenko, and Ellina V. Grigorieva
Optimization of adaptive treatment protocols for melanoma patients — 437

Index — 465

Preface

This second book, BIOKYBERNETIKA II, carries research from the first, BIOKYBERNETIKA (2025), forward through the complex behaviors that emerge with human life across scales, from molecules and cells to individuals and populations. The new subtitle, *Thermodynamics, Functional Physiology, Process Control and Society Impact*, however, reflects the intent to prioritize mathematical theory. Structure-function relations and nonlinear dynamics of competing populations are curated into a common narrative through a wide range of concepts and processes–such as tumor growth, protein architecture, biochemical regulation, and wide-sense physiology-environment interactions.

An Open Access "Overview" briefly highlights background, methods, and outcomes of each of the twenty chapters. The content architecture is again seen in its structure of parts, subparts, and finer levels, each explained in a "Rationale" on separating pages between chapters: the conceptual grid is thus announced–though only partially filled by current contributions–to stimulate research for future volumes.

Special thanks are extended to the DeGruyterBrill teams in Boston and in Berlin, to mention the encouraging enthusiasm of Ranis Ibragimov in launching the books and of Anne Sokoll of d|u|p Düsseldorf University Press, an imprint of DeGruyter, in joining for wider support, and the kind guidance in production, first by Nadja Schedensack and then by Ute Skambraks who gently helped over broken deadlines. Last, but for sure not least, special gratitude is expressed for the outstanding dedication of Vilma Vaičeliūnienė and her VTeX team to perform best in the copy-editing and technical production.

Final thanks express our appreciation for the authors, who had contributed their latest research.

Düsseldorf/Krefeld — Jochen Mau
Moscow — Sergey Mukhin
Hong Kong/Shenzhen — Guanyu Wang
Shanghai — Shuhua Xu
Belgrade — Željko Čupić

https://doi.org/10.1515/9783112216620-203

Preface

Overview

Introduction to structure and content

By *Jochen Mau (ed)*

Abstract: This second book, BIOKYBERNETIKA II, addresses Thermodynamics, Functional Physiology, Process Control, and Society Impact. For each chapter, title, pages, authors, and a brief structured content are given. Interspersed are clues to the design of parts and subparts that applies to all BIOKYBERNETIKA books, notwithstanding that several research slots may not have been covered in the present or past books: it is an invitation to contribute in the future.

Preface (p. IX)

Part I: Theories

Part I-A: Overarching theory (p. 3)

1. The role of thermodynamics in the evolution of biological complexity

Chapter by Eva Déli [pp. 5–24]

Method The grand arc of evolution is driven by the laws of thermodynamics.
Application Evolution as a perpetual cycle oscillates between two opposing strategies: mass extinction is followed by resource-rich "boom" times and life is governed by the Maximum Power Principle, favoring rapid, powerful innovation. In the crowded, competitive "bust" times of a mature ecosystem, the rules flip, and the Minimum Entropy Production Principle favors ruthless efficiency and specialization. Then mass extinctions are not failures but essential resets of this cycle.
Outcome The key to life's long-term increase in complexity is the genome—a stable, orthogonal archive that acts as a separate but connected library of successful innovations, ensuring that the lessons of each cycle are not lost but are built upon in the next cycle.

https://doi.org/10.1515/9783112216620-204

Part I-B: Systems theories (p. 27)

2. Structure-function-energy, I: Structure

Chapter by Jochen Mau [pp. 29–42]

Method The conceptual approach is borrowed from social policy: it considers energy budgets and accounting for energy expenditures, separates existential minimum from additional costs for functional needs and the overhead costs of indispensable community integration.
Application When applied to functional dynamics that evolve in hierarchical structures, the approach lends itself to a theoretical concept of scaled energy-budgets, cellular energy supplies and structural and functional energy demands.
Outcome An energy-by-(reverse internal) entropy plot is introduced; it shows structural aggregation costs as a rising energy "overhead" while the functional budgets appear as horizontal lines.

3. Seasonal forcing in rock–paper–scissors population dynamics

Chapter by Idan Sorin and Michael Zaks and Alexander Nepomnyashchy [pp. 43–69]

Method In many ecological models, quantities such as the growth rate and death rate are taken as constant parameters. In real life, this is not the case, and these quantities change through seasonality. We analyze such ecological problems by reducing the system to a discrete, season-to-season map at the transitions between seasons.
Application This is an example of a broader and less-explored class of problems, where there is a change in the system as a function of time and independent of the system's state (exogenous forcing).
Outcome We find rich behavior such as phase locking, period-doubling cascades, chaos, and a new irregular phenomenon that we call "popcorn."

Part II: Person's life-sphere

Part II-A: Person's biosphere

Part II-A.1: Body's cellular system (p. 73)

4. Three-dimensional spatiotemporal dynamics of tumor growth: A numerical study based on a nonlinear wave approach

Chapter by Natalia Levashova, Evgenii Generalov, Alexey Goltsov, and Alla Sidorova [pp. 75–91]

Method For a tumor spheroid cultivated in a Petri dish, our three-dimensional, spatiotemporal simulation of growth incorporates cell density, oxygen availability, and pH levels as key biophysical factors.
Application The dynamic interplay between these factors is captured, explicitly including the formation of necrotic cores, quiescent cell zones, and the influence of oxygen gradients on cellular behavior.
Outcome The spatial distribution of oxygen and pH significantly impacts drug penetration, drug efficacy, and the development of resistance, which suggests the approach as a promising tool for exploring therapeutic strategies, optimizing treatment protocols, and improving our understanding of tumor progression mechanisms.

5. Quantitative assessment of chirality of supersecondary protein structures

Chapter by Aleksey Lutsenko and Alla Sidorova [pp. 93–109]

Method One of the main geometric properties of proteins is the specific chirality of structures of various hierarchical levels. We present an original method for evaluating the chirality of supersecondary protein structures. Such structures have explicit functionality, which makes quantification of their chirality essential for linking geometry to biological activity.
Application With this method, it is possible to determine the sign and value of chirality of superhelices of various types, as well as small β-sheets. Some previously known regularities are confirmed and some estimated chirality values for various types of structures are obtained.
Outcome Our method for evaluating chirality in supersecondary protein structures offers a key quantitative tool for linking geometry to biological functionality. It is expected to substantially deepen understanding of protein structure both theoretically and practically, including its applications in pharmacology and artificial protein creation.

6. The relationship between chirality of secondary helical structures and the stability of lysozyme and myoglobin

Chapter by Denis Shpigun, Alla Sidorova, and Pavel Levashov [pp. 111–122]

Method Applying the results from the preceding chapter, chiral maps of regular secondary structures (α- and 3_{10}-helices) were generated to quantify the degree and sign of structural handedness.
Application To test the relationship between secondary structure chirality and thermal stability, two proteins with contrasting biological functions—lysozymes and myoglobins—were examined.
Outcome Chirality in secondary helical structures of comparable length differed due to spatial amino acid arrangement. The observed correlation between chirality and the molten-globule transition supports targeted approaches in protein engineering for enhancing enzyme stability and designing antimicrobial agents.

7. Mathematics of supercooperativity in biochemistry: Combining Michaelis–Menten, Hill, and Langmuir Equations

Chapter by Pavel Levashov [pp. 123–134]

Method Different kinetic models—Michaelis–Menten, Langmuir, Hill, and their complex variants—are unified through an equilibrium constant governing conformational transitions.
Application For oligomeric proteins that can switch between conformations with distinct kinetics, this yields a single compact “umbrella” equation.
Outcome The model captures the supercooperative behavior that underlies deviations from classical kinetics and explains how such proteins fine-tune and regulate essential biological processes.

Part II-A.2: Vital task-solving functions (p. 137)

8. Modeling circulatory substance dynamics with integrated hepatic clearance

Chapter by Alexander Khrulenko and Fedor Popko [pp. 139–152]

Method Hemodynamic equations and the convection-diffusion equation within an arbitrary closed system of elastic tubes, is implemented in the Cardiovascular Simulating System (CVSS) software package, and comparison of *in silico* results with clinical data.

Application Dynamics of indocyanine green (ICG) distribution in the human cardiovascular system, incorporating the liver's detoxifying function. Special attention was given to analyzing how liver impairment severity and portal hypertension levels influence ICG's pharmacokinetic parameters.
Outcome The results provide insights into the relationship between hepatic dysfunction and drug clearance, offering potential applications in clinical diagnostics and therapeutic monitoring, with potential applications in noninvasive liver function assessment.

9. Modeling pulsatile and diurnal rhythmicity of free and total cortisol

Chapter by Alexander N. Churilov and John G. Milton [pp. 153–175]

Method An extended system of differential equations with impulses, time delays, and feedback control on the hypothalamic-pituitary-adrenal (HPA) axis is formulated and presented with simulations in various parameter constellations.
Application Cortisol initiates glucose mobilization through glycolysis and cellular respiration to supply the energy required for wake-time physiological functions. To avoid receptor desensitization, cortisol is secreted in rhythmic pulses roughly every hour during the day. This pulsatile pattern maintains low levels of free cortisol in circulation while keeping abundant reserves bound to corticosteroid-binding globulin (CBG), ready for rapid release when physiological demand spikes.
Outcome Computational simulations of the integrated mathematical equations yield realistic hormonal profiles, offering novel insights into key features of HPA axis' role in dynamic hormone regulation.

10. Investigation of virus infection progression in tissue with reaction–diffusion models

Chapter by Anastasiia Mozokhina and Vitaly Volpert [pp. 177–198]

Method We use reaction–diffusion models that incorporate both innate and adaptive immune responses. The general model describes uninfected and infected cell populations, virus particles, interferon production, cytotoxic T lymphocytes, and antibodies, accounting for local and systemic dynamics.
Application We analyze a simplified baseline version to derive key immunological indicators, such as the virus replication number, total viral load, and infection wave speed. Analytical solutions are provided in the homogeneous case with delay, while traveling wave solutions are characterized in the spatially heterogeneous case.

Outcome The results highlight how immune mechanisms and spatial effects shape infection dynamics, offering insights into the conditions for infection clearance versus persistence and informing strategies for controlling viral spread in tissues or cell cultures.

Part II-B: Person's ecosphere exposures

Part II-B.1: Living nature cohabitation (p. 203)

11. Recent developments in epidemiological delay models

Chapter by Masoud Saade, Samiran Ghosh, Malay Banerjee, and Vitaly Volpert [pp. 205–242]

Method To move beyond unrealistic assumptions in classic epidemic models, capturing the crucial time delays that shape real outbreaks, we use Delay Differential Equations (DDEs) to explicitly model time lags, like the incubation periods.
Application In infectious diseases like COVID-19 and virus influenza, the periods between exposure, infected and infectious, and delays in reporting fundamentally change modeled epidemic dynamics, and subsequently intervention planning.
Outcome We have achieved DDE models that provide more realistic forecasts, especially of outbreak peaks, leading to better-informed and more effective public health strategies.

12. Shanghai 2022 outbreak of SARS-CoV-2 omicron revisited: Analysis of daily briefings of China National Health Commission

Chapter by Jochen Mau and Wen-Hua Tian [pp. 243–286]

Method Graphical logistic regression based on profiles of daily logistic differences, Bateman function fits to daily log-numbers of new cases for a phenomenological analysis that would not require mechanistic model parameters.
Application Comparison of daily incidences of asymptomatic infected, primary clinically symptomatic, and conversions from the former to the latter category, in Shanghai Megacity and in mainland China without Shanghai, for contrast.
Outcome Considering an epidemic outbreak as an volcano outburst does not need assumptions about unknown parameters in mechanistic epidemic models. The approach applies only with effective control as in China's 2020 and 2022 major outbreaks, that expressed in the characteristic sugarloaf shape.

Part II-B.2: Civilization (p. 289)

13. A model of territorial development of cities as autowave self-organization of conjugated natural and anthropogenic subsystems

Chapter by Alla Sidorova and Natalia Levashova [pp. 291–312]

Method This work extends the FitzHugh–Nagumo reaction–diffusion model to spatially inhomogeneous media using the method of contrast structures.
Application Spatial heterogeneity was considered with two key factors: spatial barriers (urban infrastructure and natural geobiocenoses) and the spatial distribution of population density. Predictive simulations of territorial development were performed for New Moscow and Shanghai for the period 2017–2024.
Outcome The developed model provides a framework for designing optimal conditions for the formation of *ecopolises*—urban environments that ensure comfortable human habitation while preserving the necessary level of biodiversity within urban ecosystems.

14. Game-theoretic optimization in spatial control of invasive plants: Comparison of decentralized and centralized strategies

Chapter by Emmanuelle Augeraud-Véron, Arnaud Ducrot, and Ousmane Seydi [pp. 313–343]

Method We show that the stationary equilibrium of a spatially extended invasive-species control game can be characterized as the fixed point of a globally monotone reaction operator. By using Browder's theorem, together with the spectral properties of the underlying dispersal kernel, we establish existence, uniqueness, and stability of the stationary control equilibrium under weak assumptions.
Application Spatial control of an invasive species over a finite number of discrete habitat patches. Our fixed-point algorithm converges in practice even for systems comprising hundreds of spatial patches, thereby overcoming the "curse of dimensionality" inherent in shooting-based solutions of partial-differential equation frameworks.
Outcome This offers two economically significant advantages, the full scope of inter-site externalities is captured in a single algebraic formulation, and closed-form expressions for marginal externality payments lend themselves to the design of spatially differentiated fiscal transfers that precisely internalize social costs and benefits.

15. Enhancing maximum sustainable yield in multisite single-species and predator–prey fisheries: Mathematical analysis of spatial structure, migration and species interactions

Chapter by Pierre Auger, Ali Moussaoui, Tri Nguyen-Huu, and Doanh Nguyen-Ngoc [pp. 345–362]

Method We utilize mathematical models—including single-species logistic equations, predator–prey formulations (Lotka–Volterra, Holling Type II)—alongside analytical derivations, numerical simulations, and bioeconomic models to examine fishery dynamics in networks of connected patches.
Application It is investigated how spatial connectivity, ecological heterogeneity, and economic factors influence the maximum sustainable yield (MSY) in multi-trophic fisheries systems.
Outcome It is demonstrated that connectivity does not increase MSY in single-species fisheries, but can significantly enhance predator MSY in heterogeneous prey–predator communities. Integrating spatial and economic structure yields improved long-term fishery performance and sustainability.

Part II-C: Person's sociosphere exposures

Part II-C.1: Economic factors (p. 365)

16. The effects of social support to frailty of elderly people in a longevity township of China: A community-based study

Chapter by Wenhua Tian [pp. 367–382]

Method For social support, Xiao Shuiyuan Social Support Rating Scale was applied, and for frailty, the Liu et al. (2016) frailty index. Three subscores, objective (tangible or instrumental support received), subjective (perceived emotional and interpersonal support), and utilization of support, were analyzed. Supervised one-time interviews with a standard questionnaire were conducted with 1,721 residents aged 70–84.
Application By the frailty index, one-fifth of the participants were found to be frail, and all three categories of support were statistically significantly associated with frailty. However, subgroup analyses according to gender showed different results. The three categories of support then seem to be linked differently to frailty.
Outcome Age-related differences in sensitivity imply nuanced needs for support in elderly populations: material help should be prioritized for the very old, while emotional and interpersonal support should be emphasized for younger seniors.

Part III: Technologies

Part III-B: Information sciences (p. 385)

17. Solving the magnetoencephalography inverse problem for continuously distributed source models

Chapter by Arthur I. Sabirov and Tatyana V. Zakharova [pp. 387–401]

Method We introduce a hybrid method that first separates independent activity sources (ICA) and then applies localized LORETA mapping around their centers. .
Application Mapping brain activity from MEG/EEG signals is extremely challenging: signals overlap, noise is strong, and existing algorithms often give blurred or unstable results.
Outcome The hybrid ICA–LORETA framework improves accuracy, reduces smoothing, and cuts computational cost, offering a practical tool for functional brain mapping.

18. Classification of intracytoplasmic sperm injection videos

Chapter by Alexei Shishkin and Stepan Denisov [pp. 403–415]

Method We used deep learning-based semantic segmentation, diffusion models for synthetic data generation, and gradient boosting classifiers to analyze Intracytoplasmic Sperm Injection (ICSI) videos.
Application ICSI is crucial for assisted reproduction but remains highly operator-dependent, lacking automated assessment tools.
Outcome Achieved state-of-the-art segmentation (IoU = 0.978) and outcome prediction (F1 = 0.833), provide a scalable system for clinical quality control and training.

Part III-C: Data-analytic sciences (p. 419)

19. Bioequivalence testing based on the Kullback–Leibler divergence

Chapter by Pavel V. Panov and Tatyana V. Zakharova [pp. 421–431]

Method We introduce a modified approach based on KL-divergence and kernel density estimation to calculate statistics for the bioequivalence testing.
Application Modern bioequivalence testing framework is sensitive to missing data and anomalies, as the currently used statistics are not robust.

Outcome Proposed approach yields more precise results on samples with missing data and anomalies.

Part IV: Clinical medicine (p. 435)

20. Optimization of adaptive treatment protocols for melanoma patients

Chapter by Evgenii N. Khailov and Nikolay L. Grigorenko and Ellina V. Grigorieva [pp. 437–463]

Method In a two-species Lotka–Volterra competition model over a finite time interval, optimal control minimization is carried out using the Pontryagin maximum principle and the nonoscillation theory for ordinary differential equations (ODEs). The BOCOP open-source optimal control solver was employed for numerical calculations.
Application The model describes the competition between drug-sensitive and drug-resistant cell populations during adaptive therapy. The time-dependent control function determines the optimal timing and duration of treatment periods, aiming to suspend therapy when resistant cells gain a competitive advantage and to resume it as soon as sensitive cells can still suppress a recurring expansion of resistant ones. This strategy minimizes the overall cancer cell burden throughout the entire melanoma treatment period and at its final moment.
Outcome The types of optimal control functions that solve the stated minimization problem for specific clinical data were determined analytically and numerically. The analytical results are confirmed by numerical calculations performed in the BOCOP environment, using parameter values and initial conditions from real clinical practice. We could demonstrate that our optimal adaptive protocol minimizes a patient's total cancer burden much better than either of the other protocols used in clinical practice.

Part I: Theories

Part I-A: **Overarching theory**

Rationale: 1

Part I: **Theories**

Overarching theory and systems theories

Part I-A: **Overarching theory**

Axioms and principles on which mathematically formulated systems theories can be built.

In this volume:
- The role of thermodynamics in the evolution of biological complexity

In former volume BIOKYBERNETIKA:
- Universal axioms in classical Chinese philosophy
- Exploration of category theory for structural comparisons
- Axiomatic bipolar dynamics and their control

https://doi.org/10.1515/9783112216620-001

Eva Déli

The role of thermodynamics in the evolution of biological complexity

Abstract: Charles Darwin's theory of evolution elegantly explains adaptation, yet it struggles to account for two fundamental observations: the punctuated, cyclical nature of the fossil record and life's persistent drive toward complexity in a universe governed by entropy. This work resolves these puzzles by presenting a thermodynamic foundation for evolution. We argue that evolution is not a linear march but a cyclical process, analogous to the four phases of the reversed Carnot engine, and similar to the Panarchy adaptive cycle. This cycle oscillates between two primary states. Following a mass extinction, a low-entropy "boom" phase of resource abundance and low competition favors rapid innovation and organisms that maximize energy throughput (Maximum Power Principle). As the ecosystem matures, it enters a high-entropy "bust" phase, a crowded and competitive state where selection shifts to favor organisms with maximum metabolic efficiency and specialization (Minimum Entropy Production Principle). Within this framework, mass extinctions are not failures but necessary thermodynamic resets that clear the board of hyperspecialized, brittle systems. While ecosystems may collapse, the genetic material remains a resilient archive, preserving the memory of past innovations and enabling complexity to increase over time. Current extinctions indicate that Earth's ecosystem is in the latter phase of its evolution, representing weakened resilience. This model also offers a powerful new toolkit for applied science, from guiding microbial evolution in biotechnology to designing more sustainable agricultural systems.

Keywords: Darwinian evolution, extinction, evolutionary modeling, entropy in evolution; thermodynamics of evolution, reversed Carnot cycle

MSC 2020: 80A20, 92D15, 37N25, 70K50

1 Introduction

For over a century and a half, Charles Darwin's grand narrative of evolution has served as the foundation of biology. It is based on gradual changes in a population across generations through a process where organisms with advantageous traits, such as the giraffe with a slightly longer neck and the cheetah with just a bit more speed, are more likely to survive and reproduce, passing those traits onto their offspring [8]. Yet, cracks in this el-

Eva Déli, Department of Psychology, Faculty of Health Sciences, University of Debrecen, 2-4. Sóstói Street, 4400 Nyíregyháza, Hungary, e-mail: eva.kdeli@gmail.com

https://doi.org/10.1515/9783112216620-002

egant framework reveal two profound mysteries that Darwin's original model struggled to explain.

The first mystery arises from the fossil record, our most direct account of life's past. This record does not suggest a story of slow, steady progress; instead, it reveals long periods of stability. These periods of stability—referred to by paleontologists as "stasis," where species remain relatively unchanged for millions of years, are punctuated by brief, explosive episodes of rapid transformation. This phenomenon, known as "punctuated equilibrium," proposes that evolution operates in fits and starts, often triggered by catastrophic environmental changes [11].

The second mystery goes even deeper, seemingly placing life itself in opposition to one of the universe's most fundamental laws: the Second Law of Thermodynamics. This law dictates that everything must eventually decay into disorder and dust, a process known as entropy. Yet, life does the opposite. It builds breathtaking complexity—from a single cell to a rainforest—in a universe that seems to want to tear everything down. How can this be?

This chapter posits that these are not two separate puzzles but two sides of the same coin, and the answer lies in the very law that life appears to defy: thermodynamics. As the pioneering physicist Erwin Schrödinger foresaw, life is not a magical entity evading the rules of physics; it emerges as the ultimate consequence of them [49]. Life takes advantage of the Second Law, as the increasing entropy of the environment propels biological systems to harness the continuous flow of energy from the sun.

When viewed through this thermodynamic lens, evolution is revealed not as a linear progression but as a recurring cycle between two opposing strategies. In the aftermath of a mass extinction, the destruction of the ecosystem produces massive environmental change, with opportunities for innovation. In this high-energy, low-competition environment, organisms that burn through energy rapidly gain the upper hand. This strategy, known as the Maximum Power Principle (MPP) [32], represents a "boom" phase—a period characterized by innovation and explosive diversification. The MPP is a principle in thermodynamics that states that systems, including biological ones, will evolve to maximize power, not efficiency, when the rate of energy flow is the key factor for survival.

However, this frenzy cannot last indefinitely. As life begins to reclaim the planet, ecosystems become crowded, and resources become scarce, the rules of survival shift. Now, the advantage lies with those who can extract the maximum value from every joule of energy, in line with what Ilya Prigogine described as the Minimum Entropy Production Principle (PEPP) [41]. This phase is known as the "bust" phase of stasis, favoring stability, optimization, and conservation, which is satisfied by species ideally suited to the environment. The cycle repeats.

This rhythm of boom and bust—of frantic growth followed by steady conservation—is the very pulse of nature, a pattern elegantly captured by the Panarchy adaptive cycle [16]. The Panarchy adaptive cycle is a model that describes the adaptive cycles of complex systems, including ecosystems, and how they go through phases of growth, conservation, collapse, and reorganization. In this chapter, we will map these ecological phases onto the framework of a reversed Carnot cycle, providing a thermodynamic model for how energy flow and entropy dictate the pace and direction of life. This model offers a physics foundation for Darwin's revolutionary theory.

2 Methodology

2.1 The currency of life: Energy and information

The thermodynamic foundation of life, the "order" of a rainforest or the "chaos" of a genome, is a game of energy and information. Our approach is to view the entire biosphere as a vast, self-contained system, powered by solar energy. In this model, we ignore the effects of the cosmic environment. Ecosystems consume low-entropy energy and export high-entropy waste (heat, etc.) to maintain their organization. Still, miraculously, they also build genetic complexity, which ensures the temporal continuity and evolutionary potential of life.

The entropy of the genome changed during evolution. Bacterial genomes generally exhibit higher entropy compared to eukaryotic genomes. This means bacterial genomes have a more random or disordered sequence arrangement. Thus, ancient organisms had high informational entropy. But as evolution precedes, natural selection, gene duplication, and epigenetics give rise to successful adaptations—from the machinery of photosynthesis to the architecture of a protein—lock in functional information, effectively lowering the genome's entropy [63]. How can we measure this functional information?

The Assembly Theory explains the dynamic, energy-intensive biological complexity through its causal histories and formation pathways [50]. However, genetic information primarily reflects evolutionary trajectories rather than linear advancement [55]. For example, whales transitioned to aquatic life while retaining mammalian complexity [65]. Moreover, species can accumulate or lose genes during evolution, yet maintain their evolutionary history and classification.

We can also consider the entropy of the ecosystem. We can describe interaction as a kind of "social temperature."[1] We define low social temperature as a state of "coldness". This represents a resource-rich, low-competition environment where small populations can safely explore new environments, in which genetic innovations can happen.

---

1 In large systems, individual organisms can be likened to particles, with their unique histories resembling thermal fluctuations that are proportional to their arousal levels. When resources are abundant, the system maintains a low social temperature. However, food scarcity increases social tension due to competition for resources, leading to high social temperature of erratic and risky behaviors.

Free energy within a biological system can drive genomic reorganization and the emergence of new functions. Free energy, defined thermodynamically by von Helmholtz [62], describes the maximum work a system can perform at a constant volume. In this context, free energy can drive genomic reorganization and the evolution of new functions. In far-from-equilibrium systems, organisms capable of capturing and utilizing more energy than their competitors gain a selective advantage [17, 45]. Conversely, a "hot" system is one of high arousal and frantic competition, which lacks free energy for building complexity.

According to Landauer's principle, the very act of storing information—of making a permanent decision, whether in a computer chip or a strand of DNA—has a physical cost. To lock in a choice and "erase" the other possibilities, a minimum amount of energy must be spent and dissipated as heat [28]. In biology, this means creating stable, low-entropy genetic information.

But where does the energy for this cosmic accounting come from? This brings us to life's energy budget, a fundamental equation that governs the entire evolutionary cycle:[2]

$$dE = TdS + \Delta\Omega \tag{1.1}$$

Here, the energy injected into the system, dE, can be spent in only two ways. It can be used to increase the "heat and noise"—the social temperature and chaotic interactions, in the term TdS. Or, it can be invested in building lasting structure and organizational complexity, in the term $\Delta\Omega$. Life is in a constant negotiation between these two expenditures.

After mass extinctions, the landscape is rich with the decaying organic matter of the old world, providing a massive influx of available energy (dE is high). During rapid growth, microbes, cancer cells, bacterial communities, and prey populations afford maximum power (MPP) and high entropy production, and $\Delta\Omega$ is high; see Figure 1.1, Figure 1.2, and Table 1.1. As the cycle moves toward equilibrium, $\Delta\Omega$ decreases. In close to equilibrium systems, both $\Delta\Omega$ and dS tend to a minimum. Thus, energy input increases the social temperature (arousal and competition).

With increasing genetic diversity and entropy, the system spreads to a greater number of microstates, allowing stabilization. In high-entropy, high-population-density equilibrium states, most mutations remain latent but contribute to evolutionary stability and diversity.

As evidence accumulates, uncertainty decreases, causing a decline in learning rate [22]. While the entropy and the social temperature of the ecosystem increase, evolutionary innovations reduce the informational entropy of the genome. As the ecosystem's entropy is at maximum, the genome's entropy reaches its minimum, permitted by the cycle.

2 Notation: dE for energy change, T for characteristic temperature of the system, dS for entropy change, $\Delta\Omega$ for change in organizational complexity.

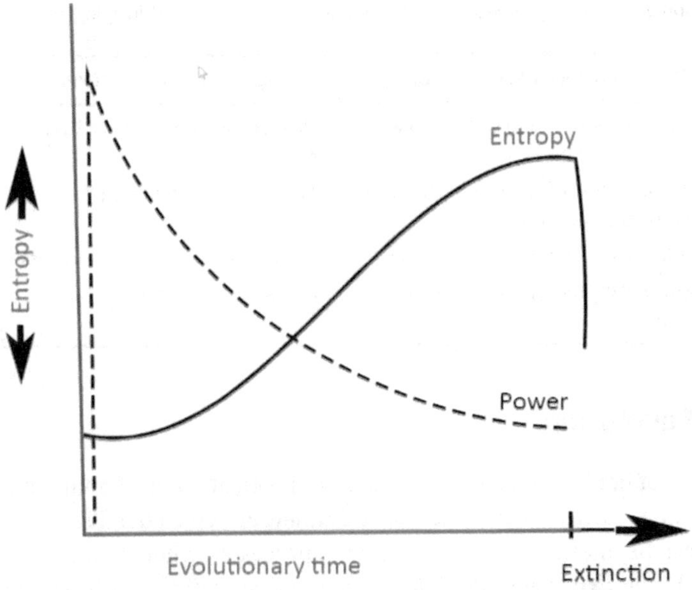

Figure 1.1: Entropy and evolutionary changes over time. The Solid Curve (Entropy): Increases gradually, with a sharp dip at an extinction event, representing an ecosystem reset. Dotted Curve (Innovation): Peaks shortly after the extinction event, representing bursts of evolutionary innovation as ecosystems reorganize in low-entropy, high-opportunity states.

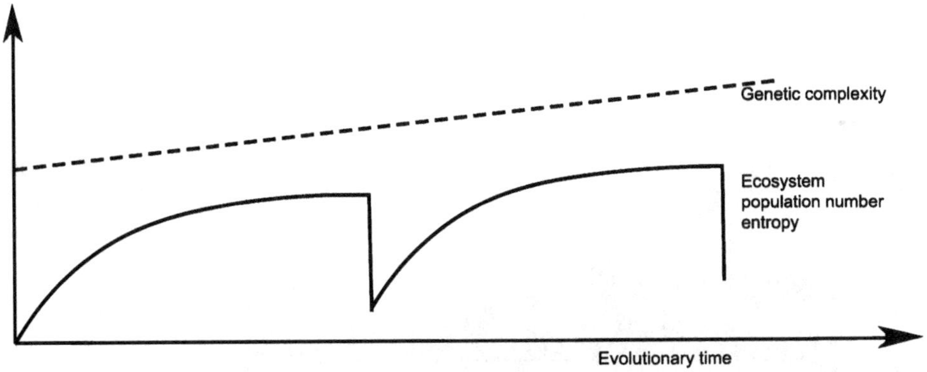

Figure 1.2: Ecosystem parameters during evolution. While the ecosystem population number and entropy repeat cyclically, genetic complexity demonstrates a steady upward trend.

Gene diversity promotes more efficient energy dissipation and minimizes entropy production [9, 48], governed by the Minimum Entropy Production Principle. For instance, evolutionary pressure selects proteins with efficient folding kinetics [12] and microbes with efficient energy or ATP utilization [23]. Adaptation is an intricate interplay between social temperature, diversity, resource availability, and other environmental factors.

Table 1.1: Analysis of evolution based on Prigogine's Theorem and the Maximum Power Principle.

Feature	Maximum Power Principle (MPP)	Prigogine's Theorem (PEPP)
The phase of evolution	Far-from-equilibrium systems: the beginning of the cycle	Near-equilibrium systems: later phase of the cycle
Regulatory Mechanism	Maximization of energy transformation (max. entropy prod.)	Minimization of entropy production
Outcome	Adaptation for optimal energy use	Stabilization of dissipative processes
Energy/Entropy Relationship	Focuses on energy flux and work output	Focuses on entropy dissipation

2.2 The engine of evolution

The evolutionary history of life is a jagged, violent diary of boom and bust. Populations explode, stabilize, and then crash, only to begin the cycle anew as depicted in Figure 1.3. These fluctuations reflect the interplay between ecological processes, where populations adapt to changing conditions, experience bottlenecks, or undergo rapid shifts in genetic makeup. These fluctuations reveal a cyclic pattern of the Panarchy cycle, a framework describing how complex systems—from forests to economies—move through four distinct phases: rapid growth, stable conservation, sudden collapse, and finally, chaotic reorganization [16].

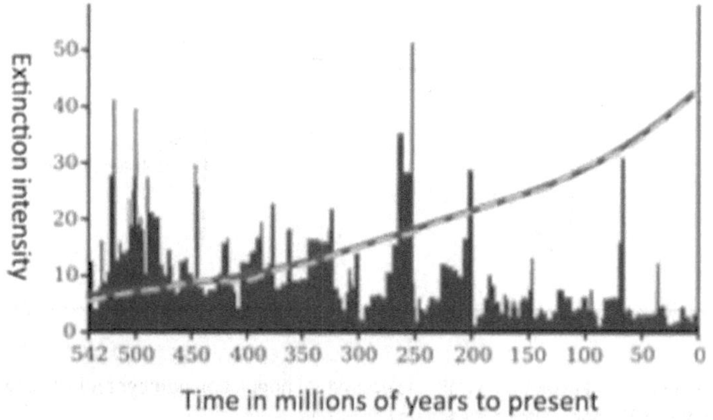

Figure 1.3: The field representation of complexity throughout evolution (Marine extinction intensity during the past five hundred million years, after Raup and Sepkoski, 1982). Genetic complexity demonstrates an upward trend (dotted line).

We can represent this ecological cycle as a refrigerator, operated as a reverse thermodynamic cycle: A refrigerator consumes energy to pump heat out of its insulated interior, creating a cool, ordered space at the cost of radiating disordered heat into the

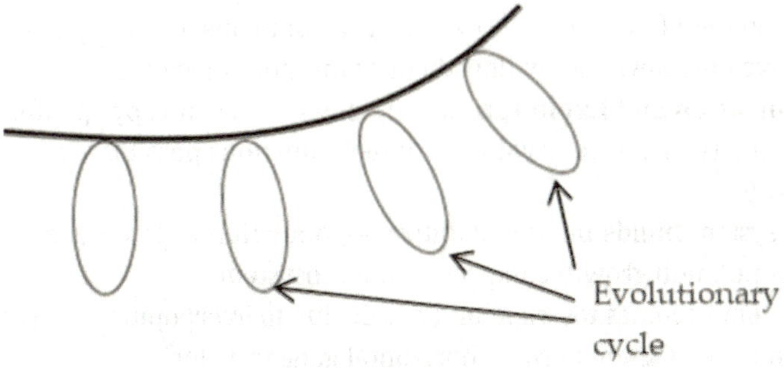

Figure 1.4: The evolutionary cycle gives rise to genetic complexity as a field fluid, chaotic, and reversible processes make up the directional evolutionary cycle (indicated by an oval). In contrast, the genetic material remains a stable field, which can only increase in complexity (solid line).

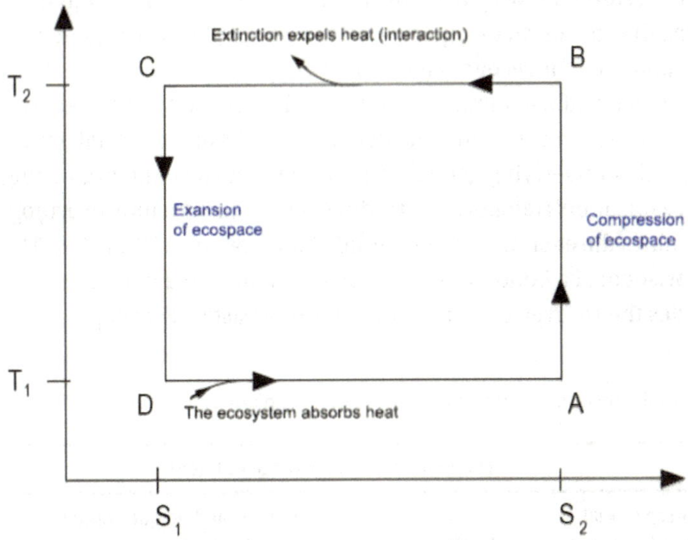

Figure 1.5: The evolutionary process as a reversed Carnot cycle: The evolutionary cycle can be divided into four discrete phases, moving between low and high social temperature (→), and low and high entropy. Phase 1 C→D: Small populations form low entropy, which allows phenotypic plasticity and the evolution of new functions. Phase 2 D→A: Population expansion gives rise to food chains. Phase 3 A→B: Overpopulation stresses ecosystems, forming high entropy and allowing horizontal gene transfer. Phase 4 B→C: Environmental shocks precipitate extinction, mainly affecting species with inflexible genomes. Each time the cycle repeats it does so with higher genetic and morphological complexity; cf. Figure 1.4.

external environment. Life relies on the reversed Carnot cycle to build organization; see Figure 1.5. It absorbs low-entropy energy from the sun to create a cool, highly ordered pocket of complexity—the biosphere—while paying its thermodynamic debt to the universe [49].

We can map the four phases of the Panarchy cycle directly onto this thermodynamic engine as in Figure 1.5, revealing how energy flow dictates the story of evolution.

Growth $C \rightarrow D$ **(Reorganization and Exploitation):** In small, low-entropy populations, phenotypic plasticity makes the evolution of new functions possible. Energy intake fuels complexity.

Expansion $D \rightarrow A$**:** The system builds up and stabilizes—population expansion gives rise to food chains. Population growth compresses the ecosystem.

Collapse $A \rightarrow B$**:** The system becomes too rigid and brittle due to overpopulation. The increasingly interconnected system permits horizontal gene transfer.

Reorganization $B \rightarrow C$**:** Mass extinction clears away inflexible organisms, creating opportunities for renewal.

After mass extinction, the ecosystem reconstitutes at a higher complexity. The secret lies in the system's memory, stored in two places: the lingering resources of the environment and, most critically, the vast library of the collective gene pool [55]. The genome's temporal stability is the potential for recovery and renewal.

We can think of this genetic information as an invisible field permeating the ecosystem, much like gravity permeates space. While the lives and deaths of individual organisms are the visible story, this underlying genetic field provides a constant, organizing blueprint. The genetic organization's temporal and hidden nature turns it into an orthogonal field,[3] which determines the evolutionary potential of the organism (Table 1.2). The genes are the inherited wisdom of a billion years of trial and error, a structural memory, and a potential that guides the recovery, even after the most devastating collapse.

Table 1.2: Comparison of the evolutionary cycle and genome representing an orthogonal field.

The evolutionary cycle	The genome as an orthogonal field
The cycle has well-defined entropy social temperature, and resource density phases.	The genome represents stable and unique energy (complexity), which can survive extinction.
The cycle's phases are chaotic, noisy, and often reversible.	Genetic complexity either increases or remains constant, decreasing only in service of greater organizational complexity.
The cycle eventually ends, forming a discrete loop.	The complexity of heritable material can increase without theoretical constraint.

3 Orthogonality refers to the independence of two quantities from each other. For example, species in the ecosystem show a spatial organization. In contrast, the genome signifies the evolutionary history and future potential of the individual and species. This independence from environmental fluctuation allows the genome to be considered an inert medium that survives extinctions.

3 The four thermodynamic phases in evolution

3.1 Phase 1: Reorganization and exploration

The fossil record of "growth phase" $C \rightarrow D$ in Figure 1.5 is a history with missing sections and chapters. Mass extinctions—moments of cataclysmic "release" of complexity as heat—represent significant gaps between geological eras (Figure 1.1). As primary productivity plunges, nutrients released from decomposed organic matter enable secondary growth [19, 21]. This resource abundance fosters high exploratory capacity and cooperation in the forming ecological vacuum [20, 26, 57]; see Table 1.3.

Table 1.3: The characterization of a typical evolutionary cycle.

	Phase One (A-B)	**Phase Two (B-C)**	**Phase Three (A-B)**	**Phase Four (B-C)**
Social Temperature	Decreasing↓ due to ecospace expansion	Low	Increasing↑ parallel to the population number	High
Entropy and Population Number	Low	Increasing↑	High	Decreasing↓
Evolutionary changes	Complexity, order, and adiabatic expansion	A growing population	Specialization, adiabatic compression	Extinction

As the relentless pressure of competition evaporates, life can take advantage of this resource-rich, low-pressure environment. It is a time for cooperation and wild experimentation [26, 57]. We saw a microcosm of this during the early COVID-19 pandemic, where a low-competition environment for the virus allowed for the rapid assembly and selection of beneficial new mutations [60].

From a thermodynamic perspective, the system is now "cold." This biological coldness is made possible by the abundance of free energy[4] $F = E - TS$—available for work. The ecosystem, then flush with potential, conspires to build innovative organizations [13]. But how does life innovate so quickly? It does so by harnessing its genetic potential. Phenotypic plasticity enables organisms to adapt to novel environments through the reuse of existing genes or the emergence of new traits that occupy previously unexplored morphospace [30, 68]. Gene duplication offers redundancies, where the "unemployed" copy can evolve a new function [24, 60].

Exaptation and the recruitment of mobile genetic elements can enhance functional complexity by shifting traits to new roles or generating entirely new functions [67]. Exap-

4 F is thermodynamic free energy, E is total energy in the system, T is absolute temperature, S is entropy, a measure of disorder.

tation is a change in function of a trait that originally evolved for a different purpose, allowing it to serve new adaptive roles without any structural changes. Perhaps the most stunning example is the evolution of antifreeze proteins in Antarctic fish. These vital proteins, which prevent ice crystals from forming in their blood, evolved from a humble digestive enzyme—a beautiful case of a gene being recruited for a new and life-saving role [5].

Initial mutations can amplify subsequent changes [40, 3]. This flurry of genetic creativity, fueled by abundant resources, allows life to rapidly fill the empty stage. It is what powered the Ediacaran and Cambrian explosions, when the blueprints for nearly all modern animal life were seemingly sketched out in a geological blink of an eye [25]. Computer simulations support the need for adaptive plasticity during rapid evolutionary change [31, 43, 37].

Other support comes from long-term single-cell experiments. In these experiments, yeast could not move beyond aggregate multicellularity [36], indicating that significant environmental changes or genetic interventions are necessary to facilitate the emergence of complexity [29]. In yeast experiments, selection for sinking speed discouraged the more difficult, long-term path toward cell differentiation and integration.

This underlines that major innovation (like integrated multicellularity) often requires a series of complex mutations that may not offer an immediate benefit. The lack of competition, abundant food source, and flexible environment enables the exploration of these complex, multistep pathways.

3.2 Phase 2: Consolidation and expansion

Once the chaotic work of reorganization is done, the survivors enter a new "expansionary phase," $D \to A$ in Figure 1.5: Thermodynamically, the ecosystem focuses on rapid growth by maximizing its energy throughput. This rush to convert environmental energy into biological complexity follows a principle known as Maximum Entropy Production; the system processes energy and creates disorder in its surroundings as fast as possible [35].

Ecologically, this unfolds as a classic succession. Pioneer species that first colonized the barren landscape steadily grow into or replaced by more sophisticated, competitive taxa [6]. Intricate food webs emerge, creating complex highways for the flow of energy and nutrients. Structures and processes stabilize, becoming more predictable and less guided by uncertain external forces.

As populations boom, however, the "social temperature" begins to rise. Competition intensifies. The advantage shifts to species with high reproductive rates and an increasingly efficient ability to utilize resources. We see this principle in the fossil record. During the late Triassic, for instance, as the climate became more humid to support lush vegetation, dinosaurs leveraged this resource boom to evolve larger body sizes, dominating the landscape and outcompeting their rivals [42].

From an information theory perspective, this entire process is a form of collective learning. Species are constantly updating their "understanding" of the environment, and we can quantify this process. The Kullback–Leibler (KL) divergence acts as a mathematical tool to measure the distance between a species' current distribution and its ideal, stable state within the ecosystem. As the system "learns" and species settle into their optimal niches, this divergence shrinks until the system reaches a temporary, high-entropy equilibrium.

With $p(x) \geq 0$ is probability distribution of species x at time 0, $q(x) > 0$ probability distribution of species x at high entropy equilibrium, the Kullback–Leibler (KL) divergence is

$$D_{\mathrm{KL}}(p(x)||q(x)) = \sum_{x \in X} p(x) \log \frac{p(x)}{q(x)} \geq 0 \tag{1.2}$$

in which $\log(p(x)/q(x))$ quantifies change. When $p = q$, the species probability distribution stabilizes, and entropy is maximized.

While a species' growth is a function of its own fitness, what truly matters in a crowded world is its fitness relative to the average. This is described by the replicator equation [56]:

$$\frac{\mathrm{d}P_i}{\mathrm{d}t} = (F_i - F)p_i \tag{1.3}$$

where the growth of a population (p_i) is determined by the gap between its fitness F_i and the mean fitness F of the entire ecosystem. It is not enough to be good; a species must be better than average. This creates a relentless evolutionary race, pushing populations to become even more efficient.

The phase is responsible for the creation of the planet's most magnificent, high-productivity ecosystems, e. g., coral reefs and rainforests. They are systems running at maximum capacity, masterpieces of energy processing. Greater speciation maximizes entropy production. The system is increasingly packed with hyperefficient specialists. The intense competition has set the stage for the cycle's next phase, where the slightest instability can bring the entire, intricate structure tumbling down.

3.3 Phase 3: Conservation and rigidity

After the expansion, the ecosystem is a mature, densely populated world characterized by high entropy overpopulation, $A \rightarrow B$ in Figure 1.5. Although it appears stable, it is a system under immense pressure and rigidity.

The ecosystem's energy budget, once dedicated to explosive growth, now shifts entirely to maintenance. The energy once used to build new structures ($\Delta\Omega$) is now consumed by the kinetic energy of a high social temperature and the cost of maintaining the existing order (TdS). Increasing social temperature prompts a transition to an asymmetrical, dominant state [1], such as apex predators.

While microstate-level fluctuations generate genetic dispersal and diversity, thermodynamic effects maintain the stability of higher-level structures [27, 44, 46]. In high population densities, energetic constraints dictate increasingly efficient energy use and redistribution. Relentless optimization drives species into even more specialized roles—commensalism, mutualism, and parasitism—as a necessity for survival [61].

This is a complexity paradox. To enhance the stability and efficiency of the whole system, the individual components must often become simpler. Parasites and endosymbionts, for instance, famously shed genes they no longer need, outsourcing functions to their hosts [38]. This reduction in freedom at the micro-level allows for a more intricate and predictable organization at the macro-level. Still, it comes at the cost of adaptability [52]. The system becomes a masterpiece of efficiency, but loses the ability to innovate.

And yet, beneath this rigid surface, the genetic undercurrents never stop flowing. Horizontal Gene Transfer (HGT) is the ability of genes to leap between species. HGT can confer adaptive advantages, such as antifreeze protein genes to Herring [14, 59] or 8 % viral genome content to the human genome [2]. This, along with "jumping genes" (transposable elements—highly dynamic components within the same genome) and hidden cryptic variations (hidden genetic differences), creates a reservoir of genetic novelty. It is why we need a new flu vaccine every year—the virus is constantly swapping parts to evade our defenses [54].

This is the phase where Darwin's "survival of the fittest" applies in its most fierce form [53]. When two species rely on the exact same resource, one will inevitably drive the other to extinction, a principle known as competitive exclusion [18]. The advantage belongs to the most ruthlessly efficient, the population that can sustain itself on the absolute minimum. The system has reached its maximum entropy; it is primed for collapse.

3.4 Phase 4: Collapse and mass extinction

The triumph of the previous phase—a world of hyperefficient specialists—contains the seeds of its own destruction entering phase $B \to C$ in Figure 1.5. In a stable environment with finite resources, the "ideal phenotype" can outcompete all generalists. The ecosystem becomes a finely tuned machine, but in doing so, it trades adaptability for perfection. It becomes brittle [47].

This fragility becomes a significant vulnerability. When the environment inevitably shifts, these dominant specialists are often the first to fall. Their very specialization becomes a liability. Extinction is often a targeted process that selectively removes the most specialized and the least adaptable species, creating a cascade that can unravel an entire ecosystem [64]. Species with low genetic diversity (saber-toothed cats and wolves) face heightened risks for extinction [39].

The most famous example is the Cretaceous–Paleogene extinction. The dinosaurs, masters of their domain for over 150 million years, were largely eliminated. Meanwhile, the survivors were often small, generalist, and living on the periphery. These generalist—mammals, birds, and other flexible genotypes—were superbly adaptable, capable of weathering the storm and inheriting a new world.

The collapse itself is often preceded by a period of mounting stress. Long-term environmental pressures, like the immense volcanic activity of the Siberian Traps that preceded the Permian extinction, can weaken an ecosystem's resilience over millennia, pushing it even closer to a tipping point [51]. Once this adaptive capacity is exceeded, the system enters a state of collapse [7]. The fact that we are witnessing staggering extinction rates today, even without a single, planet-altering catastrophe, suggests that we may be living through a similar "slow-motion catastrophe," the erosion of environmental resilience [4].

The final blow is often devastatingly simple. It shuts down the planet's energy supply. The meteorite that ended the age of dinosaurs did not kill them all directly. Instead, it threw so much debris into the atmosphere that it blocked out the sun, halting photosynthesis on a global scale. This single event starved the intricate food webs that had been built over millions of years, triggering a catastrophic energy crisis.

A mass extinction is therefore the *ultimate thermodynamic reset*. It is a violent unraveling that wipes entire lineages from the book of life, fundamentally altering the course of evolution [33]. It clears the board to prepare the stage for a new thermodynamic cycle of *R*eorganization, *C*onsolidation, *C*onservation, and *C*ollapse (*R*-triple*C*).

In the following section, we discuss three examples specifically supporting our thesis.

3.5 Evidence of thermodynamic cycles in evolution

3.5.1 The Cambrian explosion

Five-hundred and forty million years ago, the aftermath of a planetary deep freeze represented a low-entropy planet; see Table 1.4. The rising atmospheric and oceanic oxygen levels supercharged metabolism, providing a massive energy surplus for reorganization and exploration by Phase 1, $C \to D$, in the reversed Carnot cycle (Figure 1.5). This energy surplus sparked biological innovation of complex tissues, segmentation, and mobility, and significant increases in morphological and functional complexity (multicellularity). This sparked the Cambrian explosion, a frenzy of innovation where segmentation, mobility, and other multicellular complexities maximized power and conquered new niches for consolidation and expansion, by Phase 2, $D \to A$ in Figure 1.5. But as these niches filled, competition intensified, and the evolutionary pressure shifted from raw power to efficiency, expressing conservation and rigidity according to Phase 3, $A \to B$

Table 1.4: Applying the thermodynamic framework for evolution.

Event	Initial Condition	MPP Phase	PEPP Phase
Cambrian Explosion	Post-anoxia, oxygen rise (planet-scale reset)	Complex body plans, energy-rich metabolism	Ecosystem-level optimization, stable niches
Galápagos Finches	Post-drought or isolation (local entropy reset)	Adaptive radiation, niche creation	Morphological stabilization, specialization
COVID-19 Pandemic	Low competition, immune naivety (low entropy)	High replication, high mutation rates	Immune escape, lower virulence (efficiency)

in Figure 1.5. This set the stage for the next great extinction and collapse according to Phase 4, $B \rightarrow C$ in Figure 1.5.

3.5.2 Galápagos finches

The famous finches demonstrate this cycle *en miniature*; cf. Table 1.4. Fluctuating resources on the islands act as local extinction events, creating open niches (Phase 1). This triggers adaptive radiation, where beak shapes rapidly diversify to exploit every available food source [15]—a classic example of maximizing power (Phase 2). As competition rises, selection stabilizes these morphologies, optimizing the finches for their specific roles (Phase 3) until the next environmental shift resets the cycle in Phase 4. This low-innovation regime aligned with PEPP.

3.5.3 The COVID-19 pandemic

In early 2020, the novel host environment, minimal competition, and unoccupied ecological niches (human cells without immunity) provided a low-entropy state (Table 1.4). This represented Phase 1 in the reverse Carnot cycle of Figure 1.5. The virus maximized its power with early variants like Delta, which focused on rapid replication and transmission, in line with the MPP (Phase 2). Later variants like Omicron, however, emerged in a world with widespread immunity. They shifted strategy toward efficiency—reduced virulence but higher transmissibility—reflecting a system moving toward a more stable, but still competitive equilibrium (Phase 3), consistent with PEPP [34].

4 Discussion

4.1 The two arrows of time

In the vast theater of the cosmos, two great arrows of time appear to point in opposite directions. The first is the unforgiving arrow of thermodynamics, which dictates that all physical systems must decay into disorder and entropy. The second is the arrow of life, which for four billion years has pointed in the opposite direction, toward ever-increasing evolutionary complexity [66]. These contrasting arrows of time form interconnected and interdependent physical processes [10].

This chapter focused on the latter, representing the increasing complexity during biological evolution. However, unlike the growing entropy of physical systems, evolution is not a steady march but an eternal oscillation between two fundamental states. The cycle begins in the ordered, low-entropy aftermath of collapse, where small, dispersed populations in a resource-rich world are free to experiment and innovate.

This is the realm of the energy maximizers, where novelty creation is favored. But as success leads to growth, the system inevitably matures into a disordered, high-entropy state of teeming, intricate, but dangerously brittle stability. This is the realm of the entropy minimizers, where ruthless efficiency is the only path to survival. Optimal genotypes with efficient metabolic pathways and adapting to specific niches show remarkable stability [58].

This grand cycle, analogous to both the reversed Carnot engine and the Panarchy adaptive cycle, can be summarized in four acts:

1. **Reorganization and Exploration:** Mass extinction acts as a great cooling mechanism, creating a low "social temperature" where innovation can flourish.
2. **Consolidation and Expansion:** In a world of opportunity, competition drives the system toward greater complexity and energy throughput.
3. **Conservation and Rigidity:** The system reaches a state of high-entropy stability with efficient energy utilization: the hyperefficient optimal genotypes. However, the cost of this stability is an enhanced vulnerability to shocks.
4. **Collapse and Mass Extinction:** The brittle system breaks, releasing its stored complexity and clearing the board for the next cycle to begin.

4.2 Second law of intellect

This reveals a profound duality at the heart of life. On one hand, we have the dynamic, fluctuating ecosystem, vulnerable to the immediate thermodynamic laws of energy and entropy. On the other hand, we have the timeless genetic archive—a vast library of information that is shaped by the cycle but also persists through it, carrying the memory of past innovations into the future [66]. This relentless drive for genetic complexity, which builds upon itself across countless cycles, might be considered a complementary principle to cosmic decay: a "Second Law of Intellect."

4.3 Darwin's theory of evolution

In this chapter, we have laid a thermodynamic foundation for Darwin's theory of evolution, recasting it not as a linear march, but as a grand, cyclical engine powered by the fundamental laws of physics. This engine is punctuated by mass extinctions, moments of creative destruction that are not failures of the system, but essential features of its operation.

This framework does not replace Darwin; it places his engine on a more fundamental chassis. The classic "survival of the fittest" is not a universal constant but the dominant mode of selection during the high-entropy, crowded phases of the cycle. The living space's periodic compression (high entropy, energy-poor) and expansion (low entropy, energy-rich) reshape the fitness landscape, offering new evolutionary opportunities. These fluctuating thermodynamic parameters give rise to an energy pump, facilitating the transfer of energy. At equilibrium, the high-entropy ecosystem is vulnerable to external disturbances, forming Darwinian "survival of the fittest" dynamics.

By integrating the principles of thermodynamics into the Darwinian framework, we see that evolution is not a random walk. It is a directed, cyclical exploration of pathways toward greater complexity, and by understanding the fundamental rules of the game, we may finally begin to grasp not just where life has been, but where it is going.

Through the fire of these extinctions, the genetic material continues to persist, as the invisible thread connecting every cycle to the beginning of life. While dominant species and entire ecosystems may vanish, their evolutionary playbook is carried by the survivors, who often come from the fringes of the ecosystem. The ecosystem starts out and innovates from this starting point. This allows the biosphere to rebuild at even greater levels of complexity.

However, the alarming extinction rates of our current era, occurring without a singular global catastrophe, may be a sign that our biosphere's resilience is eroding from within [4].

4.4 Future directions

Evolution is a perpetual oscillation between low-entropy boom states with rapid innovation (Maximum Power Principle) and the busts of the high-entropy worlds, favoring ruthless efficiency and specialization (Minimum Entropy Production Principle). This perpetual rhythm of expansion and contraction serves as a cosmic pump, continuously reshaping the evolutionary landscape. It clears the board of hyperspecialists, creating space for new forms of life to emerge and, over time, amplifies the biosphere's capacity to retain and redistribute energy.

Understanding this cycle is a vital lens through which to view our present moment, where extinctions indicate a frailty of our global ecosystem. We must protect our planet by providing more space for natural, undisturbed environments. We can engineer microbial evolution for biotechnology, design more resilient agricultural systems, and foster a genuine transition toward planetary sustainability.

5 Conclusions

We have shown that thermodynamic forces are the invisible drivers of evolution, shifting opportunities and motivating the grand pageant of life, from its booms to its busts. While this framework requires further study, it offers a powerful perspective on why universal laws can perpetuate organization of life only in cycles.

Bibliography

[1] C. Anteneodo and N. Crokidakis. Symmetry breaking by heating in a continuous opinion model. *Physical Review E*, 95(4):042308, 2017.

[2] A. Burn, F. Roy, M. Freeman, and J. M. Coffin. Widespread expression of the ancient herv-k (hml-2) provirus group in normal human tissues. *PLoS Biology*, 20(10):e3001826, 2022.

[3] M. Burtsev, K. Anokhin, and P. Bateson. Plasticity facilitates rapid evolution. *bioRxiv*, 2022. preprint.

[4] P. Cafaro, P. Hansson, and F. Götmark. Overpopulation is a major cause of biodiversity loss and smaller human populations are necessary to preserve what is left. *Biological Conservation*, 2022.

[5] C. H. Cheng and L. Chen. Evolution of an antifreeze glycoprotein. *Nature*, 401:443–444, 1999.

[6] J. H. Connell and R. O. Slatyer. Mechanisms of succession in natural communities and their role in community stability and organization. *The American Naturalist*, 111:1119–1144, 1977.

[7] V. Dakos, B. Matthews, A. P. Hendry, J. M. Levine, N. Loeuille, J. Norberg, P. Nosil, M. Scheffer, and L. De Meester. Ecosystem tipping points in an evolving world. *Nature Ecology & Evolution*, 3(3):355–362, 2019.

[8] C. Darwin. *On the Origin of Species by Means of Natural Selection, or the Preservation of Favoured Races in the Struggle for Life*. John Murray, London, 1859.

[9] R. De la Cruz, D. Perez, and J. Serrano. Entropy production and energy efficiency in bacterial ecosystems under nutrient flux variability. *Ecological Complexity*, 36:67–75, 2018.

[10] E. Déli and J. Peters. Can dimensional anisotropy satisfy Mach's principle? A topological approach to variable dimensions of space using the Borsuk-Ulam theorem. *Journal of Astrophysics and Astronomy*, 10(7), 2022.

[11] N. Eldredge and S. J. Gould. Punctuated equilibria: An alternative to phyletic gradualism. In T. J. M. Schopf, editor, *Models in Paleobiology*, pages 82–115. Freeman Cooper, San Francisco, CA, 1972.

[12] J. L. England. Statistical physics of self-replication. *Journal of Chemical Physics*, 2013.

[13] D. H. Erwin. Novelty and innovation in the history of life. *Current Biology*, 25(19):R930–R940, 2015.

[14] L. A. Graham and P. L. Davies. Horizontal gene transfer in vertebrates: A fishy tale. *Trends in Genetics*, 37(6):501–503, 2021.

[15] P. R. Grant and B. R. Grant. Unpredictable evolution in a 30-year study of Darwin's finches. *Science*, 296(5568):707–711, 2002.
[16] L. H. Gunderson and C. S. Holling, editors. *Panarchy: Understanding Transformations in Human and Natural Systems*. Island Press, Washington, DC, 2002.
[17] C. A. S. Hall and T. McWhirter. Maximum power in evolution, ecology and economics. *Philosophical Transactions of the Royal Society A*, 381:20220290, 2023.
[18] G. Hardin. The competitive exclusion principle. *Science*, 131(3409):1292–1297, 1960.
[19] P. M. Hull. Life in the aftermath of mass extinctions. *Current Biology*, 25:R941–R952, 2015.
[20] K. Imanishi and P. J. Asquith. *A Japanese View of Nature: The World of Living Things*. Japan Anthropology Workshop Series: (JAWS). RoutledgeCurzon, 2002. https://cir.nii.ac.jp/crid/1970023484833013926.
[21] D. Jablonski and S. M. Edie. Mass extinctions and their rebounds: A macroevolutionary framework. *Paleobiology*, 51(1):83–96, 2025.
[22] M. Jepma, V. Schaaf, I. Visser, and H. M. Huizenga. Uncertainty-driven regulation of learning and exploration in adolescents: A computational account. *PLoS Computational Biology*, 16(9):e1008276, 2020.
[23] D. B. Johnson and A. Schippers. Recent advances in acidophile microbiology: Fundamentals and applications. *Frontiers in Microbiology*, 2017.
[24] M. I. Katsnelson, Y. I. Wolf, and E. V. Koonin. On the feasibility of saltational evolution. *PNAS*, 116(42):21068–21075, 2019.
[25] A. H. Knoll, E. J. Javaux, D. Hewitt, and P. Cohen. Eukaryotic organisms in Proterozoic oceans. *Philosophical Transactions of the Royal Society B: Biological Sciences*, 361(1470):1023–1038, 2006.
[26] P. A. Kropotkin. *Mutual Aid, a Factor of Evolution*. McClure, Phillips Co., New York, 1902.
[27] T. LaBar and C. Adami. Evolution of drift robustness in small populations. *Nature Communications*, 8(1):1012, 2017.
[28] R. Landauer. Irreversibility and heat generation in the computing process. *IBM Journal of Research and Development*, 5(3):183–191, 1961.
[29] D. Leon, S. D'Alton, E. M. Quandt, and J. E. Barrick. Innovation in an E. coli evolution experiment is contingent on maintaining adaptive potential until competition subsides. *PLoS Genetics*, 14, 2018.
[30] M. Levin. Bioelectric signaling: Reprogrammable circuits underlying embryogenesis, regeneration, and cancer. *Cell*, 184(8):1971–1989, 2021.
[31] N. A. Levis and D. W. Pfennig. Phenotypic plasticity, canalization, and the origins of novelty: Evidence and mechanisms from amphibians. In *Amphibian Evolutionary Biology*. Elsevier, 2019.
[32] A. J. Lotka. Contribution to the energetics of evolution. *Proceedings of the National Academy of Sciences of the United States of America*, 8(6):147–151, 1922. https://doi.org/10.1073/pnas.8.6.147.
[33] C. M. Lowery and A. J. Fraass. Morphospace expansion paces taxonomic diversification after end Cretaceous mass extinction. *Nature Ecology & Evolution*, 3(6):900–904, 2019.
[34] P. V. Markov, M. Ghafari, M. Beer, K. A. Lythgoe, P. Simmonds, N. I. Stilianakis, and A. Katzourakis. The evolution of SARS-CoV-2. *Nature Reviews. Microbiology*, 21:361–379, 2023.
[35] L. M. Martyushev and V. D. Seleznev. Maximum entropy production principle in physics, chemistry and biology. *Physics Reports*, 426(1):1–45, 2006.
[36] P. Márquez-Zacarías, P. L. Conlin, K. Tong, J. T. Pentz, and W. C. Ratcliff. Why have aggregative multicellular organisms stayed simple? *Current Genetics*, 67(6):871–876, 2021.
[37] E. T. H. Ng and A. R. Kinjo. Computational modelling of plasticity-led evolution. *Biophysical Reviews*, 14(6):1359–1367, 2022.
[38] M. A. O'Malley, J. G. Wideman, and I. Ruiz-Trillo. Losing complexity: The role of simplification in macroevolution. *Trends in Ecology & Evolution*, 31(8):608–621, 2016.
[39] M. Ørsted, A. A. Hoffmann, E. Sverrisdóttir, K. L. Nielsen, and T. N. Kristensen. Genomic variation predicts adaptive evolutionary responses better than population bottleneck history. *PLoS Genetics*, 15(6):e1008205, 2019.

[40] J. T. Pentz, J. T. Pentz, P. Márquez-Zacarías, G. O. Bozdag, A. J. Burnetti, P. J. Yunker, E. Libby, et al. Ecological advantages and evolutionary limitations of aggregative multicellular development. *Current Biology*, 30(21):4155–4164, 2020.

[41] I. Prigogine. *Étude Thermodynamique des Phénomènes Irréversibles*. Desoer, Liège, 1947. Thèse de doctorat, Université Libre de Bruxelles.

[42] M. Qvarnström, J. Vikberg Wernström, Z. Wawrzyniak, M. Barbacka, G. Pacyna, A. Górecki, J. Ziaja, A. Jarzynka, K. Owocki, T. Sulej, L. Marynowski, G. Pieńkowski, P. E. Ahlberg, and G. Niedźwiedzki. Digestive contents and food webs record the advent of dinosaur supremacy. *Nature*, 636:397–403, 2024. https://doi.org/10.1038/s41586-024-08265-4

[43] A. Rago, K. Kouvaris, T. Uller, and R. Watson. How adaptive plasticity evolves when selected against. *PLoS Computational Biology*, 15(3):e1006260, 2019.

[44] F. Ronco, M. Matschiner, A. Böhne, A. Boila, H. H. Büscher, A. El Taher, A. Indermaur, et al. Drivers and dynamics of a massive adaptive radiation in cichlid fishes. *Nature*, 589(7840):76–81, 2021.

[45] B. Sabater. Entropy perspectives of molecular and evolutionary biology. *International Journal of Molecular Sciences*, 23(8):4098, 2022.

[46] Z. R. Sailer and M. J. Harms. Molecular ensembles make evolution unpredictable. *Proceedings of the National Academy of Sciences*, 114(45):11938–11943, 2017.

[47] M. Salahshour. Evolution as a result of resource flow in ecosystems: Ecological dynamics can drive evolution. *PLoS ONE*, 18(10):e0286922, 2023.

[48] J. R. Schramski, D. K. Gattie, and J. H. Brown. Human domination of the biosphere: Rapid discharge of the Earth-space battery foretells the future of humankind. *Proceedings of the National Academy of Sciences*, 112(31):9511–9517, 2015.

[49] E. Schrödinger. *What Is Life?: The Physical Aspect of the Living Cell*. Cambridge University Press, 1946.

[50] A. Sharma, D. Czégel, M. Lachmann, C. P. Kempes, S. I. Walker, and L. Cronin. Assembly theory explains and quantifies selection and evolution. *Nature*, 622(7982):321–328, 2023.

[51] S. Shen, J. Ramezani, J. Chen, C. Cao, D. H. Erwin, H. Zhang, L. Xiang, S. D. Schoepfer, C. M. Henderson, Q. Zheng, S. A. Bowring, Y. Wang, X. Li, X. Wang, D. Yuan, Y. Zhang, L. Mu, J. Wang, and Y. Wu. A sudden end-Permian mass extinction in South China. *GSA Bulletin*, 131(1–2):205–223, 2019.

[52] A. F. Siegenfeld and Y. Bar-Yam. An introduction to complex systems science and its applications. *Complexity*, 2020:6105872:1–6105872:16, 2019.

[53] K. R. Skene. In pursuit of the framework behind the biosphere: S-curves, self-assembly and the genetic entropy paradox. *Biosystems*, 190:104101, 2020. https://doi.org/10.1016/j.biosystems.2020.104101.

[54] A. Suh. Horizontal transfer of transposons as genomic fossils of host-parasite interactions. In K. De Baets and J. Warren Huntley, editors, *The Evolution and Fossil Record of Parasitism: Coevolution and Paleoparasitological Techniques*, volume 50 of Topics in Geobiology, pages 451–463. Springer, 2021.

[55] S. M. Sundstrom and C. R. Allen. The adaptive cycle: More than a metaphor. *Ecological Complexity*, 39:100767, 2019. https://doi.org/10.1016/j.ecocom.2019.100767.

[56] P. D. Taylor and L. B. Jonker. Evolutionary stable strategies and game dynamics. *Mathematical Biosciences*, 40(1–2):145–156, 1978. https://doi.org/10.1016/0025-5564(78)90077-9.

[57] J. Tkadlec, A. Pavlogiannis, K. Chatterjee, and M. A. Nowak. Limits on amplifiers of natural selection under death-Birth updating. *PLoS Computational Biology*, 16(1):e1007494, 2020. https://doi.org/10.1371/journal.pcbi.1007494.

[58] H. Trenchard and M. Perc. Energy saving mechanisms, collective behavior and the variation range hypothesis in biological systems: A review. *Biosystems*, 147:40–66, 2016. https://doi.org/10.1016/j.biosystems.2016.07.009.

[59] J. Van Etten and D. Bhattacharya. Horizontal gene transfer in eukaryotes: Not if, but how much? *Trends in Genetics*, 36(12):915–925, 2020. https://doi.org/10.1016/j.tig.2020.09.002.

[60] R. M. Varney, D. I. Speiser, J. T. Cannon, M. A. Aguilar, D. J. Eernisse, and T. H. Oakley. A morphological basis for path-dependent evolution of visual systems. *Science*, 383(6686):983–987, 2024. https://doi.org/10.1126/science.adk1781.

[61] G. J. Vermeij. The efficiency paradox: How wasteful competitors forge thrifty ecosystems. *PNAS*, 116(36):17619–17623, 2019. https://doi.org/10.1073/pnas.1908295116.

[62] H. von Helmholtz. *Physical Memoires, Selected and Translated from Foreign Sources*. Taylor and Francis, 1888.

[63] M. M. Vopson and S. Lepadatu. Second law of information dynamics. *AIP Advances*, 12(7), 2022. https://doi.org/10.1063/5.0099243.

[64] Q. M. Webber, M. P. Laforge, M. Bonar, and E. Vander Wal. The adaptive value of density-dependent habitat specialization and social network centrality. *Nature Communications*, 15:4423, 2024.

[65] A. J. Werth. *Cetaceans as Exemplars of Evolution and Evolutionary Ecology: A Glossary*. MDPI, 2020.

[66] M. L. Wong, C. E. Cleland, D. Arend, S. J. Bartlett, H. J. Cleaves, H. Demarest, A. Prabhu, J. I. Lunine, and R. M. Hazen. On the roles of function and selection in evolving systems. *Proceedings of the National Academy of Sciences*, 120, 2023.

[67] S. Xia, N. W. VanKuren, C. Chen, L. Zhang, C. Kemkemer, Y. Shao, H. Jia, U. Lee, A. S. Advani, and A. Gschwend. Genomic analyses of new genes and their phenotypic effects reveal rapid evolution of essential functions in drosophila development. *PLoS Genetics*, 17(7):e1009654, 2021.

[68] L. Zhang, R. Reifová, Z. Halenková, and Z. Gompert. How important are structural variants for speciation? *Genes*, 12(7):2, 2021.

Part I-B: Systems theories

Rationale: 2

Part I-B: **Systems theories**

Mathematics for integrative modeling of biological, ecological, socioeconomic, and sociodynamic phenomena.

In this volume:
- Structure-function-energy, I: Structure
- Seasonal forcing in rock–paper–scissors population dynamics

In former volume BIOKYBERNETIKA:
- Stochastic formalization of agent-oriented systems
- Simplification of high-dimensional multitempo dynamic models
- Ideas of symmetry as a biophysical basis of system biomedicine
- Disorder of multi-scale control

https://doi.org/10.1515/9783112216620-003

Jochen Mau

Structure-function-energy, I: Structure

The thermodynamic concept of structure

Abstract: The triplet "structure, function, energy" serves as paradigm for physiology that understands function as enabled by energy. For "structure", the System Functional Architecture (SFA) has been formulated some years ago. For "function", an axiomatic biodynamics with bipolar control was recently added. Here, the previously considered engineering-inspired concept of human body's tri- partite functional decomposition into vital, physical, and cognitive tasks is completed with reproductive tasks solving functionality, and a central facility for coordination of the four for coherent whole-body's operations is also added. In this setting, hitherto postulated "ubiquitously available" sources of energy are found by functional drill-down to cellular levels and the bottom-up expenditure of energy budgets is theoretically explored within the SFA framework, for conceptual consistency. While budgets for basal cellular maintenance can be added up with increasing re-composition, additional costs for integration, coordination and regulation with successively expanded aggregation, from tissue via organs to functional ensembles (organ systems), emerge. Basal needs for maintenance of functional integrity (minimal energy costs of survival), additional costs for functional activity in daily living, overhead costs for structural integrity, and energy reserve can be visualized in an energy-expenditure-by-entropy plot. Though individual data currently poses a measurement issue, the present "energy" supplement shall provide grounding in cell biology to "vitalize" the theoretical suprastructure for refinement and empirical investigation.

Keywords: Human body system, functional physiology, cell biology, hierarchical structure, energy expenditure, thermodynamic entropy

MSC 2020: 92-02, 92C05, 92C30, 92C37, 80A99

Acknowledgement: Discussions during the preparation of this chapter with Perplexity AI, an advanced large language model, are acknowledged. Rationale, structure, and the content are original by the authors. Phrases were checked for fluency and clarity with Perplexity AI, and its substantial suggestions were cross-checked with Google AI. For technical support, relevant contexts in the literature were pointed out, pertinent references were provided, cross-checked, and corrected as necessary, or disregarded by the author. Still, all errors and misleading statements rest with the author.

Jochen Mau, Medical Faculty and University Hospital Düsseldorf, Heinrich Heine University Düsseldorf, Moorenstr. 5, Düsseldorf, 40225, Germany, e-mail: ismmau@hhu.de

https://doi.org/10.1515/9783112216620-004

1 Introduction

Structure is a manifestation of information [8]: as the organism emerges from randomness—maximum entropy—, it consumes free energy from environment to form structure and thus lower its own entropy, while–throughout life–dissipating metabolic heat and waste that rises environmental entropy[39].

Connecting structure and function then appears as almost a thermodynamic mandate for human physiology. A focus on function is already resonated in the field [7, 30, 9]. Understanding "function" as a capacity to change from one state to another (within set structural boundaries) suggests to call it "function enabled by energy."

A strictly hierarchical functional structure was obtained axiomatically as a top-down decomposition of functional aggregates into constituent functional units, themselves functional aggregates in next-level higher resolution (granularity) [16]. Invoking an analogy with human-machine system, an excavator in action, three canonical functional composites had been postulated for human body as top-level functional aggregate, vital tasks, physical tasks, cognitive tasks, and for a generic aggregate its unit-level dynamics were translated to aggregate-level dynamics[17]. In the first volume of BIOKYBERNETIKA, a hierarchical control concept involving a model of supply-demand match with disturbances in a stable market economy [32] was postulated[19].

The objective of this work is toadd the biological concepts of energy to previously published concepts of hierarchical functional structure, SFA, within axiomatically characterized functional dynamics, the axiomatic biodynamics, evolve

Section 2 gives an overview on physiological energy demand. Section 3 Results (p. 31) elaborates on functional structure of human body system and expands the tri-partite functional decomposition in System Functional Architecture (SFA) [19, 16] to four functional axes and a supervising center for their coordination. Discussion is found in Section 4, and the author's conclusions are drawn in Section 5.

2 Material and methods

The physiological energy demand of the human body has two fundamental categories:

1. Maintenance energy—the energy required to sustain baseline physiological functions and homeostasis under all conditions.
2. Functional-activity energy—the additional energy required to support voluntary and involuntary actions associated with a person's operational goals, including purposeful activity, reflexive response, and adaptive function. Even when no intentional activity is performed, physiological and mental processes continue to expend energy for systemic readiness and maintenance.

Cellular and tissue energy expenditure is commonly expressed in relation to the human body's basal metabolic rate (BMR) (cf. [20]) conceptually the minimal energy required to "keep the whole body alive," i. e., to maintain essential physiological functions such as respiration, circulation, cellular turnover, and basic neurological activity. In practical terms, it represents a physiological state of minimal metabolic activity: the individual must be completely rested, in a post-absorptive state (approximately 12 hours of fasting), and in a thermoneutral environment.

For a 70 kg male lying supine with minimal movement and calm cerebral activity, BMR values of approximately 60–72 $\mathrm{kcal\,h^{-1}}$ can be obtained from the Harris–Benedict equation [11, 31], the Mifflin–St Jeor equation [21], and the Schofield equation [33]; cf. [25]. A standard physiology text book has approximately 65 $\mathrm{kcal\,h^{-1}}$ [10].

All this aligns well with the average BMR for men 1600–1800 kcal/day from [27], who also have 1300–1500 kcal/day for women.

Using Watt for power, equivalent to energy in Joule per second, these values translate to:
- men 1600–1800 kcal/day: 77–87 W
- women 1200–1500 kcal/day: 58–73 W.

Systemic energy expenditure of the human body arises from the integrated energy use of all its functional units. Habitual physical activity and body weight are the main determinants for the diversity in energy requirements of adult populations with different lifestyles [13]; see [4] for tables with everyday activities and with sports, by male and female, athletes and nonathletes.

3 Results

3.1 The thermodynamic concept of structure

Proposition 3.1. *Information expresses in structure.*

Proof. Structure emerges as the organism organizes itself out of randomness, consuming free energy and releasing heat and waste to the environment [29]. According to the mass-energy-information equivalence theory [38], systems evolve toward states of lower informational entropy by compressing uncertainty and increasing internal order [22]. □

Remark 3.1. Entropy is a measure of indeterminacy or uncertainty about a system. The less the entropy, the less the indeterminacy of the system.

Assumption 1 (Concept of structure). Assume for some number n fixed

[S0] Whole human body system (wHBS), either ♂ or ♀, defines $\mathscr{B}$, the functional unit (FU) at the top, at functional level 0 (FL-0);

[S1] $\mathscr{B}$ is a functional aggregate of four canonical functional aggregates, $\mathscr{O}$, $\mathscr{P}$, $\mathscr{R}$, $\mathscr{V}$, at functional level FL-1 in ♂ and ♀ body systems;
[S2] In ♂ and ♀ body systems, there exist "High Centers" $\mathscr{C}$ for functional coordination of $\mathscr{B}$'s canonical FAs;
[S3] For $i = 0, 1, 2, \ldots, n$, every functional unit (FU) at functional level i (FL-i) is itself a functional aggregate (FA) composed of functional units (FU) at functional level $i+1$ (FL-$(i+1)$).

By its biological design, the whole human-body system, denoted by $\mathscr{B}$ in Figure 2.1, comprises the four canonical functional components for life fitness and biological mission: $\mathscr{V}$, the component for solving vital tasks, indispensable to sustain $\mathscr{B}$'s life, next, the component by which $\mathscr{B}$ solves its physical tasks, $\mathscr{P}$, primarily moving around for collecting water, food, and wood for nutrition and shelter, resp. Intimately connected is component $\mathscr{O}$, by which $\mathscr{B}$ communicates with its outside world, processes received signals for decision-making and responding, as deemed appropriate in pursuit of operational goals. While these three have to be engineered in humanoid robots, the fourth functional component, $\mathscr{R}$, is reserved for systems with reproduction—necessary, as life inevitably ends with material death, for species survival.

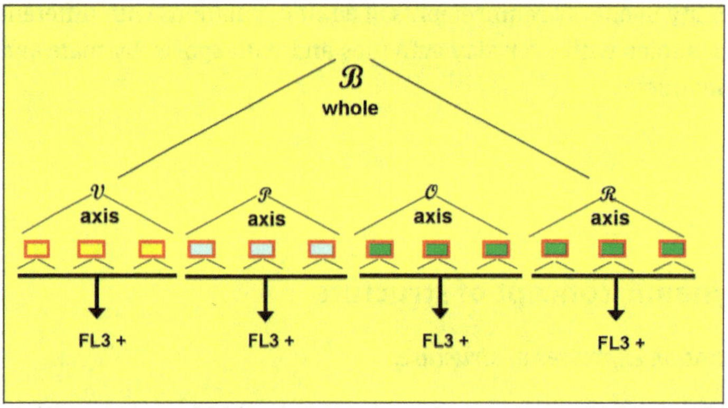

Figure 2.1: The functional breakdown of whole body system $\mathscr{B}$ at functional level FL-0 into four functional axes: $\mathscr{V}$ for vital, $\mathscr{P}$ for physical, $\mathscr{O}$ for cognitive, and $\mathscr{R}$ for reproductive functional task-solving (cf. Figure 2.2) at functional level FL-1, and each decomposed into its specific functional ensembles of organs—schematically represented by three boxes—at functional level FL-2 (above the black bar) for continuation in higher granularity, as generically shown in Figure 2.5.

To jointly fulfill $\mathscr{B}$'s mission in life, their specific actions are coordinated by a specialized "High-level Coordinating Center," say, denoted by $\mathscr{C}$ in Figure 2.2. The schematic representation in this figure shows the same functional decomposition as in Figure 2.1, however, with two nuances: the first is departure from strict hierarchy, as coordination

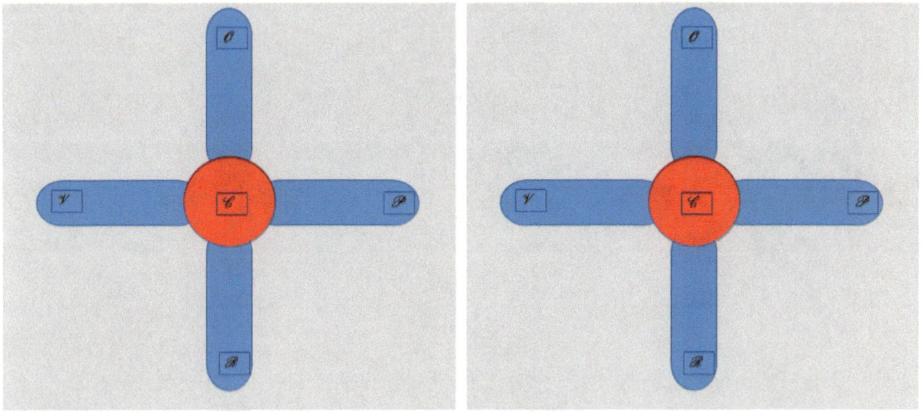

♂ ♀

Figure 2.2: Whole human body system level-1-functional decomposition into four functional axes (from North to West: $\mathcal{O}$ axis of cognitive task-solving functions, $\mathcal{P}$ axis of physical task-solving functions, $\mathcal{R}$ axis of reproductive task-solving functions, distinct for ♂ and ♀, $\mathcal{V}$ axis of vital task-solving functions) jointly coordinated by the higher level command center $\mathcal{C}$.

appears as "round table" at which the four axes "sit" for joint evaluation of all information and "committee-decisions"; the second is that male and female "round tables" will generally operate by different "consensuses" and likely reach nuanced conclusions in practically identical situations. These aspects are to be followed up in future chapters.

In further functional drill-down, each functional aggregate of $\mathcal{B}$ is itself a functional aggregate of smaller functional units.

Definition 3.1. Let $\mathcal{B}$ denote an arbitrary fixed whole human body system (wHBS) as a biological entity. Then the system functional architecture of $\mathcal{B}$, denoted by $\mathsf{SFA}(\mathcal{B})$, consists of eight functional levels (FL),

$$\mathsf{SFA}(\mathcal{B}) : \text{FL-7} \prec \text{FL-6} \prec \cdots \prec \text{FL-0} \tag{2.1}$$

where $A \prec B$ denotes "A has less structure than B," for $A, B \in \{\text{FL-0}, \ldots, \text{FL-7}\}$; see Figure 2.1 for a schematic illustration.

Remark 3.2. Different from business organizations, for which an organigram shows an arrangement of job functions with different levels of responsibilities, the diagram of SFA in Figure 2.1 includes, at every functional level, all lower-level functional units, in particular, all $\approx 10^{14}$ body cells are present at every level; only the assignments of functional roles become more elaborate.

Example 3.1. In Figure 2.3, consider FL-6, the level of different cell types; proceeding to FL-5, tissues, the set of cell types that are present in human body across all locations is then partitioned for assignment to the specific tissues in which they occur—as in a phys-

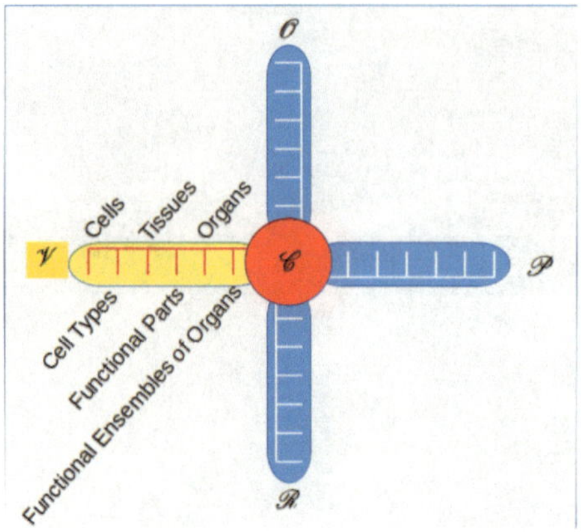

Figure 2.3: Four functional axes and central control, with ordinal scale of functional levels (FL) according to adopted System Functional Architecture (SFA) (cf. Figure 2.1), with its four functional "axes" for vital 𝒱, physical 𝒱, cognitive 𝒪, and reproductive ℛ tasks solving, coordinated by a central function 𝒞.

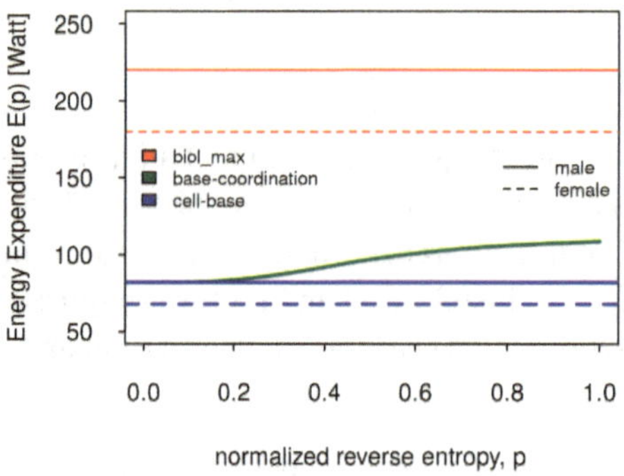

Figure 2.4: Energy expenditure by internal reverse entropy. The p axis is scaled for shrinking internal entropy due to increasingly complex structural organization, from left (maximal disorder at FL-7, cells) to right (maximal order, whole body), corresponding to an increasing aggregation from left to right on the 𝒱 axis in Figure 2.3. The vertical axis, energy expenditure, is scaled for common values of metabolic rates; see Section 2. Color codes: red for biological maximum, green for organizational "overhead," and blue for integrated body-system basal need for cellular survival, dashed for female, solid for male. The vertical difference green to red at $p = 1$ represents reserve energy, in concept.

iology text book that describes organ by organ. And again, from this level of structure, FL-5, the set of all tissues described in a physiology text book for specific locations, partitioning this set of tissues according to locations leads into the next level, of additional structuring, FL-4. In this way, all "bundles" are retained and carried upwards, level by level of increased structure, towards the completely structured whole-body level.

Scaling functional complexity in System Functional Architecture (SFA)

To scale energy expenditure continuously within the functional structure of SFA, one has to "introduce order," i. e., reduce the indeterminacy in the sense of reducing degrees of freedom for the molecules in universe by generation of molecular and then cellular structures that gradually express in the functional structures of SFA, from cells (FL-7) via tissues (FL-5) and organs (FL-3) to whole (FL-0)—in simpler words, one scales in terms of internal entropy best normed to the unit interval]0, 1[; cf. Figure 2.4.

1. Construction of horizontal axis:
   For (2.1), consider its projection onto the unit interval, with FL-7 at zero and FL-0 at 1, and assume that the entropy of $\mathsf{SFA}(\mathscr{B})$ is compressed to]0, 1[, using the reciprocal odds function of probability theory [2]

$$o'(y) = \frac{1-y}{y}, \quad 0 < y < 1, \tag{2.2}$$

   $\lim_{y\to 0} o'(y) = \infty$ and $\lim_{y\to 1} o'(y) = 0$ then. Using the inverse $\frac{1}{1+o'}$ for reciprocal odds o' to regain a linear scale, one obtains

$$p(o') = \frac{1}{1+o'(y)}, \quad 0 < y < 1, \tag{2.3}$$

   which now has $\lim_{o'\to\infty} p(o') = 0$ and $\lim_{o'\to 0} p(o') = 1$. Then $p(o')$ represent normalized reversed internal entropy on $\mathsf{SFA}(\mathscr{B})$ in (2.1), progressing from 0 of maximal disorder in (2.1) to maximal order at 1.
   In other words, p is a normalized measure, running from high entropy or disorder (dispersed, unorganized cells) to low entropy or high order (integrated, fully organized body), pseudometrically scaled from 0 (maximal entropy/disorder) to 1 (minimal entropy/maximal order), it represents an ordinal scale for SFA's increasing aggregation.
2. Scaling of vertical axis:
   This scales whole-body energy expenditure per second (Joule per second), equivalent to power (Watt), measured as the total energy output from all cellular processes—maintenance, regulatory, signaling, and coordination included. This scales minimal total energy use by all cells at rest, the rate at which the whole body's cellular processes collectively consume or transform energy.
3. The bottom horizontal (blue) line:
   Conceptually, minimal total energy use by all cells at rest—without aggregation costs.
4. Construction of the green curve:
   This curve represents the energy required to form aggregates in SFA:

$$E(p) = E_{\text{basal}} + \frac{C_{\max} p^n}{K^n + p^n} \tag{2.4}$$

As p increases (moving from left to right), organizational complexity increases, and the energetic demand for integration of lower-level functional units into upper-level aggregates and their necessary coordination to "jointly fulfill their aggregate's mission" also rises, moving the total energy expenditure upwards—until system-level efficiencies cause this increase to plateau below the body's maximum possible energy expenditure. The central position of steepest rise is only preliminarily suggested.

5. The top horizontal (red) line:
   Theoretical system maximum—the highest sustainable energy throughput physiologically possible. This line is a ceiling for short-term maximal capacity, such as during intense exercise or critical physiological stress; it greatly exceeds ordinary resting needs and daily maintenance.
6. Interpretation of points on the green curve:
   At any given point $(p, E(p))$ on the sigmoid curve, the value of $E(p)$ reflects how much energy per second the living body's cells *collectively* use, not just for their own minimal cellular survival but also including the extra costs for running whole body's integrated, organized biological systems—from the cellular system level FL-7 upwards in SFA, to a functional level in (2.1) that corresponds to the value of p.
7. Interpretation of the green-red gap at $p = 1$:
   The gap between green and red lines at high organization ($p = 1$) shows reserve capacity—the energy budget for additional vital, physical, cognitive, or reproductive activities, beyond routine homeostatic maintenance.
8. Necessary adjustments:
   The values in the plot represent approximate Basal Metabolic Rates for the entire body (cf. Section 2) and are used for illustration of the concept despite the lack of specific data. For any specific scenario, adjusted values for E_{basal} and red line values according to body mass, age, sex, and individual physiology will have to be inserted.

3.2 Scaled energy expenditure

At lowest functional level, FL-7, in (2.1), the $\approx 10^{14}$ body cell persistently generate energy for own cellular maintenance and additional functional tasks that arise from increasing aggregation in SFA; see Figure 2.5. For example, the muscle cells (myocytes) of the heart cannot perform alone and need the structures of supra-cellular organization, embedding and regulatory overhead, to jointly contract their chamber.

The bottom-up perspective in Figure 2.5 illustrates how FL-7 units of (2.1)—depicted as combustion chambers—combine in units of FL-6, labeled as cell types though functionally distinct according to their tissues of FL-5: myocytes as cell types occur in tissues of heart muscle, vascular muscle, skeletal muscle, and vessel walls throughout the circulation, e. g., to provide the same function in all places where implemented by evolution.

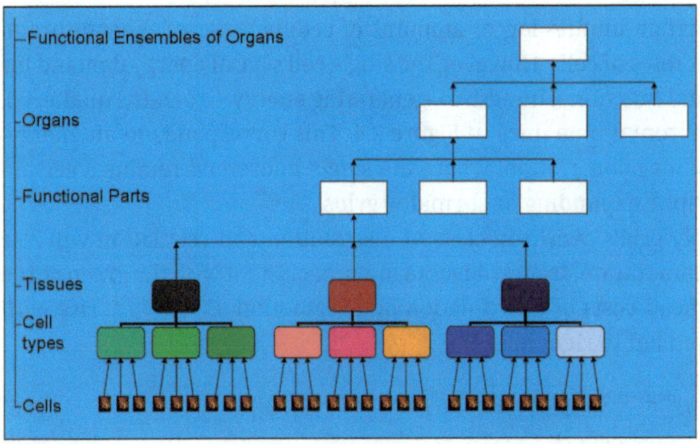

Figure 2.5: Energy confluency scheme: Extending the "tree" depicted in Figure 2.2 from FL-2 of functional ensembles of organs for increasing granularity into functional levels (FL) with higher resolution: FL-3 for individual organs, FL-4 for functional parts that constitute each organ, FL-5 for tissues in each of the functional parts of organs, FL-6 for cell types within the tissues, and last shown FL-7 for cells as the producers of energy that pervades the whole body. Different colors for cell types and for tissues are used to indicate distinction by location: same cell types in different tissues, same tissues in different organs are considered as distinct.

Hence, the concentration from $\approx 10^{14}$ body cells to one of the ≈ 400 cell types [6, 12] would be an error: one should more precisely call the FL-6 functional units "cell types *in situ*," then.

By this understanding, cells of the same type within a common tissue are functional aggregates (FAs) when seen from below, and functional units (FUs) of assigned tissues as seen in Figure 2.5 looking from above, where they contribute their specific cell-type function (e. g., contraction if tissue is muscular).

At the next step upwards, different tissues—now seen as functional units—combine their specific tissue-type functions into their assigned functional aggregate to jointly represent a specific functional part at FL-4 with a distinct functional role within an organ of FL-3. For example, blood vessel walls consist of three layers—each layer with distinct functional role—that have tissue types in common, but each tissue type remains within its anatomical territory and this way observes the strict hierarchy in the body system's functional architecture (SFA).

In transition from individual cells via tissues to fully integrated organs in the entire body, all cellular energy expenditure is bundled as aggregation progresses,

$$E_{\mathrm{FA}}^{\mathrm{cell-base}} = \sum_{\mathrm{FU} \in \mathrm{FA}} E_{\mathrm{FU}}^{\mathrm{cell-base}}, \tag{2.5}$$

for any $\mathrm{FA} = \{\mathrm{FU}_1, \dots, \mathrm{FU}_{n_{\mathrm{FA}}}\}$.

Progressive aggregation implies larger amounts of cellular-energy expenditure to sustain larger conglomerates of cells. However, the single cell's basal energy demand for its sustained survival and functional integrity—generating energy—remains unaltered, consistent with the horizontal blue lines in Figure 2.4. This corresponds to an energy-supply grid for an urban agglomeration, where the single household retains a certain base-energy budget despite expanding or shrinking grids.

However, the supply grid's overhead costs of distribution and regulation will rise with an expanding agglomeration, from villages to town, consistent with the green curve in Figure 2.4: the overhead costs of coordination and regulation, $E^{\text{base-coord}}$, rise with functional levels FL from left to right in (2.1),

$$E^{\text{base-coord}}[\text{FL-7}] < \cdots < E^{\text{base-coord}}[\text{FL-0}], \tag{2.6}$$

and accordingly shrinking internal entropy $1-p$ from left to right on the horizontal axis of Figure 2.4,

$$E^{\text{base-coord}}(0) < \cdots < E^{\text{base-coord}}(p) < \cdots < E^{\text{base-coord}}(1), \tag{2.7}$$

$0 < p < 1$. Still, the shape of the green curve needs biological grounding.

4 Discussion

As "complex systems depend on their structure to function" [8] and hierarchical structures are fundamental for organizing functional activity [1, 23, 40] the adopted System Functional Architecture (SFA) [16, 18, 19] is presented as a pivotal component in mathematical functional analysis of "how the body may work." The previous range is expanded here from three to four canonical functional "axes" by now including the body's reproductive subsystem, and thereby principally mandating parallel investigation for both sexes in the sequel.

SFA's functional decomposition of functional aggregates as "cooperatives" into assigned functional units as "permanent members" is further extended for its grounding in cell biology, i. e., explicitly integrating more levels down to cells in which the energy for body system's structural and functional integrity is generated: all functional activity sources energy from cells and cells also make the body's structure in which functional activity manifests.

Structure shapes out in cellular aggregation, from tissues via organs within four functional axes, and each step implies extra energy costs on top of base cellular maintenance and functional task-solving costs.

As the greater range of possible behaviors at lower levels (cells) becomes progressively reigned in at higher functional levels (organ, whole body), systemic indeterminacy is concurrently decreased. In thermodynamic terms, internal entropy (indeterminacy) is maximal at cellular level and minimal at whole body level. This suggests a rescaling to a normalized reverse entropy over the unit interval, as used for the energy-by-entropy diagram in Figure 2.4.

The introduced diagram seems to be new. It clarifies how entropy, energy, and organizational complexity interplay across levels of biological organization, emphasizing the shift from cell-level "freedom" to whole-body "order". In this plot, basal maintenance, functional activity-dependent energy can be separated from organization dependent costs, and the energy reserve can be seen as the difference from the body's total energy expenditure—the sum of cellular energy for basal maintenance, added cellular energy for functional task-solving across all body functions (not yet shown) and "overhead" costs for structural integration with coordination and regulation, which rise with increasing complexity to biological maximum. The implicit hypothesis is based on the observation that cells contribute energy *in situ*, do not migrate, and consequently, whether they contribute to tissue or organ function, the greater picture appears by coordination of many small localized contributions, like pixels of a screen. [24, 15, 14, 5]

The theoretical, axiomatic approach pursued here aims at the whole human body system, and is then holistic beyond an extent in empirical reach, though the Human Cell Atlas Project [28, 34, 12] and the HuMAP Project [3] have to be mentioned for similar intent.

The axiomatically presented structure is conducive to energy scaling efforts in empirical research, and lends itself to bottom-up as well as top-down approaches. Theoretically, one may take it forward to integrate behavior, motivation, and decision-making. [26, 37, 36, 35]

The next step from here is to detail the functional levels within the SFA framework with cell counts within cell types within tissues within functional parts within organs within organ systems on each functional axis for the virtual energy "piping" and parallel accounting for coordination overhead costs. A tableau for filling-in empirically validated findings or suggestions from biophysical and biochemical modeling would be helpful in a next step.

5 Conclusion

Structure organizes function, and function manifests in structure—both depend vitally on continual flow of cellular energy. It is held that the whole-body's life expresses in permanent task-solving on four functional axes, vital, physical, cognitive, and reproductive, centrally coordinated to support, at any moment, pursuit of a person's current operational goals. Top-down decomposition reveals involved tissues and cell types for functional-energy budgeting. Indispensable structural energy costs represent "overhead" and rise monotonically with increasing bottom-up recomposition. An energy-by-entropy plot visualizes the distinction.

Bibliography

[1] C. Alcocer-Cuarón, A. L. Rivera, and V. M. Castaño. Hierarchical structure of biological systems. A bioengineering approach. *Bioengineered*, 5(2):73–79, 2014. https://doi.org/10.4161/bioe.26570.

[2] J. M. Bland and D. G. Altman. The odds ratio. *British Medical Journal*, 320(7247):1468, 2000. https://doi.org/10.1136/bmj.320.7247.1468.

[3] K. Börner, P. D. Blood, J. C. Silverstein, M. Ruffalo, R. Satija, S. A. Teichmann, G. J. Pryhuber, R. S. Misra, J. M. Purkerson, J. Fan, J. W. Hickey, G. Molla, C. Xu, Y. Zhang, G. M. Weber, Y. Jain, D. Qaurooni, Y. Kong, HRA Team, A. Bueckle, and B. W. Herr II. Human BioMolecular Atlas Program (HuBMAP): 3D Human Reference Atlas construction and usage. *Nature Methods*, 22:845–860, 2025. https://doi.org/10.1038/s41592-024-02563-5.

[4] H. Braun, A. Carlsohn, M. Großhauser, D. König, A. Lampen, S. Mosler, A. Nieß, H. Oberritter, K. Schäbethal, A. Schek, P. Stehle, K. Virmani, R. Ziegenhagen, and H. Heseker. Position of the working group sports nutrition of the German Nutrition Society (DGE): Energy needs in sports. *Deutsche Zeitschrift für Sportmedizin*, 71:171–177, 2020. https://doi.org/10.5960/dzsm.2020.451.

[5] C. Cadart and R. Heald. Scaling of biosynthesis and metabolism with cell size. *Molecular Biology of the Cell*, 33(9):pe5, 2022. https://doi.org/10.1091/mbc.E21-12-0627.

[6] The Tabula Sapiens Consortium. The Tabula Sapiens: A multiple-organ, single-cell transcriptomic atlas of humans. *Science*, 376(6594), 2022. https://doi.org/10.1126/science.abl4896.

[7] A. Davies and C. Moores. *Structure of the Respiratory System, Related to Function*, page 11–28. ChurchillLivingstone, 2 edition, 2010. https://doi.org/10.1016/B978-0-7020-3370-4.00002-5.

[8] J. L. U. Doménech, J. A. Nescolarde-Selva, M. Lloret-Climent, H. Gash, and L. Segura-Abad. Structure, thermodynamics and information in complex systems. *Kybernetes*, 52(11):5307–5328, 2022. https://doi.org/10.1108/K-09-2021-0858.

[9] D. N. Granger and J. P. Granger, editors. *Colloquium Series on Integrated Systems Physiology: From Molecule to Function to Disease*. Morgan & Claypool Life Sciences, San Rafael, CA, 2009–2011.

[10] J. E. Hall. *Guyton and Hall Textbook of Medical Physiology*. Saunders, Philadelphia, 12 edition, 2011.

[11] J. A. Harris and F. G. Benedict. A biometric study of basal metabolism in man. Technical Report Publication No. 279, Carnegie Institution of Washington, Washington, DC, 1919.

[12] I. A. Hatton, E. D. Galbraith, N. S. C. Merleau, T. P. Miettinen, B. McDonald Smith, and J. A. Shander. The human cell count and size distribution. *Proceedings of the National Academy of Sciences of the United States of America*, 120(39), 2023. https://doi.org/10.1073/pnas.2303077120.

[13] W. P. T. James and E. C. Schofield. *Human Energy Requirements. A Manual for Planners and Nutritionists*. Oxford University Press, Oxford, UK, 1990. ISBN (Hardback): 978-0-19-261891-7.

[14] M. Konarzewski and A. Książek. Determinants of intra-specific variation in basal metabolic rate. *Journal of Comparative Physiology B*, 183(1):27–41, 2012. https://doi.org/10.1007/s00360-012-0698-z.

[15] R. Mateus, J. F. Fuhrmann, and N. A. Dye. Growth across scales: Dynamic signaling impacts tissue size and shape. *Current Opinion in Cell Biology*, 73:50–57, 2021. https://doi.org/10.1016/j.ceb.2021.05.002.

[16] J. Mau. Systems neuroergonomics. In R. Wang and X. Pan, editors, *Advances in Cognitive Neurodynamics (V)*, chapter 59, pages 431–437. Springer Science+Business Media, Singapore, Jan 2016.

[17] J. Mau. Translation dynamics in holistic analysis of functional human-body system. *Journal Biomedical Radioelectronics*, 7:43–46, Aug 2018.

[18] J. Mau. Functional biodynamics of human-body system. A mathematical axiomatics with functional learning and aging in life cycle. *Cognitive Neurodynamics*, 14:579–590, 2020.

[19] J. Mau. Axiomatic bipolar dynamics and their control. A holistic mathematical theory of a functional physiology. In J. Mau, S. Mukhin, G. Wang and S. Xu, editors, *BIOKYBERNETIKA. Mathematics for Theory and Control in the Human and in Society*, pages 45–75. Walter de Gruyter GmbH, Berlin/Boston, Mass. USA, 12 2024. https://doi.org/10.1515/9783111341996-004.

[20] B. K. McNab. On the utility of uniformity in the definition of basal rate of metabolism. *Physiological Zoology*, 70(6):718–720, 1997.

[21] M. D. Mifflin, S. T. St Jeor, L. A. Hill, B. J. Scott, S. A. Daugherty, and Y. O. Koh. A new predictive equation for resting energy expenditure in healthy individuals. *The American Journal of Clinical Nutrition*, 51(2):241–247, 1990.

[22] C. Mendoza Montano. Toward a thermodynamic theory of evolution: A theoretical perspective on information entropy reduction and the emergence of complexity. *Frontiers in Complex Systems*, 3:1630050, 2025. https://doi.org/10.3389/fcpxs.2025.1630050.

[23] H. Nederbrogt. Hierarchical organization of biological systems and the structure of adaptation in evolution and tumorigenesis. *Journal of Theoretical Biology*, 184(2):149–156, 1997. https://doi.org/10.1006/jtbi.1996.0266.

[24] M. Pavlicev, J. DiFrisco, A. C. Love, and G. P. Wagner. Metabolic complementation between cells drives the evolution of tissues and organs. *Biology Letters*, 20(11):20240490, 2024. https://doi.org/10.1098/rsbl.2024.0490.

[25] E. Pavlidou, S. K. Papadopoulou, K. Seroglou, and C. Giaginis. Revised Harris-Benedict equation: New human resting metabolic rate equation. *Metabolites*, 13(2):189, 2023.

[26] M. Picard. Energy transduction and the mind–mitochondria connection. *Biochemist (Lond)*, 44(4):14–18, 2022. https://doi.org/10.1042/bio_2022_118.

[27] M. S. Popson, M. Dimri, and J. Borger. Biochemistry, heat and calories. [updated 2023 jun 12]. *StatPearls [Internet]*, 2023.

[28] A. Regev, S. A. Teichmann, E. S. Lander, I. Amit, C. Benoist, E. Birney, B. Bodenmiller, P. Campbell, P. Carninci, M. Clatworthy, H. Clevers, B. Deplancke, I. Dunham, J. Eberwine, R. Eils, W. Enard, A. Farmer, L. Fugger, B. Göttgens, N. Hacohen, M. Haniffa, M. Hemberg, S. Kim, P. Klenerman, A. Kriegstein, E. Lein, S. Linnarsson, J. Lundeberg, P. Majumder, J. C. Marioni, M. Merad, M. Mhlanga, M. Nawijn, M. Netea, G. Nolan, D. Pe'er, A. Phillipakis, C. P. Ponting, S. Quake, W. Reik, O. Rozenblatt-Rosen, J. Sanes, R. Satija, T. N. Schumacher, A. Shalek, E. Shapiro, P. Sharma, J. W. Shin, O. Stegle, M. Stratton, M. J. T. Stubbington, A. van Oudenaarden, A. Wagner, F. Watt, J. Weissman, B. Wold, R. Xavier, N. Yosef, and the Human Cell Atlas Meeting Participants. The human cell atlas. *eLife*, 6:e27041, 2017. https://doi.org/10.7554/eLife.27041.

[29] T. N. F. Roach. Use and abuse of entropy in biology: A case for caliber. *Entropy (Basel)*, 22(12):1335, 2020. https://doi.org/10.3390/e22121335.

[30] E. Roux. The concept of function in modern physiology. *Journal of Physiology*, 592(11):2245–2249, 2014. https://doi.org/10.1113/jphysiol.2014.272062.

[31] A. M. Roza and H. M. Shizgal. The Harris Benedict equation reevaluated: Resting energy requirements and the body cell mass. *The American Journal of Clinical Nutrition*, 40(1):168–182, 1984.

[32] P. A. Samuelson. *Economics. An Introductory Analysis*. McGraw-Hill Book Co., New York, 8 edition, 1970.

[33] W. N. Schofield. Predicting basal metabolic rate, new standards and review of previous work. *Human Nutrition: Clinical Nutrition*, 39(Suppl 1):5–41, 1985.

[34] R. Sender, Y. Weiss, Y. Navon, I. Milo, N. Azulay, L. Keren, S. Fuchs, D. Ben-Zvi, E. Noor, and R. Milo. The total mass, number, and distribution of immune cells in the human body. *Proceedings of the National Academy of Sciences*, 120(44) 2023. https://doi.org/10.1073/pnas.2308511120.

[35] R. C. Vergara, S. Jaramillo-Riveri, A. Luarte, C. Moënne-Loccoz, R. Fuentes, A. Couve, and P. E. Maldonado. Corrigendum: The Energy Homeostasis Principle: Neuronal energy regulation drives local network dynamics generating behavior. *Frontiers in Computational Neuroscience*, 14:599670, 2020. https://doi.org/10.3389/fncom.2020.599670.

[36] R. C. Vergara, S. Jaramillo-Riveri, A. Luarte, C. Moënne-Loccoz, R. Fuentes, A. Couve, and P. E. Maldonado. The Energy Homeostasis Principle: Neuronal energy regulation drives local network dynamics generating behavior. *Frontiers in Computational Neuroscience*, 13:49, 2019. https://doi.org/10.3389/fncom.2019.00049.

[37] S. Vicencio-Jimenez, M. Villalobos, P. E. Maldonado, and R. C. Vergara. The energy homeostasis principle: A naturalistic approach to explain the emergence of behavior. *Frontiers in Systems Neuroscience*, 15:782781, 2022. https://doi.org/10.3389/fnsys.2021.782781.

[38] M. M. Vopson and S. Lepadatu. Second law of information dynamics. *AIP Advances*, 12(11):115112, 2022. https://doi.org/10.1063/5.0100358.

[39] M. Weilenmann, L. Kraemer, P. Faist, and R. Renner. Axiomatic relation between thermodynamic and information-theoretic entropies. *Physical Review Letters*, 117:260601, 2016. https://doi.org/10.1103/PhysRevLett.117.260601.

[40] C. Zardecki, S. Dutta, D. S. Goodsell, R. Lowe, M. Voigt, and S. K. Burley. Pdb-101: Educational resources supporting molecular explorations through biology and medicine. *Protein Science*, 31:129–140, 2022. https://doi.org/10.1002/pro.4200.

Idan Sorin, Michael Zaks, and Alexander Nepomnyashchy

Seasonal forcing in rock–paper–scissors population dynamics

Abstract: We study a class of cyclic dominance models with seasonal forcing, extending the classical May–Leonard competition framework. By introducing time-periodic coefficients into the growth rates and the interaction terms, we explore how environmental seasonality influences the dynamics of three-species systems. Through analytical estimates and numerical simulations, we reveal the emergence of complex oscillatory behavior, including multiyear periodic cycles, transitions to heteroclinic cycle and chaotic oscillations. Our results highlight how periodic modulation can destabilize stable periodic trajectories, generate novel attractors, and give rise to coexistence mechanisms not present in autonomous systems. These findings contribute to the understanding of biodiversity maintenance under realistic, temporally varying ecological conditions.

Keywords: May–Leonard system, Lotka–Volterra competition, seasonal forcing; heteroclinic cycle, chaos, rock-paper-scissors

MSC 2020: 37N25, 92D25, 37C60, 34C37, 34C28, 92D40

1 Introduction: Modeling of seasons in dynamical systems

1.1 Seasonal modeling of cyclic dynamical systems

Many ecological systems adapt to seasonality by changing their behavioral patterns in the course of the year [19]. One of the ways to model seasonality in dynamical systems is to introduce time-dependent system parameters; in population dynamics, these are, e. g., the prey and predator birth rates.

In [18], it was demonstrated that applying seasonal modeling to ecological dynamical systems can generate a rich variety of dynamics such as period-doubling bifurcation, multiyear cycles of different periods, parameter ranges with coexisting multiyear cycles of the same or different periods, as well as quasiperiodicity and chaos.

Since seasonality is a roughly periodic phenomenon, below we assume the parameters of the system to be periodic functions of time. For the case of long stationary seasons,

Idan Sorin, Alexander Nepomnyashchy, Department of Mathematics, Technion–Israel Institute of Technology, Haifa 32000, Israel, e-mails: idansorin@campus.technion.ac.il, nepom@technion.ac.il
Michael Zaks, Department of Physics, Humboldt Universität zu Berlin, Newtonstraße 15, 12489 Berlin, Germany, e-mail: michael.zaks@physik.hu-berlin.de

https://doi.org/10.1515/9783112216620-005

the latter can be approximated by piecewise-constant step-functions, so that throughout each season the evolution is governed by a dynamical system with the constant set of parameters. Within this approximation, transitions between the seasons are treated as instantaneous.

Step functions have been used in modeling the activity of the heart [2]; piecewise periodic functions are employed in a model of drug therapy of human immuno-deficiency virus (HIV) infection [7].

We restrict us to the case when there are just two seasons, e. g., summer and winter.

Let P denote the duration of the overall period comprising both seasons, of durations p_1 and $p_2 = P - p_1$ for the first and the second season, respectively.

For the Lotka–Volterra and May–Leonard models (3.5), modeling seasonality by using piecewise constant functions means that some of the parameters become functions of time in the form

$$\left.\begin{array}{ll} l_1, & 0 \le t \bmod P < p_1 \\ l_2, & p_1 \le t \bmod P \le P. \end{array}\right\} \tag{3.1}$$

So far, analysis of the equations of population dynamics with parameters of the form (3.1) has been carried out only in the context of "seasonal succession," when a species grows during the warmer months and dies or forms resting stages in winter. Mathematically, this means that part of the time the dynamics is governed, e. g., by the Lotka–Volterra equations whereas during the rest of the time the population of species is decaying exponentially. In such systems, the main interest was in proving the stability of equilibrium points, periodic solutions, or heteroclinic cycles [20–22].

There is a significant difference between regular autonomous dynamical systems and dynamical systems with time-dependent parameters such as in (3.1). First, the latter systems are nonsmooth, hence not all results, obtained for smooth dynamical systems, stay valid. In addition, the explicit dependence of the right-hand sides of the equations on time is equivalent to the introducing an additional cyclic variable, which can make the dynamics more involved. One can expect a significant change in the dynamics in such systems compared to their autonomous counterparts with constant parameters.

For example, below we shall see that in the considered exemplary autonomous systems chaotic behavior is impossible, but by making some of the parameters piecewise constant, we may, in principle, obtain chaos.

In the literature, the phenomenon of switching between dynamical systems usually appears in the context of control theory [10]. A typical example is a system of the form:

$$\dot{x} = \begin{cases} f_1(x), & x \in D \\ f_2(x), & x \notin D \end{cases} \tag{3.2}$$

Within this setup, switching between dynamical systems does not directly depend on time but on the instantaneous state of the system, i. e., on the values of the system variables. Systems of the type of (3.2) are autonomous while systems of the type

$$\dot{x} = \begin{cases} f_1(x,t), & t \in S \\ f_2(x,t), & t \notin S \end{cases} \tag{3.3}$$

are nonautonomous. Below, in Section 2 we present the models to be studied and summarize their properties. Section 3 describes extensions and details the results of our analysis. Section 4 provides a discussion of the findings, and Section 5 contains concluding remarks.

2 Material and methods: Multispecies competitive models

2.1 Multispecies competitive models

The general Lotka–Volterra model, which describes temporal evolution in a set of n biological species competing for common resources, is a system of ordinary differential equations (ODEs)

$$\frac{dN_i(t)}{dt} = r_i N_i(t)\left[1 - \sum_{j=1}^{n} a_{ij} N_j(t)\right]; \quad N_i \geq 0;\ i = 1, \ldots, n. \tag{3.4}$$

Here, $N_i(t)$ is the suitably normalized number of individuals of the i-th species at time t, r_i is the intrinsic growth rate of the i-th species, and a_{ij} are the interaction coefficients that measure the extent to which the j-th species affects the growth rate of the i-th species [4]. For $n > 3$, (3.4) features rich dynamics, including multiple attractors, heteroclinic networks, and chaos [16, 14, 3]. For $n = 3$, the system (3.4), after a proper rescaling, takes the form:

$$\left.\begin{aligned} \dot{N}_1 &= r_1 N_1 \left(1 - N_1 - \alpha_1 N_2 - \beta_1 N_3\right) \\ \dot{N}_2 &= r_2 N_2 \left(1 - N_2 - \alpha_2 N_1 - \beta_2 N_3\right) \\ \dot{N}_3 &= r_3 N_3 \left(1 - N_3 - \alpha_3 N_1 - \beta_3 N_2\right) \end{aligned}\right\} \tag{3.5}$$

where the dot denotes the temporal derivative. In [6] it has been shown that under nonequal growth rates r_i this system can have three kinds of attractors: a coexistence state of equilibrium (none of the variables equals zero), a limit cycle, and a heteroclinic cycle. The special case $r_1 = r_2 = r_3 = r$ is called the May–Leonard system [11]. For this system, it was proven [5] that for parameters that satisfy the relation

$$(1 - \beta_1)(1 - \beta_2)(1 - \beta_3) = (\alpha_1 - 1)(\alpha_2 - 1)(\alpha_3 - 1), \quad 0 < \beta_i < 1 < \alpha_i,$$

the attractor, can be a two-dimensional invariant manifold with a continuum of periodic solutions on it.

In the general case of (3.4) with n equal growth rates (called the *general* May–Leonard system), it has been proved that the system can be reduced to a competitive Lotka–Volterra system with $n-1$ species [6]. This implies that chaos can occur neither in (3.4) for $n = 3$ nor in the general May–Leonard system for $n = 4$, since the latter can be reduced to a competitive Lotka–Volterra system with $n = 3$.

2.2 Symmetric May–Leonard model

Here, we discuss in more detail the paradigmatic *symmetric* May–Leonard model [11] of three competing species, setting $r_1 = r_2 = r_3 = 1$, taking the circulant competition matrix

$$\begin{bmatrix} 1 & \alpha & \beta \\ \beta & 1 & \alpha \\ \alpha & \beta & 1 \end{bmatrix}$$

with $0 < \beta < 1 < \alpha$, and redenoting the variables $N_1 = u$, $N_2 = v$, and $N_3 = w$. The model becomes

$$\left.\begin{aligned} \dot{u} &= u\,(1 - u - \alpha v - \beta w) \\ \dot{v} &= v\,(1 - v - \alpha w - \beta u) \\ \dot{w} &= w\,(1 - w - \alpha u - \beta v). \end{aligned}\right\} \tag{3.6}$$

We mention in passing that the system (3.6) can also be found in the form:

$$\begin{aligned} \dot{a} &= a\big(1 - \rho - (\sigma + \zeta)b + \zeta c\big), \\ \dot{b} &= b\big(1 - \rho - (\sigma + \zeta)c + \zeta a\big), \\ \dot{c} &= c\big(1 - \rho - (\sigma + \zeta)a + \zeta b\big), \end{aligned}$$

with $\rho = a + b + c$ and σ, ζ as growth rates [15].

Dynamics are considered in the invariant octant $u, v, w \geq 0$. The system (3.6) has five equilibrium points: $(0,0,0)$, $(1,0,0)$, $(0,1,0)$, $(0,0,1)$, and the "coexistence point,"

$$u = v = w = \frac{1}{1 + \alpha + \beta}. \tag{3.7}$$

The coexistence point is stable, if $\alpha + \beta < 2$, and is oscillatory unstable otherwise. The planes $u = 0$, $v = 0$, and $w = 0$ are invariant manifolds. On the plane $w = 0$, there exists the heteroclinic trajectory leading from $(1,0,0)$ to $(0,1,0)$. Similarly, two more heteroclinic trajectories exist: from $(0, 1, 0)$ to $(0, 0, 1)$ on the plane $u = 0$ and from $(0, 0, 1)$ to $(1, 0, 0)$ on the plane $v = 0$.

Thus, for any $\beta < 1 < \alpha$, there exists the robust heteroclinic cycle

$$(1, 0, 0) \to (0, 1, 0) \to (0, 0, 1) \to (1, 0, 0)$$

Stability analysis shows that the heteroclinic cycle is attracting if $\alpha + \beta > 2$, and is repelling if $\alpha + \beta < 2$. Therefore, the coexistence point is an attractor at $\alpha + \beta < 2$, and the heteroclinic cycle is an attractor at $\alpha + \beta > 2$.

Dynamics at $\alpha + \beta = 2$ are quite nonstandard. At this combination of parameters, the plane $u + v + w = 1$ is invariant and attracts all solutions that start outside this plane. Furthermore, dynamics are conservative as the combination

$$A = \frac{uvw}{(u + v + w)^3} \tag{3.8}$$

is a constant of motion, which we define as the coexistence index.

Therefore, the attracting plane $u+v+w = 1$ is foliated into closed phase trajectories. This continuum of periodic motions is parameterized by the value of A which, on that plane, equals uvw and assumes the values from the interval $0 \le A \le 1/27$. The maximal value $A = 1/27$ corresponds to the coexistence equilibrium $u = v = w = 1/3$; the minimal value $A = 0$ is reached at all other points of equilibrium and at the heteroclinic trajectories.

Systems with cyclic competition are often called Rock-Paper-Scissors (RPS) systems due to their apparent similarity to the game: Scissors cut Paper, Paper wraps Rock, Rock blunts Scissors. In (3.6), v dominates u, w dominates v, and u, in its turn, dominates w. In such systems, there is usually a heteroclinic cycle of the kind sketched in Figure 3.1.

Systems of this kind can be extended to games with more “players” (species). For example, the game called “Rock–Paper–Scissors–Lizard–Spock” with five species has the same cyclic rules as the classical RPS game; cf. [16]. In this variant of the game, the appearance of heteroclinic networks is possible, since there are species with “nondirect interaction.”

3 Results

We first consider as specific models “ Specific aspects of dynamics” in 3.1 and “ A simple dissipative model” in 3.2, then we obtain the main result on “ Alternating May–Leonard and Lotka–Volterra dynamics” in 3.3.

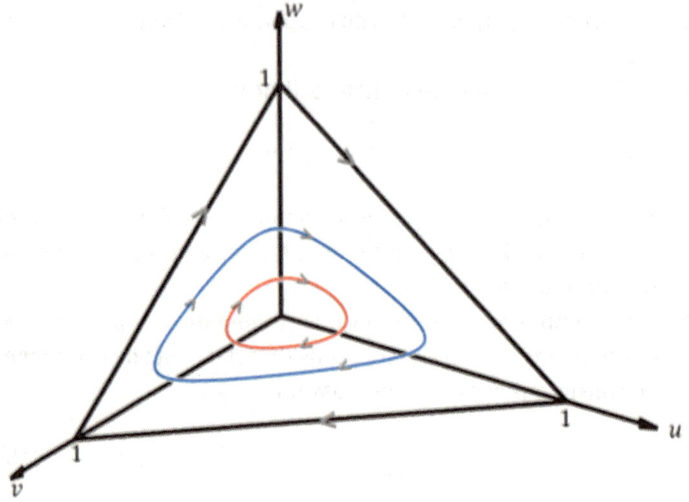

Figure 3.1: Phase portrait on the invariant simplex for $\alpha+\beta = 2$. On the plane $u+v+w = 1$ (population fractions), the boundary triangle forms a heteroclinic cycle connecting the three corner equilibria. Trajectories in the interior are closed periodic orbits; two examples with different values of the conserved first integral A are shown, illustrating that for $\alpha + \beta = 2$ the dynamics on the simplex is conservative.

3.1 Specific aspects of dynamics

Family of time-periodic solutions in the symmetric May–Leonard model
Consider (3.6) at $\alpha = 2, \beta = 0$:

$$\dot{u} = u\,(1 - u - 2v) \tag{3.9a}$$

$$\dot{v} = v(1 - v - 2w) \tag{3.9b}$$

$$\dot{w} = w(1 - w - 2u) \tag{3.9c}$$

Note that this May–Leonard model is symmetric with respect to cyclic permutation of the variables. On the attracting manifold $u + v + w = 1$, the closed trajectory $u\,v\,w = A$ is described by the equation $uv(1 - u - v) = A$, hence

$$v = \frac{u - u^2 \pm \sqrt{(u - u^2)^2 - 4Au}}{2u}. \tag{3.10}$$

On substituting expression (3.10) into (3.9), we find that

$$\dot{u} = \pm\sqrt{P(u)}, \quad P(u) = \left(u - u^2\right)^2 - 4Au, \quad 0 < u < 1. \tag{3.11}$$

In (3.11), the positive and negative signs correspond, respectively, to the epochs of the growth and decrease of $u(t)$. The solution of (3.11) with the initial condition $u(0) = u_1$, $0 < u_1 < 1$ is

$$t(u) = \frac{2}{\sqrt{u_2(u_3-u_1)}} \mathrm{F}\left(\arcsin\left(\sqrt{\frac{\frac{1}{u}-\frac{1}{u_1}}{\frac{1}{u_2}-\frac{1}{u_1}}}\right), \frac{u_3(u_2-u_1)}{u_2(u_3-u_1)}\right), \tag{3.12}$$

where F(...) denotes the elliptic amplitude. The inverse function is

$$u(t) = \frac{1}{\frac{1}{u_1} + \left(\frac{1}{u_2}-\frac{1}{u_1}\right) \cdot \mathrm{sn}^2\left[\frac{t}{2}\sqrt{u_2(u_3-u_1)}, \frac{u_3(u_2-u_1)}{u_2(u_3-u_1)}\right]} \tag{3.13}$$

where $u_1 < u_2 < u_3$ are the positive roots of the polynomial $P(u)$ (the fourth root is zero), and sn denotes the elliptic sine function. Analyzing the properties of the polynomial $P(u)$ with $0 < A < 1/27$, we conclude that $u_1, u_2 < 1$, $u_3 > 1$, so that $u(t)$ varies between u_1 and u_2, because the expression under the square root should be positive and the minimal and maximal values of u correspond to $\dot{u} = 0$. Due to the permutational symmetry of the problem, the solutions $v(t)$, $w(t)$ are the replicas of $u(t)$ shifted by $T/3$:

$$v(t) = u\left(t - \frac{T}{3}\right), \quad w(t) = u\left(t + \frac{T}{3}\right), \tag{3.14}$$

where T is the period of (3.14), which can be found as $T = 2t(u_2)$, i. e.,

$$T = \frac{4}{u_2(u_3-u_1)} K\left(\frac{u_3(u_2-u_1)}{u_2(u_3-u_1)}\right), \tag{3.15}$$

where K is the complete elliptic integral of the first kind; see Figures 3.2 and 3.3. The limits of period at the boundaries of the interval $(0, 1/27)$ are

$$\lim_{A\to 0} T = 2\pi K(1) = \infty, \qquad \lim_{A\to 1/27} T = 2\pi\sqrt{3}. \tag{3.16}$$

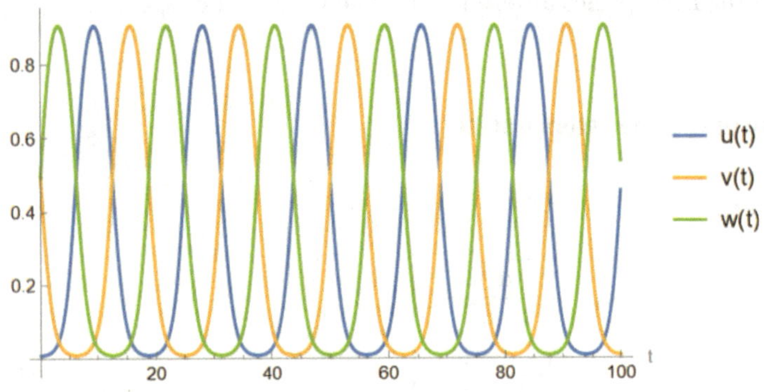

Figure 3.2: The species densities as functions of time $u(t)$ (blue), $v(t)$ (orange), and $w(t)$ (green) with the coexistence index $A = 2 \times 10^{-3}$.

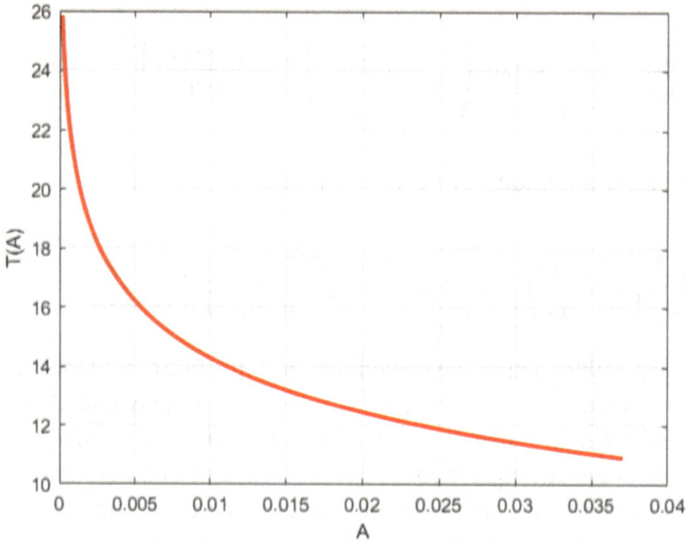

Figure 3.3: Period T versus the coexistence index A (conservative case). The oscillation period tends to infinity as $A \to 0$ (near the heteroclinic cycle) and approaches to $2\pi\sqrt{3}$ as A increases toward the coexistence point.

3.2 A simple dissipative model

Consider now the Lotka–Volterra system

$$\left.\begin{aligned} u_t &= u(\gamma - u - \alpha v), \\ v_t &= v(1 - v - \alpha w), \\ w_t &= w(1 - w - \alpha u), \end{aligned}\right\} \tag{3.17}$$

with the growth rate of the first species different from that of two other species.

3.2.1 Linear instability of the coexistence state

The coexistence solution is

$$\left.\begin{aligned} u_* &= \frac{\gamma - \alpha + \alpha^2}{1 + \alpha^3}, \\ v_* &= \frac{1 - \alpha + \alpha^2 \gamma}{1 + \alpha^3}, \\ w_* &= \frac{1 - \alpha\gamma + \alpha^2}{1 + \alpha^3}. \end{aligned}\right\} \tag{3.18}$$

We assume that

$$\alpha > 1 \quad \text{and} \quad \frac{\alpha - 1}{\alpha^2} < \gamma < \frac{\alpha^2 + 1}{\alpha}, \tag{3.19}$$

hence u_*, v_*, w_* are positive.

Stability of the coexistence state is determined by linearization of (3.17): three eigenvalues σ of the Jacobian matrix solve the equation

$$\left.\begin{aligned} \sigma^3 + (u_* + v_* + w_*)\sigma^2 \\ + (u_* v_* + v_* w_* + w_* u_*)\sigma \\ + (1 + \alpha^3) u_* v_* w_* = 0. \end{aligned}\right\} \tag{3.20}$$

At the stability boundary, the Hopf bifurcation takes place: two complex conjugate eigenvalues are purely imaginary, $\sigma = \pm i\omega$ for a real ω. This, after the substitution of the coordinate values, determines the bifurcation curve on the parameter plane (γ, α):

$$\left.\begin{aligned} \alpha(1 - \alpha - \alpha^3)\gamma^3 + (-2 + 3\alpha - 5\alpha^2 + 6\alpha^3 + \alpha^5 - \alpha^6)\gamma^2 \\ + (-4 + 7\alpha - 11\alpha^2 + 5\alpha^3 - 7\alpha^4 + \alpha^5 + \alpha^7)\gamma \\ + (1 - \alpha)^2(-2 + \alpha - 3\alpha^2 - \alpha^3 - \alpha^4) = 0. \end{aligned}\right\} \tag{3.21}$$

The curve of Hopf bifurcation splits the parameter plane into the region where the coexistence point is the attractor, and the region where it is unstable.

3.2.2 Stability of the heteroclinic cycle

System (3.17) has three saddle points $(\gamma, 0, 0)$, $(0, 1, 0)$, $(0, 0, 1)$, which are connected by heteroclinic trajectories forming a heteroclinic cycle.

Consider a trajectory that is very close to the heteroclinic cycle. The time needed to complete a "cycle" (start close to one of the saddle points and then come back to its neighborhood) can be divided into six intervals: three intervals where the solution is close to one of the saddles, with means that only one of the variables is $O(1)$, and three intervals on each of the heteroclinic trajectories, where two of the variables are $O(1)$. We will see that the major contribution comes from the time spent near the saddles.

Consider the trajectory starting between the vicinities of saddle points $(0, 0, 1)$ and $(\gamma, 0, 0)$, where u and w are $O(1)$, while $v = \epsilon \ll 1$. Near the saddle point $(\gamma, 0, 0)$, the variables are governed by the linear equations:

$$\begin{aligned} \tilde{u}_t &= -\gamma\tilde{u} - \alpha\gamma v, \\ v_t &= v, \\ w_t &= (1 - \alpha\gamma)w, \end{aligned}$$

where $\tilde{u} = u - \gamma$.

The time t_1 spent near that point depends on ϵ: during that time, the variable v grows from ϵ to $O(1)$. Since $\tilde{v} \sim \epsilon e^t$, we have $\epsilon e^{t_1} \sim 1$, so $t_1 \sim -\ln \epsilon$. During that time, w decreases from $O(1)$ to $\tilde{w}_1 \sim e^{(1-a\gamma)t_1}$.

When v becomes $O(1)$, the linear equation for v has to be corrected. Between the saddles $(\gamma, 0, 0)$ and $(0, 1, 0)$, where both v and $1 - v$ are $O(1)$, the governing equation for v is $v_t = v(1 - v)$ and the solution for v corresponds to the heteroclinic trajectory in the plane $w = 0$,

$$v(t) = \frac{1}{1 + C_1 e^{-t}}, \tag{3.22}$$

where C_1 is a constant depending on the initial conditions. Because for $v = O(1)$ and $1 - v = O(1)$, the velocity $v_t = O(1)$, the duration of the motion on that fragment of trajectory is $O(1)$. Hence, it is much smaller than t_1, and the variable w does not change its order during that interval, hence the change of the variable w during that time can be neglected. Near the point $(0, 1, 0)$, the evolution of variables is governed by the system of equations

$$u_t = (\gamma - \alpha)u, \quad \tilde{v}_t = -\tilde{v} - aw, \quad w_t = w,$$

where $\tilde{v} = v - 1$. During the time t_2 spent near that point, w growth from $\tilde{w}_1$ up to $O(1)$, hence, similar to the calculation of t_1 we get $t_2 \sim -\ln(\tilde{w}_1) = -(a\gamma - 1)\ln \epsilon$. During that time t_2, u decreases from $O(1)$ to $u_2 \sim e^{(\gamma-\alpha)t_2} = \epsilon^{(\alpha-\gamma)(a\gamma-1)}$.

Between the points $(0, 1, 0)$ and $(0, 0, 1)$, the governing equation for w is $w_t = w(1-w)$, and the heteroclinic solution on the plane $u = 0$ is

$$w(t) = \frac{1}{1 + C_2 e^{-t}}, \tag{3.23}$$

where C_2 is a constant depending on the initial conditions. Again, the duration of the time interval where $w = O(1)$ and $1 - w = O(1)$ is $O(1)$, and the variable u does not change its order during that interval.

Near the point $(0, 0, 1)$, the evolution of variables is governed by the system of equations

$$\begin{aligned} u_t &= \gamma u, \\ v_t &= (1 - \alpha)v, \\ \tilde{w}_t &= -\tilde{w} - a\tilde{u}, \end{aligned}$$

where $\tilde{w} = w - 1$. The variable u grows from $\tilde{u}_2$ until $O(1)$ during time t_3,

$$t_3 \sim -\frac{1}{\gamma}\ln(\tilde{u}_2) = -\frac{(\alpha - \gamma)(a\gamma - 1)}{\gamma}\ln \epsilon$$

During that time, v decreases from $O(1)$ to

$$\tilde{v_3} \sim e^{-(a-1)t_3} = \epsilon^{(a-1)(a-\gamma)(a\gamma-1)/\gamma}.$$

Finally, between the points $(0,0,1)$ and $(\gamma,0,0)$ the governing equation for u is $u_t = u(\gamma - u)$ and the solution for u corresponds to the heteroclinic trajectory in the plane $v = 0$,

$$u(t) = \frac{\gamma}{1 - e^{\gamma(C_3 - t)}} \tag{3.24}$$

where C_3 is a constant depending on the initial conditions. The variable v does not change its order on that interval.

We find that during the motion in the vicinity of the heteroclinic cycle $(\gamma,0,0) \to (0,1,0) \to (0,0,1) \to (\gamma,0,0)$ the order of the variable v is changed from ϵ to ϵ^c where

$$c = (a-1)(a-\gamma)(a\gamma-1)/\gamma.$$

When $c > 1$, the heteroclinic cycle is an attractor since a solution that starts close to the heteroclinic cycle becomes even closer to it. When $c < 1$, the heteroclinic cycle is a repeller. Note that if we start from a different point, we obtain the same coefficient c (with different order of multiplications). The critical case $c = 1$ divides the plane (γ, a) by the curve:

$$(a-1)(a\gamma-1)(a-\gamma) - \gamma = 0 \tag{3.25}$$

The meaning of the latter expression is as follows.

In accordance with the general theory, stability of the heteroclinic cycle requires that compression near the cycle prevails over expansion: product of all positive Jacobian eigenvalues for the participating saddles is smaller than the absolute value of the product of all closest to zero negative Jacobian eigenvalues. Equality between these two products, in terms of the coefficients of the equations, results in (3.25). Near the heteroclinic cycle, for each solution long epochs of nearly quiescent hovering over one of the saddle points alternate with relatively fast passages between the saddle points. The time of the latter passage is negligible compared to the time of slow creeping near the saddle points. Since $c > 1$, during the next "cycle," the time t_1 spent near the point $(\gamma,0,0)$ will be $-c \ln \epsilon$, so we conclude that the time that the solution spends near each saddle increases with each new turn around the cycle. This behavior can be seen in Figure 3.4.

3.2.3 Attractors in the system of ODEs

Equations (3.21) and (3.25) together with inequalities (3.19) split the parameter plane (a, γ) into three regions:

- the region A where the attractor is the equilibrium,

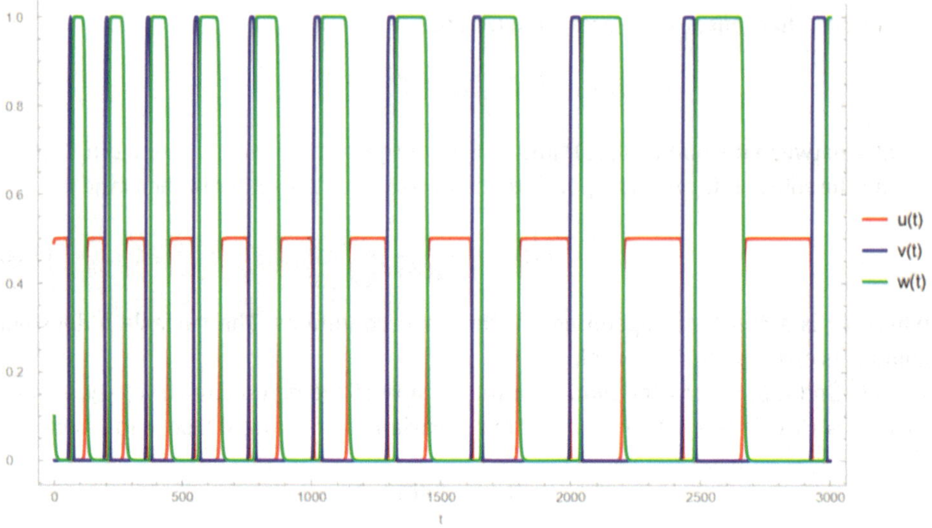

Figure 3.4: Approach to the heteroclinic cycle. An orbit on the invariant simplex spends progressively longer near the saddle equilibria and travels along the heteroclinic connections, illustrating the slow-down toward the heteroclinic cycle.

– the region B where the attractor is the heteroclinic cycle, and
– the region C between them.

In [6] it was shown that for systems in the form of (3.17) only three attractors were possible:
– a state of equilibrium,
– a heteroclinic cycle, and
– a limit cycle.

In accordance with this statement, the attractor in the region C should be the limit cycle. Indeed, on the state diagram, shown in Figure 3.5, the borders of the region C are the line of the Hopf bifurcation (in which the limit cycle is born) and the line (3.21) on which the heteroclinic cycle acquires stability. Two bifurcation curves on the parameter plane are mutually tangent in the point ($\alpha = 2, \gamma = 1$), which is the symmetric May–Leonard model (3.9).

From now on, we fix $\gamma = 0.5$ and treat α as the control parameter of the system.

For $\alpha = 2.24$, the supercritical Hopf bifurcation occurs and the stable limit cycle appears, while for $\alpha = 2.38$ the heteroclinic bifurcation takes place and the heteroclinic cycle acquires stability; hence, denote these values as $\alpha_{\text{Hopf}} = 2.24$ and $\alpha_{\text{het}} = 2.38$. Implied changes in the phase space are presented in Figure 3.6.

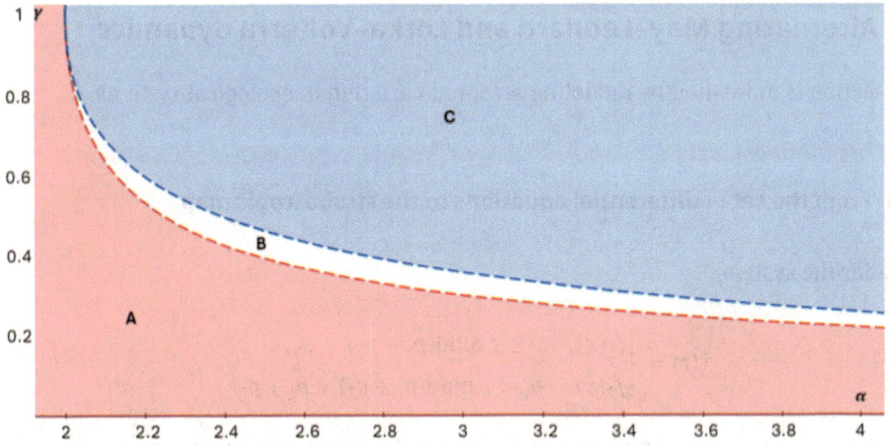

Figure 3.5: Parameters (interaction coefficient and growth rate) plane (α, γ): attractor regimes and bifurcation curves. Colored shading distinguishes three regions: **A**—stable coexistence equilibrium; **B**—stable heteroclinic cycle; **C**—stable limit cycle (between **A** and **B**). The boundary of **C** is formed by the Hopf bifurcation curve (where the limit cycle is born) and the curve on which the heteroclinic cycle becomes stable; these curves are tangent at $(\alpha, \gamma) = (2, 1)$ (the symmetric May–Leonard case).

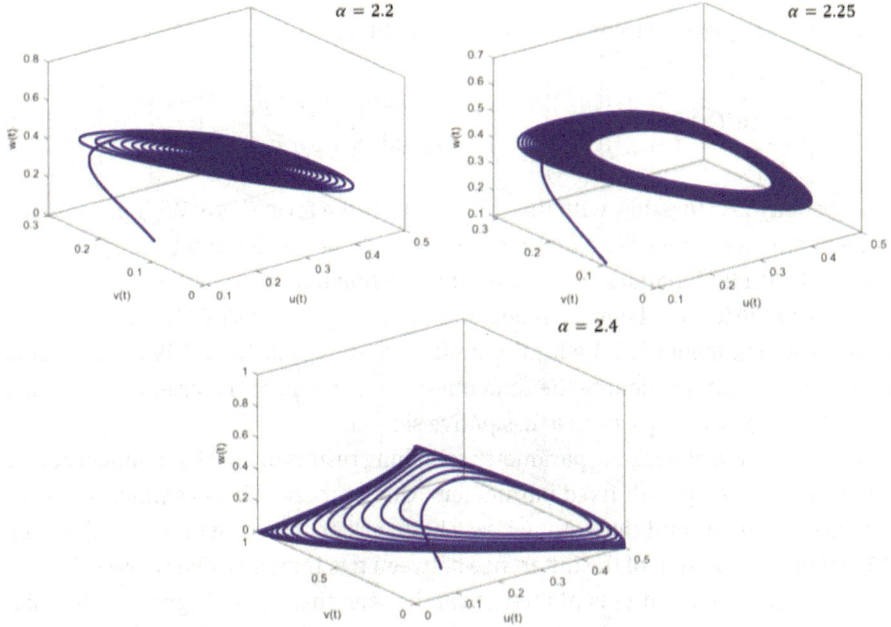

Figure 3.6: Phase trajectories at fixed growth rate γ as the interaction coefficient α varies. With the same initial condition, the flow shows (a) convergence to the coexistence equilibrium for α below the Hopf bifurcation, (b) a stable limit cycle just beyond Hopf, and (c) an attracting heteroclinic cycle once the heteroclinic threshold is crossed.

3.3 Alternating May–Leonard and Lotka–Volterra dynamics

This section is motivated by modeling seasonal variation in ecological systems.

3.3.1 From the set of differential equations to the stroboscopic map

Consider the system

$$\dot{x}(t) = \begin{cases} f_1(x), & 0 \le t \bmod(p_1 + p_2) \le p_1, \\ f_2(x), & p_1 < t \bmod(p_1 + p_2) < p_1 + p_2, \end{cases} \tag{3.26}$$

with f_1, f_2 being functions of the type (3.1), specifically, of either the Lotka–Volterra or the May–Leonard type. We choose x to be the 3-species vector $(u, v, w)^{\mathsf{T}}$ that obeys the system

$$\begin{aligned} u_t &= u(\gamma(t) - u - \alpha(t)\, v), \\ v_t &= v(1 - v - \alpha(t)\, w), \\ w_t &= w(1 - w - \alpha(t)\, u), \end{aligned} \tag{3.27}$$

where α and γ are piecewise constant functions of time:

$$(\alpha(t), \gamma(t)) = \begin{cases} (\alpha_1, \gamma_1), & 0 \le t \bmod(p_1 + p_2) \le p_1, \\ (\alpha_2, \gamma_2), & p_1 < t \bmod(p_1 + p_2) < p_1 + p_2. \end{cases} \tag{3.28}$$

The system (3.27) together with the rule (3.28) has the form (3.26). With progressing time, the parameter values are “jumping” periodically between the set (α_1, γ_1) and the set (α_2, γ_2). To study dynamics, we choose the combination of $(\alpha_1, \gamma_1) = (2, 1)$, which yields the symmetric May–Leonard model (3.9) and $(\alpha_2, \gamma_2) = (2.26, 0.5)$, which renders the Lotka–Volterra model (3.17) with the limit cycle of period 16.998849 as the global attractor. Accordingly, we denote the time interval $0 \le t \le p_1$ as a conservative season and the interval $p_1 < t < p_1 + p_2$ as a dissipative season.

The introduction of periodic parameter switching results in novel dynamical effects is impossible in a setup with fixed parameters. One of these effects reminds sensitive dependence on initial conditions, known as a hallmark of chaotic dynamics [13]. In Figure 3.7, temporal evolution of the difference between two initially rather close solutions (difference of 10^{-9} in u and v) is plotted. It can be seen that at the beginning the solutions start very close to each other, but after enough time, the difference between them reaches its maximal value ~ 0.85.

Since the solution of the conservative system is defined by its coexistence index A, it is instructive to follow the evolution of the coexistence index values. In Figure 3.8, the coexistence index levels have separated across conservative seasons, while within any

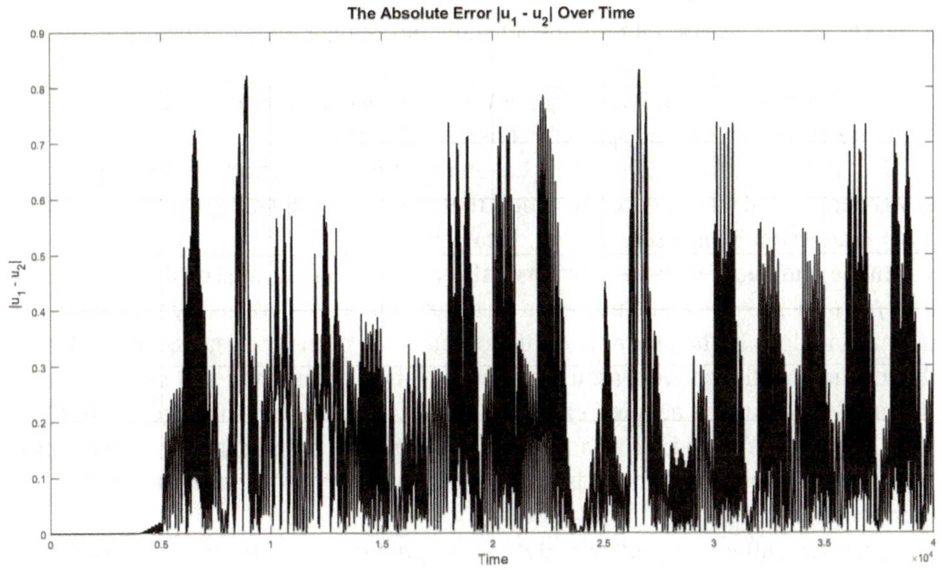

Figure 3.7: Sensitivity to initial conditions under seasonal switching. Two solutions of the species density $u(t)$ starting 10^{-9} apart separate markedly after several switching cycles with the seasons length $p_1 = p_2 = 1000$ fixed, demonstrating strong sensitivity to initial conditions.

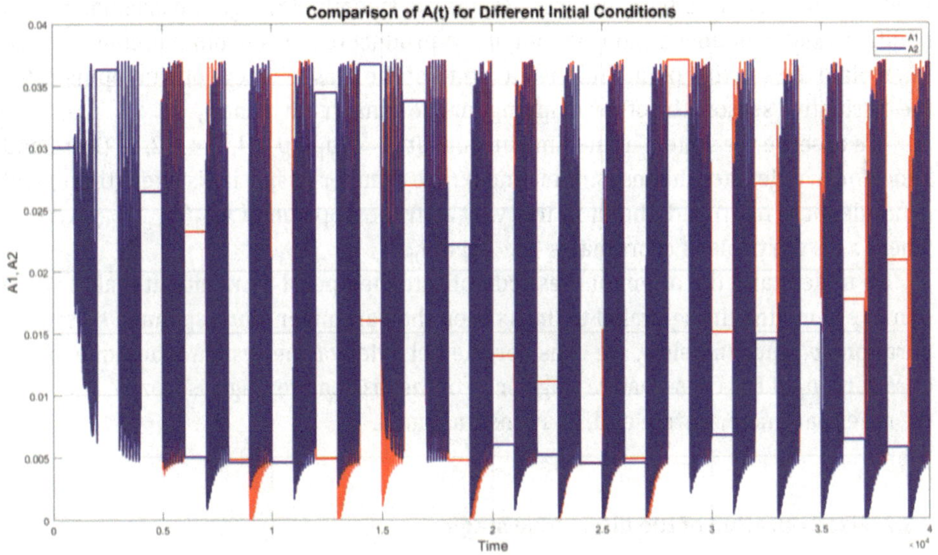

Figure 3.8: Coexistence indexes at seasonal boundaries for two nearby solutions. Within each conservative season, the stroboscopic coexistence index A is constant (plateaus). After about three full forcing cycles $(p_1 + p_2)$, the two solutions enter conservative seasons with different A (coexistence index) values, so their plateaus no longer coincide.

given conservative season A remains constant and changes only during the dissipative stage.

In order to understand the origin of this effect, we introduce a new approach, the sequence of states at certain specially chosen values of t.

If p_2 is sufficiently big, the system has enough time to approach the limit cycle from any generic initial state; hence, we expect that after the completion of every period $P = p_1 + p_2$ the system is, in a sense, close to that cycle.

In the symmetric May–Leonard case, all trajectories are located on the 2-d manifold $uvw/(u+v+w)^3 = A$, and the location on a particular closed trajectory depends on the initial conditions in the phase space: the initial value of the coexistence index A. Note that during the dissipative season the value of A is not conserved and oscillates along the limit cycle; for the adopted values of α and γ, this oscillation range is $0.00464 < A < 0.036988$. The period of the latter cycle is the "internal" characteristic of the Lotka–Volterra dynamics; in general, it is incommensurate with "externally" imposed season durations p_1 and p_2.

Therefore, after every full period $P = p_1 + p_2$ the system starts its next conservative stage with a next value of A, and is, as a result, attracted to a different periodic solution on the two-dimensional manifold. By following the sequence of endpoints of conservative seasons (or, alternatively, the sequence of endpoints of dissipative seasons) in the phase space, we obtain a mapping of the three-dimensional phase space (u, v, w) into itself. It is important to note that this mapping is smooth although the original system of differential equations is non-smooth: it is a product of two smooth mappings, which interrelate the starting point and the endpoint of the season where one mapping is for the dissipative season, the other mapping for the conservative one.

We observe the system at the time instants $\{t_n\} = \{(p_1 + p_2)n\}$, $n = 1, 2, \ldots$ (Here and henceforth, n denotes the season/time index; the number of species is fixed (three) and is not denoted by n) and obtain in this way a stroboscopic map $x_{n+1}(t_{n+1}) = f(x_n(t_n))$, where x_n is the triple of coordinates $\{u, v, w\}$ at t_n.

To understand the origin of these complicated temporal dynamics, it makes sense to move along the lines parallel to the axes on the parameter plane spanned by season durations p_1 and p_2. Below, we consider the behavior of the system when one of the durations, p_1 of the conservative stage or p_2 of the dissipative stage, is frozen, and treat the other parameter as free and, in a sense, as small.

3.3.2 Fixed duration of the dissipative stage

To start with, we fix the duration p_2 at some value that distinctly exceeds the period of the dissipative limit cycle (we use for illustrations $p_2 = 1000$), and compare the return mappings at different values of p_1. Typical examples of attracting sets are shown in Figure 3.9. At $p_1 = 0$ (i. e., under a vanishingly short conservative season) all "probing points" lie precisely on the limit cycle of the dissipative system. At $p_1 = 10$ (a nonzero,

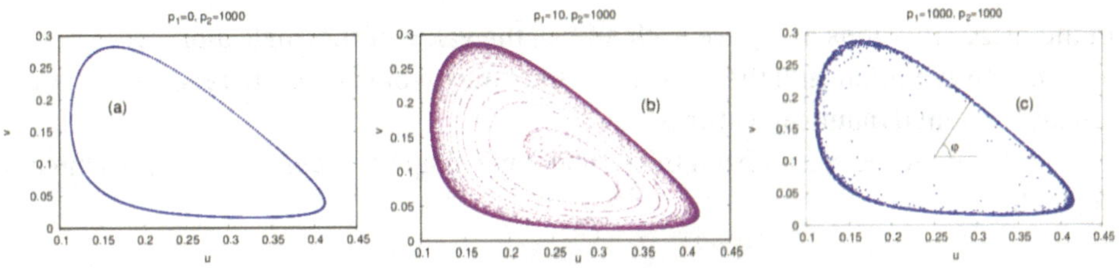

Figure 3.9: Two-dimensional projections (u vs. v) of the attractors of the stroboscopic mapping at fixed duration $p_2 = 1000$ of the dissipative season, combined with vanishing (a), moderate (b), and balanced (c), duration p_1 of the conservative season.

remarkably short conservative stage) the points fill the layered folded curve that lies inside the cycle. Finally, at $p_1 = 1000$, when both seasons have equal durations, the majority of points is located in the narrow stripe around the limit cycle; however, a closer look shows that the attracting set is "fuzzy" and disordered: there are also seldom irregular excursions inwards, far away from the curve of the limit cycle, which look like noisy outliers.

The full stroboscopic map operates in the three-dimensional space, but in the asymptotics of short conservative seasons it is well approximated by the one-dimensional circle map. In this situation, most of the time the motion is one-dimensional and occurs along the limit cycle; short conservative seasons act as a regular sequence of "kicks" that temporarily drive the system in the phase space away from the curve of the limit cycle. If the conservative stages are sufficiently short, the kicks can be viewed as weak perturbations of the periodic motion (the fact that these perturbations are conservative does not matter much). A motion along the limit cycle is adequately described in terms of the phase: the angular variable that parameterizes the closed curve in the phase space [9].

Operationally, near the limit cycle the phase can be recovered by usage of the kind of "polar coordinates": for a sufficiently large collection of the mapping points, we find the coordinates of the center of mass (e. g., $\langle u \rangle$), and for (any two out of three) coordinates calculate the angle φ_i between the line that connects the center of mass with the point (u_i, v_i) and one of the coordinate axes, e. g.,

$$\tan(\varphi_i) = \frac{v_i - \langle v \rangle}{u_i - \langle u \rangle}.$$

Reconstruction of the angular variable φ is sketched in Figure 3.9(c). By varying i, we observe the sequence of the iterations of φ_i: the desired circle mapping

$$\varphi_{i+1} = g(\varphi_i) \bmod (2\pi) \tag{3.29}$$

In the fuzzy situations of Figure 3.9(b and c), the value of the angle alone does not determine the evolution, and the circle map is not one-to-one; nevertheless, it is helpful to disclose certain dynamical features.

Circle maps, recovered from the stroboscopic maps of Figure 3.9, are presented in Figure 3.10.

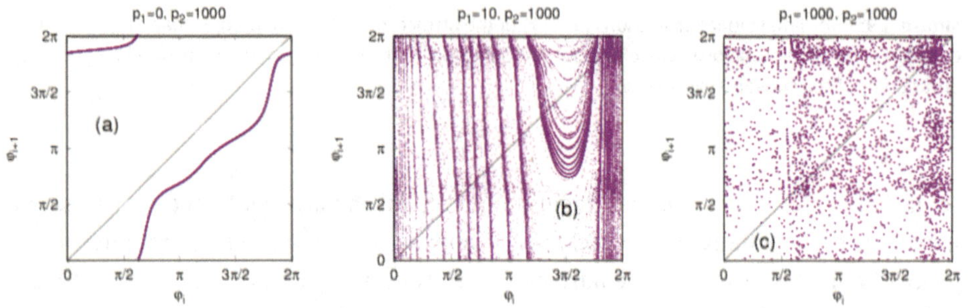

Figure 3.10: Circle maps of the angle φ_i between the line that connects the center of mass with the point (u_i, v_i) and one of the coordinate axes, constructed from the stroboscopic mapping at fixed duration p_2 of the dissipative season, and varying duration p_1 of the conservative season. The values of p_1 in the panels (a),(b),(c) coincide with respective values from the panels of Figure 3.9.

The plot of $g(\varphi)$ at $p_1 = 0$ in Figure 3.10(a) features the shape, characteristic for a circle map with irrational rotation number. The curve of the map is piecewise monotone and does not intersect the "bisecting line" $\varphi_{i+1} = \varphi_i$. In this case, there is no synchronization between two timescales of the system: the "internal" period of the limit cycle and the "externally imposed" full duration $P = p_1 + p_2$.

At $p_1 = 10$, the plot of the circle map, shown in Figure 3.10(b), is fuzzy, but still possesses a certain pattern. There are a few dozens of "branches," each of them covering the whole vertical span of the map. Each of these branches intersects the bisecting line. Every intersection delivers a fixed point $\varphi_{i+1} = \varphi_i$ of the circle map—a periodic oscillation in the underlying system of ODEs. Since the absolute value of the slope of $g(\varphi)$ in these fixed points exceeds 1, all of them are unstable. Presence of the countable set of unstable periodic orbits is an indicator of chaotic dynamics [13].

Another notable feature of the map in Figure 3.10(b) is the multibranched layered local minimum in the right part. Such locally unimodal mappings are akin to the logistic mapping and provide the framework for the sequence of period-doubling bifurcations.

Finally, the plot of the circle map at $p_1 = 1000$, shown in Figure 3.10(c), bears no particular pattern. A multitude of points in the vicinity of the line $\varphi_{i+1} = \varphi_i$ indicates strong chaoticity.

Apparently, the onset of chaos occurs during the transformation between the left and the central panel of Figure 3.10. Here, we briefly describe the sequence of events and attracting states, encountered in the course of increase of p_1.

A convenient characteristic is the observable coexistence index $A_i = uvw/(u+v+w)^3$ (or any other explicit function of the phase space coordinates) measured at the endpoint of the i-th dissipative season.

- In the range $0 < p_1 < p_{syn} = 0.091\ldots$, the shape of the map qualitatively reminds us of Figure 3.10(a). There is no synchronization between the timescales of the problem. As p_1 is increased, the curve $g(\varphi)$ comes closer to the line $\varphi_{i+1} = \varphi_i$. Observable A_i attains a range of values.
- At $p_1 = p_{syn}$, the curve $g(\varphi)$ becomes tangent to the line $\varphi_{i+1} = \varphi_i$. This is the so-called *tangent* bifurcation: the structurally unstable neutral fixed point of the circle map appears, and under the further increase of p_1, decomposes into *two* fixed points: the stable and the unstable one. The stable fixed point corresponds to the stable periodic solution in the system of ODEs. The tangent bifurcation marks the onset of synchronization: in the parameter space, the system enters the synchronization region—the so-called Arnold tongue [1]. Observable A_i converges to the single constant value.
- In the range $p_{syn} < p_1 < p_{pd} = 0.194\ldots$, the stable fixed point of the circle map persists; oscillations in the system stay synchronized with the total duration of the seasons. Close to p_{pd}, the mapping curve becomes nonmonotonic.
- At p_{pd}, the period-doubling bifurcation takes place in the map: the fixed point loses stability, and a stable periodic orbit composed of two points: $\varphi_{i+1} = g(\varphi_i)$, $\varphi_{i+2} = g(\varphi_{i+1}) = \varphi_i$ is born from it. Accordingly, the observable A_i jumps between two alternating values.
- In the range $p_{pd} < p_1 < p_{pd2} = 0.23\ldots$, the mapping orbit with period 2 is stable. At p_{pd2}, the second period-doubling bifurcation occurs: the orbit with period 2 loses stability, and the stable orbit with period 4 is born. Just beyond p_{pd2}, the observable A_i jumps between *four* alternating values.
- At $p_{pd3} = 0.236$, the further period-doubling bifurcation gives birth to the stable orbit with period 8.
- At $p_{pd4} = 0.2377$, the next period-doubling bifurcation creates the stable orbit with period 16.
- … At p_{pdj}, the period-doubling bifurcation destabilizes the map orbit with period 2^{j-1} and creates the stable orbit with period 2^j. The coexistence index A jumps between 2^j alternating values.
- At $p_1 \geq 0.238$, we observe weak chaos in the dynamics of A. The plot of the circle map consists of several filled segments, some of them folded.
- At $p_1 \approx 0.27$, the local structure of the circle map changes; now it has a shape of a narrow M-shaped band with a well-pronounced minimum, and consists of up to three thin layers.
- At $p_1 \approx 0.45$, the circle map is a narrow M-shaped band built of five thin layers.

Altogether, the evolution of the circle map follows the celebrated scenario of period-doubling bifurcations: a sequence of intervals of p_1 along which the period length is progressively doubled. Each parameter interval of p_1 between two successive transitions is smaller than the preceding one by the factor that should asymptotically tend to the universal value $\delta_F = 4.669201...$: the Feigenbaum constant [8]. This behavior is common for generic one-parameter families of unimodal mappings, the most famous representative being the logistic map [12]. Like in all those families, beyond the first threshold of chaos there is a set of windows of periodic behavior; for example, close to $p_1 = 0.25$ the mapping again has a stable orbit of period 4 (which is different from the mentioned orbit born at p_{pd2} and destabilized at p_{pd3}), but basically chaos prevails in the parameter space. For some of the numerical results, see Figure 3.11.

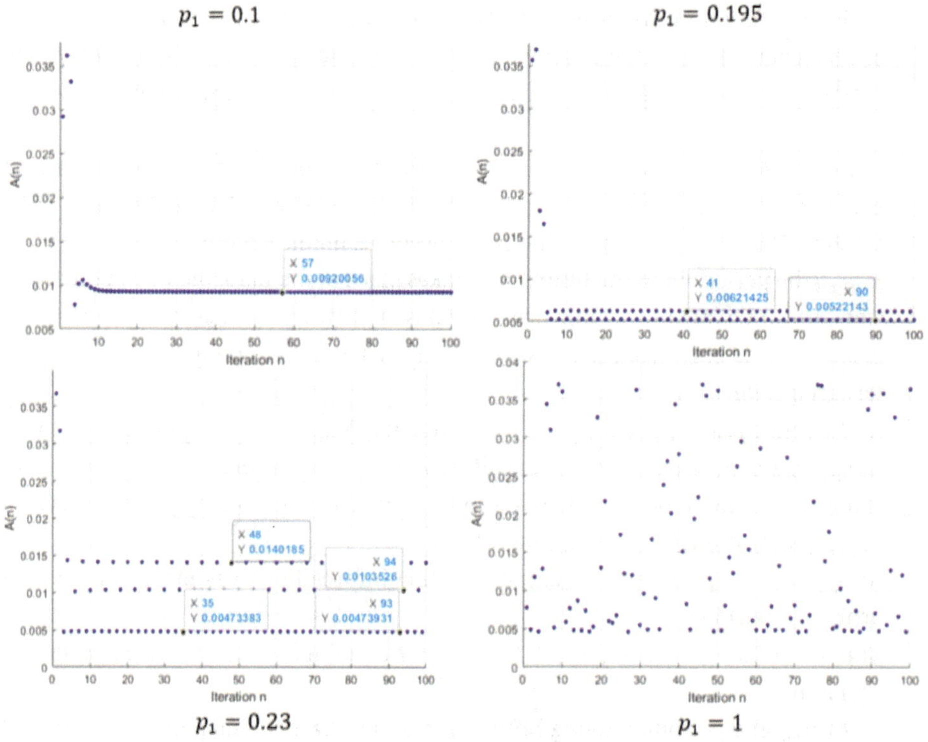

Figure 3.11: Evolution of the coexistence index A_n for various conservative season durations p_1 at fixed dissipative season duration $p_2 = 1000$. Panels illustrate locking plateaus with one, two, or four levels as the p_1 increases, followed by irregular fluctuations indicating loss of phase locking.

3.3.3 Fixed duration of the conservative stage

We consider the system (3.27) with piecewise-constant coefficients in the right-hand side of (3.28) under the assumption that the conservative season is sufficiently long, i. e., $p_1 \gg 1$. In this regime, the system has enough time to approach a periodic orbit characterized by a nearly conserved total population, so that $u+v+w \approx 1$ at the end of the conservative phase.

We first recall that for $p_2 = 0$, solutions are attracted to periodic orbits determined by the initial value of A. Here, we study the effect of a short dissipative stage $p_2 \ll 1$.

The system is governed by equations $\dot{u} = u(\gamma - u - av)$, $\dot{v} = v(1 - v - aw)$, $\dot{w} = w(1 - w - au)$; during the dissipative stage, the parameter values are $a = 2.26$, $\gamma = 0.5$. Let the endpoint of the conservative season in the phase space be the state (u_c, v_c, w_c). Since p_2 is short, expressions for the increments of the variables during the dissipative stage are well approximated by linearized equations,

$$\left.\begin{aligned} \Delta u &= p_2(\gamma - u_c - av_c) \\ \Delta v &= p_2(1 - v_c - aw_c) \\ \Delta w &= p_2(1 - w_c - au_c) \end{aligned}\right\} \tag{3.30}$$

These increments affect the quantity $A = uvw/(u+v+w)^3$. Assuming $u_c + v_c + w_c \approx 1$, we derive the evolution of $\ln A$ over the dissipative stage with size p_2:

$$\Delta(\ln A) = p_2\big[(\gamma - 1) - (a - 2) - 3(\gamma - 1)u_c + 3(a - 2)(u_c v_c + v_c w_c + w_c u_c)\big]. \tag{3.31}$$

Expression (3.31) is not sign conserving, but because the period T_{cons} of one conservative oscillation with fixed coexistence index A is generically not commensurate with the duration p_1; one can expect that the points (u_c, v_c, w_c) are uniformly distributed along the trajectory with the fixed value of A (the change of A gives the contribution of the higher order in p_2). Therefore, the total effect of dissipative stages can be estimated using the average value of $\Delta(\ln A)$ along the trajectory with the fixed A, i. e., as

$$\langle \Delta(\ln A)\rangle = p_2\big[(\gamma - 1) - (a - 2) - 3(\gamma - 1)\langle u\rangle + 3(a - 2)(\langle uv\rangle + \langle vw\rangle + \langle wu\rangle)\big]. \tag{3.32}$$

Here, we use the notation

$$\langle f\rangle = \frac{1}{T_{\text{cons}}} \int_0^{T_{\text{cons}}} f(\tau)d\tau,$$

where $T_{\text{cons}} = T_{\text{cons}}(A)$ is the period of the oscillatory solution with a coexistence index A at the conservative stage.

Now, since (3.6) implies

$$\begin{aligned} d(\ln u)/dt &= 1 - u - 2v, \\ d(\ln v)/dt &= 1 - v - 2w, \\ d(\ln w)/dt &= 1 - w - 2u, \end{aligned}$$

after taking average over the period T_{cons} from both sides, we obtain

$$\langle u \rangle = \langle v \rangle = \langle w \rangle = \frac{1}{3}. \tag{3.33}$$

Since $v(t) = u(t - T/3)$, $w(t) = u(t + T/3)$, it follows that $\langle uv \rangle = \langle vw \rangle = \langle wu \rangle$, so together with (3.33):

$$\langle \Delta \ln A \rangle = cp_2 \quad \text{and} \quad c = (\alpha - 2)(-1 + 9\langle uv \rangle) \tag{3.34}$$

Generally, c depends on A, and it can be seen that the sign of c determines the stability of the heteroclinic cycle.

Now, note that the action

$$\langle f, g \rangle = \frac{1}{T_{\text{cons}}} \int_0^{T_{\text{cons}}} f(\tau) g(\tau) d\tau$$

is an inner product. From the Cauchy–Schwarz inequality, $|\langle uv \rangle| < \sqrt{\langle u^2 \rangle} \sqrt{\langle v^2 \rangle}$ (u, v are linearly independent), and that $v(t) = u(t - T/3)$ one can show that

$$|\langle uv \rangle| < |\langle u^2 \rangle|. \tag{3.35}$$

From (3.9a), $\dot{u} = u - u^2 - 2uv$, so after taking average on both sides, from (3.33), we found that $\frac{1}{3} = \langle u^2 \rangle + 2\langle uv \rangle$. Using (3.35), we find that $\langle uv \rangle < 1/9$ (see also Figure 3.12 for numerical qualification). Therefore,

$$c < 0 \tag{3.36}$$

and:

$$\frac{\langle A_{n+1} \rangle}{\langle A_n \rangle} \approx 1 + cp_2, \tag{3.37}$$

since $p_2 \ll 1$. This implies that $A_n \to 0$ as $n \to \infty$, which means that every initial condition that is not the coexistence state, is under sufficiently small p_2 eventually attracted by the heteroclinic cycle.

Note that the analysis above has been done for general initial conditions that are not exactly on the heteroclinic cycle. The only exception is the initial condition $u = v =$

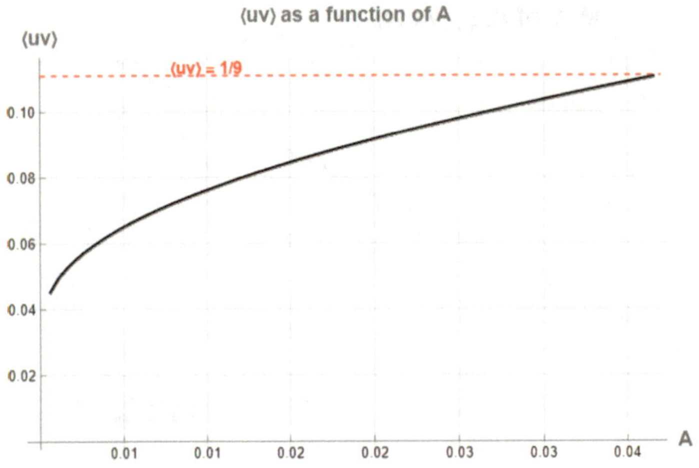

Figure 3.12: Mean product $\langle uv\rangle$ as a function of the coexistence index A. Trend of $\langle uv\rangle$ versus A consistent with the sign of the averaged update for $\ln A$, supporting attraction toward the heteroclinic cycle in the appropriate regime.

$w = 1/3$. This point is the coexistence state of the conservative stage, and not a periodic solution with nonzero T_{cons}. A coexistence point of the conservative stage is not an equilibrium state during the dissipative stage. Therefore, even if the system starts at the coexistence point, in the dissipative season it is forced to leave this state, and the same mechanism applies in subsequent cycles.

Thus, in the regime $p_2 \ll 1$, $p_1 \gg 1$, the heteroclinic cycle of the conservative dynamics becomes globally attracting.

Furthermore, as shown in [17], for states near $A = 1/27$, we can expand $\langle uv\rangle \approx 1/9 - r$, with $r = \frac{1}{27} - A \ll 1$, leading to the approximation: $\langle uv\rangle = \frac{2}{27} + A$. Substituting it in (3.34) gives

$$c = (\alpha - 2)(9\langle A_n\rangle - 1/3),$$

so that

$$\langle A_{n+1}\rangle = \langle A_n\rangle\left(1 + p_2\frac{\alpha - 2}{3}(27\langle A_n\rangle - 1)\right). \tag{3.38}$$

This matches numerical simulations even beyond the vicinity of the coexistence state; see Figure 3.13.

As p_2 increases, the system changes its behavior. While for most n, A_n is close to zero, and occasional jumps in A_n become more frequent, a phenomenon we call a “popcorn”; see Figure 3.14.

For sufficiently large p_2, the sequence A_n appears chaotic, demonstrating a type of behavior not found in classical autonomous Lotka–Volterra or May–Leonard systems.

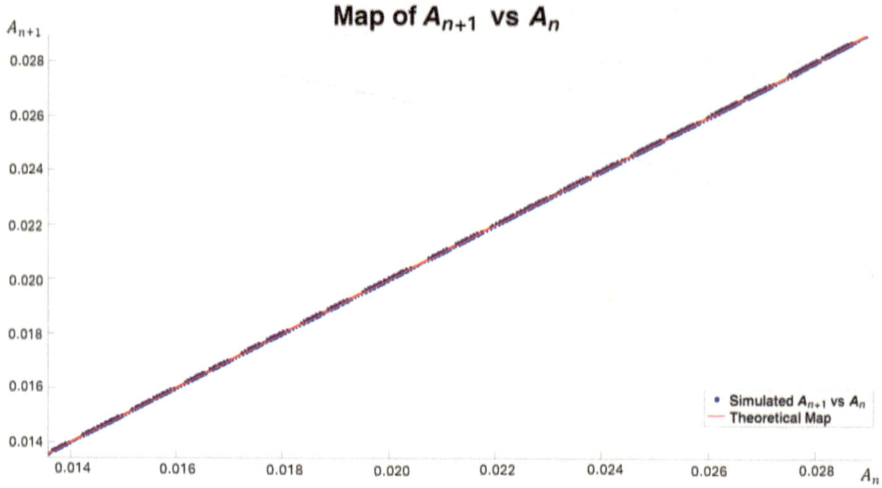

Figure 3.13: Return plot of the discrete coexistence index A_{n+1} vs. A_n: simulation versus theory. Blue dots (simulation) and a red curve (averaged theory) are compared.

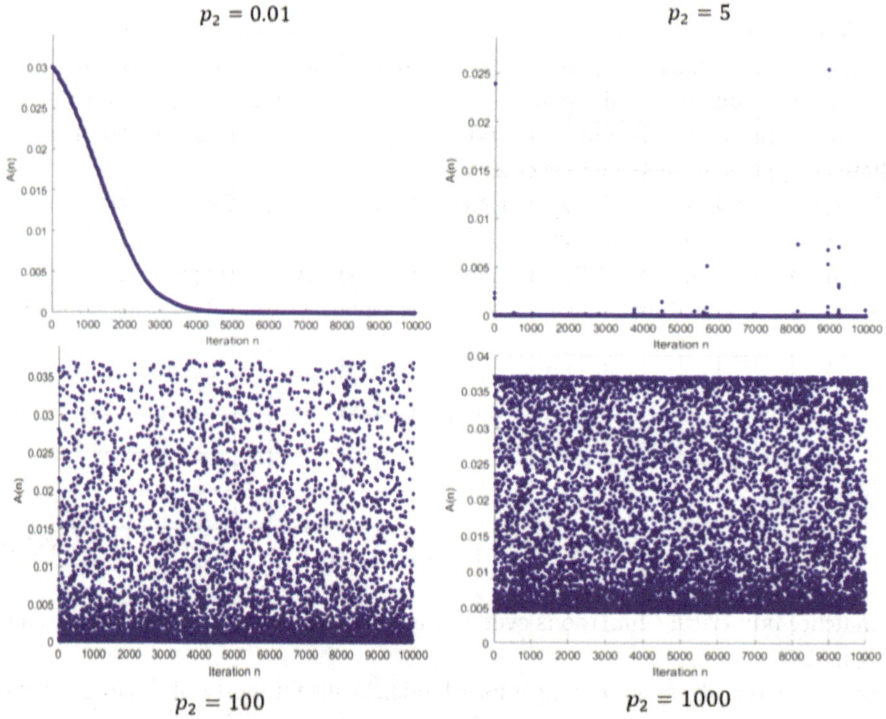

Figure 3.14: "Popcorn" dynamics of the coexistence index A_n for varying dissipative season p_2 at fixed conservative season $p_1 = 1000$. As p_2 increases, rare upward bursts in A_n become frequent and then irregular, signaling a transition from near-zero values to chaotic behavior.

4 Discussion

We have explored a class of seasonally forced competitive Lotka–Volterra models, where the coefficients at the interaction terms are modeled as piecewise constant functions of time.

Achievement

The presented approach explains how seasonal switching generates regimes that the autonomous models cannot: phase locking (synchronization windows), period-doubling cascades, irregular "popcorn" excursions of A_n, and robust chaos in the limit of long dissipative stages. The opposite limit (long conservative stage, short dissipative stage) drives A_n to zero and makes the heteroclinic cycle globally attracting.

Relevance

These findings indicate that seasonality, even when introduced as a relatively simple modulation, can fundamentally reshape ecological interactions and long-term evolutionary dynamics.

Limitations and extensions

While our analysis focused on specific model families, the methods and insights developed here are applicable to a wider class of systems, including those with spatial structure or stochastic effects. Nevertheless, we did not attempt data-based parameter identification. These are natural next steps rather than conceptual obstacles.

Compared with other works

The presented approach differs from classical "seasonal succession" studies in Lotka–Volterra/May–Leonard models, which typically alternate growth and decay and focus on equilibria or heteroclinic cycles under decay phases. Here, both seasons are fully dynamical and the alternation is conservative↔dissipative, producing new regimes (locking, cascades, chaos) and a useful stroboscopic reduction.

Perspectives

Our future research will aim to characterize the structure and stability of the nontrivial solutions identified in this study, and to generalize the framework to more complex dynamical settings, such as systems with additional species, higher-dimensional phase spaces, or alternative forms of temporal forcing.

5 Conclusion

Seasonal forcing of cyclic competition produces regimes that autonomous models cannot: synchronization windows, cascades to chaos, and when seasons are strongly imbalanced, global attraction to the heteroclinic cycle. We studied a three-species system with periodic switching between a conservative stage and a dissipative stage, and analyzed the dynamics via a stroboscopic map and a coexistence index invariant on the May–Leonard manifold. We found that season lengths select the long-term behavior: short dissipative seasons contract the invariant on average, driving the system toward the heteroclinic cycle; longer dissipative seasons introduce sensitive dependence, period-doubling, and robust chaos. The presented approach shows how realistic season lengths can stabilize coexistence, create multiyear rhythms, or trigger chaos in rock–paper–scissors population dynamics. It offers testable predictions for phase locking and irregular fluctuations, and a compact toolset for nonautonomous ecological dynamics.

Bibliography

[1] V. Aquillen. Lecture 6: Nonlinear dynamics and chaos. Lecture notes, University of Rochester, 2023.

[2] A. Atangana and S. I. Araz. Rhythmic behaviors of the human heart with piecewise derivative. *Mathematical Biosciences and Engineering*, 19:3091–3109, 2022. https://doi.org/10.3934/mbe.2022143.

[3] A. Bayliss, A. A. Nepomnyashchy, and V. A. Volpert. Beyond Rock–Paper–Scissors systems—deterministic models of cyclic ecological systems with more than three species. *Physica D: Nonlinear Phenomena*, 411:132585, 2020. https://doi.org/10.1016/j.physd.2020.132585.

[4] R. S. Cantrell and C. Cosner. *Spatial Ecology via Reaction–Diffusion Equations*. Wiley Series in Mathematical and Computational Biology. Wiley, 2003. https://doi.org/10.1002/0470871296.

[5] C.-W. Chi, S.-B. Hsu, and L.-I. Wu. On the asymmetric May–Leonard model of three competing species. *SIAM Journal on Applied Mathematics*, 58:211–226, 1998. https://doi.org/10.1137/S0036139994272060.

[6] J. Coste, J. Peyraud, and P. Coullet. Asymptotic behaviors in the dynamics of competing species. *SIAM Journal on Applied Mathematics*, 36(3):516–528, 1979. https://doi.org/10.1137/0136039.

[7] P. De Leenheer. Within-host virus models with periodic antiviral therapy. *Bulletin of Mathematical Biology*, 71(1):189–210, 2009. https://doi.org/10.1007/s11538-008-9359-5.

[8] M. J. Feigenbaum. Quantitative universality for a class of nonlinear transformations. *Journal of Statistical Physics*, 19:25–52, 1978. https://doi.org/10.1007/BF01020332.

[9] L. Glass. Cardiac arrhythmias and circle maps – a classical problem. *Chaos*, 1(1):13–19, 1991. https://doi.org/10.1063/1.165810.

[10] L. Huang, W. Qin, and S. Chen. Nonsmooth dynamics of a Filippov predator–prey ecological model with antipredator behavior. *Advances in Continuous and Discrete Models*, 2024:1–20, 2024. https://doi.org/10.1186/s13662-024-03808-5.

[11] R. M. May and W. J. Leonard. Nonlinear aspects of competition between three species. *SIAM Journal on Applied Mathematics*, 29:243–253, 1975. https://doi.org/10.1137/0129022.

[12] R. M. May. Simple mathematical models with very complicated dynamics. *Nature*, 261:459–467, 1976. https://doi.org/10.1038/261459a0.

[13] E. Ott. *Chaos in Dynamical Systems*. Cambridge University Press, 2 edition, 2002. https://doi.org/10.1017/CBO9780511803260.

[14] A. Pikovsky and A. Nepomnyashchy. Chaos in coupled heteroclinic cycles and its piecewise-constant representation. *Physica D: Nonlinear Phenomena*, 452:133772, 2023. https://doi.org/10.1016/j.physd.2023.133772.

[15] C. M. Postlethwaite and A. M. Rucklidge. Spirals and heteroclinic cycles in a spatially extended Rock–Paper–Scissors model of cyclic dominance. *Europhysics Letters*, 117:48006, 2017. https://doi.org/10.1209/0295-5075/117/48006.

[16] C. M. Postlethwaite and A. M. Rucklidge. Stability of cycling behaviour near a heteroclinic network model of Rock–Paper–Scissors–Lizard–Spock. *Nonlinearity*, 35(4):1702–1733, 2022. https://doi.org/10.1088/1361-6544/ac3560.

[17] I. Sorin, A. Nepomnyashchy, and V. Volpert. Stability of oscillations in the spatially extended May–Leonard model, 2025. https://doi.org/10.48550/arXiv.2502.18048.

[18] R. A. Taylor, J. A. Sherratt, and A. White. Seasonal forcing and multi-year cycles in interacting populations: Lessons from a predator–prey model. *Journal of Theoretical Biology*, 334:49–61, 2013. https://doi.org/10.1007/s00285-012-0612-z.

[19] C. A. Vettorazzi, D. W. Macdonald, and A. J. Dickman. Understanding the effects of seasonal variation in prey availability: A case study of large carnivores in the Maasai Mara, Kenya. *Journal of Zoology*, 312(2):92–104, 2020. https://doi.org/10.1111/jzo.13013.

[20] X. Xie and L. Niu. Global stability in a three-species Lotka–Volterra cooperation model with seasonal succession. *Mathematical Methods in the Applied Sciences*, 44(18):14807–14822, 2021. https://doi.org/10.1002/mma.7744.

[21] X. Xie and L. Niu. Heteroclinic cycles of a symmetric May–Leonard competition model with seasonal succession, 2021. https://doi.org/10.48550/arXiv.2110.12390.

[22] X. Xie and M. Chen. Global stability of a tridiagonal competition model with seasonal succession. *Mathematical Biosciences and Engineering*, 20(4):6062–6083, 2023. https://doi.org/10.3934/mbe.2023262.

Part II: Person’s life-sphere

Rationale: 3

Part II: **Person's life-sphere**

Comprehensive perspective on body's internal functional dynamics and physical behavior in pursuit of person's operational goals under permanent exposure to external factors.

Person's life-sphere is decomposed into
- biological body (biosphere)
- environmental surroundings (ecosphere)
- social setting (sociosphere)

Part II-A: **Person's biosphere**

Person's biological body as a complex system physically made from living cells and functionally understood as a nonautarkic self-steering bioautomaton that can move, plan its operations, enact decisions, and reproduce bisexually.

Biosphere is functionally decomposed into the
- cellular system
- vital task-solving functions
- physical task-solving (motor) functions
- cognitive task-solving (operational) functions
- reproductive task-solving functions
- higher-level management and control functions

Part II-A.1: **Body's cellular system**

Exploration of cellular function in formation and maintenance of the body's structures, energizing its functioning and its interactions with the body's outside world.

In this volume:
- Three-dimensional spatiotemporal dynamics of tumor growth
- The relationship between chirality of secondary helical structures and the stability of lysozyme and myoglobin
- Quantitative assessment of chirality of supersecondary protein structures
- Mathematics of supercooperativity in biochemistry

In former volume BIOKYBERNETIKA:
- Mutations as activators of biological processes
- Immunometabolic characterization of T-cells

https://doi.org/10.1515/9783112216620-006

Natalia Levashova, Evgenii Generalov, Alexey Goltsov, and Alla Sidorova

Three-dimensional spatiotemporal dynamics of tumor growth

A numerical study based on a nonlinear wave approach

Abstract: A space–time trigger model of tumor growth is presented, accounting for the concentration of hydrogen ions, oxygen, and tumor cell density at the initial stage of tumor spheroid development. The model is formulated as a system of parabolic equations with a piecewise linear right-hand side of modular type. The model captures both tumor expansion into the surrounding environment and necrotic core formation. This framework allows future extensions to heterogeneous environments and more complex tumorigenesis mechanisms. Furthermore, we investigate how the spheroid's growth dynamics depend on various system parameters, such as diffusion coefficients, glycolysis and proliferation rates, and tumor cell concentration within the proliferative zone. The modeling identified critical cellular environment parameters that govern cancer spheroid growth. The developed model and the results obtained can be used in the development of tumor control therapies.

Keywords: Proliferation, tumor cell density, hypoxia, spheroid area, necrotic core, system of parabolic equations, modular nonlinearity, small parameter

MSC 2020: 35K57, 92C50

1 Introduction

Cancer cells differ from healthy cells in several fundamental aspects:

Dysregulated growth control: Healthy cells proliferate only in response to specific physiological signals and cease division upon reaching the required population size. Cancer cells lose this regulation, dividing continuously and forming tumors.

Acknowledgement: This research was funded by Russian Science Foundation Grant 23-11-00069.The authors are very grateful to an editor who suggested significant additions to the chapter, which made it more structured, as well as for valuable comments for subsequent research on this topic.

Natalia Levashova, Department of Mathematics, Faculty of Physics, Lomonosov Moscow State University, Leninskie gory 1-2, 119991 Moscow, Russia, e-mail: levashovant@physics.msu.ru
Evgenii Generalov, Alla Sidorova, Department of Biophysics, Faculty of Physics, Lomonosov Moscow State University, Leninskie gory 1-2, 119991 Moscow, Russia, e-mails: generals1179@gmail.com, sky314bone@mail.ru
Alexey Goltsov, Systems Markers Ltd., 5 South Charlotte Str., Edinburgh, EH2 4AN, UK, e-mail: alexey.goltsov@gmail.com

https://doi.org/10.1515/9783112216620-007

Altered metabolism: Many cancer cells rely predominantly on aerobic glycolysis (the Warburg effect), even in the presence of oxygen. This adaptation enables rapid energy production and biosynthesis but also acidifies the tumor microenvironment by lowering pH.

Evasion of apoptosis: While healthy cells initiate programmed cell death (apoptosis) in response to DNA damage or cellular stress, cancer cells circumvent these self-destruct mechanisms, allowing survival despite severe genetic lesions.

Loss of differentiation: Healthy cells maintain specialized structures and functions within tissues. Cancer cells frequently lose these characteristics, reverting to a less differentiated state, which facilitates unrestricted proliferation and adaptability.

Invasiveness and metastatic potential: Cancer cells can degrade extracellular matrices, infiltrate adjacent tissues, and disseminate via the bloodstream or lymphatic system, forming secondary tumors (metastases).

Immune evasion: Through surface protein alterations and secretion of immunosuppressive factors, cancer cells can evade immune surveillance, reducing the likelihood of being detected and eliminated by the host's immune system.

Induction of angiogenesis: Unlike healthy cells, which rely on preexisting blood vessels for nutrient and oxygen supply, cancer cells can secrete proangiogenic factors such as vascular endothelial growth factor (VEGF). This stimulates the formation of new blood vessels within the tumor microenvironment, ensuring a continuous supply of nutrients and oxygen to sustain rapid growth and enabling metastasis through enhanced vascular access.

All these biological characteristics interact in complex, nonlinear ways, and make the mathematical modeling of tumor growth dynamics a challenging and multifaceted problem.

Mathematical oncology has developed a wide range of approaches for modeling tumor growth, invasion, and treatment response.

Developing models that describe tumor progression in agreement with experimental data, while allowing variation of parameters representing drug effects, or radiotherapy [30, 17] can contribute to the advancement of new and effective cancer treatment strategies [8, 10].

In silico, tumor spheroid models have emerged as a powerful complement to traditional experimental approaches, offering cost-effective and high-throughput analysis of tumor behavior [12, 24]. Existing models range from simple agent-based simulations tracking individual cell proliferation and death [28] to more complex, continuum-based models solving partial differential equations to describe nutrient diffusion and growth factor signaling [16]. The latter also include reaction–diffusion one-dimensional models that capture the wave-like propagation of tumor fronts [14, 5, 7]. In such models, the diffusion coefficient is often nonlinear and depends on the density of either healthy or tumor cells. While these models have been valuable in exploring the impact of cell-cell interactions, oxygen gradients, and drug penetration on spheroid growth dynamics, they often oversimplify biological processes. Many models rely on idealized geometries and homogenous cell populations, neglecting the spatial heterogeneity and cellular diversity inherent in real tumors [27].

Furthermore, the accuracy of these *in silico* models is heavily dependent on the availability of reliable experimental data for parameterization and validation. Existing models frequently lack detailed representations of intracellular signaling pathways, ECM remodeling, and mechanical forces, which play critical roles in tumor development and response to therapy. Often, these models are computationally expensive, limiting their application to large-scale simulations or parameter sensitivity analyses. Addressing these limitations requires developing more sophisticated models that integrate multiscale data, incorporate realistic representations of cellular and molecular processes, and leverage advanced computational techniques to improve efficiency and scalability.

Our model addresses a critical gap in current mathematical oncology by providing a three-dimensional, spatiotemporal simulation of tumor spheroid growth that incorporates key biophysical factors: cell density, oxygen availability, and pH levelstumor growth!spheroid model!pH levels. Unlike simpler models that often overlook spatial heterogeneity or rely on averaged representations of the microenvironment, our approach explicitly captures the dynamic interplay between these factors, including the formation of necrotic cores, quiescent cell zonestumor growth!spheroid model!quiescent cell zones, and the influence of oxygen gradients on cellular behavior. This level of detail is crucial for realistically simulating the response of tumors to therapeutic interventions, as the spatial distribution of oxygen and pH significantly impacts drug penetration, efficacy, and the development of resistance.

The importance of our model lies in its potential to serve as a versatile platform for exploring novel treatment strategies and optimizing existing protocols. By allowing for the manipulation of key parameters related to oxygen consumption, diffusion rates, and cellular proliferation, we can simulate a wide range of microenvironmental conditions and predict how these conditions impact tumor growth and response to treatment. This capability could accelerate the discovery of targeted therapies that exploit vulnerabilities within the tumor microenvironment and improve the effectiveness of conventional treatments by optimizing drug delivery and minimizing the development of resistance, ultimately contributing to more personalized and effective cancer therapies.

Only a 3D spatiotemporal model can accurately portray the following vital aspects of tumor biology:

1. First, the formation of the necrotic core due to oxygen and nutrient deprivation within the spheroid's center, alongside the proliferative rim exposed to the media.
2. Second, the effects of branching angiogenesis [26, 32], which is a complex three-dimensional process distributing oxygen and nutrients heterogeneously throughout the tumor, leading to differences in cell growth and drug response.
3. Third, the impact of mechanical stresses arising from cell-cell and cell-matrix interactions within the complex architecture of the spheroid.

4. Finally, this level of detail becomes crucial when modeling targeted drug delivery, as the spatial distribution of drugs can only be realistically captured in a three-dimensional framework.

All these phenomena directly impact tumor growth, treatment response, and the evolution of drug resistance, highlighting the necessity of a more complex 3D spatiotemporal approach.

In this study, we propose a three-dimensional system of parabolic reaction–diffusion equations with modular-type nonlinearity to describe the temporal and spatial evolution of tumor spheroids.

2 The mathematical model of tumor spheroid development

2.1 Model characterization

We consider a spatiotemporal trigger model of tumor growth that accounts for the dependence on

- hydrogen ion concentration $[H^+]$,
- oxygen concentration $[O_2]$,
- normalized tumor cell density u during the initial stages of tumor spheroid development

as proposed in [19].

The modeling object is a tumor spheroid cultivated in a Petri dish [12, 24].

2.2 Biological rationale

In normal conditions, a process called oxidative phosphorylation extracts energy from broken-down nutrients to generate adenosine triphosphate (ATP) in the mitochondria of human-body cells. ATP, the cell's energy currency, is essential for a cell's survival and for replication. Hence, sufficient oxygen concentrations are also essential for tumor cell proliferation. With shortage of oxygen, cells temporarily have an option to switch to energy generation by glycolysis, but cells' survival time without oxygen varies dramatically by cell type, being shortest in cell types of high energy demand, like brain cells and tumor cells in proliferation.

The formation of a necrotic core in a tumor occurs as the rapidly increasing number of tumor cells leads to inadequate blood supply inside the tumor, causing hypoxia (lack of oxygen) and nutrient deprivation in cells located at the center. These hypoxic cells shift to glycolytic metabolism for energy, producing large amounts of acidic metabolites like lactate. If the accumulating acid (lowered pH) and lack of oxygen persist, they eventually cause tumor cell death through necrosis in the center of the tumor mass [18, 11]. However, as the number of cells increases, those located within the tumor experience hypoxia, at which cells occur with a decrease in pH ultimately resulting in tumor cell death.

A distinctive feature of the model is that it captures both the outward expansion of the tumor into the surrounding medium and the formation of a necrotic core at the tumor center.

2.3 Model parameters

The model accounts for the coupled dynamics of cell proliferation, oxygen and hydrogen ion concentrations, and necrotic core formation.

- T is the characteristic time scale,
- T^{-1} is a rate factor,
- D denotes the diffusion coefficient—a parameter controlling how quickly and broadly cells disperse in space within the PDE model [13], measured in L^2 per time T, where L is the characteristic length scale.
- u denotes the tumor cell density in fluid on a Petri dish as medium surrounding cells; it is normalized to 1;
- $[H^+]$ is the hydrogen cation concentration that determines the acidity, $\mathrm{pH}_{\mathrm{medium}}$, in surrounding fluid on the Petri dish according to the formula $\mathrm{pH}([H^+]) = -\log_{10}[H^+]$; $[H^+]$ is measured in M (mol/L) for pH.
- $[O_2]$ is the oxygen concentration, measured in concentration units (often in M, μM, or mmHg).
- k_1 is the cell division rate coefficient, with dimensions T^{-1},
- k_2 is the glycolysis rate coefficient, with dimensions T^{-1},
- k_3 is proportional to the oxygen uptake rate during cell division, with dimensions T^{-1},
- τ is delay time until glycolysis begins.

At the initial moment, a Petri dish is inoculated with an agglomerate of approximately 2×10^4 cells.

2.4 Specification of the model

Further, we investigate how the spheroid's growth dynamics depend on various system parameters, such as diffusion coefficients, glycolysis and proliferation rates, and tumor cell concentration within the proliferative zone.

To accurately quantify the impact of parameter changes on tumor spheroid structure, we refined our definitions of the necrotic core and proliferative zone.

Necrotic core: The necrotic core was defined as all spatial locations within the spheroid where the cell density exceeded a predefined threshold (u-necrotic), representing nonviable cells due to crowding or oxygen deprivation, alongside an oxygen threshold.

Proliferative zone: Conversely, the proliferative zone was identified as spatial locations within the spheroid exhibiting oxygen concentrations, denoted as $[O_2]$, above the hypoxic threshold ($[O_2]$-hypoxic) and cell densities below the proliferation inhibition threshold (u_{thr}).

Implementing these definitions on our discretized simulation grid necessitated a method for determining the spheroid boundary, achieved through a cell density contour.

The precise thresholds (u-necrotic, $[O_2]$-hypoxic, and u_{thr}) were carefully selected based on biological reasoning and simulation sensitivity to thresholds, impacting regions' volumes. These detailed definitions and implementation steps are essential for rigorous and reproducible analysis of the model's spatial dynamics, allowing for more precise characterization of tumor response to microenvironmental factors.

The specific thresholds (u-necrotic, $[O_2]_{\text{hyp}}$, and u_{thr}) were meticulously chosen based on a blend of biological rationale and sensitivity analysis derived from simulations. u-necrotic, the cell density threshold for necrosis, was chosen with reference to literature [4], which indicates that cell densities exceeding X cells/mL typically induce necrosis due to limitations in nutrient diffusion, thus modeling extreme crowding, which impairs viability. $[O_2]_{\text{hyp}} = 0.01$ mM is the oxygen concentration threshold for hypoxia and was selected with regard to existing literature that defines hypoxia in similar cell types as oxygen concentrations below X mmHg (or Y μM), initiating the cells' transition to anaerobic metabolism and consequent cell death. u_{thr} is the threshold for proliferation, reflected the point at which cell-cell contact inhibition and competition begin to limit growth [20].

To evaluate each threshold, we performed a parameter sweep analysis, varying each threshold across plausible ranges while maintaining other parameters at fixed values, and calculating and measuring the ensuing changes in necrotic core volume, proliferative zone volume, and spheroid volume. The model showed a particular sensitivity to hypoxic conditions concerning $[O_2]$ levels, which influenced the size of the necrotic core, underscoring the critical need for an appropriate balance between oxygen availability and consumption. Furthermore, an increase in the threshold for proliferation (u_{thr}) was found to reduce the size of the proliferative zone, which in turn slowed the expansion of the spheroid volume.

To describe the processes of tumor expansion into the surrounding medium and cell necrosis within the spheroid, we employ a system of parabolic equations with a piecewise-linear right-hand side of the so-called modular type [19], namely equations of the form

$$\frac{\partial u}{\partial t} - D_u \Delta u = -\begin{cases} T^{-1}u, & [H^+] \geq [H^+]_{\text{thr}} \\ \begin{cases} T^{-1}(u - u_{\max}), & u \geq u_{\text{thr}} \\ -k_1 u[O_2], & u < u_{\text{thr}} \end{cases}, & [H^+] < [H^+]_{\text{thr}} \end{cases} \tag{4.1}$$

$$\frac{\partial [H^+]}{\partial t} - D_{\text{H}} \Delta [H^+] = -\begin{cases} T^{-1}([H^+] - [H^+]_{\max}), & [H^+] \geq [H^+]_{\text{thr}} \\ \begin{cases} -k_2 u(t-\tau), & [O_2] < [O_2]_{\text{hyp}} \\ T^{-1}[H^+], & [O_2] \geq [O_2]_{\text{hyp}} \end{cases}, & [H^+] < [H^+]_{\text{thr}} \end{cases} \tag{4.2}$$

$$\frac{\partial [O_2]}{\partial t} - D_{O_2} \Delta [O_2] = -\begin{cases} T^{-1}([O_2] - [O_2]_{\min}), & [O_2] \leq [O_2]_{\text{thr}} \\ T^{-1}([O_2] - S[O_2]_{\max}) + k_3 u[O_2], & [O_2] > [O_2]_{\text{thr}} \end{cases} \tag{4.3}$$

for dynamics of tumor cell density, pH concentrations, and oxygen concentrations in the medium, respectively.

Equation (4.1) governs the dynamics of tumor cell density. The sink term, $-T^{-1}u$ in the first line of the right-hand side of Equation (4.1) models a function identically equal to zero in the region where $[H^+] > [H^+]_{\text{thr}}$, meaning that the pH level is unsuitable for cell survival. This situation is characteristic of the necrotic core of a tumor. If the oxygen concentration is sufficient and the cell density is below a certain threshold value u_{thr}, active cell proliferation occurs [12] that is described by the term $k_1 u[O_2]$, where k_1 is the cell division rate coefficient. When the cell density reaches the threshold value ($u = u_{\text{thr}}$), cell division ceases, and the region $u \geq u_{\text{thr}}$ attains the maximum density $u_{\max} = 1$. This saturation state is modeled by the source term $T^{-1}(u - u_{\max})$ in the second line of the right-hand side of Equation (4.1).

Equation (4.2) describes pH dynamics. In cells inoculated at the initial moment, anaerobic glycolysis is not immediately activated, but after a short delay, which is accounted for in the equation by the time-delayed argument $u(t-\tau)$ in the term $k_2 u(t-\tau)$, responsible for glycolysis. Here, k_2 is the glycolysis rate coefficient with dimensions T^{-1}. The delay time τ represents the time before glycolysis begins in inoculated cells and is inversely proportional to the glycolysis constant, i. e., $\tau = 1/k_2$. Anaerobic glycolysis occurs under hypoxic conditions [29], i. e., when the oxygen concentration is below the threshold $[O_2]_{\text{hyp}}$. If the oxygen concentration exceeds this threshold, glycolysis does not occur, and the ion concentration $[H^+]$ remains minimal ($\approx 10^{-7}$ M), as modeled by the third line in the right-hand side of Equation (4.2), which represents a zero solution. The first line in the right-hand side of Equation (4.2) models the attainment of the maximum $[H^+]$ concentration in the necrotic region (pH $\approx$ 3).

Equation (4.3) describes oxygen concentration dynamics. The term $-T^{-1}([O_2] - S[O_2]_{\max}) - k_3 u[O_2]$ in the second line of the right-hand side of Equation (4.3) models oxygen consumption by proliferating cells when the oxygen concentration is sufficiently high ($[O_2] > [O_2]_{\text{thr}}$). The multiplier S is introduced to adjust the volume of oxygen supply from the outside. The first line in the right-hand side of Equation (4.3) models the attainment of the minimum possible concentration $[O_2]_{min}$ in regions of maximum cell density.

3 Results

3.1 Model calibration and parameter selection

To determine the characteristic spatial and temporal scales, we used data from [12], where an experiment was conducted on the growth of neuroblastoma cells in a Petri dish. For the numerical experiment based on Equations (4.1)–(4.3), the computational domain was chosen as a cube with edge length $L = 0.1$ mm.

The initial cell distribution was modeled as a homogeneous spheroid of diameter 0.014 mm at the maximum possible cell density ($u_{max} = 1$). The initial oxygen concentration was set to $[O_2]_{max} = 0.0556$ mM [12] in the cell-free region and $[O_2]_{min} = 0.05[O_2]_{max}$ in the initial cell agglomeration region [1]. The initial proton concentration $[H^+]$ was set to zero, corresponding approximately to pH = 7.2–7.4. Homogeneous Neumann boundary conditions were imposed on all boundaries of the computational domain.

The pH threshold below which tumor cell necrosis occurs was set to 3.022, and the minimal possible pH value was set to 3, corresponding to a proton concentration of 1 mM in a saturated aqueous solution. The oxygen concentration threshold defining hypoxic conditions was $[O_2]_{hyp} = 0.2[O_2]_{max}$ [3]. The oxygen concentration threshold for active cell proliferation was set to $[O_2]_{thr} = 0.036$ mM. This value was chosen based on formula for the front propagation speed from [25]: to model an oxygen concentration profile with a step-like distribution (minimum at the tumor center, maximum far from the cell cluster, and a steep gradient at the tumor edge), the threshold must exceed the average of the minimum and maximum oxygen concentrations. The threshold value u_{thr} in this paper is a variable parameter.

Experimental data from [12] show that the initial tumor diameter was 0.01 mm, and after 7 days, the spheroid surface area exceeded 0.01 mm^2. By matching the tumor growth time from experiments with the dimensionless time in the model under similar initial conditions, we obtain $T = 19$ hours for the model's dimensionless time unit.

As noted in [25], traveling-front-type solutions in equations with modular nonlinearity occur only for sufficiently small coefficients of the second spatial derivative. Numerical experiments yielded the following coefficients: $D_{O_2} \sim 0.0001\,\mathrm{L^2/T} \approx 1.5\times10^{-11}\,\mathrm{mm^2/s}$, $D_H \sim 0.0001\,\mathrm{L^2/T} \approx 1.5\times10^{-11}\,\mathrm{mm^2/s}$, $D_u \sim 0.00001\,\mathrm{L^2/T} \approx 1.5\times10^{-12}\,\mathrm{mm^2/s}$. It should be noted that, in models employing a nonlinear diffusion coefficient for the cells, its value becomes negligibly small at the boundary of the growth zone. This is consistent with the low value of the cell diffusion coefficient, D_u chosen in our model. In many models commonly found in the literature [31, 12], the diffusion equations typically employ oxygen and cation diffusion coefficients on the order of $10^{-2}\,\mathrm{mm^2/s}$.

In [22], the experimentally determined oxygen diffusion coefficient in multicellular spheroids was reported to be 10^{-3} mm^2/s. This value was also adopted in the models described in [21, 3]. The difference of eight orders of magnitude between these values and the diffusion coefficients used in our model arises from the normalization of the cell density to its maximum attainable value, which can reach 3×10^8 cells/mL [9].

Numerical experiments have shown that the model is very sensitive to variations in the diffusion coefficients (see Section 4).

The coefficient k_1, representing the cell division rate, was chosen based on the doubling time of the cell population (12–24 hours [12]). Since one model time unit equals 19 hours, $k_1 = 1$ corresponds to one cell division every 19 hours.

The coefficient k_2 represents the glycolysis rate. Aerobic glycolysis occurs within 2–3 hours [6]; given that one model time unit equals 19 hours, k_2 should be in the range $6\,T^{-1} - 8\,T^{-1}$. Numerical simulations used $k_2 = 6$.

The coefficient k_3 represents cellular oxygen homeostasis. Physically, it is proportional to the oxygen uptake rate during cell division. In the model, k_3 was chosen based on the ratio between oxygen levels in the organoid corresponding to $p_{O_2} = 20$–60 mmHg (typical in the culture medium) and $p_{O_2} = 5$–20 mmHg (hypoxic state) [1]. Numerical simulations were implemented for various values of parameter k_3 within the physically admissible range for this ratio.

3.2 The results of model calculations for different parameter values

In this study, we present a mathematical model of tumor spheroid development aimed at investigating the effects of various parameters on its growth dynamics and structural organization. Particular emphasis is placed on examining the role of oxygen consumption, diffusion coefficients, and the threshold cell concentration for proliferation.

The simulation results indicate that the parameters k_1 and k_2, varied within the ranges $0.6 \le k_1 \le 1.2$ and $6 \le k_2 \le 8$, respectively, have negligible impact on the overall tumor growth dynamics. This demonstrates the robustness of the model with respect to variations in these parameters.

In contrast, parameters related to oxygen concentration—most notably k_3 ($1 \le k_3 \le 12$), which defines the oxygen consumption rate during cell proliferation—exert the greatest influence on spheroid dynamics. An increase in k_3 leads to an expansion of the quiescent cell region (see Figure 4.1), with the width of this region being particularly sensitive to k_3 values.

In the numerical experiments, the results of which were presented below, we used the following set of fixed parameters.

Figure 4.1(a)–(c), (d)–(f), and (g)–(i) illustrate changes in the central cross-section of the spheroid at 2 days post-inoculation ((a), (d), (g)), at 5 days ((b), (e), (h)) and at 7 days

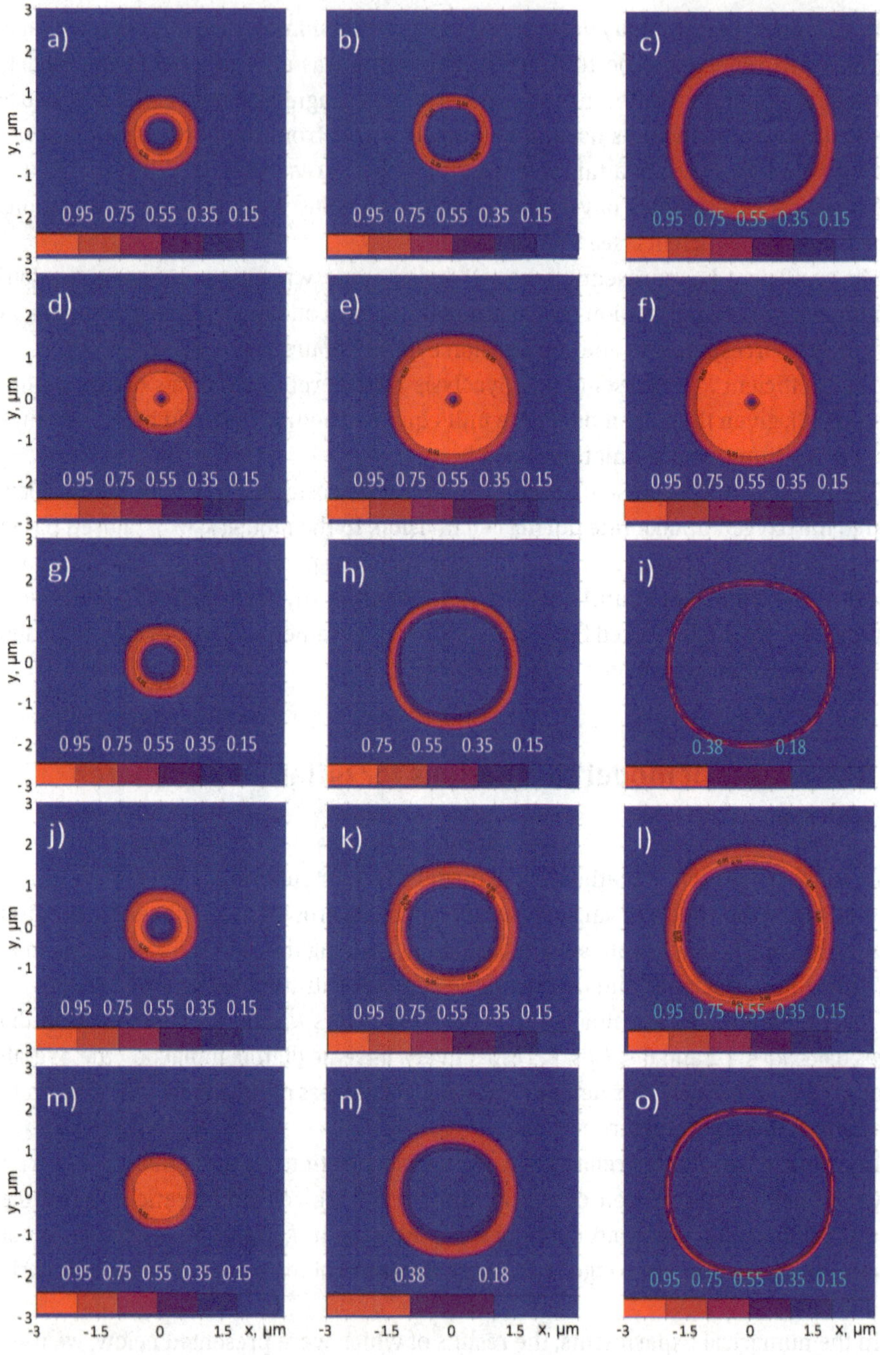

Figure 4.1: The dynamics of the tumor spheroid growth in the central cross-section at 2 days post-inoculation in displays a), d), g), j), m), at 5 days post-inoculation in displays b), e), h), k), n), and at 7 days post-inoculation in displays c), f), i), l), o) – all with the fixed set of parameters in Tab. 4.1, with fixed $D_u = 0.00001\,L^2/T$, $u_{thr} = 0.05$ and variation of other parameters according to Tab. 4.2.

Table 4.1: Fixed parameter values in calculations of the 3D spheroid growth model in (4.1)–(4.3).

u_{max}	k_1	k_2	L	D_H
1	1	6	0.1 mm	0.0001 L^2/T
$[O_2]_{max}$	$[O_2]_{min}$	$[O_2]_{hyp}$	$[O_2]_{thr}$	$[H]_{thr}$
0.0556 mM	0.00278 mM	0.0112 mM	0.036 mM	0.6 mM

Table 4.2: Parameter variation in calculations of the 3D spheroid growth model in (4.1)–(4.3).

Parameter	Units	Number of Days After Inoculation			Figure 4.1
		2 days	5 days	7 days	Displays
S		1	1	1	a, b, c
k_3		4	4	4	a, b, c
D_{O_2}	L^2/T	0.0002	0.0002	0.0002	a, b, c
S		1	1	1	d, e, f
k_3		1	1	1	d, e, f
D_{O_2}	L^2/T	0.0002	0.0002	0.0002	d, e, f
S		1	1	1	g, h, i
k_3		5	5	5	g, h, i
D_{O_2}	L^2/T	0.0002	0.0002	0.0002	g, h, i
S		2	2	2	j, k, l
k_3		6	6	6	j, k, l
D_{O_2}	L^2/T	0.0002	0.0002	0.0002	j, k, l
S		1	1	1	m, n, o
k_3		4	4	4	m, n, o
D_{O_2}	L^2/T	0.0004	0.0004	0.0004	m, n, o

((c), (f), (i)) for the set of fixed parameters in Table 4.1, $S = 1$, $D_{O_2} = 0.0002\,L^2/T$, $D_u = 0.00001\,L^2/T$, $u_{thr} = 0.05$, and different k_3 values.

The plots in Figure 4.1(a)–(c) correspond to $k_3 = 4$. These results highlight the critical role of the oxygen gradient in determining spheroid structure and growth dynamics. The model reveals that k_3, representing the oxygen consumption rate, has a substantial effect on necrotic core formation and the extent of the quiescent zone. This finding aligns with fundamental concepts of tumor heterogeneity, whereby hypoxic processes in the tumor core trigger metabolic adaptations, including a shift to anaerobic glycolysis and the emergence of therapy resistance.

The observed increase in necrotic area at higher k_3 values reflects real tumor behavior, where highly proliferative cells consume oxygen more rapidly than can be replenished by diffusion, leading to central necrosis (Figure 4.1(a)–(c): $k_3 = 4$). Conversely, low k_3 values $k_3 = 1$ correspond to enhanced hypoxia tolerance, allowing cells to survive under extreme conditions (see Figure 4.1(d)–(f): $k_3 = 1$). A decrease in the oxygen consumption coefficient from $k_3 = 4$ to $k_3 = 1$ induces a different traveling wave regime, corresponding to spheroid growth without necrotic core formation (nonnecrotic tumor regime) [2].

Furthermore, the introduction of an additional parameter S into the model enables the simulation of angiogenesis, which supplies oxygen and nutrients required for sustained tumor growth; cf. Figure 4.1(j)–(l): $S = 2$, $k_3 = 6$, $D_{O_2} = 0.0002\,L^2/T$, $D_u = 0.00001\,L^2/T$, and $u_{thr} = 0.05$, while the remaining parameters take values from the fixed set in Table 4.1.

Simulations show that without an additional oxygen source ($S = 1$), spheroid growth is limited, and cell death begins after approximately 5 days (Figure 4.1(g)–(i): $S = 1$, $k_3 = 5$, $D_{O_2} = 0.0002\,L^2/T$, $D_u = 0.00001\,L^2/T$, $u_{thr} = 0.05$) are consistent with *in vivo* angiogenesis data [15].

Increasing oxygen diffusion in the model alters spheroid structure: at sufficiently high diffusion rates, no necrotic core forms; however, as spheroid volume increases, hypoxia emerges, eventually leading to necrosis (Figure 4.1(m)–(o): $S = 1$, $k_3 = 4$, $D_{O_2} = 0.0004\,L^2/T$, $D_u = 0.00001\,L^2/T$, and $u_{thr} = 0.05$).

This underscores the importance not only of oxygen availability but also of its effective delivery to tumor cells. High oxygen diffusion may act as either a favorable or unfavorable factor for tumor growth, depending on tumor size, cell density, and cellular oxygen consumption rates. At elevated concentrations, oxygen can also inhibit cell growth by oxidizing molecular structures, participating in oxidative stress, and initiating other cytotoxic processes.

Doubling diffusion coefficient D_u causes an increase in spheroid growth rate from 2.5 μm/day to 1 μm/day (Figure 4.2 a)–c): $S = 1$, $k_3 = 4$, $D_{O_2} = 0.0002\,L^2/T$, $D_u = 0.00002\,L^2/T$, and $u_{thr} = 0.05$). This dependence of spheroid growth rate is consistence with other results of tumor spheroid growth [23].

The modeling further reveals that the threshold cell concentration required for active proliferation is an important determinant of tumor growth rate (Figure 4.2 d)–f): $S = 2$, $k_3 = 4$, $D_{O_2} = 0.0002\,L^2/T$, $D_u = 0.00001\,L^2/T$, and $u_{thr} = 0.1$, g)–i): $S = 2$, $k_3 = 4$, $D_{O_2} = 0.0002\,L^2/T$, $D_u = 0.00001\,L^2/T$, and $u_{thr} = 0.4$).

High cell density in the proliferative zone slows tumor expansion, likely due to competition for resources (oxygen and nutrients) and accumulation of metabolic byproducts.

This result reflects the complex interplay between cell proliferation, metabolism, and the tumor microenvironment. High cell density may also increase intercellular contact and mechanical stress, further restricting proliferation (Figure 4.2(g)–(i)).

Notably, cells adapted to hypoxia (low k_3) can continue to proliferate even at high cell densities without supplemental oxygen, highlighting the adaptive capacity of tumor cells under adverse conditions (Figure 4.2(j)–(l): $S = 1$, $k_3 = 1$, $D_{O_2} = 0.0002\,L^2/T$, $D_u = 0.00001\,L^2/T$, and $u_{thr} = 0.1$).

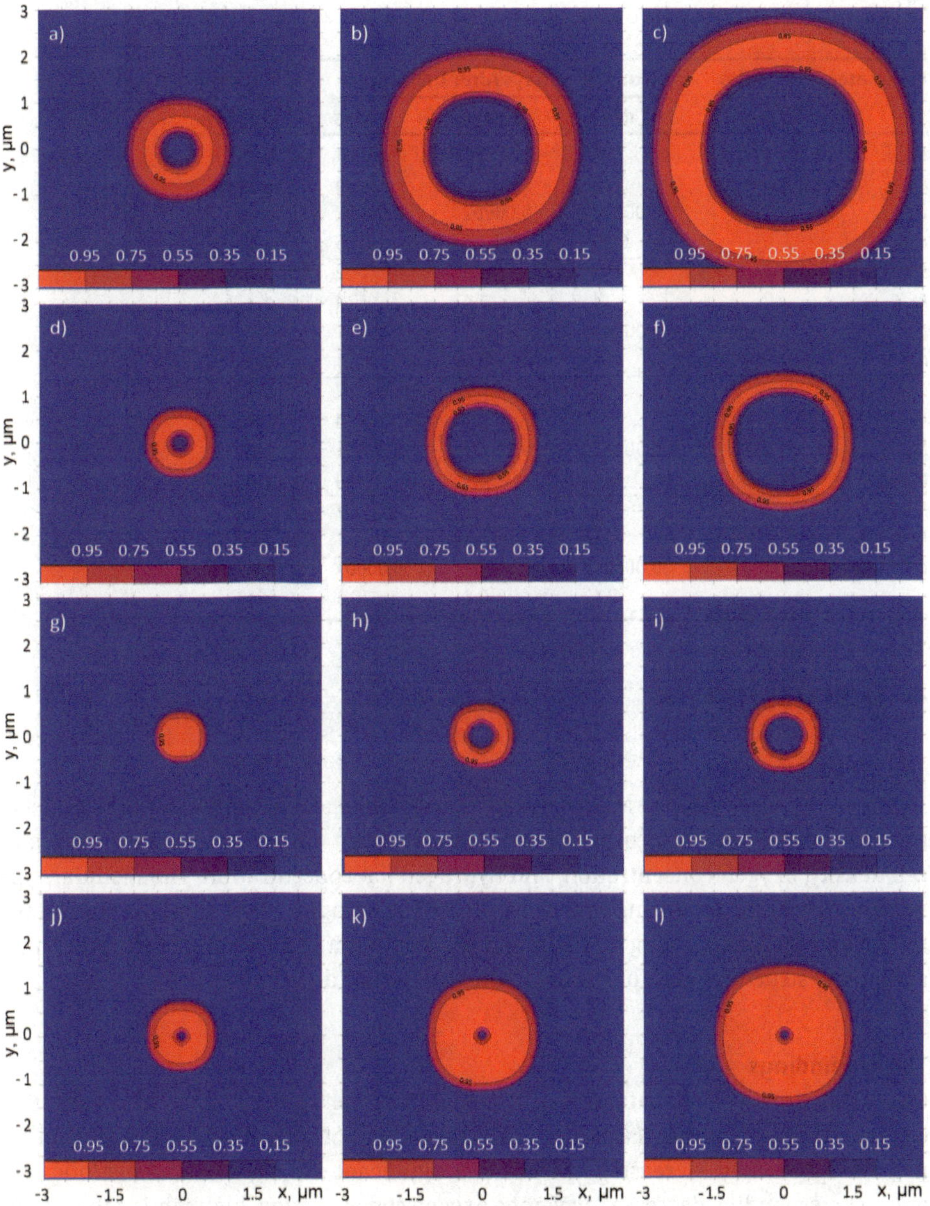

Figure 4.2: The dynamics of the tumor spheroid growth in the central cross-section at 2 days post-inoculation in displays a), d), g), j), at 5 days post-inoculation in displays b), e), h), k) and at 7 days post-inoculation in displays c), f), i), l) – all with the fixed set of parameters in Tab. 4.1, with fixed D_{O_2} = 0.0002 L^2/T and variation of other parameters according to Tab. 4.3.

Table 4.3: Parameter variation in calculations of the 3D spheroid growth model in (4.1)–(4.3).

Parameter	Units	Number of Days After Inoculation			Figure 4.2
		2 days	5 days	7 days	Displays
S		1	1	1	a, b, c
k_3		4	4	4	a, b, c
D_u	L^2/T	0.00002	0.00002	0.00002	a, b, c
u_{thr}		0.05	0.05	0.05	a, b, c
S		2	2	2	d, e, f
k_3		4	4	4	d, e, f
D_u	L^2/T	0.00001	0.00001	0.00001	d, e, f
u_{thr}		0.1	0.1	0.1	d, e, f
S		2	2	2	g, h, i
k_3		4	4	4	g, h, i
D_u	L^2/T	0.00001	0.00001	0.00001	g, h, i
u_{thr}		0.4	0.4	0.4	g, h, i
S		1	1	1	j, k, l
k_3		1	1	1	j, k, l
D_u	L^2/T	0.00001	0.00001	0.00001	j, k, l
u_{thr}		0.1	0.1	0.1	j, k, l

4 Discussion

The proposed model

In this study, we developed and analyzed a three-dimensional spatiotemporal model of tumor spheroid growth incorporating coupled reaction–diffusion equations for tumor cell density, oxygen concentration, and hydrogen ion concentration. The modular-type nonlinearity in the governing equations allowed us to capture both the expansion of the tumor into the surrounding medium and the formation of a necrotic core, thus reproducing key structural and functional features of tumor development.

The key findings

Our three-dimensional spatiotemporal model of tumor spheroid growth reveals that oxygen consumption rate (k_3) is a critical determinant of spheroid architecture and dynamics. Specifically, increasing k_3 leads to a significant expansion of the necrotic core and a corresponding decrease in average oxygen concentration, highlighting the importance of balancing oxygen supply and demand. Furthermore, reducing the threshold cell concentration for proliferation (u_{thr}) enhances cell division in the outer regions, influencing overall spheroid growth rate and highlighting the role of cell density-dependent inhibition.

These findings emphasize the sensitivity of spheroid growth to parameters governing oxygen metabolism and cell density and underscore their significance for understanding tumor heterogeneity and therapeutic response. By quantitatively linking these parameters to specific structural features of the spheroid, our model provides a valuable tool for investigating targeted treatment strategies, optimizing drug delivery, and predicting the impact of microenvironmental factors on tumor evolution and drug resistance.

Model limitations

The tumor spheroid model presented here, while valuable, remains simplified and requires further development to incorporate additional parameters that are realized in the tumorigenic process *in vivo*, such as angiogenesis, immune response, drug treatment effects, and hormonal regulation. It does not account for the heterogeneity of the tumor cell population, the interactions between different cell types (e. g., immune cells and fibroblasts), or the influence of the extracellular matrix and other components of the tumor microenvironment. Moreover, the model does not consider genetic instability and evolutionary variability of tumor cells, which may lead to further transformations of malignant cells, alterations in their metabolic requirements, and changes in their sensitivity to hypoxia. The absence of these factors limits the direct applicability of the simulation results to more complex *in vivo* systems and clinical contexts.

5 Conclusion

The three-dimensional spatiotemporal model of tumor spheroid growth provides a flexible framework for investigating tumor growth dynamics under varying microenvironmental conditions. Its ability to account for oxygen and pH heterogeneity, necrosis, and angiogenesis-related effects makes it a promising tool for exploring therapeutic strategies, optimizing treatment protocols, and improving our understanding of tumor progression mechanisms.

Bibliography

[1] J. A. Bertout, S. A. Patel, and M. C. Simon. The impact of o_2 availability on human cancer. *Nature Reviews. Cancer*, 8(12):967–975, 2008.

[2] H. M. Byrne and M. A. J. Chaplain. Growth of nonnecrotic tumors in the presence and absence of inhibitors. *Mathematical Biosciences*, 130:151–181, 1995. https://doi.org/10.1016/0025-5564(94)00117-3.

[3] A. Caiazzo and I. Ramis-Conde. Multiscale modeling of palisade formation in glioblastoma multiforme. *Journal of Theoretical Biology*, 383:145–156, 2015. https://doi.org/10.1016/j.jtbi.2015.07.021.

[4] C. Carmona-Fontaine, V. Bucci, L. Akkari, M. Deforet, J. A. Joyce, and J. B. Xavier. Emergence of spatial structure in the tumor microenvironment due to the Warburg effect. *Proceedings of the National Academy of Sciences of the United States of America*, 110(48):19402–19407, 2013. https://doi.org/10.1073/pnas.1311939110.

[5] C. Colson, F. Sánchez-Garduño, H. M. Byrne, P. K. Maini, and T. Lorenzi. Travelling-wave analysis of a model of tumour invasion with degenerate, cross-dependent diffusion. *Proceedings of the Royal Society A. Mathematical, Physical and Engineering Sciences*, 477(2256), 2021. https://doi.org/10.1098/rspa.2021.0593.

[6] DB DifferenceBetrween.net. Difference between aerobic and anaerobic glycolysis. https://www.differencebetween.net/science/health/difference-between-aerobic-and-anaerobic-glycolysis/.

[7] M. El-Hachem, S. W. McCue, and M. J. Simpson. Travelling wave analysis of cellular invasion into surrounding tissues. *Physica D: Nonlinear Phenomena*, 428:133026, 2021. https://doi.org/10.1016/j.physd.2021.133026.

[8] C. Fornari, F. Cordero, D. Manini, G. Balbo, and R. A. Calogero. Mathematical approach to predict the drug effects on cancer stem cell models. *Electronic Notes in Theoretical Computer Science*, 277:29–39, 2011. https://doi.org/10.1016/j.entcs.2011.09.033.

[9] J. P. Freyer and R. M. Sutherland. A reduction in the in situ rates of oxygen and glucose consumption of cells in emt6/ro spheroids during growth. *Journal of Cellular Physiology*, 124(3):516–524, 1985.

[10] M. Frieler, C. Pho, B. H. Lee, H. Dobrovolny, G. Akkaraju, and A. Naumov. Effects of doxorubicin delivery by nitrogen-doped graphene quantum dots on cancer cell growth: Experimental study and mathematical modeling. *Nanomaterials*, 11, 01 2021. https://doi.org/10.3390/nano11010140.

[11] Z. gang Liu and D. Jiao. Necroptosis, tumor necrosis and tumorigenesis. *Cell Stress*, 4(1):1–8, 2019. https://doi.org/10.15698/cst2020.01.208.

[12] S. Hervas-Raluy, B. Wirthl, P. E. Guerrero, G. R. Rei, et al. Tumour growth: An approach to calibrate parameters of a multiphase porous media model based on in vitro observations of neuroblastoma spheroid growth in a hydrogel microenvironment. *Computers in Biology and Medicine*, 159:106895, 2023. https://doi.org/10.1016/j.compbiomed.2023.106895.

[13] A. M. Jarrett, E. ABF Lima, D. A. Hormuth II., M. T. McKenna, X. Feng, D. A. Ekrut, A. C. M. Resende, A. Brock, and T. E. Yankeelov. Mathematical models of tumor cell proliferation: A review of the literature. *Expert Review of Anticancer Therapy*, 18(12):1271–1286, 2018. https://doi.org/10.1080/14737140.2018.1527689.

[14] J. M. Jepson, R. D. O'Dea, J. Billingham, and N. T. Fadai. Pattern formation and travelling waves in a multiphase moving boundary model of tumour growth. *Mathematical Medicine and Biology: A Journal of the IMA*, 40(4):327–347, 2023. https://doi.org/10.1093/imammb/dqad008.

[15] M. P. Kiseleva, I. S. Golubeva, V. P. Deryagina, A. V. Lantsova, L. V. Ektova, E. A. Kornyushenkov, L. M. Borisova, and E. A. Generalov. Influence of polysaccharide from *Helianthus tuberosus* l. on antiproliferative activity of n-glycoside indolo[2,3-a]carbazole derivative lcs-1269. *Biofizika*, 69:1109, 2024. https://doi.org/10.31857/S0006302924050164.

[16] M. Kuznetsov and A. Kolobov. Investigation of solid tumor progression with account of proliferation/migration dichotomy via Darwinian mathematical model. *Journal of Mathematical Biology*, 80(3):601–626, 2020. https://doi.org/10.1007/s00285-019-01434-4.

[17] M. B. Kuznetsov, V. V. Gubernov, and A. V. Kolobov. Analysis of anticancer efficiency of combined fractionated radiotherapy and antiangiogenic therapy via mathematical modelling. *Russian Journal of Numerical Analysis and Mathematical Modelling*, 33(4):225–242, 2018. https://doi.org/10.1515/rnam-2018-0019.

[18] S. Y. Lee, M. K. Ju, H. M. Jeon, E. K. Jeong, and Y. J. Lee. H. G. Park, C. H. Kim, S. I. Han, and H. S. Kang. Regulation of tumor progression by programmed necrosis. *Oxidative Medicine and Cellular Longevity*, 2018:3537471, 2018. https://doi.org/10.1155/2018/3537471.

[19] N. T. Levashova, E. A. Generalov, A. E. Sidorova, and A. N. Goltsov. Modular-type nonlinearity in the modeling of tumor spheroid growth. *Theoretical and Mathematical Physics*, 224(1):1202–1210, 2025. https://doi.org/10.1134/S0040577925070086.

[20] S. Mangani, S. Kremmydas, and N. K. Karamanos. Mimicking the complexity of solid tumors: How spheroids could advance cancer preclinical transformative approaches. *Cancers (Basel)*, 17(7):1161, 2025. https://doi.org/10.3390/cancers17071161.

[21] A. Martínez-González, G. F. Calvo, L. A. Pérez-Romasanta, and V. M. Pérez-García. Hypoxic cell waves around necrotic cores in glioblastoma: A biomathematical model and its therapeutic implications. *Bulletin of Mathematical Biology*, 74(12):2875–2896, 2012. https://doi.org/10.1007/s11538-012-9786-y.

[22] W. F. Mueller-Klieser and R. M. Sutherland. Oxygen consumption and oxygen diffusion properties of multicellular spheroids from two different cell lines. *Advances in Experimental Medicine and Biology*, 180:311–321, 1984. https://doi.org/10.1007/978-1-4684-4895-5_30.

[23] H. M. Narayana et al. In silico estimates of the free energy rates in growing tumor spheroids. *Journal of Physics. Condensed Matter*, 22(19):194122, 2010. https://doi.org/10.1088/0953-8984/22/19/194122.

[24] H. Narayanan. In silico estimates of the free energy rates in growing tumor spheroids. *Journal of Theoretical Biology*, 524:110739, 2021. https://doi.org/10.1016/j.jtbi.2021.110739.

[25] N. N. Nefedov, E. I. Nikulin, and A. O. Orlov. Front motion in a problem with weak advection in the case of a continuous source and a modular-type source. *Differential Equations*, 58(6):757–770, 2022.

[26] S. M. Peirce, E. J. Van Gieson, and T. C. Skalak. Multicellular simulation predicts microvascular patterning and in silico tissue assembly. *The FASEB Journal*, 18(6):731–733, 2004. https://doi.org/10.1096/fj.03-0933fje.

[27] I. Ramis-Conde, M. A. J. Chaplain, and A. R. A. Anderson. Mathematical modelling of cancer cell invasion of tissue. *Mathematical and Computer Modelling*, 47(5–6):533–545, 2008. https://doi.org/10.1016/j.mcm.2007.02.034.

[28] R. Sulimanov, K. Koshelev, V. Makarov, A. Mezentsev, M. Durymanov, L. Ismail, K. Zahid, Y. Rumyantsev, and I. Laskov. Mathematical modeling of non-small-cell lung cancer biology through the experimental data on cell composition and growth of patient-derived organoids. *Life*, 13(11):2228, 2023. https://doi.org/10.3390/life13112228.

[29] C. Ward, J. Meehan, M. E. Gray, A. F. Murray, et al. The impact of tumour pH on cancer progression: Strategies for clinical intervention. *Exploratory Target Antitumor Therapy*, 1:71–100, 2020. https://doi.org/10.37349/etat.2020.00005.

[30] Y. Watanabe, E. L. Dahlman, K. Z. Leder, and S. K. Hui. A mathematical model of tumor growth and its response to single irradiation. *Theoretical Biology and Medical Modelling*, 13:6, 2016. https://doi.org/10.1186/s12976-016-0032-7.

[31] J. West, F. Rentzeperis, C. Adam, R. Bravo, K. A. Luddy, M. Robertson-Tessi, and A. R. A. Anderson. Tumor-immune metaphenotypes orchestrate an evolutionary bottleneck that promotes metabolic transformation. *Frontiers in Immunology*, 15:1323319, 2024. https://doi.org/10.3389/fimmu.2024.1323319.

[32] H. Yanagisawa, M. Sugimoto, and T. Miyashita. Mathematical simulation of tumour angiogenesis: Angiopoietin balance is a key factor in vessel growth and regression. *Scientific Reports*, 11(1):419, 2021. https://doi.org/10.1038/s41598-020-79824-8.

Aleksey Lutsenko and Alla Sidorova

Quantitative assessment of chirality of supersecondary protein structures

Abstract: Despite the rapid development of neural network mechanisms for determining the eventual protein structure by amino acid sequence, the crucial physical and biophysical relationships between protein structure and function remain generally unknown. It can be extremely difficult to produce the optimal structure to achieve the desired functionality. To simplify this task, it is necessary to study the various biophysical characteristics of proteins in general, as a class of molecules. In particular, this paper proposes a universal technique for estimating the sign and value of chirality for several types of supersecondary protein structures. Its applicability is not limited by any features of the protein, except for the actual presence of the type of structure in question. The obtained results of chirality calculations for various protein structures demonstrate convincing compliance with established and widely known rules.

Keywords: Chirality in proteins, supersecondary protein structure, coiled coil, β-sheet, chirality assessment

MSC 2020: 14J33, 92C05, 92D20, 92E10, 92-08, 92-10

1 Introduction

1.1 Background

1.1.1 Protein structure and function relationship problem

The discovery of a comprehensive relationship between an amino acid sequence and a complete three-dimensional structure, as well as the relationship between the structure and function of a protein, is still one of the most important major tasks in structural biophysics. In recent years, the problem of determining a three-dimensional structure based on an amino acid sequence has been partially solved using artificial intelligence technologies [22, 51], which have made it possible to formally answer this question, although with some limitations [37, 2, 44, 7]. These restrictions may be circumvented in the future.

Acknowledgement: This research was supported by the Theoretical Physics and Mathematics Advancement Foundation "BASIS" to Aleksey Lutsenko (project 23-2-10-25-1).

Aleksey Lutsenko, Alla Sidorova, Faculty of Physics, Lomonosov Moscow State University, Leninskie gory 1-2, 119234 Moscow, Russia, e-mails: aleksluchrus@yandex.ru, sky314bone@mail.ru

https://doi.org/10.1515/9783112216620-008

However, neural networks operate on a "black box" principle. They provide information about the structure of a protein with a given amino acid sequence, but do not answer the question of why it is correct. Actual physical principles responsible for this phenomenon remain known only very conditionally and to a limited extent. The relationship between the structure and functions of proteins is even more poorly established in general. Thus, in order to improve understanding of the processes of protein formation and functioning, it is necessary to study their structure at different levels and from the perspective of various scientific disciplines. The regularities in the structure of proteins should be identified in terms of their geometric, physical, chemical and biophysical properties, and characteristics. Then, perhaps, it will be possible not only to achieve an understanding of specific well-studied proteins features, but also to discover much more general relationships between the structure and functions of proteins.

1.1.2 Chirality as general protein feature

One of the most important geometric characteristics of proteins is chirality [4, 3, 49, 14, 40]. An object is chiral if it cannot be superimposed with its mirror image. Such an object and its reflection are called enantiomers in relation to each other. Chirality as such should be distinguished from pseudochirality [14]. Pseudochirality imitates chirality, but only at the highest hierarchical structural level. For example, the two hands of a person are not enantiomers in the full sense of the word, since they consist of identical molecules while their mirror images contain mirror-reflected objects.

It should be noted that both concepts of chirality and pseudochirality are relevant for proteins. However, since proteins consist only of the left amino acids, at all levels of the structure, except for the amino acid level itself, chirality is understood in a sense of pseudochirality. So, when it comes to the right- or left-handed helix, it is not meant that they are complete mirror images of each other, but only that at the highest hierarchical level their structure has a corresponding chirality. In general, chirality at any hierarchical level is naturally considered only in the context of the relative arrangement of the elements of the previous level. In the following text, we will also use the term chirality in that sense.

For a number of objects, there are generally accepted rules according to which one enantiomer is considered right-handed (D) and the other left-handed (L). For example, for molecules that are chiral relative to a carbon atom, there are rules that make it possible to determine the precedence of the radicals of this atom, and depending on their relative position, to designate the chirality of the molecule relative to that carbon atom. This rule has been slightly adjusted for amino acids in order to universalize it, taking into account their structural features. For three-dimensional helices, there is also a rule that defines their chirality. However, in general case, there is no such agreement for an abstract chiral geometric construction.

In our work, we propose to extend the concept of chirality to the category of quantity. We propose to determine not only the sign of chirality (D or L), but also its quantitative value for chiral objects. This approach allows not only to divide structures into two categories (right-handed and left-handed), but also to compare them, obtaining information about their relative characteristics. In particular, in our previous papers [39, 27, 50] we proposed a quantitative characterization of chirality for the secondary protein structure and a qualitative definition of chirality value for superhelical structures.

1.2 Subject of research

The structure of proteins is usually divided into several hierarchical levels. Primary structure is the amino acid sequence itself. Secondary structure represents local regular and irregular structures. Tertiary structure is the complete structure of the polypeptide chain. Finally, in protein complexes the quaternary structure can be distinguished. It represents the mutual arrangement of several chains within one complete protein complex. Nevertheless, there are many rather explicit structural motifs that do not fit into this sequence. These include, first of all, supersecondary structures: stable assemblies of several secondary structures held together in a specific position. Such motifs are clearly higher in the hierarchy than secondary structures, but cannot be classified as either tertiary or quaternary structures, since they are most often only part of but not the whole molecule.

Supersecondary structures are quite diverse [15, 8, 23, 24, 36]. These include, in particular, β-sheets, coiled coil superhelices, collagen superhelices, α- and β-hairpins, solenoids, and barrels of various types. The coiled coil superhelices, collagen superhelices, and β-sheets are the easiest to identify and separate from the rest of the molecule, as they all have a characteristic and well-distinguishable structure.

1.2.1 Coiled coil and collagen superhelices

Coiled coil superhelices or simply coiled coils consist of several intertwined secondary α-helices. There is a wide variety of superhelices of this type [21, 28, 48]. Nevertheless, the classical coiled coil, called also coiled coil 7/2, is characterized by an original inner structure with a special arrangement of amino acid residues. This type of coiled coils is the most abundant (more than 70 % among all coiled coils from the PDB [6] according to CCdb [46] as of July 2025).

A classical coiled coil consists of two or more α-helices. These helices can be located both parallel to each other and antiparallel [26]. The structure of each α-helix follows a certain pattern: the relative positions of amino acid residues are repeated every seven positions. Such a set of seven residues is called a heptad. Individual α-helices have about 3.6 amino acid residues per turn, but in a coiled coil one heptad is distributed in two turns, and the number of amino acid residues per turn is reduced to about 3.5 [26]. For this exact reason, that type of coiled coils is called 7/2.

The coiled coils stability is provided in a first place by the hydrophobic core of the coiled coil. The first and fourth positions of each α-helix heptad (or, as they are commonly referred to, positions a and d) are occupied as a rule by hydrophobic amino acid residues, which form a dense package due to hydrophobic interactions [26]. Thus, they are protected from the solvent, which is energetically beneficial for the molecule. Due to these interactions, coiled coils form a stable and durable structure in the protein. Such a characteristic superhelical arrangement is called "knobs-into-holes" and was predicted by F. H. C. Crick based on calculations in 1952–1953 [10–12].

Amino acid residues in the fifth and seventh positions (positions e and g) can also stabilize the coiled coil due to ionic interactions or due to the formation of salt bridges between α-helices [52, 25]. These positions, as well as positions a and d, are especially important for the coiled coil oligomeric state determination [25]. These exact residues mainly determine whether a coiled coil will consist of two, three, four, or five helices.

There are many studies showing which amino acid residues are more suitable for certain heptad positions [52, 47, 1, 20]. In general, it has been shown that leucine, isoleucine, and meteonin are among the most preferred residues in positions a and d. All of them belong to nonpolar, i. e., hydrophobic, aliphatic amino acids. Proline and glycine are the least preferred. This amino acid composition ensures maximum stability of the coiled coil.

Collagen superhelices consist of three polyproline helices [38]. These helices are held together by hydrogen bonds. The amino acid composition of these helices is rather monotonous and follows a peculiar pattern. Every third residue in the chain is glycine, and the gaps are occupied, most often, by proline or hydroxyproline. Although such superhelices are found only in a small set of different proteins, they play an important role because collagen is a key structural protein in many organisms, including humans, where it makes up about 30 % of the mass of all body proteins.

1.2.2 β-sheets and β-hairpins

The structure of β-sheets was shown by Pauling and Corey in a number of articles in the middle of the last century [32, 31, 33]. β-sheets are relatively flat structures consisting of several β-strands. A β-sheet of two strands is called a β-hairpin. β-sheets are often described as part of the secondary structure of a protein, but since each β-strand already has a specific characteristic spatial structure, it itself belongs to the secondary level of the protein structure [16]. Accordingly, β-sheets consisting of several such structures should be attributed to the supersecondary level of the structure.

The β-strand is an almost flat and only slightly twisted in terms of the main atomic chain fragment of the protein chain. Its structural repeating unit is two amino acid residues [54, 16]. In order to unify the consideration of different elements of supersecondary protein structure, we will also call this unit a "turn." These turns form a zigzag structure that is peculiar for the β-strand. A single strand is not fixed by hydrogen bonds, and is generally quite unstable. Instead, hydrogen bonds connect several strands together and hold the structure of the β-sheet, which already acquire a fairly significant overall strength [54]. Accordingly, due to the sinuous structure of the individual strands, the sheet formed from them is pleated.

Strands of one β-sheet can be strongly spaced along the primary amino acid sequence, but close to each other in space. Adjacent strands can be positioned either parallel or antiparallel with respect to each other.

1.2.3 Chirality on different protein structural levels

Different classes of elements of the protein structure of various hierarchical levels are often characterized by a certain sign of chirality. Amino acids in proteins have almost exclusively the left stereo form. The right amino acid residues are not included in proteins on ribosomes during matrix protein synthesis [17, 18, 30, 16, 19] and seem to be used in proteins in exceptional cases. The fundamental reasons for the selective choice of one enantiomer, as well as the reasons why the left but not right amino acids were selected are an issue under strong discussion.

There are several types of secondary helical elements. These include α-helices, polyproline helices, 3_{10} helices, and π-helices. All listed above, except for polyproline helices, have as a rule, right chirality (with the exception of short, deformed or artificial structures). Polyproline helices have left chirality. This configuration is due to the possible mutual arrangement of the side chains of the residues in the helices. It should ensure, on the one hand, that neighboring amino acid residues will not interfere with each other to occupy the corresponding positions in the helix; and, on the other hand, that all the necessary stabilizing links will be formed.

Coiled coils, as a rule, have left chirality. Apparently, this is due to the heptadic structure of the helices. The heptadic repeat of α-helices causes a characteristic dense packing of amino acid residues. Due to this, the residues are arranged so that the protruding side chains from one helix fit exactly into the gaps between the side chains of the other helix. The heptadic repeat ensures the distribution of 7 residues into 2 turns. This is a lower density of turns than in a separate α-helix, so the α-helix has to be a little less twisted in coiled coil than without it. The left direction of the coiled coil rotation is achieved due to this circumstance [26]. Similarly, the collagen superhelix formed from the left-handed polyproline helices is right-twisted.

Individual β-strands are quite unstable, but as a rule, they are left-handed. The chirality of the β-sheet as a whole can be investigated from different perspectives. One can look on the structure perpendicular to the strands or along them. If viewed perpendicular to the strands, the sheet twists to the left, and if viewed parallel it twists to the right

[9]. For uniformity with the consideration of superhelices, we propose to consider the chirality of the sheet in the direction along the strands.

From the described properties of supersecondary structures, it follows that several of their most typical classes are characterized by a twist in the direction opposite to that in which the individual elements of the structure are twisted. However, it is desirable to support this rule with the possibility to mathematically determine the sign and value of chirality for certain real protein fragments. This will make it possible to study the geometric properties of not only supersecondary structures in general, but also their particular implementations, as well as compare the characteristics of structural elements of different existing proteins.

The general features of coiled coils, collagen superhelices, and β-sheets should be emphasized. All these structural motifs belong to the supersecondary level of protein organization and consist of several separate secondary structures, each having a helical (or, in the case of β-strands, close to a helical) structure. Individual elements can be located in parallel or (for β-strands and coiled coils) antiparallel direction relative to neighboring secondary structures. Combining several secondary structures into a supersecondary one leads to increased stability for coiled coils [54] as well as collagen superhelices [38] and β-sheets [54, 16]. Due to all these features, we assumed that they should be considered in a similar way when creating a model for calculating chirality, and their common characteristics should be identified.

1.3 The current state of affairs

From the very beginning, the researchers noted the characteristic direction of supersecondary protein structures chirality. Thus, in the works of Crick [10–12], in which the structure of the coiled coils was theoretically predicted for the first time, it was shown that the coiled coil and its α-helices are twisted in opposite directions. Similarly, in one of the first works on the structure of collagen, [34] it was shown that the polyproline helices should be left-handed and the collagen superhelix should be right-handed. These assumptions were later definitively confirmed.

Although the general structure of the β-sheet was described as early as the 1950s [32, 31, 33], the characteristic direction of sheet twist was shown only 20 years later in [9]. Subsequently, the sign of chirality was identified for some other supersecondary protein motifs consisting of α-helices and β-strands [35, 42, 43, 53].

However, almost all models, both the earlier ones mentioned above and the more modern ones, either consider superhelices as a class without taking into account their individual characteristics, or, conversely, concentrate on a particular protein. Of the works discussing the definition of chirality of individual supersecondary structures, the work [29] is particularly noteworthy. In this work, it is suggested to evaluate the chirality of the "interface curves" of individual α-helices in a coiled coil. The interface curve is a curve passing between the curve connecting the residues in heptadic positions a and the curve connecting the residues in heptadic positions *d*. In general, the interface curve in this model fairly well reflects the direction of movement of the α-helix axis. The results of this work showed that the interface curves have left chirality, which corresponds to the expected sign of chirality for coiled coils. However, this approach has its limitations. The authors note that the sign of chirality of the interface curves does not necessarily coincide with the sign of chirality of the entire superhelix. In addition, this method requires at least two heptadic repeats in α-helices. Finally, it is adapted to determine the sign, but not the value of chirality.

No other works suggesting ways to determine the sign and, moreover, the chirality value for supersecondary protein structures could be found. Thus, the question of determining the chirality of individual, real, rather than ideal supersecondary protein structures remains open.

2 Data and methods

2.1 Data

The information needed for calculations can be obtained from any suitable source, but for our research group it is convenient to take data from the Protein Data Bank (PDB) [6]. It contains data on the structure of many proteins. In addition, for a coiled coil it is necessary to know which exact amino acid residues are included in the superhelical structure. We use the CCdb database for this [46]. It should be noted that when examining the CCdb database, we did not encounter superhelices in which the constituent helices would be of different lengths.

2.2 Methods

We have previously suggested methods for calculating the sign [39, 50] and the value [27] of chirality of superhelical protein structures. The method described in this chapter is an improved method from [27] and a natural development of the method from [39, 50]. It allows one to calculate the chirality of coiled coils, collagen superhelices, and β-hairpins.

Our proposition is to understand the chirality value of a superhelical structure as a scalar numerical indicator that characterizes how much each individual element of the superhelix is twisted around the axis of the entire superhelix as a whole. As will be shown later, this qualitative definition allows one to calculate the chirality of β-hairpins as well.

Let us first consider the chirality calculation method for coiled coil, and then specify the points that should be taken into account when using the method for collagen superhelices and β-hairpins.

2.2.1 The method application to coiled coil

Let us consider the application of our method to a coiled coil from a structure with PDB ID 5A7D [13] as an example. This structure consists of two antiparallel α-helices. Each α-helix consists of 15 amino acid residues.

Let us define the *center* of a set of points $A_1, A_2, \ldots, A_n$ as such point O that its coordinates are calculated as the arithmetic mean of the corresponding coordinates of all points from the set.

At the first stage of our algorithm, it is required to construct the axes of the α-helices that make up the coiled coil. In the classical coiled coil, the α-helix has a noninteger number of amino acid residues (3.5) per turn, unlike the polyproline helices in collagen (about 3 residues per turn) and β-strands (about 2 residues per "turn"). In the previous version of our method, [27] it was proposed to construct the axes of helices and superhelices as straight lines. This introduced distortions into the calculations and made the method inapplicable for more complex structures such as β-hairpins. Therefore, we now propose to build axes in the form of polygonal chains. In particular, for the coiled coil we apply the technique described in [45]. In this work, the positions of the points of α-helical axis can be calculated for each residue of the helix, except for the two terminal points at each end. Thus, for an α-helix consisting of n amino acid residues, the axis is plotted at $n-4$ points (Figure 5.1b). Since our method requires that the helix axis consist of at least 2 points, our calculation method is not applicable to superhelices with less than 6 amino acid residues in at least one helix. The axis points of the entire superhelix are calculated as the centers of the corresponding points of all axes of the helices (Figure 5.1c).

Let us consider a further calculation on a small fragment of a superhelix (Figure 5.1d–e). For the first segment of the helical axes, the vectors perpendicular to the corresponding segment of the superhelical axis are drawn to both end points (Figure 5.1d). It can be seen from the point of view perpendicular to the axis of the helix that there is a small angle between these vectors (Figure 5.1e). In this example, the angle is about 1.8° for both α-helices. It is important to take into account the direction of the angle. We use the sign of the mixed product of the two vectors directed to the helical axis and the vector directed along the superhelical axis for this purpose. We chose the sign in such a way that the positive value corresponds to the right direction of rotation, and the negative value corresponds to the left direction. Since when the helix advances towards the observer, the right helix turns counterclockwise, and its direction of rotation corresponds to the positive direction of the angle in trigonometry. Accordingly, in this case, the left helix turns clockwise, which corresponds to a negative angle in trigonometry.

The twist angle is calculated similarly for the second segment of the axes (Figure 5.1f), the third, and so on. For each helix, the calculated angles are summed up taking into account the sign. Then the arithmetic mean of the results for all helices is

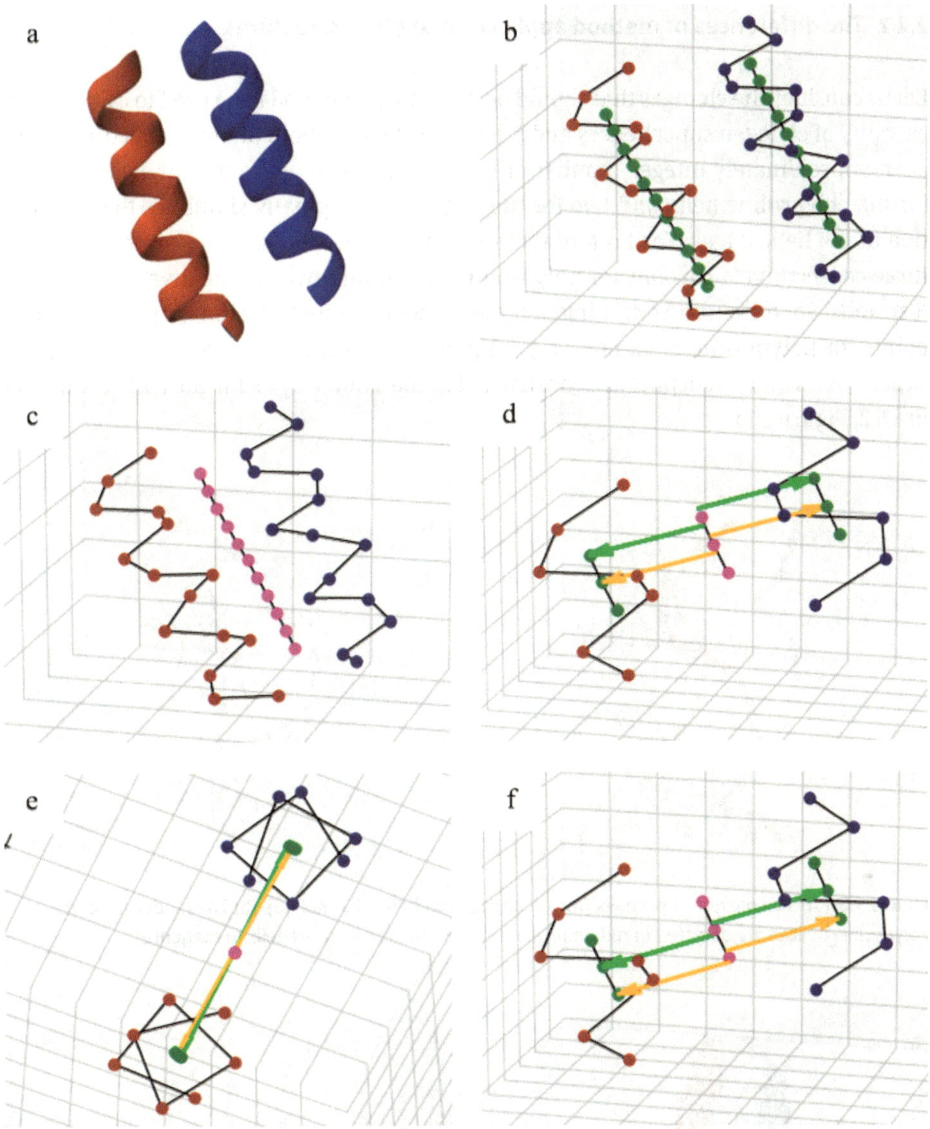

Figure 5.1: (a) The coiled coil from a structure with PDB ID 5A7D [13]. (b–f) The chirality calculation steps. α-helices are depicted in red and blue, helices axes in green and the axis of superhelix in magenta. Pairs of vectors from axis of superhelix to helical axes are depicted in light green and orange.

calculated. The resulting number is considered as the chirality value of the superhelix. The sign of this number indicates the direction of rotation of the superhelix: positive values indicate right superhelices, negative values indicate the left ones.

2.2.2 The differences of method application to other structures

Let us consider the changes that should be made to the method to be used to calculate the chirality of collagen superhelices and β-hairpins. First, in both of these structures, there is an approximately integer number of residues per turn of the helical element: three for the polyproline helix and two for the β-strand. This greatly simplifies the construction of the helical axis. For both of these structures, we calculate the centers for every three consecutive α-carbons for polyproline helices and two consecutive α-carbons for β-strands. So, for a unit consisting of n amino acid residues, the axis is plotted at $n-2$ points for polyproline helix and at $n-1$ points for β-strand. The axes of the collagen superhelices and β-hairpins are constructed in the same way as for the coiled coils (Figure 5.2, Figure 5.3).

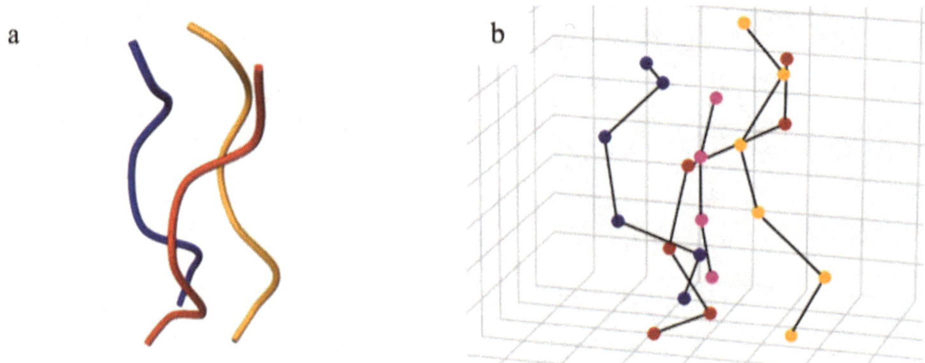

Figure 5.2: (a) The collagen superhelix from a structure with PDB ID 1K6F [5]. (b) The model structure. Polyproline helices are depicted in red, blue, and orange; the axis of superhelix in magenta.

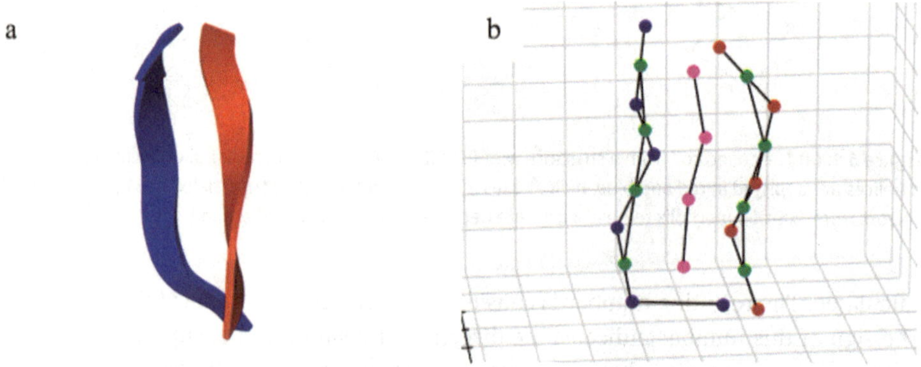

Figure 5.3: (a) The β-hairpin from a structure with PDB ID 1A4Z [41]. (b) The model structure. β-strands are depicted in red and blue; the axes of strands in green and the axis of superhelix in magenta.

The PDB files keep information about which amino acid residues form particular β-sheets. Therefore, a database similar to CCdb is not needed for such structures. However, in a β-hairpin, the strands can have different lengths. Since the method is applicable only to supersecondary structures with the same element length, it is necessary to consider only a part of the residues from a longer β-strand. In order to automatically select the necessary residues, we calculate the sum of the distances from the α-carbons of one strand to respective α-carbons of another one, with different selections of residues from a longer strand. In the end, we select an option with the smallest sum.

3 Results

We performed calculations for 990 coiled coils from the CCdb database [46], 12 collagen superhelices and 116 β-hairpins. Our results show that 969 of these superhelices have a left-handed twist, and that is about 98 % of total calculated structures. All the collagen superhelices turned out to be right. There were 107 right-handed β-hairpins, which corresponds to about 92 % of the total number of β-hairpins examined.

Since the vast majority of calculated structures have predicted chirality sign, the results are highly consistent with the data available in the literature. About 98 % of coiled coils, all collagen superhelices and about 92 % β-hairpins have the expected sign of chirality. That strongly confirms the correctness of our approach and calculation method.

There are structures with a chirality sign that does not correspond to the expected one, i. e., right-handed coiled coils and left-handed β-hairpins. It can be assumed that there are more such irregular structures in β-hairpins than in coiled coils because of some reasons. First, β-strands are not stabilized by hydrogen bonds and are more loose structures, than α-helices. In β-sheets, such stabilization is provided only in supersecondary level of structure. Second, β-hairpins are as a rule much shorter than coiled coils. In our study, no β-hairpins with more than 9 amino acid residues in a strand were examined. Third, as it was mentioned above, there can be many similar structures in our selection due to absence of methodology of duplicate structures. Finally, it is natural for β-sheets to have more than two strands. Such structures are more stable and it can be expected that their chirality values are more regular. Still the particular structures as well as the physical principles that cause their chirality are the subject of future studies. Tables 5.1, 5.2, 5.3 provide a selection of calculation results for structures with same number of units for coiled coils, collagen superhelices, and β-hairpins, respectively. The tables show the order of chirality values as well as the correspondence between chirality value and the length of individual units in amino acids.

The results show that there is no clear dependency between chirality value and helices or strands length in amino acid residues. It seems logical that the amino acid composition and the environment of superhelices play crucial role in their inner structure,

Table 5.1: The calculation results for sample data on coiled coils.

Superhelix	Helices number	Helices length	Chirality
6BB8_2	2	9, 9	−7,310 (L)
1BSM_1	2	9, 9	−16,649 (L)
4A0C_12	2	10, 10	−9,124 (L)
5AIQ_0	2	10, 10	−40,077 (L)
1AU1_3	2	11, 11	−28,990 (L)
6AYS_0	2	12, 12	33,228 (D)
3BJA_2	2	13, 13	−62,983 (L)
4A1S_3	2	14, 14	−19,852 (L)
4B0F_5	2	15, 15	−45,800 (L)
5AHZ_10	2	17, 17	−18,393 (L)
1AY2_0	2	19, 19	−109,341 (L)
6BK8_14	2	25, 25	−80,432 (L)
6BL6_6	2	35, 35	−4,851 (L)
3A7P_0	2	64, 64	−217,647 (L)
4A7F_0	2	134, 134	−504,411 (L)

Table 5.2: The calculation results for sample data on collagen superhelices.

Superhelix	Helices number	Helices length	Chirality
1BKV_0	3	7, 7, 7	75,851 (D)
1BKV_1	3	11, 11, 11	145,993 (D)
4AU3_0	3	15, 15, 15	204,143 (D)
3WN8_0	3	22, 22, 22	341,289 (D)
1X1K_0	3	25, 25, 25	380,290 (D)
1NAY_0	3	26, 26, 26	380,171 (D)
1X1K_1	3	26, 26, 26	390,530 (D)
1CAG_0	3	27, 27, 27	425,615 (D)
1CGD_0	3	27, 27, 27	426,154 (D)
2DRT_0	3	28, 28, 28	445,917 (D)
2DRX_0	3	28, 28, 28	428,016 (D)
1K6F_0	3	29, 29, 29	413,291 (D)

and consequently, in the sign and value of chirality, but a detailed and clear overview and explanation of this matter is not provided yet.

The calculations also show that the chirality value of different supersecondary structures have the same order of magnitude. It seems that coiled coils have lesser absolute chirality value per amino acid residue, so they are less twisted than collagen superhelices or β-hairpins. This observation should be validated with bigger structure samples and by the means of statistical analysis.

Table 5.3: The calculation results for sample data on β-hairpins.

Hairpin	Strands number	Strands length	Chirality
1BOU_C	2	3, 3	28,525 (D)
1BE3_I	2	3, 3	3,434 (D)
1BMF_T	2	3, 3	46,821 (D)
3GCF_AF	2	3, 4	−4,370 (L)
13PK_I	2	3, 4	36,443 (D)
1A07_B	2	4, 3	14,408 (D)
3GCF_AL	2	4, 4	41,670 (D)
1A42_A	2	4, 4	25,680 (D)
1A31_A	2	4, 4	59,595 (D)
1AOI_D	2	4, 4	44,447 (D)
1A2V_I	2	5, 4	48,698 (D)
1A4Z_F	2	5, 6	67,678 (D)
3LKF_E	2	6, 6	58,492 (D)
1DFA_C	2	7, 7	103,063 (D)
1A4Z_B	2	9, 9	122,773 (D)

4 Discussion

Purpose

Proteins have really complex inner structure, and chirality is one of its main geometric properties. Different elements of protein structure have their specific chirality sign on various levels. The more comprehensive understanding of this feature can significantly improve the overall knowledge about proteins, both theoretical and practical. These ideas can be useful in such areas as pharmacology and artificial protein creation.

Findings and research gaps

This study presents the original definition of chirality of superhelical structures and the method for evaluating the chirality of supersecondary protein structures. Such structures as a rule have explicit functionality, and that makes them a reasonable research target. The presented method allows the calculation of the sign and value of chirality of coiled coils, collagen superhelices, and β-hairpins. It is shown that coiled coils have predominantly left-handed twist, and collagen superhelices and β-hairpins—right-handed twist. These results correspond well with the available data. The exceptions to this rule still should be studied properly.

This approach also makes it possible to compare different structures of the same or of different types. It is shown that different types of structures have chirality value of same order of magnitude. But the physical principles responsible for the chirality value of particular structures are known only in a general sense. The details of these principles remain unknown and should be clarified.

Advantages of the improved method

The previous version of this method was described in [27]. In that version, the axes of helices and superhelices were approximated as straight lines. The improved method showed a significant difference with it. The representation of axes of helices and superhelices as polygonal chains, as well as the change of the database to the library of coiled-coil domains (CCdb), led to a significant decrease in the calculated chirality value for many superhelices. For some entries, the chirality value decreased by about 30 %.

Another advantage of the improved method is the possibility of the chirality calculation for coiled coils, the α-helices of which consist of only 6 amino acid residues, as opposed to 12 residues in the previous version of the method.

Difficulties with the improved method

The difficulty of using this method for collagen superhelices lies in the fact that in order to correctly calculate the chirality value it is necessary to identify helices that make up the superhelix, as well as establish an exact correspondence between amino acid residues in different polyproline helices. This information is not contained in the PDB files, and an identifying and matching system should be developed. A similar problem may arise for coiled coils containing α-helices of different lengths.

It can be assumed that in some PDB structures having several similar subunits β-sheets have the same structure in each subunit. Then it turns out that the calculation is performed several times for almost equal structures. Thus, 6 of the 9 left-handed β-hairpins were detected in the same oligomeric protein with PDB ID 3GCF. This nuance may, of course, spoil the statistics. A methodology to eliminate duplicate structures should be developed.

Finally, β-sheets often consist of more than two strands. The method of assessment the chirality value for such cases is one of the goals of our further work.

5 Conclusion

This work establishes a universal method for calculation of chirality of supersecondary protein structures. The approach confirms that coiled coils are left-handed, while collagen superhelices and β-hairpins are right-handed. It can be assumed that the chirality value is primarily affected not by helix or strand length, but by amino acid composition and the structure surroundings. Still the testing of this hypothesis is the matter of further research.

Bibliography

[1] A. Acharya, S. B. Ruvinov, J. Gal, J. R. Moll, and C. Vinson. A heterodimerizing leucine zipper coiled coil system for examining the specificity of a position interactions: Amino acids I, V, L, N, A, and K. *Biochemistry*, 41(48):14122–14131, 2002. https://doi.org/10.1021/bi020486r.

[2] F. Azzaz, N. Yahi, H. Chahinian, and J. Fantini. The epigenetic dimension of protein structure is an intrinsic weakness of the alphafold program. *Biomolecules*, 12(10):1527, 2022. https://doi.org/10.3390/biom12101527.

[3] S. D. Banik and N. Nandi. Chirality and protein biosynthesis. In P. Cintas, editor, *Biochirality: Origins, Evolution and Molecular Recognition*, pages 255–305. Springer Berlin Heidelberg, Berlin, Heidelberg, 2013. https://doi.org/10.1007/128_2012_369.

[4] L. D. Barron. From cosmic chirality to protein structure: Lord Kelvin's legacy. *Chirality*, 24(11):879–893, 2012. https://doi.org/10.1002/chir.22017.

[5] R. Berisio, L. Vitagliano, L. Mazzarella, and A. Zagari. Crystal structure of the collagen triple helix model [(pro-pro-gly)10]3. *Protein Science*, 11(2):262–270, 2002. https://doi.org/10.1110/ps.32602.

[6] H. M. Berman, J. Westbrook, Z. Feng, G. Gilliland, T. N. Bhat, H. Weissig, I. N. Shindyalov, and P. E. Bourne. The protein data bank. *Nucleic Acids Research*, 28(1):235–242, 01 2000. https://doi.org/10.1093/nar/28.1.235.

[7] L. M. F. Bertoline, A. N. Lima, J. E. Krieger, and S. K. Teixeira. Before and after alphafold2: An overview of protein structure prediction. *Frontiers in Bioinformatics*, 3, 2023. https://doi.org/10.3389/fbinf.2023.1120370.

[8] C. I. Branden and J. Tooze. *Introduction to Protein Structure*. Garland Publishing, New York, 1999. https://doi.org/10.1201/9781136969898.

[9] C. Chothia. Conformation of twisted β-pleated sheets in proteins. *Journal of Molecular Biology*, 75(2):295–302, 1973. https://doi.org/10.1016/0022-2836(73)90022-3.

[10] F. H. C. Crick. Is α-keratin a coiled coil? *Nature*, 170(4334):882–883, 1952. https://doi.org/10.1038/170882b0.

[11] F. H. C. Crick. The Fourier transform of a coiled-coil. *Acta Crystallographica*, 6(8–9):685–689, 1953. https://doi.org/10.1107/S0365110X53001952.

[12] F. H. C. Crick. The packing of α-helices: Simple coiled-coils. *Acta Crystallographica*, 6(8–9):689–697, 1953. https://doi.org/10.1107/S0365110X53001964.

[13] S. Culurgioni, S. Mari, P. Bonetti, S. Gallini, G. Bonetto, M. Brennich, A. Round, F. Nicassio, and M. Mapelli. Insc: lgn tetramers promote asymmetric divisions of mammary stem cells. *Nature Communications*, 9(1):1025, Mar 2018. https://doi.org/10.1038/s41467-018-03343-4.

[14] A. V. Efimov. Chirality and handedness of protein structures. *Biochemistry (Moscow)*, 83(1):S103–S110, 2018. https://doi.org/10.1134/S0006297918140092.

[15] A. V. Efimov. Standard structures in proteins. *Progress in Biophysics and Molecular Biology*, 60(3):201–239, 1993. https://doi.org/10.1016/0079-6107(93)90015-C.

[16] A. V. Finkelstein and O. B. Ptitsyn. *Protein Physics. A Course of Lectures*. Academic Press, Cambridge, Massachusetts, 2016.

[17] M. Friedman. Chemistry, nutrition, and microbiology of d-amino acids. *Journal of Agricultural and Food Chemistry*, 47(9):3457–3479, 1999. https://doi.org/10.1021/jf990080u.

[18] N. Fujii. D-amino acids in living higher organisms. *Origins of Life and Evolution of the Biosphere*, 32(2):103–127, 2002. https://doi.org/10.1023/A:1016031014871.

[19] N. Fujii, T. Takata, N. Fujii, K. Aki, and H. Sakaue. D-amino acids in protein: The mirror of life as a molecular index of aging. *Biochimica Et Biophysica Acta. Proteins and Proteomics*, 1866(7):840–847, 2018. https://doi.org/10.1016/j.bbapap.2018.03.001.

[20] M. M. Gromiha and D. A. D. Parry. Characteristic features of amino acid residues in coiled-coil protein structures. *Biophysical Chemistry*, 111(2):95–103, 2004. https://doi.org/10.1016/j.bpc.2004.05.001.

[21] M. Gruber and A. N. Lupas. Historical review: Another 50th anniversary – new periodicities in coiled coils. *Trends in Biochemical Sciences*, 28(12):679–685, 2003. https://doi.org/10.1016/j.tibs.2003.10.008.

[22] J. Jumper, R. Evans, A. Pritzel, T. Green, M. Figurnov, O. Ronneberger, K. Tunyasuvunakool, R. Bates, A. Žídek, A. Potapenko, A. Bridgland, C. Meyer, S. A. A. Kohl, A. J. Ballard, A. Cowie, B. Romera-Paredes, S. Nikolov, R. Jain, J. Adler, T. Back, S. Petersen, D. Reiman, E. Clancy, M. Zielinski, M. Steinegger, M. Pacholska, T. Berghammer, S. Bodenstein, D. Silver, O. Vinyals, A. W. Senior, K. Kavukcuoglu, P. Kohli, and D. Hassabis. Highly accurate protein structure prediction with alphafold. *Nature*, 596(7873):583–589, Aug 2021. https://doi.org/10.1038/s41586-021-03819-2.

[23] B. Kobe and A. V. Kajava. When protein folding is simplified to protein coiling: The continuum of solenoid protein structures. *Trends in Biochemical Sciences*, 25(10):509–515, 2000. https://doi.org/10.1016/S0968-0004(00)01667-4.

[24] I. Koch and T. Schäfer. Protein super-secondary structure and quaternary structure topology: Theoretical description and application. *Current Opinion in Structural Biology*, 50:134–143, 2018. https://doi.org/10.1016/j.sbi.2018.02.005.

[25] A. N. Lupas and J. Bassler. Coiled coils – a model system for the 21st century. *Trends in Biochemical Sciences*, 42(2):130–140, Feb 2017. https://doi.org/10.1016/j.tibs.2016.10.007.

[26] A. N. Lupas and M. Gruber. The structure of α-helical coiled coils. In *Fibrous Proteins: Coiled-Coils, Collagen and Elastomers*, volume 70 of Advances in Protein Chemistry, pages 37–38. Academic Press, 2005. https://doi.org/10.1016/S0065-3233(05)70003-6.

[27] A. Lutsenko, A. Sidorova, D. Shpigun, E. Belova, and V. Tverdislov. A method for calculating the sign and degree of chirality of supercoiled protein structures. *Symmetry*, 15(11), 2023. https://doi.org/10.3390/sym15112051.

[28] E. Moutevelis and D. N. Woolfson. A periodic table of coiled-coil protein structures. *Journal of Molecular Biology*, 385(3):726–732, 2009. https://doi.org/10.1016/j.jmb.2008.11.028.

[29] S. Neukirch, A. Goriely, and A. C. Hausrath. Chirality of coiled coils: Elasticity matters. *Physical Review Letters*, 100:038105, Jan 2008. https://doi.org/10.1103/PhysRevLett.100.038105.

[30] C. Ollivaux, D. Soyez, and J.-Y. Toullec. Biogenesis of d-amino acid containing peptides/proteins: Where, when and how? *Journal of Peptide Science*, 20(8):595–612, 2014. https://doi.org/10.1002/psc.2637.

[31] L. Pauling and R. B. Corey. Configurations of polypeptide chains with favored orientations around single bonds. *Proceedings of the National Academy of Sciences*, 37(11):729–740, 1951. https://doi.org/10.1073/pnas.37.11.729.

[32] L. Pauling and R. B. Corey. The pleated sheet, a new layer configuration of polypeptide chains. *Proceedings of the National Academy of Sciences*, 37(5):251–256, 1951. https://doi.org/10.1073/pnas.37.5.251.

[33] L. Pauling and R. B. Corey. Two rippled-sheet configurations of polypeptide chains, and a note about the pleated sheets. *Proceedings of the National Academy of Sciences*, 39(4):253–256, 1953. https://doi.org/10.1073/pnas.39.4.253.

[34] G. N. Ramachandran and G. Kartha. Structure of collagen. *Nature*, 176(4482):593–595, Sep 1955. https://doi.org/10.1038/176593a0.

[35] J. S. Richardson. Handedness of crossover connections in beta sheets. *Proceedings of the National Academy of Sciences*, 73(8):2619–2623, 1976. https://doi.org/10.1073/pnas.73.8.2619.

[36] V. R. Rudnev, L. I. Kulikova, K. S. Nikolsky, K. A. Malsagova, A. T. Kopylov, and A. L. Kaysheva. Current approaches in supersecondary structures investigation. *International Journal of Molecular Sciences*, 22(21), 2021. https://doi.org/10.3390/ijms222111879.

[37] K. M. Ruff and R. V. Pappu. Alphafold and implications for intrinsically disordered proteins. *Journal of Molecular Biology*, 433(20):167208, 2021. https://doi.org/10.1016/j.jmb.2021.167208.

[38] M. D. Shoulders and R. T. Raines. Collagen structure and stability. *Annual Review of Biochemistry*, 78:929–958, 2009. https://doi.org/10.1146/annurev.biochem.77.032207.120833.

[39] A. E. Sidorova, E. V. Malyshko, A. O. Lutsenko, D. K. Shpigun, and O. E. Bagrova. Protein helical structures: Defining handedness and localization features. *Symmetry*, 13(5), 2021. https://doi.org/10.3390/sym13050879.

[40] J. Skolnick, H. Zhou, and M. Gao. On the possible origin of protein homochirality, structure, and biochemical function. *Proceedings of the National Academy of Sciences*, 116(52):26571–26579, 2019. https://doi.org/10.1073/pnas.1908241116.

[41] C. G. Steinmetz, P. Xie, H. Weiner, and T. D. Hurley. Structure of mitochondrial aldehyde dehydrogenase: The genetic component of ethanol aversion. *Structure*, 5(5):701–711, May 1997. https://doi.org/10.1016/S0969-2126(97)00224-4.

[42] M. J. E. Sternberg and J. M. Thornton. On the conformation of proteins: The handedness of the β-strand-α-helix-β-strand unit. *Journal of Molecular Biology*, 105(3):367–382, 1976. https://doi.org/10.1016/0022-2836(76)90099-1.

[43] M. J. E. Sternberg and J. M. Thornton. On the conformation of proteins: The handedness of the connection between parallel β-strands. *Journal of Molecular Biology*, 110(2):269–283, 1977. https://doi.org/10.1016/S0022-2836(77)80072-7.

[44] A. O. Stevens and Y. He. Benchmarking the accuracy of alphafold 2 in loop structure prediction. *Biomolecules*, 12(7), 2022. https://doi.org/10.3390/biom12070985.

[45] S. V. Strelkov and P. Burkhard. Analysis of α-helical coiled coils with the program TWISTER reveals a structural mechanism for stutter compensation. *Journal of Structural Biology*, 137(1):54–64, 2002. https://doi.org/10.1006/jsbi.2002.4454.

[46] K. Szczepaniak, A. Bukala, A. M. da Silva Neto, J. Ludwiczak, and S. Dunin-Horkawicz. A library of coiled-coil domains: From regular bundles to peculiar twists. *Bioinformatics*, 36(22–23):5368–5376, 12 2020. https://doi.org/10.1093/bioinformatics/btaa1041.

[47] B. Tripet, K. Wagschal, P. Lavigne, C. T. Mant, and R. S. Hodges. Effects of side-chain characteristics on stability and oligomerization state of a de novo-designed model coiled-coil: 20 amino acid substitutions in position “d”. *Journal of Molecular Biology*, 300(2):377–402, 2000. https://doi.org/10.1006/jmbi.2000.3866.

[48] L. Truebestein and T. A. Leonard. Coiled-coils: The long and short of it. *BioEssays*, 38(9):903–916, 2016. https://doi.org/10.1002/bies.201600062.

[49] V. A. Tverdislov. Chirality as a primary switch of hierarchical levels in molecular biological systems. *Biophysics*, 58(1):128–132, Jan 2013. https://doi.org/10.1134/S0006350913010156.

[50] V. Tverdislov, A. Sidorova, O. Bagrova, E. Belova, A. Lutsenko, E. Semenova, D. Shpigun, and N. Levashova. Ideas of symmetry as a biophysical basis of system biomedicine. In J. Mau, S. Mukhin, G. Wang, and S. Xu, editors, *BIOKYBERNETIKA. Mathematics for Theory and Control in the Human and in Society*, pages 123–148. De Gruyter, Berlin, Boston, 2025. https://doi.org/10.1515/9783111341996-008.

[51] M. Varadi, S. Anyango, M. Deshpande, S. Nair, C. Natassia, G. Yordanova, D. Yuan, O. Stroe, G. Wood, A. Laydon, A. Žídek, T. Green, K. Tunyasuvunakool, S. Petersen, J. Jumper, E. Clancy, R. Green, A. Vora, M. Lutfi, M. Figurnov, A. Cowie, N. Hobbs, P. Kohli, G. Kleywegt, E. Birney, D. Hassabis, and S. Velankar. Alphafold protein structure database: Massively expanding the structural coverage of protein-sequence space with high-accuracy models. *Nucleic Acids Research*, 50(D1):D439–D444, 11 2021. https://doi.org/10.1093/nar/gkab1061.

[52] K. Wagschal, B. Tripet, P. Lavigne, C. Mant, and R. S. Hodges. The role of position a in determining the stability and oligomerization state of α-helical coiled coils: 20 amino acid stability coefficients in the hydrophobic core of proteins. *Protein Science*, 8(11):2312–2329, 1999. https://doi.org/10.1110/ps.8.11.2312.

[53] D. W. Weatherford and F. R. Salemme. Conformations of twisted parallel beta-sheets and the origin of chirality in protein structures. *Proceedings of the National Academy of Sciences*, 76(1):19–23, 1979. https://doi.org/10.1073/pnas.76.1.19.

[54] D. Whitford. *Proteins: Structure and Function*. John Wiley & Sons Ltd, Chichester, West Sussex, 2005.

Denis Shpigun, Alla Sidorova, and Pavel Levashov

The relationship between chirality of secondary helical structures and the stability of lysozyme and myoglobin

Abstract: This study describes the structural features of lysozymes and myoglobins from various organisms. To elucidate the correlation between chirality and stability, comparatively diminutive single-subunit proteins were chosen, given their extensive representation in experimental literature. Of particular scientific interest is the determination of the relationship between protein's structural properties and their thermal stability, quantifiable via the melting temperature of the protein globule (or melting temperatures of individual domains). Chirality constitutes one such critical property, defined as the inherent incompatibility of an entity with its mirror image. Individual amino acid residues additionally contribute to the stabilization of lysozyme and myoglobin. Based on the structural analysis of lysozymes and myoglobins, chirality maps of regular secondary structures (α- and 3_{10}-helices) were generated for these proteins derived from various organisms, employing a method for calculating the sign and degree of chirality. The findings demonstrate that the chirality degree within helical structures is contingent upon the spatial arrangement of constituent amino acids; consequently, secondary structures possessing helices of equivalent length can exhibit variations in their degree of chirality. This investigation explores the association between the phase transition temperature to the "molten globule" state and the chirality degree exhibited by secondary helical structures. These insights are valuable for advancing protein engineering efforts aimed at enhancing enzyme stability and designing novel therapeutics. Overall, understanding the structural determinants of lysozyme and myoglobin stability allows to develop more effective antimicrobial agents and biotechnological applications.

Keywords: Chirality in proteins, secondary protein structure, lysozyme, myoglobin, stability, α-helix, 3_{10}-helix, temperature

MSC 2020: 14J33, 92C05, 92D20, 92E10, 92-08, 92-10

Acknowledgement: This research was supported by the Theoretical Physics and Mathematics Advancement Foundation "BASIS" to Denis Shpigun (project 23-2-10-8-1).

Denis Shpigun, Alla Sidorova, Faculty of Physics, Lomonosov Moscow State University, Leninskie gory 1-2, 119234 Moscow, Russia, e-mails: denish.den@mail.ru, sky314bone@mail.ru
Pavel Levashov, Faculty of Chemistry, Lomonosov Moscow State University, Leninskie gory 1-3, 119234 Moscow, Russia, e-mail: levashov@yahoo.com

https://doi.org/10.1515/9783112216620-009

1 Introduction

The enzyme lysozyme is a protein with a molecular weight of approximately 14–15 kDa, comprising 125–129 amino acid residues. The globular structure of lysozyme incorporates two domains that delineate a cleft, wherein the active site of this enzyme binds a peptide molecule between the two domains.

Lysozyme is found in saliva, tears, milk of mammals, blood serum, lymphatic tissues and possesses immunomodulatory, antihistamine, and antitumor properties [4]. In connection with the above properties of lysozyme, the study of its thermal stability seems relevant. There are three main types of lysozymes: c-type (hen or normal type), g-type (goose type), and i-type (invertebrate type). Hen egg white is the richest source of this enzyme and contains about 0.3 % lysozyme by weight [16]. Milk lysozymes belong to the c-type family with wide variability in structure, physicochemical properties, and concentrations [1]. An important role in ensuring the structural stability and, therefore, the thermal stability, of lysozymes is played by salt bridges [3], hydrophobic and electrostatic interactions [21], irregular secondary structures, as well as the value of the hydrogen index (pH). The structural and thermal stability of lysozyme are significantly influenced by specific amino acid residues, including phenylalanine, lysine, aspartic acid, glutamic acid, arginine [15], and proline [6]. Aspartic and glutamic acids are the main amino acid residues that make up the catalytic center of lysozyme. Substitution of just a few amino acids in the helices can significantly affect the stability of the entire structure.

Myoglobin is a member of the globin superfamily, which also includes hemoglobin, and is a single polypeptide chain of 153–154 amino acids with an oxygen binding site [20]. It is produced in muscle and cardiac cells in response to increased oxygen demand during exercise. It transports oxygen from the sarcolemma to the mitochondria of vertebrate cardiac and muscle cells [24]. In the study of myoglobin, special attention is paid to its stability. Thus, a high pH can increase the thermal stability of myoglobin [25] and with a decrease in the number of hydrogen bonds and hydrophobicity, the stability of myoglobin also decreases [11]. Of particular interest is the establishment of the relationship between the structural properties of proteins and their thermal stability. One of these important properties is chirality. This paper examines the structural features of lysozymes and myoglobins from various organisms and the relationship between the stability of lysozymes and myoglobins and their chirality.

2 Methods

According to our method, a sufficient condition for the method of assessing the chirality of regular and irregular protein structures is the mutual arrangement of α-carbons [17–19]. The determinant of any three vectors $\vec{v}_i = (x_i, y_i, z_i)'$, $i = 1, 2, 3$, constructed between consecutive reference points $C\alpha$, with regard to the $C\alpha$ coordinates presented in the PDB database, is calculated as "mixed product,"

$$\left.\begin{aligned} \begin{vmatrix} x_1 & y_1 & z_1 \\ x_2 & y_2 & z_2 \\ x_3 & y_3 & z_3 \end{vmatrix} &= ([\vec{v}_1, \vec{v}_2], \vec{v}_3) \\ &= (y_1 z_2 - y_2 z_1) x_3 + (z_1 x_2 - z_2 x_1) y_3 + (x_1 y_2 - x_2 y_1) z_3, \end{aligned}\right\} \tag{6.1}$$

mathematically a scalar triple product.

For the right three vectors, the sign of the mixed product is positive and the helix is right-handed (D); for the left, it is negative and the helix is left-handed (L). The total value of the chirality of protein helical structures is calculated as

$$\left.\begin{aligned} \chi_{\text{real}} &= \sum_{i=1}^{n-3} \frac{([\vec{v}_i, \vec{v}_{i+1}], \vec{v}_{i+2})}{C_i}, \\ C_i &= \left(\frac{1}{3} \sum_{j=0}^{2} |\vec{v}_{i+j}| \right)^5. \end{aligned}\right\} \tag{6.2}$$

$\epsilon \equiv$ Atomic coordinate error (input uncertainty)

$\delta_v \equiv$ Propagation of ϵ to vector length

$\delta_\chi \equiv$ Propagation of ϵ to chirality degree

$m \equiv$ Scalar triple product (mixed product), $\quad m_{x_0} = \dfrac{\partial m}{\partial x_0}$

$|\vec{v}| \equiv$ Norm (length) of vector $\vec{v}$, $|\vec{v}|_{x_0} = \dfrac{\partial |\vec{v}|}{\partial x_0}$

The main error formulae can then be concisely written as

$$\delta_v = \epsilon \sqrt{\frac{2(x_0 - x_1)^2 + 2(y_0 - y_1)^2 + 2(z_0 - z_1)^2}{(x_0 - x_1)^2 + (y_0 - y_1)^2 + (z_0 - z_1)^2}} \tag{6.3}$$

$$\delta_\chi \approx \epsilon \sqrt{\frac{3(m_{x_0} |\vec{v}|^5 - 5m|\vec{v}|^4 |\vec{v}|_{x_0})^2 + 3m_{x_2}}{[(x_0 - x_1)^2 + (y_0 - y_1)^2 + (z_0 - z_1)^2]^5}} \tag{6.4}$$

where ϵ is the atomic coordinate error, δ_v the propagated error in vector length, δ_χ the propagated error in degree of chirality.

Thus, small deviations in the coordinates of atoms and the spatial arrangement of amino acids do not affect the sign of chirality, but can affect the value of the degree of chirality of the helices.

Differences in the values of the degree of chirality can be caused by various reasons: due to the use of different methods for determining structures in proteins, errors in determining the coordinates of atoms and the sequence of amino acids in the helices. Since all the helices in lysozymes were determined by X-ray structural analysis, possible differences in the values of the chirality degree of the helices are associated with the amino acid composition of the polypeptide chain. Amino acids differ in the sizes of side radicals, which affect the arrangement of amino acids relative to each other. In this case, the values of the mixed products can change, since the angles between the vectors connecting the reference points can change.

As an example, let us consider two 3_{10}-helices in turkey lysozyme that have the same length (5 amino acid residues), but different values of the chirality degree. Both helices, like the entire 1XFT protein [12], were obtained using the X-ray structural analysis method. According to formulas (6.3)–(6.4), for 3_{10}-helices in turkey lysozyme, the error in determining the length of the vectors $\delta_v = 0.008$ and the error in the degree of chirality $\delta_\chi = 0.008$. The obtained results demonstrate an insignificant effect of the method for determining the structure of these helices on the value of their chirality degree. At the same time, when analyzing the values of mixed products of vectors and the chirality degree of these structures, one can note differences due to the mutual arrangement of amino acids and, accordingly, their spatial coordinates. Thus, it can be argued that not only the chirality sign of secondary structures, but also the value of the chirality degree adequately assess the spatial structure of helices and, therefore, are associated with the stability of helices and the proteins that include them.

3 Results

3.1 Lysozymes

In this work, the structure of lysozymes of 6 organisms (echidna, camel, hen, human, bull, and house mouse) was studied, most of which belong to the c-type lysozyme and were obtained using X-ray structural analysis. Lysozymes of camel, echidna, and human belong to milk lysozymes. These lysozymes were selected for the study based on the availability of publicly available information on their structure.

For the considered organisms chirality maps of helical structures (Figure 6.1a), amino acids distribution of lysozymes (Figure 6.1b), alignment by amino acid composition in lysozyme helices (Figure 6.2) were constructed. The chirality maps (Figure 6.1a) show the dependence of the chirality degree of regular secondary helical structures (α-helices and 3_{10}-helices) of lysozymes on the number of amino acid residues in the polypeptide chain.

The value of the chirality degree of helical structures depends on the spatial position of individual amino acids that form the helix. Therefore, for secondary structures with

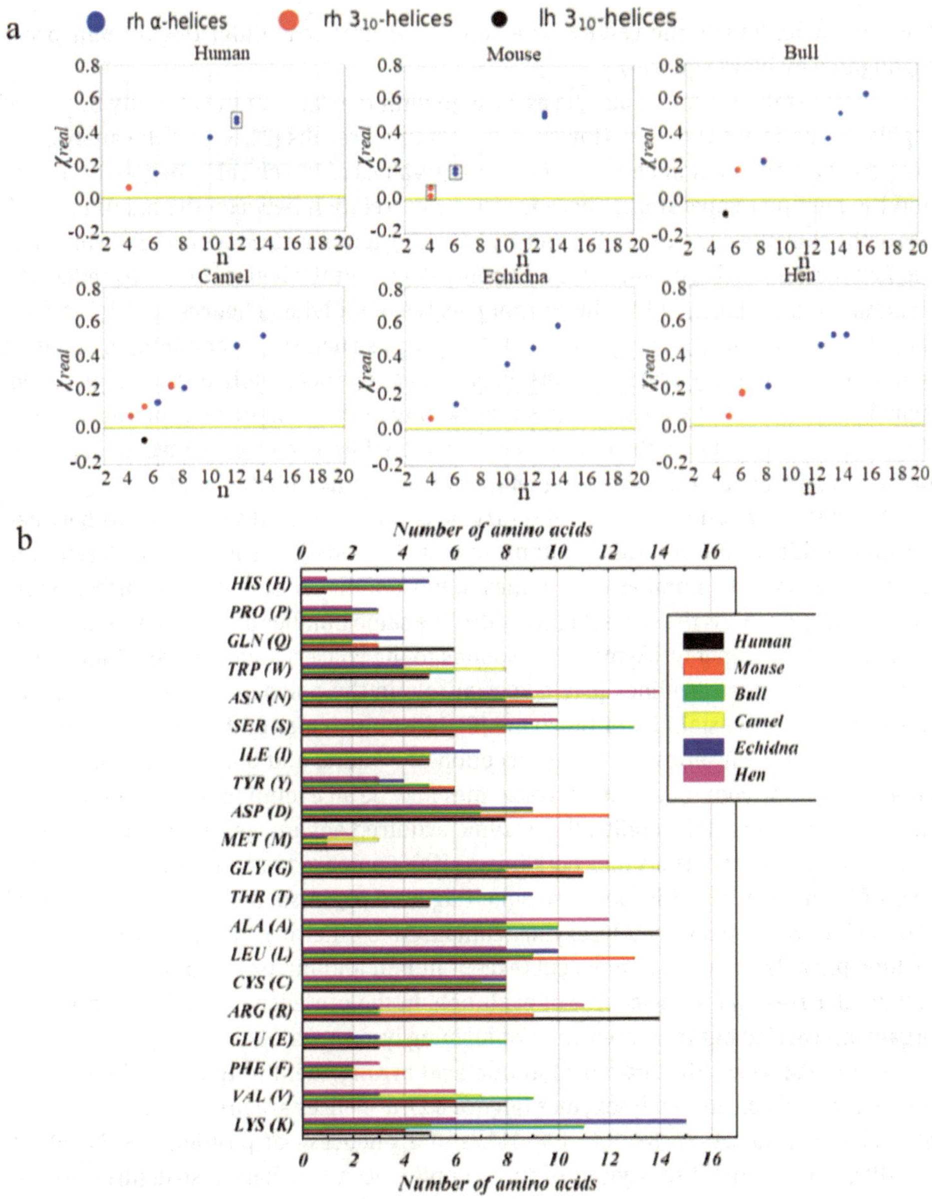

Figure 6.1: (a) Chirality maps of helical structures of lysozymes. The horizontal axis shows the length of helical structures in amino acid residues, the vertical axis shows the chirality degree (rh – right-handed, lh – left-handed). In the graphs, right-handed *α*-helices are shown in blue, right-handed 3_{10}-helices are shown in red, and left-handed 3_{10}-helices are shown in black. Regions of helices of the same length, but with different chirality degrees, are marked with black squares. (b) Distribution of amino acids in lysozymes of various organisms. In the figure, amino acids belonging to human lysozyme are highlighted in black, mouse lysozyme – in red, bull lysozyme – in green, camel lysozyme – in yellow, echidna lysozyme – in blue, and hen lysozyme – in crimson.

the same helix length, the value of the chirality degree may differ (Figure 6.1a, areas highlighted by black squares).

Protein stability should also be associated with the value of the chirality degree of regular secondary structures. However, the obtained results [22, 5, 7] of the analysis do not provide a correlation between the average values of the chirality degree of helices and the melting temperature. There is also no correlation between the maximum values of the chirality degree of helices and the melting temperature. We associate this with the fact that the average value of the chirality degree of the helices does not reflect the structure of the helices, unlike the chirality degree of individual helices, which is associated with the amino acid composition of the helical structures. In addition, the melting temperatures (one or more) reflect the phase transitions of an individual domain (or domains), the state of which determines the stability of the protein structure as a whole. Therefore, the question of the relationship between the chirality of the secondary structures of lysozymes and their melting temperature requires further study.

It is known that the structure of lysozyme is stabilized by disulfide bonds between cysteine residues and, as a rule, contains four such bonds [13]. For the considered sample of organisms, the number of cysteines in most of the considered structures, except for camel lysozyme, is the same (Figure 6.1b). The same number of cysteines in the structure (except for camel lysozyme) corresponds to the conservative four disulfide bonds in most c-lysozymes. Prolines play a special role in the polypeptide chain, which significantly affect the stability of lysozyme [6]. Most lysozymes of the considered organisms contain two prolines (with the exception of echidna and bull lysozymes). The increased arginine content in human lysozymes may be an evolutionary feature, when at the expense of thermal stability, the enzyme acquired advantages in terms of antibacterial action against various microorganisms [14]. Compared to other organisms, an excess of lysine was noted in bull lysozyme (Figure 6.1b). Also, an increased content of asparagine was noted in hen lysozyme (compared to other organisms). These residues include phenylalanine, glutamic acid (except in bull and mouse), and proline. This result emphasizes that in the course of evolution, in the presence of a common trend, all organisms carried out their own ways of forming lysozymes.

Figure 6.2 shows the regions of amino acid arrangement in lysozyme helices. For the majority of considered lysozymes (Figure 6.2), α–helices are predominantly located at positions 4–15, 24–37, 89–100, 108–114, and 3_{10}-helices—at positions 19–23, 79–84, 103–107, 119–124. In all lysozymes of this sample, secondary helical structures are not observed at positions 38–74. This region is probably necessary for flexible adjustment of the gap geometry between domains upon substrate binding. Lysozymes contain a large number of charged amino acids (lysine, arginine, aspartate). This is likely due to the fact that these amino acids can improve the binding of lysozyme to the negatively charged surface of the bacterium, enhancing bacterial lysis and can promote the penetration of lysozyme into the cell periplasm [10].

As noted above, the average value of the chirality degree is not directly related to the thermal stability of proteins. However, it turns out that the number of charged amino

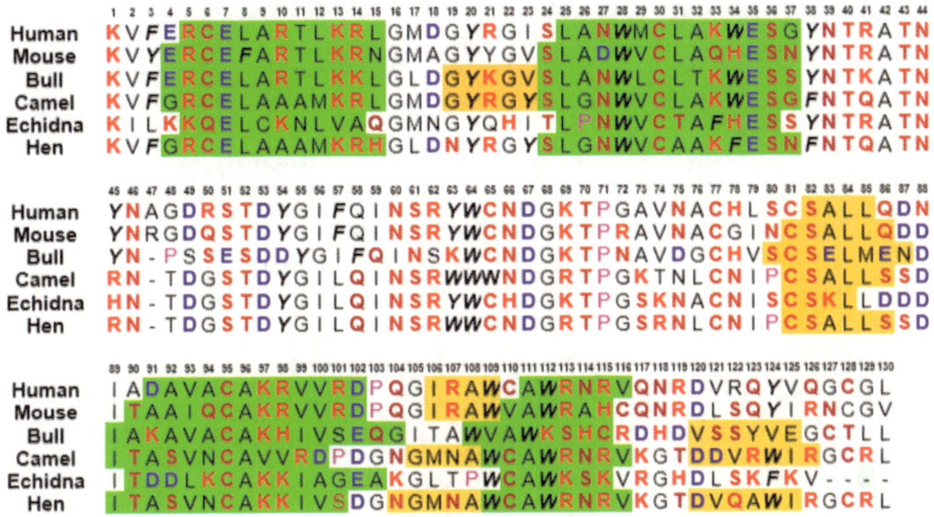

Figure 6.2: Alignment of the primary structure and arrangement of secondary structures in lysozymes of the studied organisms. Residues corresponding to the *α*-helix are colored green; those corresponding to the 3_{10}-helices are colored orange. Polar amino acids are shown in bold: positively charged are shown in red (K, R, H), negatively charged - in purple (D, E), uncharged amino acids - in brown (S, T, C, N, Q). Aromatic amino acids are shown in bold oblique font in black (F,W,Y). Nonpolar amino acids are shown in regular font in black (G, A, V, 1, L, M). Proline (P), due to its important role in the stability of lysozymes, is shown in pink.

acids responsible for the stability of the helices is also not related to the average value of chirality of the helical structures. Analysis of the number of charged amino acids (Figure 6.2) showed that the greatest number of positively charged amino acids in *α*-helices ($n = 8$) are in the lysozymes of human, mouse, echidna, and hen, and the minimum number is in camel – 7. The greatest number of negatively charged amino acids in 3_{10}-helices ($n = 5$) are in the lysozymes of mouse and echidna, and the minimum number is in the lysozyme of hen ($n = 2$). These results confirm our conclusion: the stability of lysozymes, as well as the chirality degree of their helical structures, depend significantly on the amino acid composition of the polypeptide chain. But the average value of the chirality degree does not reflect the features of the amino acid structure, therefore it cannot be a characteristic of protein stability, unlike the chirality degree of individual secondary structures.

3.2 Myoglobins

Myoglobins of five organisms were analyzed: boar, whale, horse, elephant, and human. The chirality maps of *α*-helices and 3_{10}-helices of myoglobins are shown in Figure 6.3a. Amino acid distributions of myoglobins are shown in Figure 6.3b, and data on the amino acid composition of helices are shown in Figure 6.4.

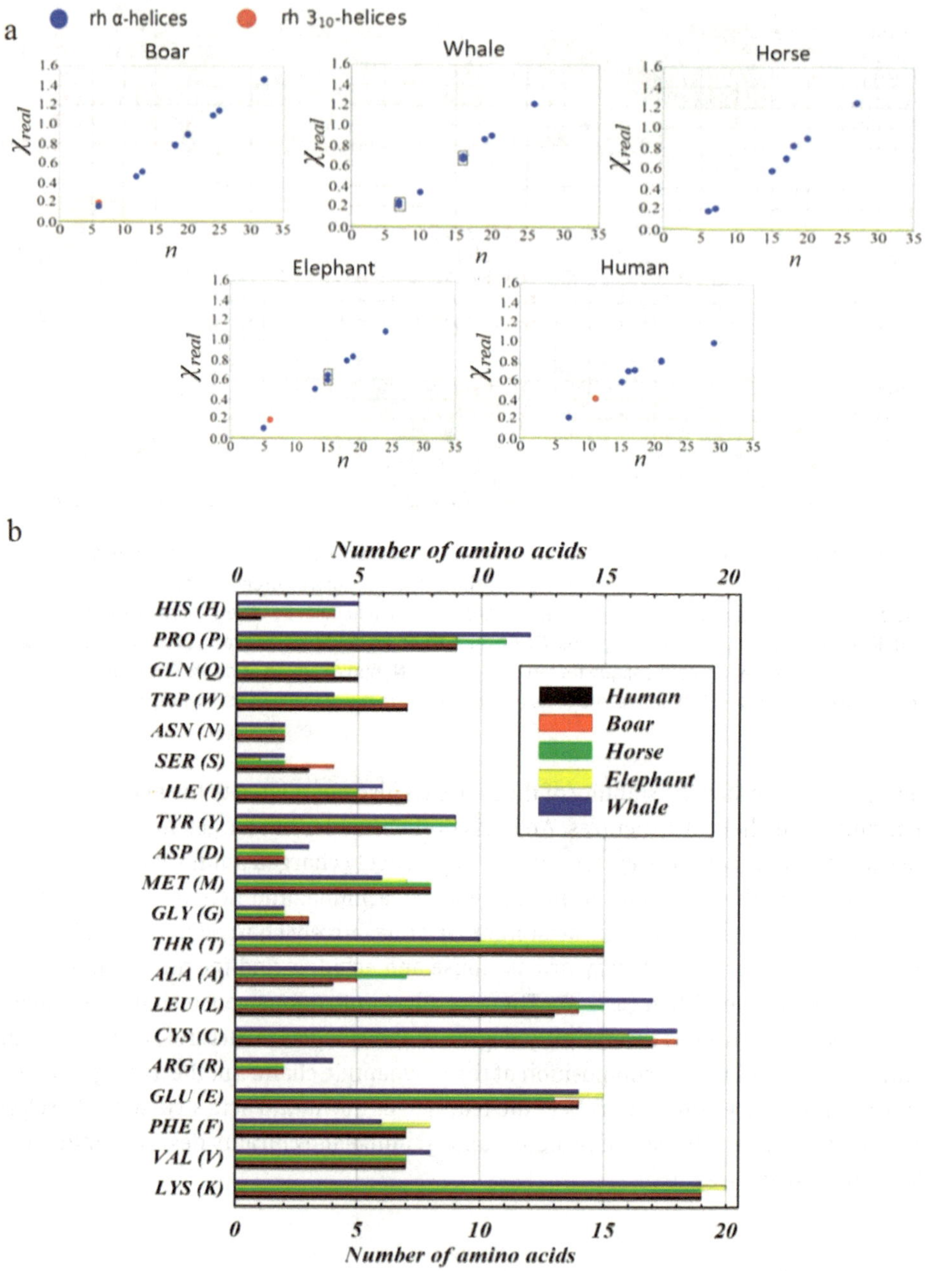

Figure 6.3: (a) Chirality maps of helical structures of myoglobins. The horizontal axis shows the length of helical structures in amino acid residues, the vertical axis shows the chirality degree (rh – right-handed, lh – left-handed). In the graphs, right-handed *α*-helices are shown in blue, right-handed 3_{10}-helices are shown in red. Regions of helices of the same length, but with different chirality degrees, are marked with black squares. (b) Distribution of amino acids in myoglobins of various organisms. In the figure, amino acids belonging to human myoglobin are highlighted in black, boar myoglobin – in red, horse myoglobin – in green, elephant myoglobin – in yellow, and whale myoglobin – in blue.

The chirality maps of myoglobins (Figure 6.3a) and the data obtained [23, 9, 2] demonstrate that this sample of organisms does not contain left-handed helices, which are sterically strained structures. Similar to lysozymes, the results obtained do not provide a correlation between the average values of the chirality degree of helical structures and the melting temperature of myoglobins. There is also no correlation between the maximum values of the chirality degree of the helices and the melting temperature. The differences in the observed stability temperatures are associated with differences in the amino acid composition of the myoglobins under study. Therefore, the question of the relationship between the chirality of the secondary structures of myoglobins and their melting temperature requires further study. In this sample of myoglobins, the average length of the right-handed α-helices is 13.6 amino acid residues, and the average length of right-handed 3_{10}-helices is 8 amino acid residues. In addition, the absence of left-handed helices is noted (Figure 6.3a). These results indicate the stability of the myoglobins under study.

Figure 6.4 shows the distribution of amino acids in myoglobin helices. According to studies, the regular structures of myoglobins form a high-density region (mainly of α-helices), interspersed with small regions of irregular structures [8]. In the considered

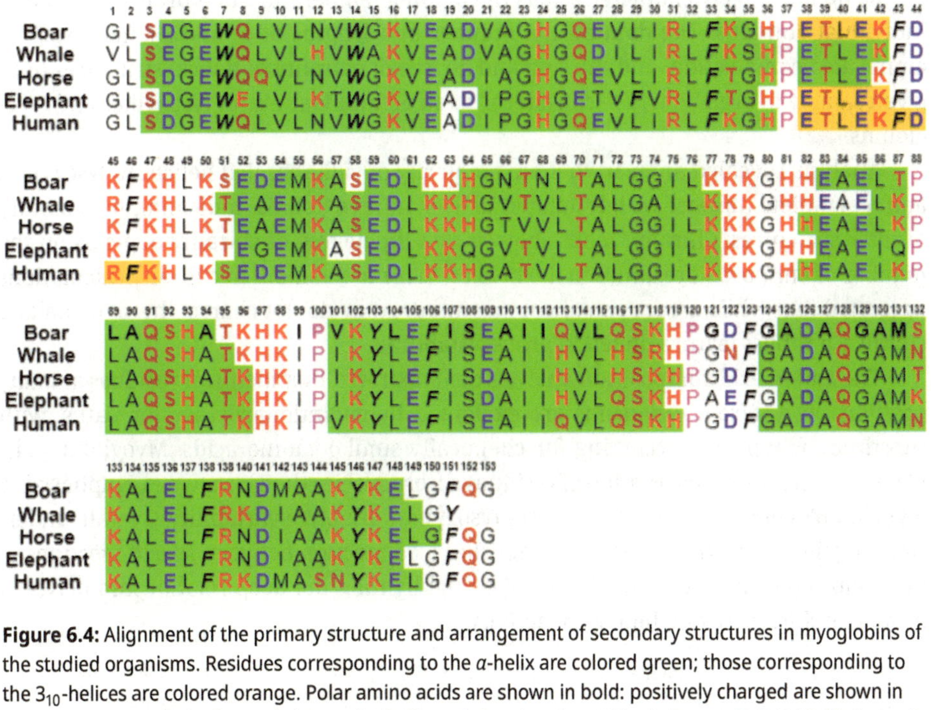

Figure 6.4: Alignment of the primary structure and arrangement of secondary structures in myoglobins of the studied organisms. Residues corresponding to the α-helix are colored green; those corresponding to the 3_{10}-helices are colored orange. Polar amino acids are shown in bold: positively charged are shown in red (K, R, H), negatively charged - in purple (D, E), uncharged amino acids - in brown (S, T, C, N, Q). Aromatic amino acids are shown in bold oblique font in black (F,W,Y). Nonpolar amino acids are shown in regular font in black (G, A, V, 1, L, M). Proline (P) is shown in pink.

sample of myoglobins, the overwhelming majority of regular structures belong to right-handed α-helices. Due to this, myoglobin forms a stable protective framework of heme.

For this sample of myoglobins in α-helices, similar to the sample of lysozymes, the presence of identical amino acids is noted. For example, for positively charged amino acids: lysine (K); histidine His (H); glutamic acid (E); aspartic acid (D). And, similar to the lysozyme sample, substitutions usually occur for amino acids that are similar in radical. Unlike lysozymes, in the myoglobin sample, many more matches can be noted for different organisms for amino acids in the same positions, which indicates a more conservative nature of myoglobins, compared to lysozymes.

Similar to lysozymes, the number of charged amino acids responsible for the stability of myoglobin helices is also not associated with the average chirality of the helical structures.

4 Discussion

We present the results of a study of the relationship between the structural features of lysozymes and myoglobins of various organisms and their thermal stability, which can be quantitatively characterized as the melting temperature of a protein globule or individual domains. The data on the number and lengths of α- and 3_{10}-helices, their amino acid composition, and the sign and degree of chirality of helical structures for a given sample of organisms are considered as structural features of lysozymes and myoglobins.

Analysis revealed an overwhelming majority of right-handed helices in lysozymes and a complete absence of left-handed helices in myoglobins, indicating the stability of their structures. Chirality maps showed that the degree of chirality depends on the number of amino acid residues in the helices. Error assessment in chirality calculations suggested that while errors can affect magnitude, they do not change the sign, making chirality a robust feature linked to protein stability.

The amino acid composition analysis demonstrated that identical residues at many positions in different organisms are explained by evolutionarily similar paths, with substitutions typically occurring for chemically similar amino acids. Myoglobins displayed much more conservation of residues across species than lysozymes, emphasizing myoglobin's conservative nature verses lysozyme's adaptive variability. Overall, understanding these structural features establishes a connection between specific sections of these enzymes and their functional roles, highlighting the deep relationship between protein architecture and biological activity.

5 Conclusion

Understanding the relationship between the structural properties of lysozymes and myoglobins and their thermal stability is crucial for elucidating their functional stability under diverse conditions. One key aspect of this investigation is the chirality of protein structural elements. This study demonstrates that key secondary structure features—chirality and conserved amino acid positions—are fundamental determinants of enzyme thermal stability and function, directly linking protein architecture with evolutionary adaptation and molecular resilience. The discovery of a correlation between the chirality degree of helices in lysozymes and myoglobins and the amino acid composition of their polypeptide chains, as well as between the sign and degree of chirality of helices and thermal stability, requires further exploration.

Bibliography

[1] N. Benkerroum. Antimicrobial activity of lysozyme with special relevance to milk. *African Journal of Biotechnology*, 7:4856–4867, 2008.

[2] J. Coria-Hernández, R. Meléndez-Pérez, A. Méndez-Albores, and J. L. Arjona-Román. Changes in myoglobin content in pork longissimus thoracis muscle during freezing storage. *Revista Mexicana de Ciencias Pecuarias*, 11(3):651–668, 2020.

[3] R. Das and M. Gerstein. The stability of thermophilic proteins: A study based on comprehensive genome comparison. *Functional & Integrative Genomics*, 1:76–88, 2000.

[4] S. R. Feroz, S. B. Mohamad, Z. S. Bakri, S. N. Malek, and S. Tayyab. Interaction of flavokawain b with lysozyme: A photophysical and molecular simulation study. *Journal of Luminescence*, 160:101–109, 2015.

[5] J. Funahashi, K. Takano, K. Ogasahara, Y. Yamagata, and K. Yutani. The structure, stability, and folding process of amyloidogenic mutant human lysozymel. *The Journal of Biochemistry*, 120(6):1216–1223, 1996.

[6] T. Herning, K. Yutani, K. Inaka, R. Kuroki, M. Matsushima, and M. Kikuchi. Role of proline residues in human lysozyme stability: A scanning calorimetric study combined with X-ray structure analysis of proline mutants. *Biochemistry*, 31(31):7077–7085, 1992.

[7] M. T. Ho, Z. Zou, B. Bhandari, and N. Bansal. Properties of functional camel milk powder. In C. L. Floch-Fouéré, P. Schuck, G. Tanguy, L. Lanotte and R. Jeantet, editors, *Drying in the Dairy Industry - From Established Technologies to Advanced Innovations*, pages 226–240. CRC Press, Taylor & Francis Group, Boca Raton, 2020.

[8] J. C. Kendrew, R. G. Parrish, J. R. Marrack, and E. S. Orlans. The species specificity of myoglobin. *Nature*, 174(4438):946–949, 1954.

[9] C. Küng, O. Protsenko, R. Vanella, and M. Nash. Decoding stability and epistasis in human myoglobin by deep mutational scanning and codon-level machine learning. *bioRxiv preprint*, 2024.

[10] P. A. Levashov, D. A. Matolygina, E. D. Ovchinnikova, I. Y. Adamova, D. A. Gasanova, S. A. Smirnov, and A. V. Levashov. The bacteriolytic activity of native and covalently immobilized lysozyme against gram-positive and gram-negative bacteria is differentially affected by charged amino acids and glycine. *FEBS Open Bio*, 2019.

[11] L. Lin, R. J. Pinker, and N. R. Kallenbach. .alpha.-Helix stability and the native state of myoglobin. *Biochemistry*, 32(47):12638–12643, 1993.

[12] I. Margiolaki and J. P. Wright. Synchrotron X-ray powder diffraction study of hexagonal Turkey egg-white lysozyme. *Acta Crystallographica*, D61:423–432, 2005. https://doi.org/10.2210/pdb1xft/pdb.

[13] A. N. Muttathukattil, P. C. Singh, and G. Reddy. Role of disulfide bonds and topological frustration in the kinetic partitioning of lysozyme folding pathways. *The Journal of Physical Chemistry B*, 2019.

[14] N. V. Rastriga, D. A. Gasanova, and P. A. Levashov. Adsorption of lysozyme on living cells of escherichia coli and its bacteriolytic activity in the presence of glycine and charged amino acids. *Moscow University Chemistry Bulletin*, 78:89–94, 2023.

[15] N. Saadati-Eskandari, L. Navidpour, P. Yaghmaei, and A. Ebrahim-Habibi. Amino acids as additives against amorphous aggregation: In vitro and in silico study on human lysozyme. *Applied Biochemistry and Biotechnology*, 2019.

[16] E. A. Sauter and J. E. Montoure. The relationship of lysozyme content of egg white to volume and stability of foams. *Journal of Food Science*, 37:918–920, 1972.

[17] A. Sidorova, V. Bystrov, A. Lutsenko, D. Shpigun, E. Belova, and I. Likhachev. Quantitative assessment of chirality of protein secondary structures and phenylalanine peptide nanotubes. *Nanomaterials*, 11(12):3299, 2021.

[18] A. E. Sidorova, A. O. Lutsenko, D. K. Shpigun, E. V. Malyshko, and V. A. Tverdislov. Methods to determine the chirality sign for helical and superhelical protein structures. *Biophysics*, 66(3):357–363, 2021.

[19] A. E. Sidorova, E. V. Malyshko, A. O. Lutsenko, D. K. Shpigun, and O. E. Bagrova. Protein helical structures: Defining handedness and localization features. *Symmetry*, 13(5):879, 2021.

[20] T. P. Silverstein, S. R. Kirk, S. C. Meyer, and K. L. M. Holman. Myoglobin structure and function: A multiweek biochemistry laboratory project. *Biochemistry and Molecular Biology Education*, 43(3):181–188, 2015.

[21] A. Tigerström. Thermostability of proteins. *BIOS*, 76:22–27, 2005.

[22] S. Venkataramani, J. Truntzer, and D. R. Coleman. Thermal stability of high concentration lysozyme across varying pH: A Fourier transform infrared study. *Journal of Pharmacy and Bioallied Sciences*, 5:148, 2013.

[23] L. Wan, M. B. Twitchett, L. D. Eltis, A. G. Mauk, and M. Smith. In vitro evolution of horse heart myoglobin to increase peroxidase activity. *Proceedings of the National Academy of Sciences*, 95(22):12825–12831, 1998.

[24] J. B. Wittenberg and B. A. Wittenberg. Myoglobin function reassessed. *Journal of Experimental Biology*, 206(12):2011–2020, 2003.

[25] M. Yao, J. Qi, X. Yang, Y. Mao, B. Xu, J. Hao, Y. Gu, Y. Zhang, and R. Liang. Effect of different pH on the structure and thermal stability of myoglobin. *Meat Research*, 38(6):1–8, 2024.

Pavel Levashov

Mathematics of supercooperativity in biochemistry

Combining Michaelis–Menten, Hill, and Langmuir Equations

Abstract: We present a universal mathematical model for oligomeric proteins and multiprotein complexes (enzymes, transport proteins, and receptors) exhibiting "supercooperative" behavior characterized by deviations from classical Michaelis–Menten kinetics, or the simple Langmuir adsorption isotherm equation, with exhibiting complex dependences of the enzymatic reaction rate on substrate concentration, or adsorption on ligand concentration, including variants of non-monotonic functions and the presence of multiple linear regions in double reciprocal plot. Based on an extensive set of experimental data from the literature, we show that this can be consistently described by a single, mathematically compact expression. The model represents the protein as an oligomer capable of adopting two (or more) interconvertible conformational states: a basal form, and a form saturated with substrate (or ligand). Each state is characterized by its own kinetic or adsorption parameters, and their interconversion is regulated by an equilibrium constant reflecting the cooperative transition during multisite substrate binding, where the fractions of different conformational states are expressed similar to the Hill equation. The Michaelis–Menten, Langmuir, and Hill equations are special cases of the new universal equation. The model encompasses a wide range of non-Michaelis (non-Langmuir) behavior within a general theoretical framework. Its predictive power is confirmed by the excellent agreement between the calculated curves and various experimental data, confirming the suitability of this approach as a general model for the analysis of supercooperative protein systems involved in the regulation of biological processes.

Keywords: non-Michaelis kinetics, cooperative regulation of enzymes, supercooperativity, nicotinamide adenine dinucleotide (NAD^+)-kinase, lysozyme

MSC 2020: 92C45

Acknowledgement: The section of theoretical analysis of lysozyme kinetics was carried out within the framework of the project "Activation of antibacterial bacteriolytic proteins of milk by low-molecular additives." The research was funded by the Russian Science Foundation, Grant 25-26-00144.

Pavel Levashov, Faculty of Chemistry, Lomonosov Moscow State University, Leninskie gory 1-11B, 119234 Moscow, Russia, e-mail: levashov@yahoo.com

https://doi.org/10.1515/9783112216620-010

1 Introduction

Supercooperativity refers to the phenomenon in which the binding of several molecules of a ligand or substrate to an oligomeric protein (transport protein or enzyme) significantly alters the adsorption or enzymatic characteristics of this protein complex. In recent years, biochemistry has accumulated many experimental examples of special cases of "non-Michaelis" enzyme kinetics, when very unusual dependencies of the initial reaction rate on the substrate concentration are observed, which cannot be described by equations developed for enzymes with positive or negative cooperativity. Such "unusual" dependencies often have the form of curves with minima and maxima, or the form of two or three "pyramid steps."

Such unusual kinetics are often associated with mechanisms of fine regulation of metabolic processes that have a dramatic effect on the state of a living cell. For example, for the key enzyme regulating the ratio of anabolic and catabolic processes in the cell, namely NAD^+-kinase, an "anomalous" dependence of activity on the substrate concentration (NAD^+) is known with two clearly discernible linear sections in the Lineweaver–Burk coordinates [1].

Sometimes, enzymes with complex kinetics function in the cell as part of multiprotein complexes. For example, NAD^+-kinase can be included in a multienzyme complex together with glutamate dehydrogenase [13, 2]. Glutamate dehydrogenase is an enzyme that plays a key role in nitrogen metabolic processes in the body.

A complex dependence called *supercooperativity* has also been described for the kinetics of platelet aggregation [14]. A complex "stepwise" dependence of activity on substrate concentration is demonstrated by alkaline phosphatase from the olfactory mucosa of rats [9]. This alkaline phosphatase is part of the receptor complexes of the animal's olfactory system, allowing the animal, like the most perfect physical device, to analyze odors in a wide range of concentrations (differing by seven or more orders of magnitude).

The well-known bacteriolytic enzyme lysozyme also demonstrates "stepwise supercooperativity" under several conditions [11], which may be associated with a feature of the regulation of this enzyme in the case of assembly into a membrane-attacking antibacterial oligomer [10].

"Stepwise kinetics" is also demonstrated by ATPase, which is part of complexly organized protein complexes of ion channels embedded in the cell membrane [8]. Complex "two-phase stepwise" kinetics are also demonstrated by some plant transport systems responsible for the assimilation of inorganic nitrogen [3].

This work examines the selection of a universal mathematical model for describing complexly regulated supercooperative transport proteins and enzymes for which the dependence of the equilibrium adsorption of a ligand on its free concentration (in the case of an enzyme, on the initial rate of the enzymatic reaction and the substrate concentration) has several linear sections in double reciprocal plot (in the case of an enzyme, in the Lineweaver–Burk plot).

2 Methods

For simplicity of presentation in general questions of mathematical transformations, below we write about enzymes and rates of enzymatic reactions, meaning that the mathematical formulas for the adsorption of ligands on transport proteins have the same form, except that the formulas will contain adsorption capacities of, instead of maximum reaction rates, and instead of the Michaelis constants there will be ligand dissociation constants—the substrate concentration changes to the ligand concentration in this case.

According to our assumption, a "sharp" conformational transition occurs when several n molecules of the substrate are bound by an n-oligomeric enzyme molecule, accompanied by a change in the properties of the enzyme, according to the chemical equation:

$$E_n + nS \xrightarrow{\mathrm{K}} E_nS_n, \tag{7.1}$$

where E is the enzyme, E_n is the oligomeric enzyme, S is the substrate, E_nS_n is the oligomeric enzyme after the binding of n substrate molecules, and K is the equilibrium constant. The binding of n substrate molecules is apparently the mechanism of physiological regulation of the enzyme, so it is logical to assume a rapid establishment of equilibrium between these two forms of the enzyme. Equilibrium is probably achieved by the time the initial rate of the enzymatic reaction begins to be measured experimentally. We can write the expression for K:

$$K = \frac{[E_n][S]^n}{E_nS} \tag{7.2}$$

From formula (7.2) follows the expression for the proportion of the enzyme in the state E_n (this equation can be easily derived under the assumption $[S] \gg [E]$, which is usually the case in real biological systems):

$$g1([S]) = \frac{K}{K + [S]^n} \tag{7.3}$$

The value of $g1$ changes from 1 to 0, and accordingly, the proportion of the enzyme in the state E_nS_n is expressed by the equation:

$$g2([S]) = 1 - g1([S]) = 1 - \frac{K}{K + [S]^n} = \frac{[S]^n}{K + [S]^n}. \tag{7.4}$$

The last expression in formula (7.4) is actually a variant of the Hill equation [6], which is sometimes used to describe the activity of oligomeric cooperative enzymes. Hill proposed such an equation for an approximate description of the isotherm of oxygen binding by hemoglobin. In its physical meaning, $\sqrt[n]{K}$ numerically corresponds to the concentration $[S]_{0.5}$ (the middle of the transition region) of the half-height of the curve $g1([S])$ when the relative shares of the enzyme in the two states are equal to

$$g1_{0.5} = g2_{0.5} = 0.5 = \frac{[S]_{0.5}^n}{K + [S]_{0.5}^n}. \tag{7.5}$$

As was said earlier, the dependence of the enzymatic reaction rate on the substrate concentration has two "Michaelis" segments; for each of the two enzyme states, the enzymatic reaction rate is described by the Michaelis–Menten equation. Knowing the Michaelis parameters of each of the two enzyme forms and the fraction of the enzyme converted from state E_n to state E_nS_n, we obtain the formula for the resulting enzymatic reaction rate:

$$\left.\begin{aligned} v([S]) &= g1([S]) \cdot \frac{v_1 \cdot [S]}{K_{m1} + [S]} + g2([S]) \cdot \frac{v_2 \cdot [S]}{K_{m2} + [S]} \\ &= g1([S]) \cdot \frac{v_1 \cdot [S]}{K_{m1} + [S]} + (1 - g1([S])) \cdot \frac{v_2 \cdot [S]}{K_{m2} + [S]} \end{aligned}\right\} \tag{7.6}$$

from where we derive the expression:

$$g1([S]) = \frac{v([S]) - \dfrac{v_2 \cdot [S]}{K_{m2} + [S]}}{\dfrac{v_1 \cdot [S]}{K_{m1} + [S]} - \dfrac{v_2 \cdot [S]}{K_{m2} + [S]}} \tag{7.7}$$

Substituting into formula (7.7), the experimental values $v([S])$ and the previously found v_1, v_2, K_{m1}, K_{m2}, we now calculate $g1([S])$. The curve $g1([S])$ can be calculated on the basis of experimental data. Now we transform expression (7.3) and obtain the formula:

$$\ln\left(\frac{1 - g1}{g1}\right) = n \ln([S]) - \ln(K) \tag{7.8}$$

Using the values of $g1$ calculated from the experimental data, we can plot the linear dependence of $\ln((1 - g1)/g1)$ on $\ln([S])$ to determine n and K.

Now, let us consider what form the experimental adsorption isotherms of transport proteins might take if Equation like (7.6) is satisfied for the biological system. We will write it as follows: where Am1 and K1 are the sorption capacity and ligand dissociation constant for the protein complex at low ligand concentration, the sorption capacity of the protein complex not saturated with ligand (L), and Am2 and K2 are the sorption capacity and ligand dissociation constant for the protein complex at high ligand concentration. For adsorption equilibrium in the Langmuir isotherm approximation, for each individual conformational state of the transport protein, we obtain the following Equation (7.9).

$$
\left.
\begin{aligned}
A([L]) &= g1([L]) \cdot \frac{A_m1 \cdot [L]}{K1 + [L]} + g2([L]) \cdot \frac{A_m2 \cdot [L]}{K2 + [L]} \\
g2([L]) &= 1 - g1([L]) = 1 - \frac{K}{K + [L]^n} = \frac{[L]^n}{K + [L]^n}.
\end{aligned}
\right\}
\tag{7.9}
$$

Figure 7.1A shows the typical shape of the oxygen adsorption isotherms for myoglobin and hemoglobin [4] in arbitrary units along the scale axes. To compare the shapes

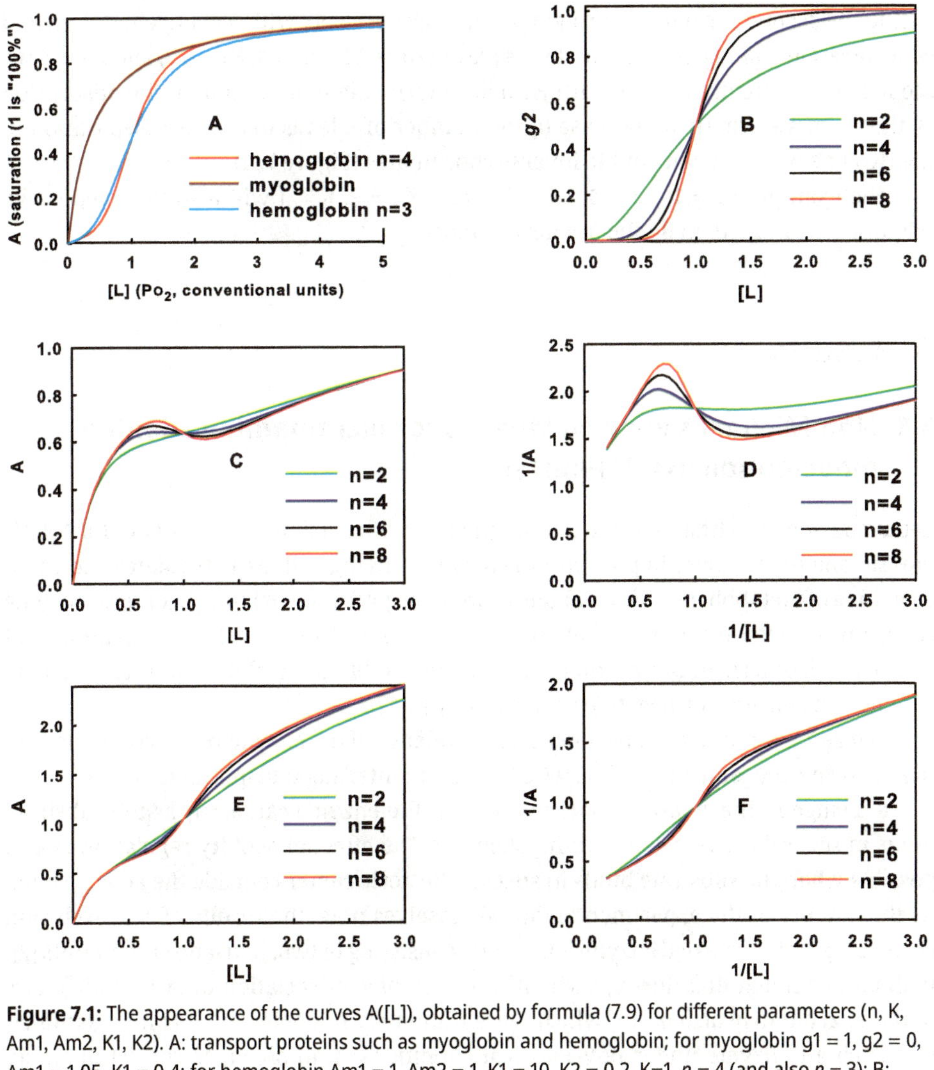

Figure 7.1: The appearance of the curves A([L]), obtained by formula (7.9) for different parameters (n, K, Am1, Am2, K1, K2). A: transport proteins such as myoglobin and hemoglobin; for myoglobin g1 = 1, g2 = 0, Am1 = 1.05, K1 = 0.4; for hemoglobin Am1 = 1, Am2 = 1, K1 = 10, K2 = 0.2, K=1, n = 4 (and also n = 3); B: type of dependence g2([L]) for different values of n (2, 4, 6, 8, oligomeric proteins in biological systems often have an even number of subunits and ligand binding centers); C: A([L]) for the case A1 = A2 = 1, K1 = 0.3, and K2 = 2, K=1, different values of n (2, 4, 6, 8); D: A([L]) from the example of graph C in coordinates $A^{-1}([L]^{-1})$; E: A([L]) for the case A1 = 1, A2 = 4, K1 = 0.3, and K2 = 2, K=1, different values of n (2, 4, 6, 8); F: A([L]) from the example of graph E in coordinates $A^{-1}([L]^{-1})$.

of the characteristic curves, we did not use exact constants for these proteins from a particular animal; we used various parameter ratios. For myoglobin, we take G1 = 1, G2 = 0, Am1 = 1.05, K1 = 0.4. For hemoglobin, we take Am1 = 1, Am2 = 1, K1 = 10, K2 = 0.2, K=1, n = 4 (and also n = 3). We see that with a slight decrease in the ligand concentration (partial pressure of oxygen), hemoglobin gives up almost all of its oxygen to myoglobin (supplies the body's tissues with oxygen).

Figure 7.1B shows $g2([L])$ at different n. As we can see, with increasing n, the transition to the second form of the protein occurs more sharply with a change in the ligand concentration. Taking as an example the parameters A1 = A2 = 1, K1 = 0.3, and K2 = 2, we see a complex picture (Figure 7.1C) with an intermediate minimum on the adsorption isotherm curve, where an increase in the number of n leads to a clearer separation of the two linear sections in double inverse coordinates (Figure 7.1D).

The example A1 = 1, A2 = 4, K1 = 0.3, and K2 = 2, K = 1 will lead to a step curve (Figures 7.1E and 7.1F in double reciprocal plot).

3 Results

3.1 Non-Michaelis enzyme kinetics for nicotinamide adenine dinucleotide (NAD^+)-kinase

Let us consider as a first example the complex non-Michaelis enzyme kinetics for NAD^+-kinase, one of the most important enzymes in a living cell, as it regulates the rapid switching of metabolism from catabolic to anabolic processes. It is an oligomeric enzyme consisting of several subunits, that converts NAD^+ into $NADP^+$ with the expenditure of ATP. The oligomeric organization of the enzyme is due to its ability to finely regulate activity, e. g., positive or negative cooperativity [5, 12].

Figure 7.2A shows the experimental dependence of the enzymatic reaction rate v on the substrate concentration $[S]$ as ($v([S])$) with an intermediate minimum [2].

A change in the catalytic characteristics of the enzyme can occur as the substrate binds in several active centers of the oligomer. The effect of activity regulation is also possible when the substrate binds in special allosteric centers outside the active center. As the authors of the experimental data themselves note, the results of size-exclusion chromatography revealed enzyme oligomers consisting of two, four, and six subunits [2]. It should be noted that during such chromatography, dissociation of larger oligomers of the enzyme may also occur, which were ultimately not detected. It can be assumed that such an enzyme under physiological conditions is an oligomer consisting of six subunits, and possibly eight to twelve subunits. After multistage purification of the enzyme by different methods, it was already represented by one subunit and had a simple Michaelis–Menten dependence of the rate of the enzymatic reaction on the concentration of the substrate [2].

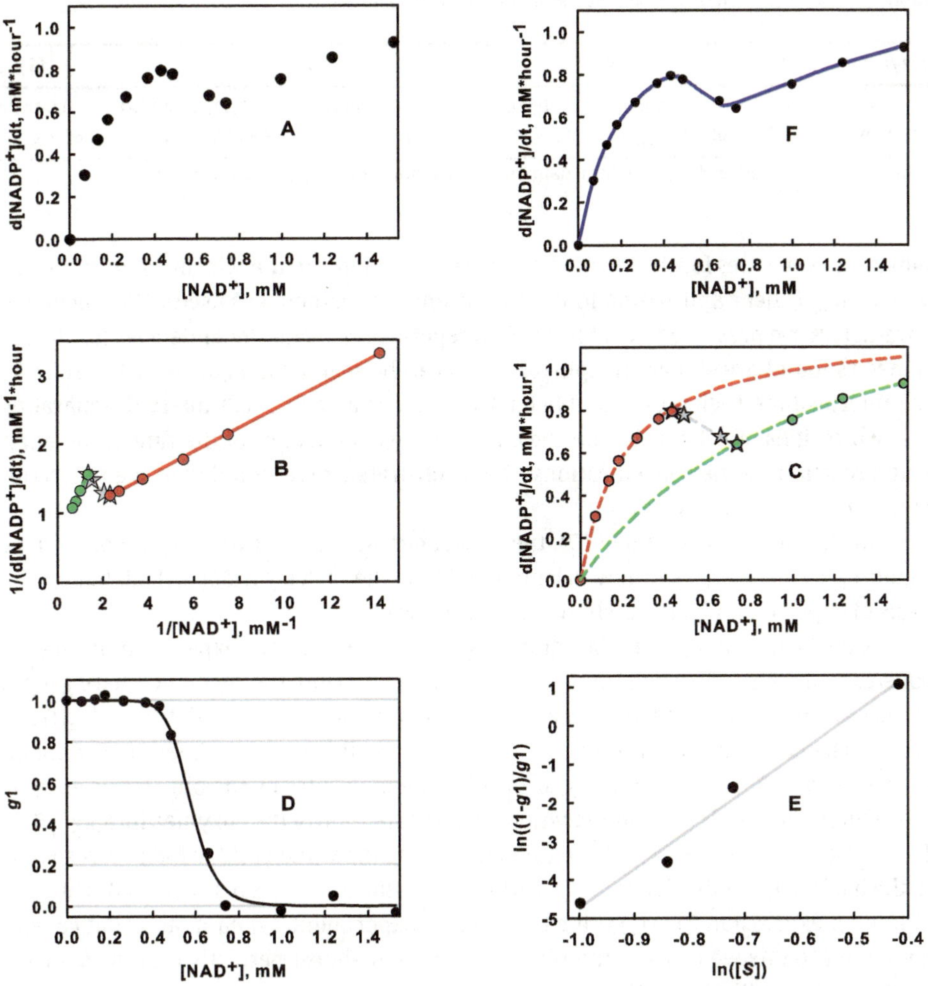

Figure 7.2: Dependence of the enzymatic reaction rate for NAD$^+$-kinase on the concentration of the substrate (NAD$^+$) [2]. A: original experimental data [2]; B: experimental data in double reciprocal plot (Lineweaver–Burk plot); C: two "Michaelis" curves (calculated from linear segments in double inverse coordinates); D: calculated from the experimental data values $g1([S])$, (curve $g1([S])$ according to Equation (7.7) based on the calculated parameters); E: Dependence of $\ln((1-g1)/g1)$ on $\ln([S])$ to determine the parameters n and K according to Equation (7.8); F: initial experimental data and a curve constructed according to Equation (7.6) based on parameters determined from experimental data (7.10).

In double reciprocal plot (Lineweaver–Burk plots, Figure 7.2B), two linear segments are observed (for low substrate concentrations and for high substrate concentrations). In each of these two segments, the curve corresponds to the Michaelis–Menten equation. The parameters of the Michaelis–Menten equation are different for each of the two segments. From the linear segment in Lineweaver–Burk plots, we determine the

Table 7.1: Kinetic parameters of enzymes estimated in both applications.

Enzyme	Ref.	v_1	K_{m1}	v_2	K_{m2}	n	$[S]_{0.5}$
NAD^+-kinase	[2]	1.191 mM·h^{-1}	0.206 mM	1.607 mM·h^{-1}	1.112 mM	10	0.59 mM
Lysozyme	[11]	1.004 $A_{650\,nm}$·s^{-1}	0.302 $A_{650\,nm}$	2.947 $A_{650\,nm}$·s^{-1}	0.994 $A_{650\,nm}$	16	0.65 $A_{650\,nm}$

Legend: Ref. = reference of experimental data; mM = millimol per L; $A_{650\,nm}$ = Optical density.

maximum velocities for the Michaelis–Menten equation and the Michaelis constants: v_1, v_2, K_{m1}, K_{m2} (linear approximation; the least squares method was used). The calculated parameters are presented in Table 7.1. Two separate Michaelis dependencies for the parameters found are shown in Figure 7.2C. As can be seen from Figures 7.2A, 7.2B, 7.2C, the intermediate region of "non-Michaelis" has a relatively small interval, separating the two regions of "Michaelis" on the curve of the dependence of the rate of the enzymatic reaction on the concentrations of the substrate (corresponding to the two states of the enzyme, respectively).

Substituting into formula (7.7), the experimental values $v([S])$ and the previously found v_1, v_2, K_{m1}, K_{m2}, we now calculate $g1([S])$. The curve $g1([S])$, calculated on the basis of experimental data, is shown in Figure 7.2D.

Using the formula (8) and the values of $g1$ calculated from the experimental data, we construct the dependence of $\ln((1-g1)/g1)$ on $\ln([S])$. Only the segment where $g1([S])$ changes from 1 to 0 is used as this dependence. In this case, the condition $1 > g1([S]) > 0$ is met. The curve is shown in Figure 7.2E. Using the linear approximation, we obtain the values $n = 10.05 \approx 10$ (binding of 10 substrate molecules by the oligomeric enzyme molecule, which leads to a conformational transformation of the enzyme), $\ln(K) = -5.31$, $K = 0.00494 \approx 0.005$, $[S]_{0.5} = \sqrt[n]{k} = 0.588$mM. Using the value of [S], which corresponds to the half-height of the dependence $g1([S])$, the value of $[S]_{0.5}$ is also 0.59 mM. Thus, we obtain an expression for $v([S])$ for the original experimental data in accordance with equation (7.6). As we can see, the resulting equation almost perfectly describes the experimental data (Figure 7.2F):

$$\left.\begin{aligned} v([S]) &= \frac{K}{K+[S]^n}\frac{v_1[S]}{K_{m1}+[S]} + \frac{[S]^n}{K+[S]^n}\frac{v_2[S]}{K_{m2}+[S]} \\ &= \frac{0.005}{0.005+[S]^{10}}\frac{1.191[S]}{0.206+[S]} + \frac{[S]^{10}}{0.005+[S]^{10}}\frac{1.607[S]}{1.112+[S]} \end{aligned}\right\} \tag{7.10}$$

3.2 Non-Michaelis enzyme kinetics for lysozyme

Figure 7.3A shows another type of experimental dependence of the initial rate of the enzymatic reaction on the substrate concentration [11]. Figure 7.3 also shows the results of processing this data according to the scheme used in the previous example with NAD^+-kinase. In this case, the enzyme is lysozyme, a bacteriolytic enzyme that lyses bacterial

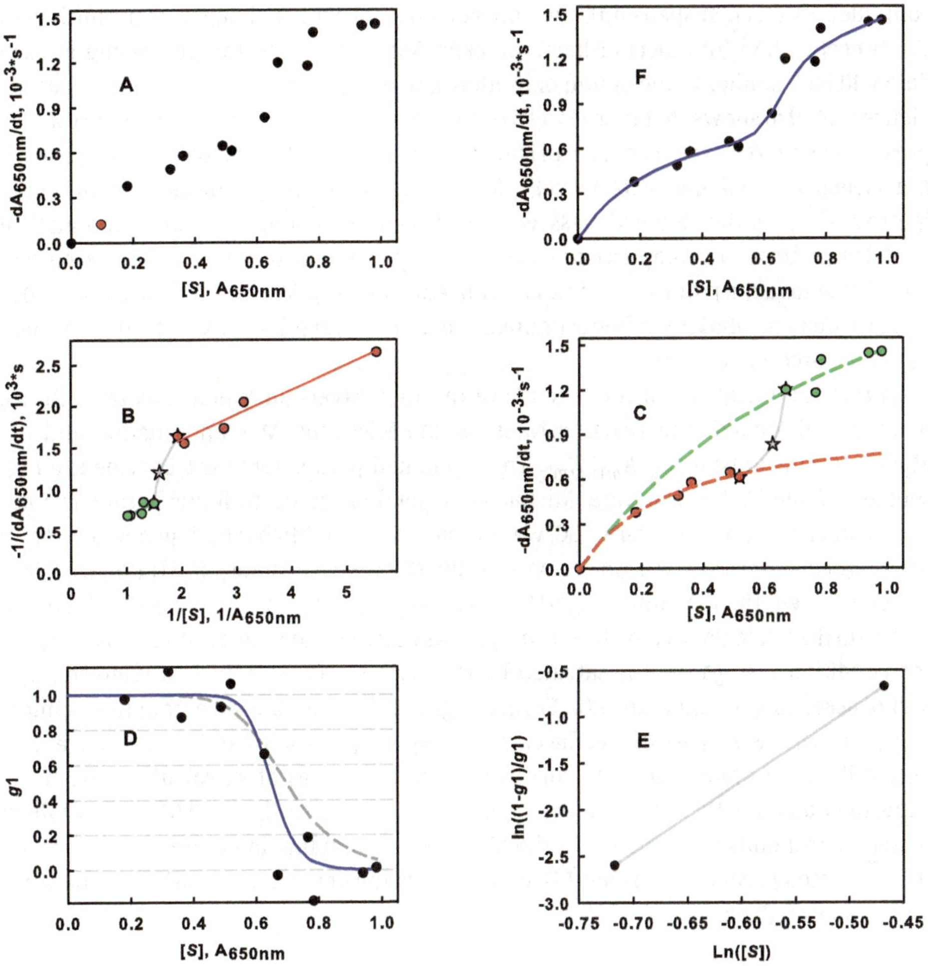

Figure 7.3: Dependence of the rate of the enzymatic reaction of lysozyme (rate of decrease in bacterial concentration) on the concentration of the substrate (bacteria) [11]. A: original experimental data [11]; B: experimental data in double inverse coordinates (Lineweaver–Burk plot); C: two "Michaelis" curves (calculated from linear segments in double reciprocal plot); D: calculated from the experimental data values $g1([S])$, (curve $g1([S])$ according to Equation (7.7) based on the calculated parameters); E: Dependence of $\ln((1-g1)/g1)$ on $\ln([S])$ to determine the parameters n and K according to Equation (7.8) (dashed line) and solid line for g1([S]0.5) ≈ 0.65, n ≈ 16; F: initial experimental data and a curve constructed according to Equations (7.6) and (7.11) based on parameters determined from experimental data.

cells. It is known that under certain conditions, lysozyme can form oligomers [10, 7]. The "giant" substrate in this example is whole bacterial cells and a cell wall copolymer. In this example, the phenomenon of the "non-Michaelis" shape of the curves, meaning a conformational transition during "binding of n ligand molecules" can, at the chemical level, mean the binding of some unknown low-molecular ligand (e. g., an oligosaccharide contained in a cell suspension), the concentration of which changes as a multiple of the concentration (number) of bacterial cells. Mathematically, the processing of such data will look similar to the option of binding n molecules of a low-molecular substrate. Figure 7.3A also shows that there is a strong scatter of points and a significant error. The presence of a large error is not surprising for this complex system, where the substrate is a suspension of living bacterial cells. Increasing the number of measurements could improve the situation. Nevertheless, we will check the capabilities of our mathematical model even for such a complex example of real experimental data with limited accuracy. We will remove the point with a low cell concentration (optical absorption less than 0.1) from further calculations as being limitedly reliable and significantly distorting the data in double inverse coordinates.

From the linear sections of the curve in Lineweaver–Burk plot (Figure 7.3B), we determine the maximum reaction rates for the Michaelis–Menten equation and the Michaelis constants: v_1, v_2, K_{m1}, K_{m2}. The calculated parameters for lysozyme are presented in Table 7.1. Two separate "Michaelis" dependencies for the found parameters are shown in Figure 7.3C. In Figure 7.3C, we compare the two "Michaelis dependencies" and the original experimental data. Using formula (7.7), we calculate $g1([S])$ (Figure 7.3D). As we can see, the dependence $g1([S])$ has a significant scatter of points, which can make further calculations of K and n approximate. For the «inter-Michaelis» region, the condition $1 > g1 > 0$ is satisfied by only two points, by which we determine K and n, according to Equation (7.8). From the graph (Figure 7.2E) we obtain the estimate $[S]_{0.5} \approx 0.7, n \approx 8$. As we can see, the curve $g1([S])$ for $[S]_{0.5} \approx 0.7$ and $n \approx 8$ (Figure 7.3D, dotted line) very approximately corresponds to the values of $g1$ calculated from the experimental data. We select another approximation of $g1([S]_{0.5}) \approx 0.65, n \approx 16$, which better corresponds to the points in the figure (blue solid line in Figure 7.3D). Based on the calculated parameters (Table 7.1), we construct the dependence $v([S])$ in accordance with Equation (7.6):

$$\left.\begin{aligned} v([S]) &= \frac{K}{K+[S]^n}\cdot\frac{v_1\cdot[S]}{K_{m1}+[S]}+\frac{[S]^n}{K+[S]^n}\cdot\frac{v_2\cdot[S]}{K_{m2}+[S]} \\ &= \frac{0.001}{0.001+[S]^{16}}\cdot\frac{1.004\cdot[S]}{0.302+[S]}+\frac{[S]^{16}}{0.001+[S]^{16}}\cdot\frac{2.947\cdot[S]}{0.994+[S]} \end{aligned}\right\} \tag{7.11}$$

As we can see, the curve $v([S])$, constructed according to formula (7.11) (the solid line in Figure 7.3F) describes the experimental points relatively well. The difficulties of calculating n can be overcome by a large number of experimental points in the transition region. Literary experimental data from different authors usually suffer from a small number of points near the substrate concentration region corresponding to $[S]_{0.5}$.

4 Discussion

It is particularly valuable that the equations described in this work are not empirical or simply convenient approximations of complex experimental curves that ignore the real physical meaning of individual parameters.

The general equation is derived directly from well-founded formulas describing the thermodynamics and kinetics of individual stages of physicochemical processes in multiprotein biological systems. Individual equations such as Langmuir, Michaelis–Menten, and Hill are monotonic functions of ligand concentration.

In the case of a mixture of protein complexes with different Langmuir, Michaelis, or Hill parameters in a biological system, the general equation may have a more complex form.

Curves with both intermediate minimum and step-type are excellently described and explained by the new general equation, which can significantly facilitate the study and interpretation of the functioning of complex macromolecular complexes in biological systems, e. g., for the case of Na^+/K^+-ATPase [8].

5 Conclusion

The model encompasses a wide range of non-Michaelis (non-Langmuir) behavior within a general theoretical framework. Its predictive power is confirmed by the excellent agreement between the calculated curves and various experimental data. It is then suitable as a general model for the analysis of supercooperative protein systems involved in the regulation of biological processes

Bibliography

[1] D. K. Apps. Pigeon-liver NAD kinase: The structural and kinetic basis of regulation of NADPH. *European Journal of Biochemistry*, 55(2):475–483, 1975. https://doi.org/10.1111/j.1432-1033.1975.tb02184.x.

[2] E. R. Bulygina, O. G. Busygina, and V. I. Telepneva. Change in the kinetic properties of NAD-kinase from pigeon heart muscle during purification (in Russian). *Biokhimiya (Moscow)*, 48(6):906–913, 1983. https://biochemistrymoscow.com/ru/archive/1983/48-06-0906/. Abstract: https://pubmed.ncbi.nlm.nih.gov/6309257/.

[3] T. Enomoto, N. Tone, N. Ishii, H. Hirono, A. Oi, Y. Hirono, N. Ikka, and H. Yamashita. Effect of ammonium: Nitrate application ratios on growth and nitrogen metabolism of tea plants (*Camellia sinensis* l.). *Plant Direct*, 2025. https://doi.org/10.1002/pld3.70084.

[4] J. Feher. Oxygen and carbon dioxide transport. In *Quantitative Human Physiology*, pages 586–594, 2012. https://doi.org/10.1016/B978-0-12-382163-8.00064-5.

[5] S. Garavaglia, A. Galizzi, and M. Rizzi. Allosteric regulation of bacillus subtilis NAD kinase by quinolinic acid. *Journal of Bacteriology*, 185(16):4844–4850, 2003. https://doi.org/10.1128/JB.185.16.4844-4850. 2003.

[6] A. V. Hill. The possible effects of the aggregation of the molecules of haemoglobin on its dissociation curves. *The Journal of Physiology*, 40(supplement):iv–vii, 1910. https://doi.org/10.1113/jphysiol.1910.sp001386.

[7] M. Matwiejczyk, A. Zambrowicz, and M. Besman. Lysozyme monomer, dimer, and oligomers: A review with a focus on immunological potential. *International Journal of Peptide Research and Therapeutics*, 31:79, 2025. https://doi.org/10.1007/s10989-025-10742-x.

[8] J. L. E. Monti, M. R. Montes, and R. C. Rossi. Steady-state analysis of enzymes with non-Michaelis-Menten kinetics: The transport mechanism of Na^+/K^+-ATPase. *The Journal of Biological Chemistry*, 293(4):1373–1385, 2018. https://doi.org/10.1074/jbc.M117.799536.

[9] O. M. Poltorak, P. A. Levashov, and E. S. Chukhray. Hill generalized equation and kinetics of reactions performed by protein complexes. Obobshchennoe uravnenie Khilla i kinetika reaktsiy, osushchestvlyaemykh belkovymi kompleksami. *Zhurnal Fizicheskoi Khimii (Russian Journal of Physical Chemistry A)*, 69(3):511–514, 1995 (in Russian).

[10] S. A. Ragland and A. K. Criss. From bacterial killing to immune modulation: Recent insights into the functions of lysozyme. *PLoS Pathogens*, 13(9):e1006512, 2017. https://doi.org/10.1371/journal.ppat.1006512.

[11] A. V. Shnitko, M. G. Chernysheva, P. A. Levashov, and G. A. Badun. Kinetic analysis as an approach to studying specific features of lysozyme–pluronic complexes. *Russian Chemical Bulletin*, 70(7):1400–1403, 2021. https://doi.org/10.1007/s11172-021-3230-3.

[12] B. Srinivasan. Explicit treatment of non-Michaelis-Menten and atypical kinetics in early drug discovery. *ChemMedChem*, 16(6):899–918, 2021. https://doi.org/10.1002/cmdc.202000791.

[13] V. I. Telepneva, O. L. Voronina, and E. R. Bulygina. NAD kinase interaction with the activator–glutamate dehydrogenase. *Biochemical International*, 14(5):903–910, 1987. Abstract: https://pubmed.ncbi.nlm.nih.gov/2841941/.

[14] P. V. Vrzheshch, A. V. Tatarintsev, D. E. Yershov, and S. D. Varfolomeyev. Cell response kinetics: The phenomenon of supercooperativity in aggregation of human platelets. *Thrombosis Research*, 66(5):537–547, 1992. https://doi.org/10.1016/0049-3848(92)90308-W.

Part II-A: Person’s biosphere

Part II-A.2: **Vital task-solving functions**

Rationale: 4

Part II-A.2: **Vital task-solving functions**

Body system's core functions that keep its cellular system intact to empower all internal physiological and external interaction functions.

Internal physiological vital task-solving functions primarily serve to provide oxygen and nutrients to cells for mitochondrial energy generation, to remove residuals, and to support homeostasis. This implies highly interwoven multiscale dynamics and controls.

In this volume:
- Modeling circulatory substance dynamics with integrated hepatic clearance
- Modeling pulsatile and diurnal rhythmicity of free and total cortisol
- Investigation of virus infection progression in tissue with reaction–diffusion models

In former volume BIOKYBERNETIKA:
- Structural modeling of vascular networks
- AI application for diagnosis of blood-fow disorders
- Hemodynamics of glucose / insulin regulation
- Dynamics of lymph flow
- Rhythmicity in the neuro-endocrine system as an integrate-and-fire mechanism
- Kinetic network model of the neuroendocrine of HPA axis dynamics
- Inflammation and immune response in atherosclerosis

https://doi.org/10.1515/9783112216620-011

Alexander Khrulenko and Fedor Popko

Modeling circulatory substance dynamics with integrated hepatic clearance

Abstract: The ability to quantitatively model substance transport in the cardiovascular system is important for drug development and for assessing hepatic disorders, where cirrhosis and portal hypertension alter both hemodynamics and metabolic clearance. This study presents algorithms for the numerical solution of hemodynamic equations and the convection-diffusion equation within an arbitrary closed system of elastic tubes, implemented using the Cardiovascular Simulating System (CVSS) software package. Using this computational framework, based on a quasi-one-dimensional approximation of the hemodynamic equations, we conducted numerical experiments to investigate the dynamics of indocyanine-green (ICG) distribution in the human cardiovascular system, incorporating the liver's detoxifying function. Special attention was given to analyzing how liver impairment severity and portal hypertension levels influence ICG's pharmacokinetic parameters. The results provide insights into the relationship between hepatic dysfunction and drug clearance, offering potential applications in clinical diagnostics and therapeutic monitoring, with potential applications in noninvasive liver function assessment, surgical and transplant planning, and prediction of pharmacokinetics for hepatically cleared drugs.

Keywords: Mathematical modeling, substance transfer, hemodynamics, Child–Pugh classification

MSC 2020: 65N21, 92C50

1 Introduction

This study presents a computational approach to modeling the pharmacokinetics of indocyanine green (ICG)—a diagnostic tracer whose clearance serves as a sensitive marker of hepatic function. The development of novel pharmaceuticals necessitates a thorough understanding of substance distribution dynamics in human blood circulation. These processes are critically dependent on hepatic function and hemodynamic parameters. Liver diseases, particularly fibrosis and cirrhosis accompanied by portal hypertension, induce significant hemodynamic alterations that directly affect drug metabolism. According to the World Health Organization [5], hepatic disorders remain

Alexander Khrulenko, Fedor Popko, Faculty of Computational Mathematics and Cybernetics, Lomonosov Moscow State University, Ulitsa Kolmogorova, 1, Moscow, 119991, Russia, e-mails: khrulenko@cs.msu.ru, popkofd@my.msu.ru

https://doi.org/10.1515/9783112216620-012

among the leading global causes of mortality, underscoring the importance of such investigations.

The Cardiovascular Simulating System (CVSS) [3], implementing quasi-one-dimensional hemodynamic equations, is employed to simulate human circulatory system hemodynamics [7], quantify blood substance concentrations, and determine ICG half-life under varying physiological conditions.

The model is validated against clinical trial data [4], with simulations conducted for three distinct scenarios corresponding to Child–Pugh [6] classification stages A, B, and C of liver cirrhosis.

This computational framework enables precise determination of significant ICG pharmacokinetic parameters, baseline values for healthy hepatic function and progressive increases in half-life corresponding to Child–Pugh classes A, B, C cirrhosis.

2 Material and methods

2.1 Cardiovascular System Modeling framework

2.1.1 Graph-based hemodynamic model

The cardiovascular system is represented as a directed graph where:

- *Edges* correspond to blood vessels.
- *Nodes* represent:
  - Branching points of the vascular network.
  - Major organs and tissues.
  - The heart chambers (left ventricle and right atrium).

Blood flow dynamics are modeled using a quasi-one-dimensional approximation governed by the following system of equations:

1. Mass conservation (continuity equation):

$$\frac{\partial S}{\partial t} + \frac{\partial (SU)}{\partial x} = 0.$$

2. Momentum conservation (Navier–Stokes simplified):

$$\frac{\partial U}{\partial t} + \frac{\partial}{\partial x}\left(\frac{U^2}{2}\right) + \frac{1}{\rho}\frac{\partial P}{\partial x} = F_T + F_{TP},$$

where: $S(x,t)$ – vessel cross-sectional area (luminal area); $U(x,t)$ – blood flow velocity; $P(x,t)$ – intravascular pressure; ρ – blood density (assumed constant); F_T – gravitational force density; F_{TP} – viscous friction force (wall shear stress).

3. Constitutive relation (pressure-area relationship)
   The system is closed by a pressure-area constitutive equation that describes vessel compliance. A piecewise-linear model is used:

$$S(P) = \begin{cases} S_{\min}, & P \leq P_{\min}; \\ S_{\min} + \frac{S_{\max}-S_{\min}}{P_{\max}-P_{\min}}(P - P_{\min}), & P_{\min} \leq P \leq P_{\max}; \\ S_{\max}, & P \geq P_{\max}. \end{cases}$$

   Each vessel has unique parameters ($S_{\min}, S_{\max}, P_{\min}, P_{\max}$) defining its distensibility. More sophisticated nonlinear or viscoelastic models can be incorporated as needed.

2.1.2 Boundary and coupling conditions

1. At internal nodes (branching points):
   - *Mass conservation:* Total inflow equals total outflow.
   - *Pressure continuity:* Uniform pressure is maintained across connected vessels.
2. At tissue/organ nodes:
   - *Mass conservation* is enforced.
   - *Darcy's law* governs perfusion through capillary beds.
3. Heart model (closed-loop circulation): The heart is represented by two lumped-parameter chambers:
   - *Left ventricle* (systemic circulation pump)
   - *Right atrium* (venous return receiver)

   This model takes into account that the duration of systole/diastole depends on venous return and ventricular filling. Self-consistent coupling ensures physiologically realistic pressure-volume dynamics.

2.2 Transport of substances in a closed vascular system

2.2.1 Transport equations on a vascular graph

The transport of substances through a graph representing the cardiovascular system is analyzed using a quasi-one-dimensional approach. We assume that the solution to the hemodynamic equations described in Section 2 is known, and at each point of any vessel with index i, there exists a certain substance dissolved in blood with a mass concentration $c_i(x,t)$. The transport of the substance occurs due to convection and diffusion, and the flux $W_i(x,t)$ in the vessel can be expressed as the sum of convective $W_{c,i}$ and diffusive $W_{d,i}$ components [2]:

$$W_i = W_{c,i} + W_{d,i} = (cSu)_i - DS_i\frac{\partial c_i}{\partial x}, \tag{8.1}$$

where D is the diffusion coefficient, assumed constant along the vessel and uniform across all edges of the graph.

The integral equation describing mass conservation of the substance over a spatial segment $[x_1, x_2]$ of a vessel and a time interval $[t_1, t_2]$ is given by

$$\int_{x_1}^{x_2} (Sc)_i\Big|_{t_1}^{t_2} dx = \int_{t_1}^{t_2} (W_i(x_1, t) - W_i(x_2, t))\, dt. \tag{8.2}$$

2.2.2 Boundary conditions: Coupling at internal graph vertices

We introduce a topological orientation parameter $z_{i,k} \in \{-1, +1\}$ that characterizes the incidence relationship between edge i and vertex k in the vascular network:

$$z_{i,k} = \begin{cases} +1 & \text{if edge } i \text{ originates from vertex } k \text{ (outgoing);} \\ -1 & \text{if edge } i \text{ terminates at vertex } k \text{ (incoming).} \end{cases}$$

The boundary conditions are defined by coupling conditions at the internal vertices of the graph: continuity of mass concentration and volumetric substance flux. Let $c_{i_j,k}$ denote the boundary concentration of the substance on edge i_j at vertex k, which connects edges i_j ($j = 1, \ldots, n_{\text{deg}}(k)$ – number of edges at the branch vertex k). These conditions can be expressed as:

1. Concentration continuity:

$$c_{i_j,k} = c_{i_l,k}, \quad j \neq l;\ j, l = 1, \ldots, n_{\text{deg}}(k). \tag{8.3}$$

2. Flux balance (mass conservation):

$$\sum_{j=1}^{n_{\text{deg}}(k)} z_{i_j,k} \int_{x_1^{(i_j)}}^{x_0^{(i_j)}} (Sc)_{i_j}\Big|_{t_1}^{t_2} dx = \int_{t_1}^{t_2} \left(G_k(t) + \sum_{j=1}^{n_{\text{deg}}(k)} z_{i_j,k} W_{i_j}(x_1^{(i_j)}, t) \right) dt. \tag{8.4}$$

Here, $G_k(t)$ represents the source (or sink) of the substance at the internal vertex k. The points $x_0^{(i_j)}$ and $x_1^{(i_j)}$ lie on edge i_j, where $x_0^{(i_j)}$ corresponds to vertex k, and $x_1^{(i_j)}$ is located in close proximity to this vertex.

Alternatively, instead of conditions (8.3)–(8.4), other coupling conditions can be imposed, e. g., continuity of volumetric concentration under flux through an internal vertex.

2.2.3 Model closure: Heart balance equation

The transport model on the vascular graph is closed by a balance equation over the time interval $[t_1, t_2]$ for the self-consistent heart model described in Section 2.1.2:

$$C_k(t_2)V_k(t_2) = C_k(t_1)V_k(t_1) + \int_{t_1}^{t_2} (z_{V,k}W_V(t) + z_{A,k}W_A(t))\, dt, \tag{8.5}$$

where V_k and C_k are the blood volume and substance concentration in the heart, respectively; W_A and W_V represent the substance fluxes from the aorta into the heart and from the venous system into the heart, respectively; the coefficients $z_{V,k}$ and $z_{A,k}$ are defined analogously to Section 2.2.2 for the atrium and aorta.

This formulation of the balance equation ensures a self-consistent heart model with respect to flow dynamics [2].

2.3 Approximation of transport equations and coupling conditions

2.3.1 Discretization scheme

Let the number of edges in the graph be n_{link}, and the number of vertices be $m_{\text{node}} + 1$. Consider a vessel labeled i with length l_i. On the interval $[0, l_i]$, which serves as the domain of the independent variable x, we introduce a uniform spatial grid with stepsize $h = l_i/J_{i,\max}$ and a temporal grid with time step τ.

The discrete grid is defined as

$$\omega_{\tau h} = \{x_j = x_{j-1} + h,\ j = 0, 1, 2, \ldots, J_{i,\max};\ t_n = \tau n,\ n = 0, 1, 2, \ldots\}.$$

On this grid, we define the following discrete functions:

$$s = s_{i,j}^n = s_j^n = S_i(x_j, t_n), \quad u = u_{i,j}^n = u_j^n = u_i(x_j, t_n),$$
$$c = c_{i,j}^n = c_j^n = c_i(x_j, t_n).$$

2.3.2 Finite-difference approximation

For each internal spatial node ($j = 1, 2, \ldots, J_{i,\max} - 1$) of every vessel ($i = 1, \ldots, n_{\text{link}}$), the transport equations (8.1)–(8.2) are approximated using the following finite-difference scheme:

$$(sc)_t = \frac{1}{h_i}(\widehat{w}_{j-1/2} - \widehat{w}_{j+1/2}), \tag{8.6}$$

where

$$w_{j+1/2} = \overline{(csu)} - D\bar{s}c_x. \tag{8.7}$$

Here, we use the following notation: $y_j = y(x_j, t_n), \hat{y}_j = y(x_j, t_{n+1}), y_x = (y_{j+1} - y_j)/h_i$, $\bar{y} = (y_j + y_{j+1})/2$.

Since this scheme is three-point in space, boundary conditions must be specified at each edge endpoint.

2.3.3 Coupling conditions at internal vertices

The coupling conditions (8.3)–(8.4) are discretized as follows. For each internal vertex k ($k = 0, \ldots, m_{\text{node}}$), we define the difference equations:

1. Concentration continuity:

$$\hat{c}_{i_j, j(i_j)} = \hat{c}_{i_l, j(i_l)}, \quad j \neq l,\ j, l = 1, \ldots, n_{\deg}(k), \tag{8.8}$$

2. Mass conservation:

$$\sum_{l=1}^{n_{\deg}(k)} \frac{h_{i_l}}{2}\left((sc)_{i_l, j(i_l)}\right)_t = G_k + \sum_{l=1}^{n_{\deg}(k)} \left(z_{i_l,k} \widehat{w}_{i_l, j(i_l) - 0.5 z_{i_l,k}}\right). \tag{8.9}$$

Here, $j(i_l)$ denotes the boundary grid point on edge i_l, and $j(i_l) - 0.5z_{i_l,k}$ corresponds to the midpoint of the segment adjacent to vertex k.

2.3.4 Heart balance equation

The integral balance equation (8.5) is approximated for the vessel l connected to the heart:

$$\frac{h_l}{2}\left((sc)_{l,j(l)}\right)_t = z_{l,k}\left(\widehat{w}_{l,j(l)-0.5z_{l,k}} - \widehat{w}_{l,j(l)}\right), \tag{8.10}$$

$$(C_k V_k)_t = z_{l,k} \widehat{w}_{l,j(l)}, \tag{8.11}$$

$$\frac{h_l}{2}\left((sc)_{l,j(l)}\right)_t = z_{l,k} \widehat{w}_{l,j(l)-0.5z_{l,k}}, \tag{8.12}$$

where $(V_k)_t = z_{l,k}\widehat{(su)}_{l,j(l)}$.

The flux $w_{l,j(l)}$ at the node adjacent to the heart is computed as

$$w_{l,j(l)} = (su)_{l,j(l)}\underline{C} - Ds_{l,j(l)} z_{l,k} \frac{C_k - c_{l,j(l)}}{h_l}. \tag{8.13}$$

Here, $\underline{C}$ is defined as: C_k if blood is ejected from the heart ($z_{l,k} u_{l,j(l)} < 0$) or $c_{l,j(l)}$ otherwise ($z_{l,k} u_{l,j(l)} \geq 0$).

2.3.5 Conservation properties

As shown in [2], the proposed finite-difference scheme (8.6)–(8.13) is conservative. This property ensures numerical stability, allowing long-term simulations necessary for studying the dynamics of blood-borne substance transport in the cardiovascular system.

2.4 Cardiovascular simulating system software package (CVSS)

For simulating transport phenomena in the large-scale circulatory network comprising approximately two thousand vessels, we employed the specialized CVSS (Cardiovascular Simulating System) software package. The vascular graph incorporates:

1. Anatomically detailed submodels:
   - Hepatic circulation system
   - Renal vascular networks
   - Coronary circulation
   - Cerebral vasculature
2. Physiological basis:
   - Constructed using established human physiological data
   - Incorporates representative averaged parameters
   - Maintains characteristic vascular properties
3. Validation and applications: The model has demonstrated its research capability through multiple successful applications:
   - Reproduction of periodic cardiovascular dynamics
   - Simulation of portal hypertension
   - Investigation of hemodynamic disturbances
4. Key features:
   - Parameters adjustable across physiological ranges
   - Capable of patient-specific customization
   - Compatible with clinical data integration
   - Supported by CVSS toolkit functionality

The model's flexibility in parameter variation within physiological bounds enables its adaptation for personalized medicine applications when patient-specific data becomes available through the CVSS framework.

Note: The CVSS package provides comprehensive tools for both network parameterization and simulation analysis, making it particularly suitable for fundamental hemodynamic research and clinical applications.

3 Results

3.1 Substance extraction by organs and tissues

For a detailed description of mass transport in user-defined organs and tissues, each graph node is endowed with *sources* and/or *sinks*. The rate at which an organ releases a substance into the bloodstream is prescribed by the source flux Q_s (mass per unit time). Removal of circulating substance is characterized by the *extraction coefficient* $\alpha \in [0, 1]$.

Let Q_{in} be the incoming mass flux from all vessels connected to the node, and Q_{out} the outgoing flux. Mass balance in the node then reads

$$Q_{\text{out}} = (1 - \alpha)\, Q_{\text{in}} + Q_s. \tag{8.14}$$

Limiting cases of α

1. $\alpha = 0$ (*no extraction*). The node does not remove the substance: $Q_{\text{out}} = Q_{\text{in}} + Q_s$.
2. $\alpha = 1$ (*complete extraction*). The incoming substance is entirely absorbed: $Q_{\text{out}} = Q_s$.

Equation (8.14) is imposed at every organ node and couples the convective–diffusive transport in adjacent vessels with metabolic uptake or release inside the tissue.

3.2 Liver modeling

The liver receives a *dual* inflow: roughly 20 % of hepatic blood arrives via the hepatic artery, while the remaining 80 % is delivered by the portal vein draining the splanchnic circulation. Inside the organ both vessels divide into left and right branches and further into *segmental* branches. These supply independent functional units whose venous blood converges into hepatic veins and finally into the inferior vena cava.

We adopt Couinaud's eight-segment scheme (Figure 8.1), which partitions the liver into two lobes (right: segments 5–8; left: segments 1–4). Segment 1 (caudate lobe) contributes only a minor share of total flow and is therefore omitted. Segments 2–8 are represented as individual graph nodes in the cardiovascular model, each endowed with:

- two inflow edges (a portal–venous branch and a hepatic–arterial branch);
- one outflow edge (a segmental hepatic-vein branch).

Within a segment, microvascular exchange with parenchyma is described by the extraction formalism introduced in Section 3.1. Adjusting the segment-specific diffusion coefficient and extraction factor allows us to mimic regional perfusion loss or shunting that accompanies cirrhosis.

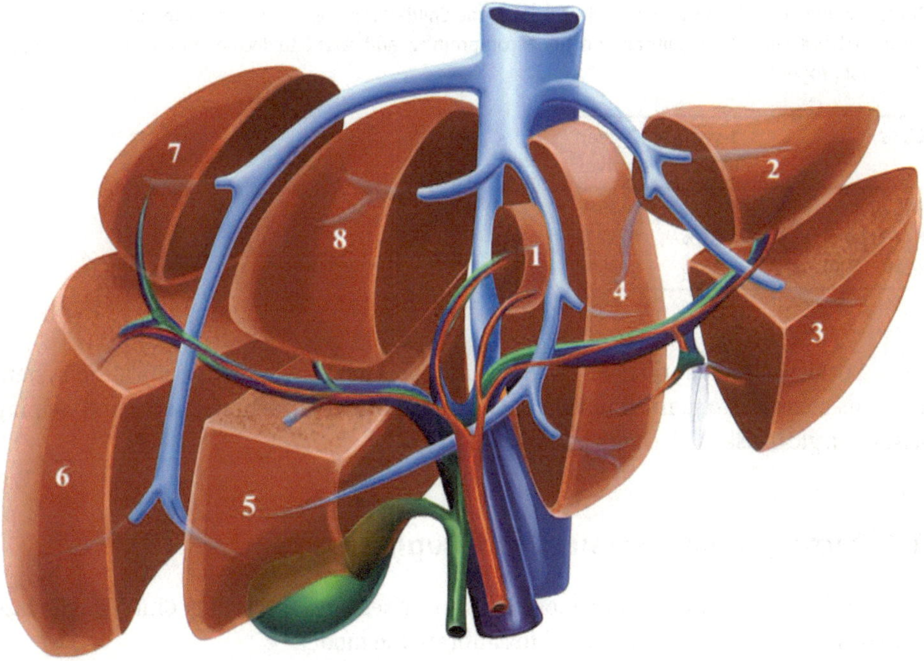

Figure 8.1: Graph of hepatic circulation with the location of regions in which substances are extracted.

3.3 Child–Pugh classification of cirrhosis severity

The Child–Pugh score is a widely used clinical tool that combines five clinical variables to grade the severity of hepatic dysfunction and predict prognosis in patients with chronic liver disease. The score evaluates total bilirubin, serum albumin, prothrombin time/INR, ascites severity, and hepatic encephalopathy grade. Each parameter is scored from 1 to 3 points, with higher scores indicating worse function. The sum of these points (ranging from 5 to 15) determines the Child–Pugh class: A (5–6 points), B (7–9 points), or C (10–15 points). Table 8.1 lists the detailed scoring criteria, while Table 8.2 shows

Table 8.1: Child–Pugh scoring criteria for assessing the severity of chronic liver disease. Each parameter is scored from 1 to 3 points, with higher scores indicating greater impairment. The total score (5–15 points) determines the Child–Pugh class (see Table 8.2).

Parameter	1 point	2 points	3 points
Ascites	none	mild, controlled	tense, refractory
Total bilirubin (μmol L^{-1})	< 34	34–50	> 50
Serum albumin (g dL^{-1})	> 3.5	2.8–3.5	< 2.8
Hepatic encephalopathy	none	grade I–II	grade III–IV
Prothrombin time prolongation (s) / INR	1–4 / < 1.70	4–6 / 1.71–2.20	> 6 / > 2.20

Table 8.2: Predicted 1- and 2-year survival rates by the Child–Pugh class. Class A corresponds to compensated cirrhosis, class B to significant functional compromise, and class C to decompensated cirrhosis with the worst prognosis.

Class	1-year survival	2-year survival
A (5–6 points)	100 %	85 %
B (7–9 points)	81 %	57 %
C (10–15 points)	45 %	35 %

the associated survival rates at 1 and 2 years for each class. This classification system remains a cornerstone in clinical hepatology, guiding treatment decisions and helping assess surgical risk in cirrhotic patients.

3.4 Portal pressure versus Child–Pugh class

Clinical data show that portal venous pressure rises with worsening Child–Pugh class [1]. The following averages are used to calibrate the model.

Values from Table 8.3 guide the adjustment of segmental parameters in the disease scenarios described next.

Table 8.3: Mean portal venous pressure (mmHg) reported for patients in different Child–Pugh classes of cirrhosis. Values are given as mean ± standard deviation.

Class	Pressure (mmHg)
A	10.2 ± 5.6
B	12.1 ± 7.6
C	14.6 ± 7.2

3.5 Modeling indocyanine-green dynamics in the cardiovascular system

3.5.1 Indocyanine green as a diagnostic marker of liver function

Indocyanine green (ICG) is commonly used during medical investigations for assessing the functionality of the liver. Another significant property of ICG is that it is completely removed from the body by the liver, without being metabolized anywhere else. This makes ICG considerably valuable for assessing how effectively the liver is cleaning the blood. Amounts of ICG in the blood directly reflect how effectively liver is functioning.

This helps clinicians obtain clear evidence about the health of the liver and how liver disease is likely to progress.

Because of this, regularly measuring ICG levels in the blood has become an important method for evaluating liver health, especially for patients with serious liver conditions like cirrhosis and portal hypertension. Monitoring ICG helps doctors not only measure how severely liver function is impaired but also predict potential health complications, assess the risks involved in surgeries, and plan individualized treatments. In medical practice, data from ICG tests is very useful when planning liver surgeries such as resections or transplants. It also helps doctors determine how well medications are working and monitor changes in the patient's liver condition over time. Overall, using ICG greatly improves the accuracy of medical assessments and helps doctors make timely and effective decisions.

3.5.2 Disease scenarios: Cirrhosis classes A, B, and C

To investigate how hepatic dysfunction alters indocyanine-green (ICG) kinetics, three disease scenarios were constructed in accordance with the Child–Pugh classification (classes A, B, and C). Disease severity is introduced *segmentwise* by reducing the diffusion coefficient in selected portal and arterial branches, thereby (i) lowering segmental perfusion and (ii) elevating the modeled portal pressure to typical clinical values (13 mmHg for A, 15 mmHg for B, 17 mmHg for C). Complete flow shutdown is represented by assigning a near-zero diffusion coefficient to the affected segment. The healthy model (*baseline*) retains physiological parameters throughout the hepatic graph.

3.6 Calibration of the hepatic extraction coefficient

In the initial stage of the study, the extraction parameter of indocyanine green (ICG), which characterizes the liver segments' capability to eliminate the substance, was calibrated for the "healthy liver" model. The goal of this stage was to adjust the parameter so that the calculated elimination half-life closely matched the known experimental half-life values of ICG in healthy individuals (approximately 180 seconds).

A series of computational experiments were conducted to test various extraction coefficient values. The results are presented in Table 8.4.

As shown in Table 8.4, an extraction coefficient of $\alpha = 0.2$ resulted in a significantly prolonged half-life of 443 seconds, inconsistent with clinical data. Increasing the extraction coefficient to $\alpha = 0.4$ provided a calculated half-life of 173 seconds, closely aligning with experimental values for healthy subjects. Consequently, $\alpha = 0.4$ was identified as the optimal extraction coefficient and was subsequently used in further calculations.

Table 8.4: Dependence of indocyanine-green (ICG) half-life on the extraction coefficient α used in the model calibration. A higher extraction coefficient corresponds to faster hepatic clearance.

Extraction coefficient α	Half-life (s)
0.1	567
0.2	443
0.3	204
0.4	173
0.5	131

3.6.1 Clinical reference data for validation

Validation of the mathematical model relied on published clinical data that report indocyanine-green (ICG) half-lives under different degrees of hepatic impairment. In particular, we employed the study conducted at the S. M. Kirov Military Medical Academy, Saint Petersburg [4], which included 149 patients with liver cirrhosis and 20 control subjects without evidence of liver disease. All participants underwent a standard ICG clearance test. The plasma elimination rate (PER) of ICG was markedly reduced in cirrhotic patients (7.6 ± 3.4 %/min) compared with controls (22.9 ± 4.2 %/min). Furthermore, PER progressively declined with worsening Child–Pugh class:

- class A: 11 ± 3.5 %/min;
- class B: 7.6 ± 2.9 %/min;
- class C: 5.3 ± 2.0 %/min.

Half-life values used for comparison were obtained from $t_{1/2} = \frac{\ln 2}{k_e}$, $k_e = \frac{\mathrm{PER}}{100}$. The resulting half-life values obtained from clinical studies are shown in Table 8.5.

Table 8.5: Clinical reference data for indocyanine-green (ICG) plasma elimination rate (PER) and half-life $t_{1/2}$ by the Child–Pugh class, adapted from [4]. Values are given as mean ± standard deviation. PER = plasma elimination rate in % min^{-1}.

Child–Pugh class	PER (% min^{-1})	Half-life $t_{1/2}$ (min)
A	11 ± 3.5	7.0 ± 2.2
B	7.6 ± 2.9	10.7 ± 4.1
C	5.3 ± 2.0	15.2 ± 5.8

3.6.2 Modeled half-lives versus clinical observations

After fixing the extraction coefficient at $\alpha = 0.4$ for the *healthy-liver* baseline, three simulations were run that reflect Child–Pugh classes A, B, and C. For each case, the indocyanine-green (ICG) half-life $t_{1/2}$ was obtained from the numerically computed concentration curve as the time required for the circulating amount of ICG to decrease to 50 % of its initial value.

The results on picture 8.2 show a monotonic prolongation of $t_{1/2}$ with increasing cirrhosis severity: from 328s in class A to 668s in class C. This trend aligns with the progressive decline in plasma elimination rate reported in [4], thereby supporting the validity of the coupled cardiovascular-liver model across the full Child–Pugh spectrum.

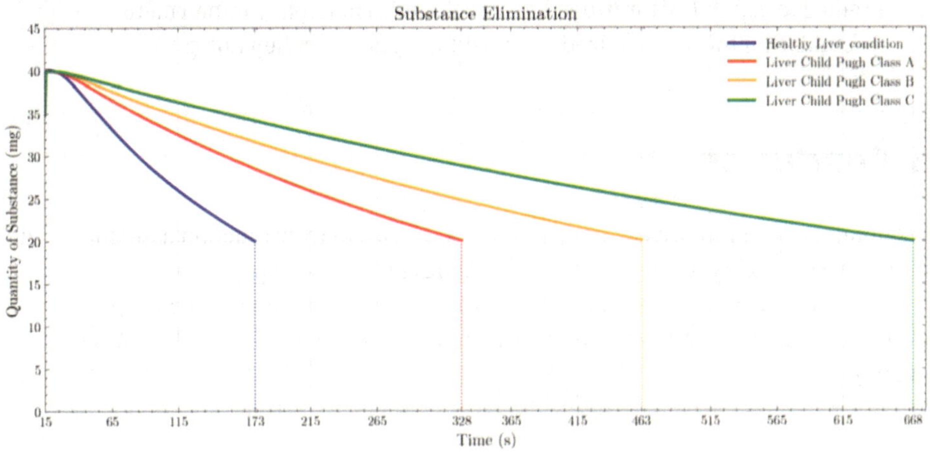

Figure 8.2: Substance elimination curves for indocyanine green (ICG) under different liver conditions. The y-axis shows the remaining quantity of ICG in blood (mg), and the x-axis shows elapsed time (s). Colored lines indicate: blue – healthy liver; red – Child–Pugh class A; orange – Child–Pugh class B; green – Child–Pugh class C. Vertical dotted lines mark the half-life ($t_{1/2}$) for each case.

4 Discussion

In this study, numerical experiments were carried out using the CVSS software package to investigate the dynamics of indocyanine-green distribution within the human cardiovascular system, taking into account the detoxifying function of the liver. Particular emphasis was placed on analyzing the impact of liver impairment severity and portal hypertension levels on the pharmacokinetic parameters of the substance.

The calculations yielded numerical values for the ICG half-life periods under various liver conditions: healthy liver and liver with varying degrees of cirrhosis severity according to the Child–Pugh classification (classes A, B, and C). To ensure accurate computations, an optimal extraction parameter was preliminarily determined, reflecting the hepatic segments' ability to eliminate the substance based on real experimental data. The numerical results obtained were consistent with published clinical data [4], confirming the validity of the proposed model and the adopted approach.

The calculations demonstrated that increasing severity of liver cirrhosis according to the Child–Pugh classification significantly prolongs the ICG half-life periods, indicating reduced liver functional activity. These findings can be valuable for future research in predicting cirrhosis progression dynamics, which in turn can assist physicians in making decisions regarding individualized therapies for patients with liver diseases.

The demonstrated correlation between ICG clearance kinetics and cirrhosis severity (Child–Pugh classification) provides:
1. A quantitative framework for noninvasive liver function assessment.
2. Predictive capability for drug pharmacokinetics in hepatic impairment.
3. A foundation for personalized therapeutic regimens in hepatology.

5 Conclusion

We obtained a quantitative framework for noninvasive liver function assessment with predictive capability for drug pharmacokinetics in hepatic impairment.

These findings advance computational approaches to modeling hepatic pathophysiology and drug disposition, suggest direct application in individualized clinical decision making.

Bibliography

[1] M. Al Mahtab, S. M. Noor E. Alam, M. A. Rahim, M. A. Alam, F. A. Khondaker, A. L. Moben, M. Mohsena, and S. M. F. Akbar. Hepatic venous pressure gradient measurement in Bangladeshi cirrhotic patients: A correlation with child's status, variceal size, and bleeding. *Euroasian Journal of Hepato-Gastroenterology*, 7(2):142–145, 2017.

[2] A. G. Borzov, A. V. Dreval, and S. I. Mukhin. Modeling blood glucose dynamics considering the topology of the systemic circulation. *Mathematical Modeling*, 27(2):3–24, 2015.

[3] A. G. Borzov, S. I. Mukhin, and N. V. Sosnin. Conservative substance transfer schemes in vascular systems closed through the heart. *Differential Equations*, 48(7):935–944, 2012.

[4] I. I. Dzidzava, B. N. Kotiv, D. P. Kashkin, A. A. Kochatkova, S. A. Bugaev, A. V. Smorodskiy, and A. V. Slobodyanik. Quantitative assessment of liver function using indocyanine green clearance test. *Transplantology*, 29–36, 2010.

[5] World Health Organization. Sdr, chronic liver disease and cirrhosis, all ages per 100 000, 2023. https://gateway.euro.who.int/ru/indicators/hfa_236-1860-sdr-chronic-liver-disease-and-cirrhosis-all-ages-per-100-000.

[6] R. N. H. Pugh, I. M. Murray-Lyon, J. L. Dawson, M. C. Pietroni, and R. Williams. Transection of the oesophagus for bleeding oesophageal varices. *British Journal of Surgery*, 60(8):646–649, 1973.

[7] T. R. Zhaleev, V. A. Kubyshkin, S. I. Mukhin, A. F. Rubina, and A. B. Khrulenko. Mathematical modeling of blood flow in the hepatic vascular system. *Applied Mathematics and Informatics: Proceedings of the Faculty of Computational Mathematics and Cybernetics, Lomonosov Moscow State University*, 61:42–56, 2019.

Alexander N. Churilov and John G. Milton

Modeling pulsatile and diurnal rhythmicity of free and total cortisol

Abstract: Cortisol (CORT) is a steroid hormone essential for stress and immune responses, which also plays an important role in metabolism of glucose, proteins, and lipids. CORT is secreted in adrenal glands and then released into the bloodstream, where its small fraction (about 5 %) circulates in a free form. The most of CORT is bound to carrier proteins, mainly to corticosteroid-binding globulin (CBG) and albumin, forming bound CORT complexes. The free CORT is physiologically active, its molecules can penetrate the cells and bind to specific intracellular glucocorticoid receptors (GR), exerting a number of responses. On the contrary, the molecules of bound CORT cannot cross the cell membranes and remain in plasma circulation. In response to physiological challenges, such as fever, the CORT-protein complex can dissociate releasing CORT in a free form. In this way, the CORT regulatory system is adapted to varying clinical conditions. Here, we study these adaptive mechanisms by extending our previously proposed mathematical model of the hypothalamus-pituitary-adrenal (HPA) hormonal axis. Our model uses a system of impulsive differential equations to describe the rhythmicity of HPA hormones. We explore performance of this extended system with the help of computer simulations under clinically relevant parameter conditions.

Keywords: Hormonal regulation, ultradian rhythm, circadian rhythm, free cortisol, protein-bound cortisol

MSC 2020: 34K11, 34K45, 37N25, 92C20

1 Introduction

The hypothalamic-pituitary-adrenal axis (HPA) maintains hormone balance through a negative feedback loop in which the adrenal gland communicates with the brain. This loop is mediated by cortisol (CORT), a steroid hormone produced in the adrenal cortex, which suppresses the release of peptide hormones produced in the hypothalamus and in the pituitary gland. Only a small fraction of cortisol, known as free cortisol, is biologically active because it can freely diffuse into cells. The majority, however, circulates in an inactive form, bound primarily to corticosteroid-binding globulin (CBG). This bound

Alexander N. Churilov, Independent researcher, St. Petersburg, Russia, e-mail: alexander.n.churilov@gmail.com
John G. Milton, Department of Neurology, Dell Medical School, The University of Texas at Austin, Red River Street 1501, Austin, TX 78712, USA, e-mail: john.milton@austin.utexas.edu

https://doi.org/10.1515/9783112216620-013

fraction forms a large reservoir in the plasma. If free cortisol levels drop, cortisol can dissociate from the binding proteins to restore its concentration in blood and tissues.

In this way, the feedback signal is flexible rather than fixed: binding affinities and protein capacities can subtly adjust the availability of free cortisol. From a systems or cybernetic perspective, this mechanism provides adaptive control under changing physiological conditions, such as elevated body temperature or fever.

2 Methods

2.1 Background in physiology

2.1.1 General regulation circuit of the hypothalamic-pituitary-adrenal (HPA) axis

We consider dynamics of the HPA hormonal axis, which is influenced by an external circadian drive. The mathematical model employs a scheme of regulated exocytosis previously proposed in our earlier publications to describe the episodic release of peptide hormones involved in the axis [10, 11, 50].

The HPA axis represents a classic example of a biological regulatory system with a negative feedback, involving the hypothalamus, pituitary and adrenal gland. The axis comprises corticotropin-releasing hormone (CRH, corticoliberin), adrenocorticotropic hormone (ACTH, corticotropin) and cortisol (CORT, also known as hydrocortisone). The first two hormones are polypeptides released in the hypothalamus and the anterior pituitary (adenohypophysis), respectively, while CORT is a steroid hormone secreted in the adrenal cortex.

The paraventricular nucleus of the hypothalamus releases CRH through the median eminence into the hypophyseal portal system, where it is transported to the anterior pituitary and stimulates pituitary corticotrophs to release ACTH. In turn, ACTH reaches the adrenal cortex through the systematic circulation, where it stimulates production of CORT released into the bloodstream. Excessive release of CORT is compensated by negative feedback loops that diminish the secretion rates of CRH and ACTH [16, 36, 63, 71]. The corresponding schematic diagram is presented in Figure 9.1. The binding unit will be discussed in detail in the next section.

Concentration of a hormone in the blood plasma is determined by two competing processes: clearance and secretion [59]. Hormonal secretion is a diverse process that greatly depends on the hormone's water or fat solubility. The main distinction should be made between peptides, which are hydrophilic and lipophobic, and steroids, which are lipophilic and hydrophobic.

Neuropeptides such as CRH and ACTH are synthesized from precursor proteins by the endoplasmatic reticulum and subsequently packed into dense-core secretory vesicles [64], some of which are organized into release pools. Since a vesicle is enclosed by

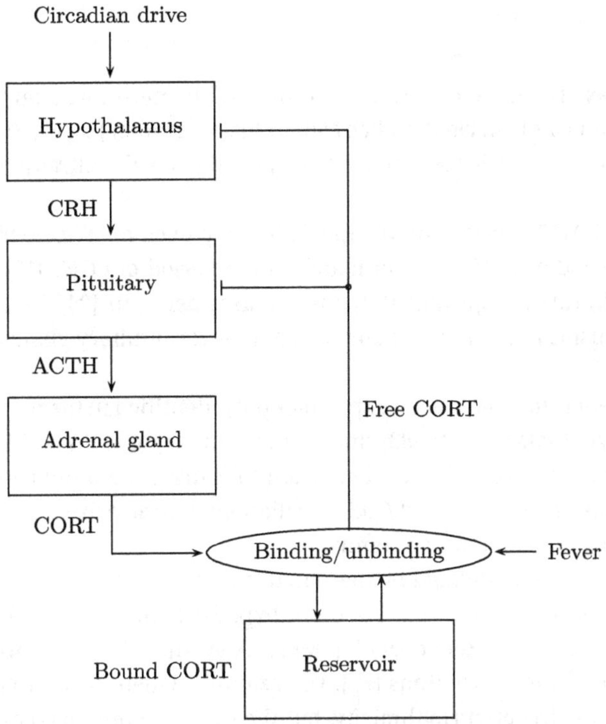

Figure 9.1: Schematic representation of the hypothalamic-pituitary-adrenal (HPA) hormonal axis with adaptively adjusted negative feedbacks. CRH: corticotropin-releasing hormone; ACTH: adrenocorticotropic hormone; CORT: cortisol. Bar-headed lines depict inhibitory connections.

a lipid bilayer, a lipophobic peptide cannot simply cross it. The release of these vesicles into the extracellular environment is achieved via a calcium-regulated exocytosis mechanism, see Chap. 6 [30]; [46, 49, 54]. As a result, peptides tend to be released episodically, that is their release is pulsatile [44, 59, 77].

In contrast, cortisol, a steroid hormone, is synthesized from cholesterol by adrenal cortical cells. Being lipophilic, steroids are released immediately into the circulation by diffusing through the lipid plasma membranes of the secretory cells. Therefore, cortisol secretion occurs continuously over time; however, the pulsatility of its release can be influenced by the upstream hormones within the regulatory circuit.

Both peptides and steroids are released in response to external inputs that can be stimulatory or inhibitory. Nevertheless, hormones often exhibit a basal secretion rate that is independent on external stimuli. This minimal secretion rate may persist even when an endocrine gland is isolated from external regulatory influences.

2.1.2 Hormonal oscillations

The oscillatory hormonal release is characterized by rhythmicity. The most important hormonal rhythms are ultradian (short period) and circadian (day periodic) [28, 63, 71, 76]. Further, we assume that the exocytosis mechanism is responsible for the ultradian rhythmicity.

The impulses of CORT and ACTH release occur approximately once or twice per hour. At the same time, the period of CRH does not match to the period of CORT [16]. For example, the CRH period in rats is approximately three pulses per hour [24]. This suggests that the pulse mechanisms in the hypothalamus and anterior pituitary glands are different [70].

Intrinsic ultradian pulse generators have been experimentally identified in the paraventricular nucleus of the hypothalamus [48, 83] and in the pituitary gland [17, 34, 70]. In particular, [70] states that for certain levels of constant CRH drive the pituitary-adrenal system supports self-sustained ACTH and CORT oscillations characterized by a physiological ultradian frequency (about once per hour).

The hypothalamic CRH release is modulated by the signal from the circadian clock master, which lies in the suprachiasmatic nucleus (SCN) of the hypothalamus and is modulated by the light/dark entrainer (called Zeitgeber) [2]. Most likely, the SCN is not involved in the generation of the ultradian rhythms [63]. For example, when the SCN is lesioned in rats, there is a loss of circadian rhythmicity, but the CORT rhythm persists [80]. The human CORT and ACTH levels reach their daily nadir between about 11:00 PM and 03:00 AM. The rise starts at about 3:00–4:00 AM and peaks around 8:00–9:00 AM [37].

Thus, there are three rhythms in the HPA axis, two of which are pulsatile. Together they combine to form an oscillation having a rather complicated frequency spectrum. The CORT release mainly follows the ACTH rhythmicity pattern with some time delay, due to time which the adrenal needs to synthesize CORT [82]. Besides that, CORT and ACTH concentrations may rise irregularly, in stress situations [12].

2.1.3 Cortisol binding to proteins

Unlike neuropeptides CRH and ACTH, CORT is hydrophobic (poorly soluble in blood plasma), so it cannot exist in plasma in high concentrations but needs a protein transporter to travel through the general circulation. Most of CORT is bound (associated) to hydrophilic proteins, mainly to CBG (also known as transcortin); however, much smaller fractions of CORT can be bound to albumin or orosomucoid [8, 22]. The unbound parts of CORT and CBG remain intact (free). The free CORT is also able to travel in blood plasma, but in smaller concentrations.

CBG is secreted from hepatocytes [22]. It is hydrophilic and easily enters the bloodstream, where it binds CORT in a 1:1 molecular ratio. The obtained complex (the CBG-bound cortisol) is also hydrophilic, it helps to deliver CORT to target organs and tissues, e. g., to sites of inflammation.

The widely accepted "free hormone hypothesis" [5, 6, 23] suggests that only the free (fat-soluble) fraction of CORT is able to diffuse across lipid membranes into target tissue cells, where the free CORT binds intracellular glucocorticoid receptors (GR). After their binding, the GR become activated and exert various biological effects [57]. Microdialysis experiments in rats demonstrated that levels of subcutaneous tissue CORT mirror plasma-free CORT, but differ significantly from the plasma total CORT, which is usually measured in clinical practice [6, 56].

Approximately 4–6 % of CORT is free, while its 80–90 % is bound to CBG and about 9–15 % is bound to other binding proteins, mainly albumin [33]. The clearance rate of free CORT is much greater than that of bound CORT (see further Table 9.1).

The so-called "reservoir hypothesis" [6] claims that another function of the bound cortisol is to serve a buffered source (a reservoir) of the free CORT, which can be rapidly released by a dissociation of the CORT-CBG complex, see [8, 52, 78]. Hence, the free and bound fractions of CORT allow partial interconvertions,

$$\text{CORT} + \text{CBG} \leftrightharpoons \text{CBG-bound CORT}.$$

Our modeling focuses on the binding of CORT to CBG. The binding of CORT to albumin and orosomucoid can be considered similarly; however, they have less impact on the system's dynamics. We also neglect CORT binding to some versions of CBG with special properties, namely to glycosylated CBG [8] and elastase-cleaved CBG [53].

There are two important aspects of the binding of CORT to CBG [6]: (1) the binding affinity and (2) the binding capacity. The binding affinity refers to the strength of connection between CORT and CBG (in more general chemical terms, they can be called ligand and receptor). Let us take a certain amount of free (unoccupied) CBG. The greater the affinity between CORT and CBG, the more CORT can be bound to this amount of CBG, and thus the less CORT remains free.

The protein's binding capacity is the maximal concentration of the binding protein in plasma. The capacity of CBG is strongly limited, so not all of the available CORT is usually bound to CBG. CBG remains close to constant throughout adult life [53], it is saturated at about 400–500 nmol/L [23]. In model simulations in [53], the total CBG level was taken 800 nmol/L.

The CBG plasma concentration follows a weakly expressed circadian rhythm with a peak in the morning and nadir in the evening [47, 78]. The CBG concentration is increased in pregnancy and when using oral contraceptive pills. Decreases in CBG can be caused, e. g., by ethanol, cirrhosis, insulin, hyperthyroidism, and SERPINA6 gene mutations. The affinity of CORT for CBG may also vary and is increased by glycosylation and decreased by inflammation, high body temperature, ethanol and SERPINA 6 mutations [78].

The affinity of CORT for albumin is much less than that for CBG [53]. At the same time, the binding capacity of plasma albumin is much greater (it can be over 1000-fold more than that of CBG [6] or 670000 nM [53]), so CORT never saturates the albumin [78]. Albumin and orosomucoid take over the role of CBG in the case of the congenital or induced absence or reduction of CBG [22].

Let us denote by C_f, C_b, B_f, B_b the plasma concentrations of the free CORT, CBG-bound CORT, free CBG, and CORT-bound (occupied) CBG, respectively. Neglecting binding of CORT to albumin and of CBG to some other glucocorticoids [6], the total CORT and CBG concentrations C, B can be represented as

$$C = C_f + C_b, \tag{9.1}$$

$$B = B_f + B_b. \tag{9.2}$$

Since CORT and CBG are bound in 1:1 ratio, one has $B_b = C_b$ (actually, they refer to the same CORT-CBG complex).

For simplicity, assume that the term B in (9.2) is constant and is equal to CBG binding capacity $B \equiv B^0 = \text{const}$. This implies

$$B_f = \text{Sat}(C_b) = \begin{cases} B^0 - C_b, & C_b \le B^0, \\ 0, & C_b > B^0. \end{cases} \tag{9.3}$$

This observation suggests that we can neglect the fraction of CBG bound to other steroids and the fraction of CORT bound by other proteins.

The most common numerical characteristic of binding CORT for CBG is the equilibrium dissociation constant [29]

$$K_d^0 = \frac{C_f^0\, B_f^0}{C_b^0}, \tag{9.4}$$

where the concentrations are measured at equilibrium. The smaller the K_d^0, the higher is the binding affinity, and the more tightly cortisol is bound to CBG. In particular, if $B_f^0 = K_d^0$, then

$$C_f^0 = C_b^0 = \frac{1}{2}(C_f^0 + C_b^0),$$

i. e., exactly the half of the total CORT is bound to CBG in this case. The K_d^0 is usually measured in nanomoles per liter (nmol/L, nM) or in milligrams per liter (mg/L) [79].

When analyzing hormonal rhythms, we are interested in concentrations C_f, B_f and C_b that are not constant, but time dependent.

The dynamics of binding/unbinding processes was mathematically modeled in [14, 15, 27, 42, 53, 55, 74] with the help of the Michaelis–Menten rate law. The latter assumes that the binding rate is proportional to the concentration product $C_f(t)B_f(t)$ [38].

Following [27], the dynamics of the reversible binding of the free cortisol $C_f(t)$ and of the CBG-bound cortisol $C_b(t)$ can be described by

$$\left.\begin{aligned}\frac{\mathrm{d}C_f}{\mathrm{d}t} &= -k_c C_f - k_{\mathrm{on}} C_f B_f + k_{\mathrm{off}} C_b + Z,\\ \frac{\mathrm{d}C_b}{\mathrm{d}t} &= -k_b C_b - k_{\mathrm{off}} C_b + k_{\mathrm{on}} C_f B_f.\end{aligned}\right\} \quad (9.5)$$

Here, k_c and k_b are the clearance coefficients of CORT and bound CORT, respectively, $Z(t)$ is the free CORT secretion rate, k_{on} and k_{off} are the cortisol kinetic parameters, namely the association rate constant (on-rate) and dissociation rate constant (off-rate), respectively [72]. The coefficient k_b is often neglected in mathematical models, since the half-life of C_b is rather large, it ranges 60–90 min [22].

Combining (9.3) and (9.5), one obtains

$$\left.\begin{aligned}\frac{\mathrm{d}C_f}{\mathrm{d}t} &= -\Big(k_c + k_{\mathrm{on}}\mathrm{Sat}(C_b)\Big) C_f + k_{\mathrm{off}} C_b + Z,\\ \frac{\mathrm{d}C_b}{\mathrm{d}t} &= -(k_b + k_{\mathrm{off}}) C_b + k_{\mathrm{on}} C_f \mathrm{Sat}(C_b).\end{aligned}\right\} \quad (9.6)$$

From (9.6), it is seen that the metabolic decay coefficient for $C_f(t)$ is time dependent and is equal to

$$W(t) = k_c + k_{\mathrm{on}}\mathrm{Sat}(C_b(t)). \quad (9.7)$$

Similar to the binding dissociation constant (9.4), we introduce a time-dependent binding dissociation function

$$K_d(t) = \frac{C_f(t)\mathrm{Sat}\,(C_b(t))}{C_b(t)}. \quad (9.8)$$

In the case when the secretion rate $Z(t)$ is constant, i. e., $Z(t) \equiv Z^0$, the equilibrium state of (9.6) is asymptotically stable and satisfies the relationships

$$K_d^0 = \frac{k_{\mathrm{off}} + k_b}{k_{\mathrm{on}}}, \quad k_c C_f^0 + k_b C_b^0 = Z^0.$$

2.2 Modeling impulsive hormonal release: A review

An impulse can be thought of as a short-time abrupt change in a level or an action (more generally, in a process). In nature, impulses are usually released in trains, so that periods of a fast change are interspersed with periods of a slow change. An impulse train has specific characteristics such as shapes, amplitudes (peaks), and timing (including frequencies and phases); see [18].

The impulsive nature of hormonal release requires a specific mathematical technique. Surveys of the existing mathematical models in endocrinology are given in [4, 31, 85]. The models can be divided into several main classes: ordinary differential equations (ODE), delay differential equations, partial differential equations, and stochastic models.

The most advanced representative of models based on ODE is the celebrated Hodgkin–Huxley model [25], which takes into account ion currents in different types of voltage gated ion channels. The Hodgkin–Huxley model has a number of modifications and simplifications. The appearance and disappearance of oscillations in this model is associated with classical bifurcations and with a division of its dynamics into fast and slow motions [69].

Additional effects can be reached by including time delays into differential equations. In particular, delays play a key role in oscillatility of the pituitary-adrenal ultradian generator introduced in [73, 82].

As an alternative, one can consider models with instant spikes (delta impulses), which can be mathematically classified as ODE with impulses [3]. In mathematical modeling, they are called ODE with event location [67] (the event occurs when an impulse is fired). Their solutions are piecewise continuous, but at some time points they exhibit discontinuities of the first kind (also called jump discontinuities or just jumps). A prominent biological representative of this class is the integrate-and-fire (IAF) model of the electrical nerve excitation, which has a number of modifications [25]. IAF system fires an impulse when its certain coordinate (a potential) reaches a given threshold. The IAF mechanism refers only to pulse timing, while pulse amplitude modulation should be considered separately.

In endocrinology, the IAF formalism was previously used to model vasopressin and oxytocin release [32, 43].

Following the classification of [4, 31], the mathematical model that we consider here refers to the class of functional models, which are much reduced in complexity compared with physiological models of the Hodgkin–Huxley type. Functional models attempt to minimize the physiological details and focus on the "big picture" in order to capture the main functionality of the endocrine system.

Our previous research on HPA [10] was inspired by a significant interest to the HPA ultradian pulsatility that arose in the recent decades [58, 71, 81, 82], in which the role of the anterior pituitary was emphasized. In that work, we put forward a new parsimonious impulsive model based on the exocytosis scheme. It was suggested that the adenohypophyseal hormone ACTH is produced by a pulse generator whose timing is defined by the IAF mechanism. As for pulse amplitudes, they were defined from the same membrane potential and from some amplitude-modulating function, which determined the shape of an ACTH pulse. The secretion process of the hypothalamic hormone CRH was somewhat simplified, it was modeled with a periodic sawtooth function. In our further paper [11], the control scheme was augmented to a system with two IAF pulse generators located in the hypothalamus and in the anterior pituitary glands.

In the present work, the ideas of [9–11] are further developed, in particular, the model [11] was adapted to the HPA hormonal axis. As in [11], we suggest the existence of two centers of the HPA pulsatility, namely CRH and ACTH pulse generators, whose oscillations are superimposed. The first novelty is that a traditional circadian sine-like SCN input was replaced by an asymmetric Möbius wave [60, 61], which leads to more realistic hormonal profiles. The second novelty is that we extend this model to include CORT binding to CBG.

2.2.1 Impulsive model of a peptide hormone's release

The release of a peptide hormone by a neuroendocrine cell is a complex Ca^{2+}-mediated process referred to as exocytosis [46, 68]. This is a regulated Ca^{2+}-dependent fusion of secretory granules with cell membranes. The triggering event is the calcium entry, which causes the cell membrane to rapidly depolarize and trigger an influx of Ca^{2+} ions [68]. During this secretory process, the secretory granules move to the plasma membrane where they dock and then release their contents. The fusion between the secretory granules and plasma membrane is also mediated by SNARE protein complexes [46], which we considered in more detail in [50].

Exocytosis is applicable to both hormones and neurotransmitters. Unlike the latter, the hormone contents is accumulated in dense-core secretory vesicles (DCVs) and then released into the bloodstream for the distribution throughout the body [46, 68].

We modeled hormonal exocytosis using an integrate-and-fire mechanism. There are two ways this model can be developed. From the point of view of the membrane potential, the fast step is the hormone-triggered depolarization of membrane potential and the slow step is the subsequent recharging. From the point of view of hormonal release, the fast step is the release of hormone and the slow step is the Ca^{2+}-triggered movements of the DCVs. It should be noted that the slow and fast stages are opposite in these two approaches. For simplicity, we modeled the timing of hormone release from the point of view of the membrane potential changes using an IAF scheme. We also use an IAF-based scheme to describe the shape of the hormone pulse.

Assume that between the firing times the hormone's concentration $H(t)$ and the potential $V(t)$ satisfy ordinary differential equations (see [10, 11])

$$\frac{\mathrm{d}H}{\mathrm{d}t} = -kH + F(V)\,S(t), \tag{9.9}$$

$$\frac{\mathrm{d}V}{\mathrm{d}t} = -\mu(V - V^0) + S_v(t). \tag{9.10}$$

The potential $V(t)$ refers not to a single cell or a single release vesicle, but to a community of cells and a pool of vesicles that release simultaneously. Here, k, μ, and V^0 are positive parameters. The functions $S(t)$, $S_v(t)$ present modulating inputs from other organs, $F(V)$ is a shaping function, whose meaning will be explained further.

The firing times t_n are found from the IAF relationships

$$t_0 = 0, \quad t_{n+1} = \min\{t : t > t_n,\ V(t) = \Delta\}, \tag{9.11}$$

where Δ is a given positive threshold. That is, an impulse occurs when the potential $V(t)$ reaches the threshold Δ. After the firing, the potential resets to zero, i. e.,

$$V(t_n^+) = 0, \quad n \geq 0, \tag{9.12}$$

where the superscript plus denotes the right-sided limit.

The function $H(t)$ describes a generated pulse train, (9.9) relates to its amplitudes and shapes, and equations (9.10)–(9.12) define its timing. From (9.10), the greater is $S_v(t)$, the faster $V(t)$ reaches the threshold, and hence the smaller is the time interval between impulses. On the contrary, when $S_v(t)$ decreases, impulses become sparser. That is the function $S_v(t)$ determines pulse timing. If Δ is too large compared with $S_v(t) + \mu V^0$, the threshold is never reached and the behavior of $H(t)$, $V(t)$ is not oscillatory.

Equation (9.9) determines a pulse between firings. The modulation input $S(t)$ defines the rate of $H(t)$, i. e., pulse amplitudes. A shaping function was previously introduced in [10] to model pulse shape of a pulse. This function allows to provide a description of pulse-amplitude modulation based on the same potential. The shaping functions transform a sawtooth potential functions $V(t)$ into a function $F(V(t))$, which may be of reverse sawtooth or of more complex shape. In particular, the shaping function can be chosen in the form of an exponential function

$$F(V) = \exp(-\lambda V) \tag{9.13}$$

where λ is a constant parameter; see Figure 9.2. An alpha function [10, 51] can be taken instead.

The exocytosis is regulated by the input functions $S(t)$, $S_v(t)$, which combine excitatory and inhibitory inputs from the other organs. The excitatory components of $S_v(t)$ contribute to reaching the firing threshold, while inhibitory components slow down this process. Thus, the firing frequencies either increase, or decrease. Similarly, the excitatory and inhibitory components of $S(t)$ influence pulse peaks (amplitudes), which also either increase, or decrease. From the control perspective, these pulse-frequency and pulse-amplitude modulation mechanisms allow for positive and negative feedbacks.

In the HPA axis, the most important negative feedbacks are from the adrenal gland to the hypothalamus and to the pituitary. In this case, the main inhibitory signal is the blood concentration of the free CORT, $C_f(t)$. When the level of free CORT grows, the pulse amplitudes and frequencies of CRH and ACTH are suppressed.

An inhibitory input can be modeled by an expression $L(C_f(t))$ described with the help of a function $L(C_f)$, decreasing for $C_f \geq 0$. Assume that $L(C_f)$ is normalized so that

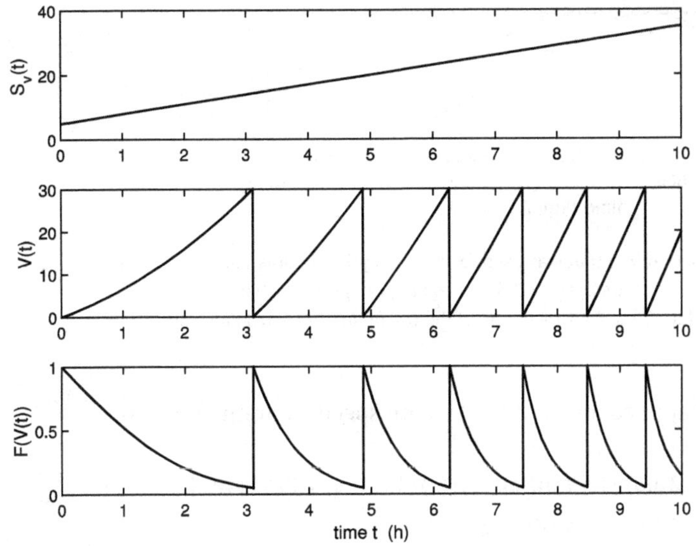

Figure 9.2: Linearly increasing modulation input $S_v(t)$, membrane potential function $V(t)$, and its transform, a shaped modulating coefficient $F(V(t))$, from top to bottom, evolving in time t. It is seen that the firing frequency increases with the increase of $S_v(t)$. The shaping function $F(V)$ is taken exponential (normalized to have peak equal to one). Time 0 corresponds to the beginning of an ultradian pulse train.

$L(0) = 1$. In the following simulations, a repressor form of the Michaelis–Menten function will be taken [1]. Namely,

$$L(C_f) = \frac{1}{1 + C_f/h}, \tag{9.14}$$

where the parameter h corresponds to the value C_f of the half-maximal repression.

2.2.2 Modeling the SCN circadian modulator

The 24-h periodic input from the SCN to the HPA periodically modulates the ultradian HPA pulses in their amplitudes. In the simplest case, the circadian input can be represented as a harmonic function or a sum of harmonic-like functions; see, e. g., [45].

In our modeling of the SCN drive, we used a scheme called Frequency Modulated Möbius (FMM) model [60, 61]. FMM provides a useful tool to deal with oscillatory processes, especially biomedical, including chronobiology, electocardiography, neuroscience, and corticotropin levels. In particular, this method allows to introduce realistically looking asymmetries [20, 40, 84] into standard sinusoidal curves.

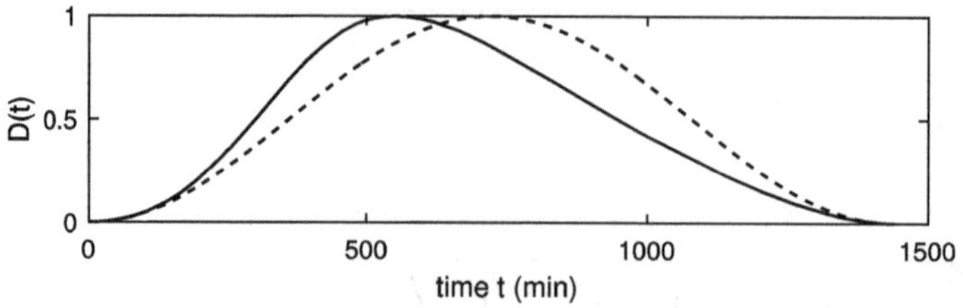

Figure 9.3: Graph of the function $D(t)$ that was used for modeling the circadian drive (solid line). For comparison, the graph of a cosine function $D_0(t) = 0.5\,(1 - \cos(\omega t))$ is given (dashed line). The time is taken in minutes (1440 minutes per day). Time 0 corresponds to the beginning of a circadian wave.

In our simulations, we applied an FMM model with specific parameters:

$$\left.\begin{aligned} &\varphi(t) = 2\,\mathrm{atan}(1.5\,\tan((\omega t - \pi/2)/2)), \quad \omega = 2\pi/1440, \\ &D(t) = 0.5\,(1 - \cos(\varphi(t) - \varphi(0))). \end{aligned}\right\} \tag{9.15}$$

The function $D(t)$ is periodic with the circadian period 1440 min (24 h); see Figure 9.3. The shape of the graph is asymmetric (cf. [13]), it can be modified by fitting required FMM parameters [62].

3 Results: Mathematical model of the entire HPA axis

3.1 Mathematical description

The following basic simplifications were made for composing our model:

(i) The model is limited to CORT binding for CBG. The albumin and orosomucoid bindings can be considered similarly, but they have less impact on the system's dynamics. We will also neglect a change in binding affinities of some special fractions of CORT and CBG [8, 22, 53]. For example, the CBG that was cleaved from a complex, has binding affinity ten times lower than that of the intact CBG [22].

(ii) The binding of some small fractions of CBG to steroids other than CORT (e. g., for corticosterone and progesterone) is neglected.

(iii) It is assumed that the plasma concentration of CBG is constant during the simulation. Actually, the plasma CBG concentration slightly varies about a baseline following a circadian rhythm, but the impact of circadian rhythmicity is relatively small [47].

(iv) The consideration is limited to CORT stored in the blood plasma. The dynamics of an extravascular fraction of CORT stored in tissues is neglected. Notice that dynamic models including tissue CORT exist [15, 55, 74], and augmenting our equations with them is not a mathematical problem. However, such an extension requires a knowledge of additional parameters that are difficult to identify (see the Supplementary Information [15]).

Let $R(t)$, $A(t)$, $C_f(t)$, and $C_b(t)$ be plasma concentrations at a time t of CRH, ACTH, and of free and bound CORT, respectively.

We consider a model of the HPA axis under unstressed (basal) conditions. The model contains three main compartments: the hypothalamic pulse generator, the anterior pituitary pulse generator, and the adrenal glands hormonal release. Let $V_r(t)$ and $V_a(t)$ be normalized and consolidated membrane potentials of the hypothalamic and pituitary secretory cells involved in the regulation process.

Assume that between the firing times the hormones concentrations $R(t)$, $A(t)$, $C_f(t)$, $C_b(t)$, and the potentials $V_r(t)$, $V_a(t)$ satisfy ordinary differential equations

$$\frac{dR}{dt} = -k_r R + g_r\, F_r(V_r)\, L_r(C_f)\,(D + D_0), \tag{9.16}$$

$$\frac{dV_r}{dt} = -\mu_r(V_r - V_r^0) + \nu_r\, L_r(C_f)\, D, \tag{9.17}$$

$$\frac{dA}{dt} = -k_a A + g_a\, F_a(V_a)\, L_a(C_f)\,(R + A_0), \tag{9.18}$$

$$\frac{dV_a}{dt} = -\mu_a(V_a - V_a^0) + \nu_a\, L_a(C_f)\, R, \tag{9.19}$$

$$\frac{dC_f}{dt} = -k_c C_f - k_{\text{on}} C_f \text{Sat}(C_b) + k_{\text{off}} C_b + g_c A + C_0, \tag{9.20}$$

$$\frac{dC_b}{dt} = -(k_b + k_{\text{off}}) C_b + k_{\text{on}} C_f \text{Sat}(C_b). \tag{9.21}$$

Equations (9.16), (9.17) describe the hypothalamic pulse generator, equations (9.18), (9.19) are related to the pituitary pulse generator, and (9.20), (9.21) correspond to adrenal release (see (9.6)). The saturation function $\text{Sat}(C_b)$ is defined by (9.3).

The firing times $t_n^{(r)}$ correspond to the CRH generator and $t_n^{(a)}$ correspond to the pituitary generator, as well as the potentials' resets, are found from the relationships (9.11), (9.12) with $V = V_r$, $\Delta = \Delta_r$ and $V = V_a$, $\Delta = \Delta_a$, respectively. Here, Δ_r, Δ_a are given positive thresholds. That is impulses occur when either the potential $V_r(t)$ reaches the threshold Δ_r or the potential $V_a(t)$ reaches the threshold Δ_a (at certain times two impulses can be fired simultaneously).

In (9.16)–(9.21), the parameters k_r, k_a, k_b, k_c, g_r, g_a, g_c, D_0, A_0, C_0, μ_r, ν_r, μ_a, ν_a, V_r^0, V_a^0, k_{on}, k_{off} are positive constants.

The modulating circadian input $D(t)$ is described with a 24-h periodic function, e. g., (9.15). The shaping functions $F_r(V_r)$, $F_a(V_a)$ and the inhibitory functions $L_r(C_f)$, $L_a(C_f)$ can be taken by formulas (9.13) and (9.14) as it was explained previously.

3.2 Simulations

System (9.16)–(9.21) was applied to simulations.

For dynamics, the most critical are clearance coefficients. In the simplest case, a clearance coefficient k can be estimated with the help of the measured hormone's half-life $t_{1/2}$ using the formula [41]

$$k = \ln(2)/t_{1/2}.$$

The values of half-lives and clearance rate coefficients that we used in modeling are shown in Table 9.1. It is seen that the clearance rates of free and bound CORT differ by ≈ 23 times. Hence, the decrease rate of the total CORT is biexponential; cf. [27].

Table 9.1: Clearance coefficients and half-lives of the HPA hormones. CRH stands for corticotropin-releasing hormone, ACTH is adrenocorticotropic hormone, CORT is cortisol, and CBG is corticosteroid-binding globulin.

Hormone	Coefficient	Half-life	Value	References
CRH	k_r	25 min	0.0277	[65, 66]
ACTH	k_a	10 min	0.0693	[19]
free CORT	k_c	4 min	0.1733	[27]
CBG-bound CORT	k_b	90 min	0.0077	[22]
CBG		5 days		[78]

The other parameters were chosen to provide the best fit between actual and simulated data. Reference values for CRH levels were found in [21] and for ACTH and CORT levels in [7]. They can be varied depending on clinical observations.

The following system was taken:

$$\frac{dR}{dt} = -0.0277\,R + \exp(-0.2\,V_r)\,\frac{2000}{1 + C_f/0.001}\,(D + 0.05), \tag{9.22}$$

$$\frac{dV_r}{dt} = -0.005(V_r - 100) + \frac{100}{1 + C_f/0.001}\,D, \tag{9.23}$$

$$\frac{dA}{dt} = -0.0693\,A + \exp(-0.04\,V_a)\,\frac{100}{1 + C_f/0.1}\,(R + 0.2), \tag{9.24}$$

$$\frac{dV_a}{dt} = -0.005(V_a - 100) + \frac{10}{1 + C_f/0.1}\,R, \tag{9.25}$$

$$\frac{dC_f}{dt} = -0.1733\,C_f + 0.04\,C_b - 0.002\,C_f\mathrm{Sat}(C_b) + 3.0\,A + 0.01\,, \tag{9.26}$$

$$\frac{dC_b}{dt} = -0.0477\,C_b + 0.002\,C_f\mathrm{Sat}(C_b). \tag{9.27}$$

The upper CBG limit was taken $B^0 = 700$, $k_{\mathrm{on}} = 0.002$, $k_{\mathrm{off}} = 0.04$. The firing thresholds were chosen $\Delta_r = 10$, $\Delta_a = 55$.

Assume that the hypothalamic generator's input is modulated by the circadian drive $D(t)$ that follows the frequency modulated Möbius model (9.15).

From the control perspective, the decay linear parts of equations (9.26), (9.27) present a low pass-filter, which responds to an input signal modifying its amplitudes and phases depending on the signal's frequency. In particular, a small phase shift is revealed with a delay of the CORT profile with respect to the ACTH profile. Thus, we can observe a small delay even without its explicit introduction into equation (9.26); cf. [82].

The system with the parameter values given in (9.22)–(9.27) clearly demonstrates three rhythms, the two of them (the hypothalamic and the pituitary) are ultradian and the third one is circadian (24-h periodic). The rhythms are superimposed to exhibit an oscillation with a complex frequency spectrum (see Figure 9.4). By a suitable choice of the firing thresholds, one can assign any desirable periods to the ultradian rhythms. For the present simulations, we have chosen the CRH firing frequency of 3 pulses per hour and the ACTH firing frequency of 9 pulses per day (cf. [24, 84]). The pattern of CORT generally follows that of ACTH. In this paper, we have not introduced any time delay into the equation of CORT release (9.27), but we did it in our previous publication [10]. In that work, a time delay of 15 min was included into the CORT's secretion rate, however, unlike [82], it had little effect on dynamics other than to introduce a small phase shift into the CORT pattern.

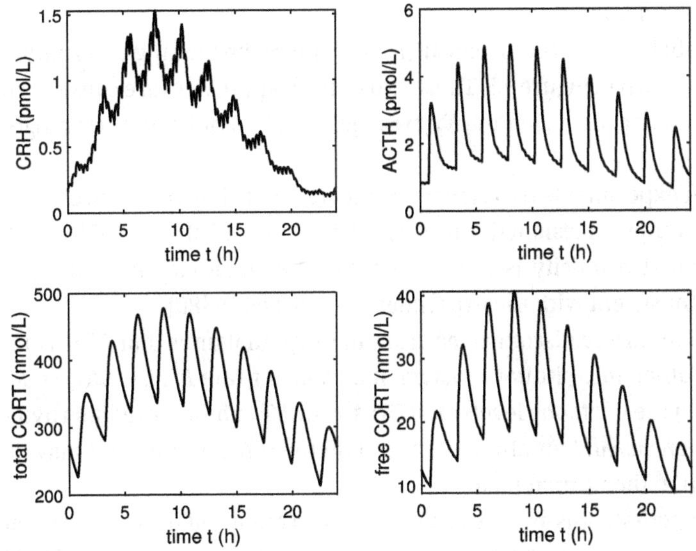

Figure 9.4: 24-h profiles for human plasma CRH (top, left), ACTH (top, right), total CORT (bottom, left), and free CORT (bottom, right). Here, CRH is corticotropin-releasing hormone, ACTH is adrenocorticotropic hormone, and CORT is cortisol. The firings occur with periods ≈ 21 min for CRH and ≈ 150 min for ACTH. The superimposed ultradian rhythms are clearly seen in the CRH profile. Time 0 corresponds approximately to the CORT nadir at 23:00 clock time.

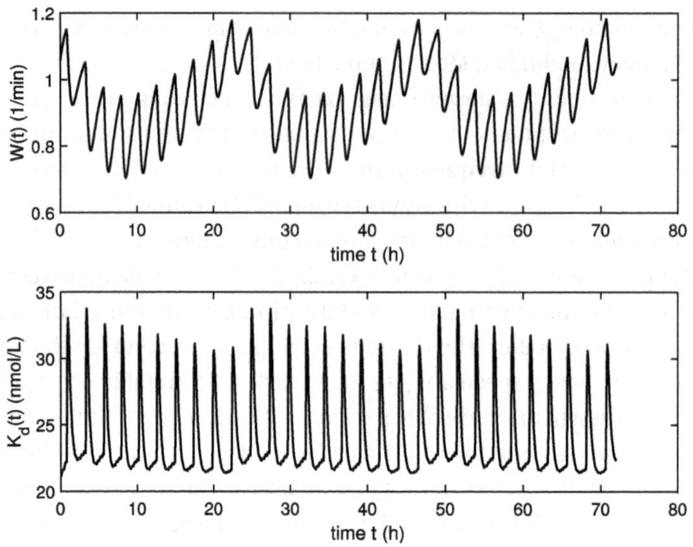

Figure 9.5: 72-h profiles for the metabolic decay coefficient $W(t)$ given by (9.7) and for the binding dissociation function $K_d(t)$ given by (9.8). Time 0 corresponds approximately to 23:00 clock time.

Some simulated profiles are shown in Figure 9.4. The total CORT profile mainly follows the bound CORT, but not the free CORT.

Graphs of the metabolic decay coefficient $W(t)$ (9.7) and of the binding dissociation function $K_d(t)$ (9.8) are shown in Figure 9.5. These two crucial parameters are considered not constant, but follow a circadian rhythm, $K_d(t)$ ranges 22–34 nmol/L and $W(t)$ ranges 0.7–1.2 min^{-1}.

Further simulation experiments (i)–(v) were conducted (not shown in figures).

(i) If the circadian SCN input is lesioned (i. e., $D(t) \equiv 0$ in (9.22), (9.23)), the CRH, ACTH, and CORT ultradian rhythmicity is maintained, but the circadian rhythmicity is abolished. This is consistent with the experiments described in [80].

(ii) Assume that the pituitary is disconnected from the hypothalamus, but CRH is infused into the hypothalamic-pituitary system at a constant rate R^0, i. e., $R(t) \equiv R^0$ in (9.24), (9.25). Then the hormone levels of ACTH and CORT exhibit ultradian rhythmicity. Their frequencies and amplitudes depend on the infusion rate and may be less or greater than in the normal mode.

(iii) Suppose that the hypothalamus and the pituitary are disconnected from the adrenal gland, i. e., $C_f(t) \equiv 0$ in (9.22)–(9.25). Then ultradian and circadian rhythms of CRH and ACTH are maintained. Their amplitudes and frequencies are greater than in the normal mode, since the suppressive effect of the free CORT is eliminated.

(iv) Suppose that the adrenal gland is disconnected from the pituitary, i. e., $A(t) \equiv 0$ in (9.20). Then the cortisol rhythmicity fails and $C_f(t)$, $C_b(t)$ tend to constant values.

(v) Let the pituitary be entirely isolated from the hypothalamus and from the adrenal gland. Then $R(t) \equiv 0$ and $C_f(t) \equiv 0$ in (9.24), (9.25). In this case, an ultradian ACTH rhythm is maintained, namely system (9.24), (9.25) has a periodic solution, but its amplitude and frequency are damped when compared to the normal mode. This result is consistent with the in vitro experiment described in [17]. Notice that a popular hypothesis of the pituitary-adrenal oscillator [81, 82] does not explain this phenomenon.

4 Discussion

The main cybernetic connections of the HPA axis are presently well understood and recognized. They include a number of feedback and feedforward interactions. However, there is no generally accepted mathematical model for ultradian oscillatility in this system. In the last decade, the ultradian oscillation model based on a pituitary-adrenal generator with delay gained a great popularity [82, 85]. Nevertheless, this model cannot explain some phenomena observed in experiments, such as the presence of ultradian oscillations in the pituitary gland disconnected from the adrenal gland. In our recent publications [10, 11], we proposed an alternative way to explain the origin of ultradian oscillations in the HPA. Our modeling supports the idea that regulated exocytosis plays the key role in understanding the HPA short-period pulsatility; cf. [46]. We suggest that our parsimonious model of impulsive equations may be sufficient to capture the main dynamics observed in practice.

Spectrum disorders and other abnormalities in hormonal rhythms may signal health problems [16, 26]. For example, in [26] it was stated that a lack of CORT ultradian pulsatility is associated with poorer working memory performance, reduced activity, and a poorer quality of sleep.

This paper focuses on modeling dynamics of plasma CORT (free and CBG-bound). Changing in binding parameters under some kinds of stress manifests itself in creating an additional control mechanism. For instance, an increase in body temperature above normal levels leads to a decrease in affinity of CORT to binding proteins. In turn, the decrease of affinity increases the concentration of free (unbound) CORT in blood. Hence, the negative feedbacks that CORT exerts on the hypothalamus and the pituitary become amplified, even if the production of CORT in the adrenal cortex remains the same. In effect, a variable binding affinity provides an adaptation of the HPA system to high temperature and fever stress.

An alternative mechanism may increase an amount of free CORT (and also amplify feedbacks) by a dissociation of the bound CORT, which can be considered as an emergency reservoir. A number of adaptations of the HPA axis to sepsis and hyperinflammation was discussed in [75]. Among them are mentioned the liberation of circulating CORT from the plasma binding proteins and a variable cortisol half-life. Thus, the homeostatic control looks very flexible and can include both association and dissociation.

It is known that the free CORT concentration in plasma may increase in emergency 6- to 12-fold [14]. In Figure 2, [22] it is seen that the rise in body temperature from 37 °C to 42 °C infers a near fivefold increase of the dissociation constant K_d^0. Experiments made for 10 healthy volunteers [79] showed that an increase in the body temperature from 37 °C to 41 °C resulted in an increase of about 80 % free CORT. In this way, an adjustment of the binding mechanism is directed to maintenance of the body temperature at a physiological level 37 °C [22].

The individual variations in the CBG production and normal mode affinity can be explained genetically (depending on polymorphisms in serpin family genes, especially in the SERPINA6 gene) [6, 39]. About ten known SERPINA6 mutations lead to reduced CBG capacity or affinity and have clinical significance. For example, a rare SERPINA6 null mutation (with no CBG in blood) implies fatigue, hypotension, chronic pain, and overweight [39].

The Beijing Centers for Diseases Control and Prevention collected blood samples from a massive cohort of 2287 anonymous young Chinese workers undergoing medical tests for employment health certification (more than 95 % of them were of Han ethnicity) [39]. When screening these samples, it was discovered that about 3 % of these persons (1:35 of men and 1:37 of women) had a SERPINA6 mutation leading to a 50 % reduction in CBG release. Interestingly that although the CBG levels were much reduced, the free CORT concentrations and the CBG affinities for CORT were found in normal ranges [39]. Thus, the HPA regulation circuit successfully compensates the CBG deficiency. However, it seems that the ability of bound CORT to serve an emergency reservoir was much reduced (see [52]), which makes patients more sensitive to fever. If the mutation percentage of 3 % can be extended to the entire Han population, one can assume that this CBG deficiency mutation affects more than 40 million people.

In [35], it was emphasized that in the last decade endocrinologists came to understand the clinical importance of patterns of hormone levels varying across the day. It was suggested that a new role of an endocrinologist is to bring these patterns to normality. This requires a deep understanding of the control and signaling mechanisms that govern the hormonal axis. We hope that our work will contribute to a better insight into the HPA functionality. Of course, as more observations and medical details become available, we can expect that further updates to our model may be required.

5 Conclusion

We have enhanced our previously proposed mathematical model, which incorporates two coupled impulsive generators, to more accurately capture the rhythmic dynamics of the hypothalamic-pituitary-adrenal (HPA) hormonal axis. Particular focus was given to the kinetic modeling of cortisol's association and dissociation with glucocorticoid-binding globulin (CBG). These binding and unbinding processes constitute a crucial component of an adaptive feedback mechanism that modulates plasma cortisol levels under diverse clinical conditions. Our integrated mathematical and computational simulations yield realistic hormonal profiles, offering novel insights into key features of HPA axis functioning and its dynamic regulation.

Bibliography

[1] U. Alon. *Introduction to Systems Biology: Design Principles of Biological Circuits*. CRC Press, Boca Raton, FL, 2 edition, 2019.

[2] I. P. Androulakis. Circadian rhythms and the HPA axis: A systems view. *WIREs Mechanisms of Desease*, 13(4):e1518, 2021.

[3] D. Bainov and P. Simeonov. *Impulsive Differential Equations: Periodic Solutions and Applications*. Longman, Harlow, UK, 1993.

[4] R. Bertram. Mathematical modeling in neuroendocrinology. *Comprehensive Physiology*, 5(2):911–927, 2015.

[5] D. D. Bikle. The Free Hormone Hypothesis: When, why, and how to measure the free hormone levels to assess vitamin D, thyroid, sex hormone, and cortisol status. *Journal of Bone and Mineral Research Plus*, 5(1):e10418, 2021.

[6] C. W. Breuner, H. E. Beyl, and J. L. Malisch. Corticosteroid-binding globulins: Lessons from biomedical research. *Molecular and Cellular Endocrinology*, 514:110857, 2020.

[7] B. J. Carroll, J. C. Ritchie, H. Rogers, and D. K. Kim. Fast feedback inhibition of adrenocorticotropic hormone secretion by endogenous cortisol in humans. *Neuroendocrinology*, 109(4):299–309, 2019.

[8] W. L. Chan, R. W. Carrell, A. Zhou, and R. J. Read. How changes in affinity of corticosteroid-binding globulin modulate free cortisol concentration. *The Journal of Clinical Endocrinology and Metabolism*, 98(8):3315–3322, 2013.

[9] A. N. Churilov, J. Milton, and E. R. Salakhova. An integrate-and-fire model for pulsatility in the neuroendocrine system. *Chaos (AIP journal)*, 30(8):083132, 2020.

[10] A. N. Churilov and J. G. Milton. Modeling pulsativity in the hypothalamic-pituitary-adrenal hormonal axis. *Scientific Reports*, 12:8480, 2022.

[11] A. N. Churilov and J. G. Milton. An integrate-and-fire mechanism for modeling rhythmicity in the neuroendocrine system. In J. Mau, S. Mukhin, G. Wang and S. Xu, editors, *BIOKYBERNETIKA: Mathematics for Theory and Control in the Human and in Society*, pages 365–376. De Gruyter, Berlin, 2025.

[12] M. Conrad, C. Hubold, B. Fischer, and A. Peters. Modeling the hypothalamus-pituitary-adrenal system: Homeostasis by interacting positive and negative feedback. *Journal of Biological Physics*, 35(2):149–162, 2009.

[13] M. Debono, C. Ghobadi, A. Rostami-Hodjegan, H. Huatan, M. J. Campbell, J. Newell-Price, K. Darzy, D. P. Merke, W. Arlt, and R. J. Ross. Modified-release hydrocortisone to provide circadian cortisol profiles. *The Journal of Clinical Endocrinology and Metabolism*, 94(5):1548–1554, 2009.

[14] R. I. Dorin, Z. Qiao, C. R. Qualls, and F. K. Urban. III. Estimation of maximal cortisol secretion rate in healthy humans. *The Journal of Clinical Endocrinology and Metabolism*, 97(4):1285–1293, 2012.

[15] R. I. Dorin, F. K. Urban III., I. Perogamvros, and C. R. Qualls. Four-compartment diffusion model of cortisol disposition: Comparison with 3 alternative models in current clinical use. *Journal of the Endocrine Society*, 7(2):1–13, 2023.

[16] C. M. B. Focke and K. J. Iremonger. Rhythmicity matters: Circadian and ultradian patterns of HPA axis activity. *Molecular and Cellular Endocrinology*, 501:110652, 2020.

[17] M. Gambacciani, J. H. Liu, W. H. Schwartz, V. S. Tueros, D. D. Rasmussen, and S. S. C. Yen. Intrinsic pulsatility of ACTH release from the human pituitary in vitro. *Clinical Endocrinology*, 26(5):557–563, 1987.

[18] A. Kh. Gelig and A. N. Churilov. *Stability and Oscillations of Nonlinear Pulse-Modulated Systems*. Birkhäuser, Boston, 1998.

[19] A. E. Gonzalez-Iglesias and M. E. Freeman. Brain control over pituitary gland hormones. In D. W. Pfaff, N. D. Volkow, and J. L. Rubenstein, editors, *Neuroscience in the 21st Century: From Basic to Clinical*, pages 2295–2344. Springer, New York, NY, 3 edition, 2022.

[20] A. Hartmann, J. D. Veldhuis, M. Deuschle, H. Standhardt, and I. Heuser. Twenty-four hour cortisol release profiles in patients with Alzheimer's and Parkinson's disease compared to normal controls: Ultradian secretory pulsatility and diurnal variation. *Neurobiology of Aging*, 18(3):285–289, 1987.

[21] K. Hashimoto, T. Nishioka, Y. Numata, T. Ogasa, J. Kageyama, and S. Suemaru. Plasma levels of corticotropin-releasing hormone in hypothalamic-pituitary-adrenal disorders and chronic renal failure. *Acta Endocrinologica*, 128(6):503–507, 1993.

[22] D. Henley, S. Lightman, and R. Carrell. Cortisol and CBG – Getting cortisol to the right place at the right time. *Pharmacology & Therapeutics*, 166:118–135, 2016.

[23] D. E. Henley and S. L. Lightman. New insights into corticosteroid-binding globulin and glucocorticoid delivery. *Neuroscience*, 180:1–8, 2011.

[24] G. Ixart, G. Barbanel, J. Nouguier-Soulé, and I. Assenmacher. A quantitative study of the pulsatile parameters of CRH-41 secretion in unanesthetized free-moving rats. *Experimental Brain Research*, 87:153–158, 1991.

[25] E. M. Izhikevich. *Dynamical Systems in Neuroscience: The Geometry of Excitability and Bursting*. MIT Press, Cambridge, MA, 2007.

[26] K. Kalafatakis, G. M. Russell, C. J. Harmer, M. R. Munafo, N. Marchant, A. Wilson, J. C. Brooks, C. Durant, J. Thakrar, P. Murphy, N. J. Thai, and S. L. Lightman. Ultradian rhythmicity of plasma cortisol is necessary for normal emotional and cognitive responses in man. *Proceedings of the National Academy of Sciences of the United States of America*, 115(17):E4091–E4100, 2018.

[27] D. M. Keenan, F. Roelfsema, and J. D. Veldhuis. Endogenous ACTH concentration-dependent drive of pulsatile cortisol secretion in the human. *American Journal of Physiology: Endocrinology and Metabolism*, 287(4):E652–E661, 2004.

[28] L. J. Kriegsfeld, J. LeSauter, T. Hamada, S. M. Pitts, and R. Silver. Circadian rhythms in the endocrine system. In D. W. Pfaff, A. P. Arnold, and A. M. Etgen, editors, *Hormones, Brain and Behavior*, volume 2, pages 33–91. Academic Press, 1 edition, 2002.

[29] D. Lee, J. Kim, and G. Lee. Simple methods to determine the dissociation constant, K_d. *Molecules and Cells*, 47(10):100112, 2024.

[30] G. Leng. *The Heart of the Brain: The Hypothalamus and Its Hormones*. MIT Press, Cambridge, MA, 2018.

[31] G. Leng and D. J. MacGregor. Mathematical modelling in neuroendocrinology. *Journal of Neuroendocrinology*, 20(6):713–718, 2008.

[32] G. Leng and D. J. MacGregor. Models in neuroendocrinology. *Mathematical Biosciences*, 305:29–41, 2018.

[33] J. G. Lewis, C. J. Bagley, P. A. Elder, A. W. Bachmann, and D. J. Torpy. Plasma free cortisol fraction reflects levels of functioning corticosteroid-binding globulin. *Clinica Chimica Acta*, 359:189–194, 2005.

[34] S. Lightman. Rhythms within rhythms: The importance of oscillations for glucocorticoid hormones. In P. Sassone-Corsi and Y. Christen, editors, *A Time for Metabolism and Hormones: Research and Perspectives in Endocrine Interactions*, pages 87–99. Springer, Cham, CH, 2016.

[35] S. Lightman. Clinical endocrinology – time for a reset? *Journal of the Endocrine Society*, 8(4):1–4, 2024.

[36] S. L. Lightman, M. T. Birnie, and B. L. Conway-Campbell. Dynamics of ACTH and cortisol secretion and implications for disease. *Endocrine Reviews*, 41(3):470–490, 2020.

[37] S. L. Lightman and B. L. Conway-Campbell. Circadian and ultradian rhythms: Clinical implications. *Journal of Internal Medicine*, 296:121–138, 2024.

[38] R. Lim, N. L. P. Martin, J. Chae, W. J. Kim, C.-M. Ghim, and P.-J. Kim. Generalized Michaelis-Menten rate law with time-varying molecular concentrations. *PLoS Computational Biology*, 19(12):e1011711, 2023.

[39] H.-Y. Lin, C. Underhill, J.-H. Lei, A. Helander-Claesson, H.-Y. Lee, B. R. Gardill, Y. A. Muller, H. Wang, and G. L. Hammond. High frequency of SERPINA6 polymorphisms that reduce plasma corticosteroid-binding globulin activity in Chinese subjects. *The Journal of Clinical Endocrinology and Metabolism*, 97(4):E678–E686, 2012.

[40] P. Linkowski, J. Mendlewicz, R. Leclercq, M. Brasseur, P. Hubain, J. Golstein, G. Copinschi, and E. Van Cauter. The 24-hour profile of adrenocorticotropin and cortisol in major depressive illness. *The Journal of Clinical Endocrinology and Metabolism*, 61(3):429–438, 1985.

[41] B. Liu, J. Peng, and Y. Liu. The concept of half-life of hormone. *Medical & Biological Engineering & Computing*, 33:822–826, 1995.

[42] Y.-W. Liu, Z.-Y. Hu, J.-H. Peng, and B.-Z. Liu. A dynamical model for the pulsatile secretion of the hypothalamo-pituitary-adrenal axis. *Mathematical and Computer Modelling*, 29(4):103–110, 1999.

[43] D. J. MacGregor and G. Leng. Phasic firing in vasopressin cells: Understanding its functional significance through computational models. *PLoS Computational Biology*, 8(10):e1002740, 2012.

[44] N. Malandrino and R. J. Smith. Synthesis, secretion, and transport of peptide hormones. In A. Belfiore and D. LeRoith, editors, *Principles of Endocrinology and Hormone Action*, pages 29–42. Springer, Cham, CH, 2018.

[45] V. M. Marković, Ž. Čupić, V. Vukojević, and L. Kolar-Anić. Predictive modeling of the hypothalamic-pituitary-adrenal (HPA) axis response to acute and chronic stress. *Endocrine Journal*, 58(10):889–904, 2011.

[46] T. F. J. Martin. Chapter 3 – Control of hormone secretion. In J. L. Jameson and L. J. De Groot, editors, *Endocrinology: Adult and Pediatric*, pages 26–37. W. B. Saunders, 6 edition, 2010.

[47] J. Melin, N. Hartung, Z. P. Parra-Guillen, M. J. Whitaker, R. J. Ross, and C. Kloft. The circadian rhythm of corticosteroid-binding globulin has little impact on cortisol exposure after hydrocortisone dosing. *Clinical Endocrinology*, 91(1):33–40, 2019.

[48] J. L. Mershon, C. S. Sehlhorst, R. W. Rebar, and J. H. Liu. Evidence of a corticotropin-releasing hormone pulse generator in the macaque hypothalamus. *Endocrinology*, 130(5):2991–2996, 1992.

[49] D. J. Michael, H. Cai, W. Xiong, J. Ouyang, and R. H. Chow. Mechanisms of peptide hormone secretion. *Trends in Endocrinology and Metabolism*, 17(10):408–415, 2006.

[50] J. Milton and A. Churilov. Hormone oscillations in the HPA axis: Dynamical diseases and beyond. *Exploration of Endocrine and Metabolic Diseases*, 2:101434, 2025.

[51] J. Milton and T. Ohira. *Mathematics as a Laboratory Tool*. Springer, Cham, CH, 2 edition, 2021.

[52] M.-P. Moisan. CBG: A cortisol reservoir rather than a transporter. *Nature Reviews. Endocrinology*, 9(2):78, 2013.

[53] P. T. T. Nguyen, J. G. Lewis, J. Sneyd, R. S. F. Lee, D. J. Torpy, and P. R. Shorten. Development of a formula for estimating plasma free cortisol concentration from a measured total cortisol concentration when elastase-cleaved and intact corticosteroid binding globulin coexist. *Journal of Steroid Biochemistry and Molecular Biology*, 141:16–25, 2014.

[54] J. J. Park and Y. P. Loh. How peptide hormone vesicles are transported to the secretion site for exocytosis. *Molecular Endocrinology*, 22(12):2583–2595, 2008.

[55] N. Picard-Hagen, V. Gayrard, M. Alvinerie, H. Smeyers, R. Ricou, A. Bousquet-Mélou, and P.-L. Toutain. A nonlabeled method to evaluate cortisol production rate by modeling plasma CBG-free cortisol disposition. *American Journal of Physiology: Endocrinology and Metabolism*, 281(5):E946–E956, 2001.

[56] X. Qian, S. K. Droste, M. Gutièrrez-Mecinas, A. Collins, F. Kersanté, J. M. H. M. Reul, and A. C. E. Linthorst. A rapid release of corticosteroid-binding globulin from the liver restrains the glucocorticoid hormone response to acute stress. *Endocrinology*, 152(10):3738–3748, 2011.

[57] S. Ramamoorthy and J. A. C. Cidlowski. Corticosteroids: Mechanisms of action in health and disease. *Rheumatic Diseases Clinics of North America*, 42(1):15–31, 2015.

[58] J. Rankin, J. J. Walker, R. Windle, S. L. Lightman, and J. R. Terry. Characterizing dynamic interactions between ultradian glucocorticoid rhythmicity and acute stress using the phase response curve. *PLoS ONE*, 7(2), 2012.

[59] R. S. Robeva, J. R. Kirkwood, R. L. Davies, L. S. Farhy, M. L. Johnson, B. P. Kovatchev, and M. Straume. *An Invitation to Biomathematics*. Academic Press, Amsterdam, 2007.

[60] C. Rueda, I. Fernández, Y. Larriba, and A. Rodríguez-Collado. The FMM approach to analyze biomedical signals: Theory, software, applications and future. *Mathematics (MDPI)*, 9:1145, 2021.

[61] C. Rueda, Y. Larriba, and S. D. Peddada. Frequency modulated Möbius model accurately predicts rhythmic signals in biological and physical sciences. *Scientific Reports*, 9:18701, 2019.

[62] C. Rueda Sabater, A. Rodríguez-Collado, and Y. Larriba. A novel wave decomposition for oscillatory signals. *IEEE Transactions on Signal Processing*, 69:960–972, 2021.

[63] G. M. Russell and S. L. Lightman. The human stress response. *Nature Reviews. Endocrinology*, 15(9):525–534, 2019.

[64] A. F. Russo. Overview of neuropeptides: Awakening the senses? *Headache*, 57(Suppl. 2):37–46, 2017.

[65] P. W. Saphier, M. Faria, A. Grossman, D. H. Coy, G. M. Besser, B. Hodson, M. Parkes, E. A. Linton, and P. J. Lowry. A comparison of the clearance of ovine and human corticotrophin-releasing hormone (CRH) in man and sheep: A possible role for CRH-binding protein. *Journal of Endocrinology*, 133:487–495, 1992.

[66] T. H. Schürmeyer, P. C. Avgerinos, P. W. Gold, W. T. Gallucci, T. P. Tomai, G. B. Cutler Jr., D. L. Loriaux, and G. P. Chrousos. Human corticotropin-releasing factor in man: Pharmacokinetic properties and dose-response of plasma adrenocorticotropin and cortisol secretion. *The Journal of Clinical Endocrinology and Metabolism*, 59(3):1103–1108, 1984.

[67] L. F. Shampine, I. Gladwell and S. Thompson. *Solving ODEs with MATLAB*. Cambridge University Press, New York, 2003.

[68] D. U. Silverthorn. *Physiology: An Integrated Approach*. Pearson, London, UK, 8 edition, 2019.

[69] J. Song, C. R. Laing, and S. Liu. Ion channels and the diversity of spontaneous firing in anterior pituitary corticotrophs: A dynamical analysis. *Communications in Nonlinear Science and Numerical Simulation*, 131:107789, 2024.

[70] F. Spiga, J. J. Walker, R. Gupta, J. R. Terry, and S. L. Lightman. Glucocorticoid dynamics: Insights from mathematical, experimental and clinical studies. *Journal of Endocrinology*, 226(2):T55–T66, 2015.

[71] F. Spiga, J. J. Walker, J. R. Terry, and S. L. Lightman. HPA axis-rhythms. *Comprehensive Physiology*, 4:1273–1298, 2014.

[72] D. A. Sykes, L. A. Stoddart, L. E. Kilpatrick, and S. J. Hill. Binding kinetics of ligands acting at GPCRs. *Molecular and Cellular Endocrinology*, 485:9–19, 2019.

[73] J. R. Terry, J. J. Walker, F. Spiga, and S. L. Lightman. Dynamics of the HPA axis: A systems modeling approach. In D. J. MacGregor and G. Leng, editors, *Computational Neuroendocrinology*, pages 252–283. Wiley, 2016.

[74] F. K. Urban, R. I. Dorin, I. Perogamvros, and C. R. Qualls. Numerical endocrinology: Portable solutions of cortisol on small edge devices using AI. *Computers in Biology and Medicine*, 200:111379, 2026.

[75] G. Van den Berghe, A. Téblick, L. Langouche, and J. Gunst. The hypothalamus-pituitary-adrenal axis in sepsis- and hyperinflammation-induced critical illness: Gaps in current knowledge and future translational research directions. *eBioMedicine*, 84:104284, 2022.

[76] J. D. Veldhuis. Pulsatile hormone secretion: Mechanisms, significance and evaluation. In D. Lloyd and E. L. Rossi, editors, *Ultradian Rhythms from Molecules to Mind*, pages 229–248. Springer, 2008.

[77] J. D. Veldhuis, D. M. Keenan, and S. M. Pincus. Motivations and methods for analyzing pulsatile hormone secretion. *Endocrine Reviews*, 29(7):823–864, 2008.

[78] K. C. Verbeeten and A. H. Ahmet. The role of corticosteroid-binding globulin in the evaluation of adrenal insufficiency. *Journal of Pediatric Endocrinology and Metabolism*, 31(2):105–115, 2018.

[79] M. Vogeser and J. Briegel. Effect of temperature on protein binding of cortisol. *Clinical Biochemistry*, 40(9–10):724–727, 2007.

[80] E. J. Waite, M. McKenna, Y. Kershaw, J. J. Walker, K. Cho, H. D. Piggins, and S. L. Lightman. Ultradian corticosterone secretion is maintained in the absence of circadian cues. *European Journal of Neuroscience*, 36(8):3142–3150, 2012.

[81] J. J. Walker, F. Spiga, E. Waite, Z. Zhao, Y. Kershaw, J. R. Terry, and S. L. Lightman. The origin of glucocorticoid hormone oscillations. *PLoS Biology*, 10(6):e1001341, 2012.

[82] J. J. Walker, J. R. Terry, and S. L. Lightman. Origin of ultradian pulsatility in the hypothalamic-pituitary-adrenal axis. *Proceedings of the Royal Society B*, 277(1688):1627–1633, 2010.

[83] Y-E. Wu, R. Enoki, Y. Oda, Z-L. Huang, K. Honma, and S. Honma. Ultradian calcium rhythms in the paraventricular and subparaventricular zone in the hypothalamus. *Proceedings of the National Academy of Sciences of the United States of America*, 115:E9469–E9478, 2018.

[84] E. A. Young, N. E. Carlson, and M. B. Brown. Twenty-four-hour ACTH and cortisol pulsatility in depressed women. *Neuropsychopharmacology*, 25(2):267–276, 2001.

[85] E. Zavala, K. C. A. Wedgwood, M. Voliotis, J. Tabak, F. Spiga, S. L. Lightman, and K. Tsaneva-Atanasova. Mathematical modelling of endocrine systems. *Trends in Endocrinology and Metabolism*, 30(4):244–257, 2019.

Anastasiia Mozokhina and Vitaly Volpert

Investigation of virus infection progression in tissue with reaction–diffusion models

Abstract: Mathematical modeling is widely used now for support physiological experiments, check hypothesis, and predict targets for pharmacological therapy in treating virus infections. In this work, we investigate the progression of viral infections in tissue using reaction–diffusion models that incorporate both innate and adaptive immune responses. The general model describes uninfected and infected cell populations, virus particles, interferon production, cytotoxic T lymphocytes, and antibodies, accounting for local and systemic dynamics. We analyze a simplified baseline version to derive key immunological indicators, such as the virus replication number, total viral load, and infection wave speed. Analytical solutions are provided in the homogeneous case with delay, while traveling wave solutions are characterized in the spatially heterogeneous case. The results highlight how immune mechanisms and spatial effects shape infection dynamics, offering insights into the conditions for infection clearance versus persistence and informing strategies for controlling viral spread in tissues or cell cultures.

Keywords: Reaction–diffusion equations, delayed equations, wave solution, virus infection, immune response

MSC 2020: 92C50, 35R10, 35K57

1 Introduction

Understanding the spread of viral infections within tissues is crucial for controlling disease progression, limiting transmission, and informing treatment strategies—particularly given the limited availability of effective antiviral drugs. Despite significant advances in experimental virology and immunology, translating this knowledge into predictive frameworks remains challenging due to the differences between *in vitro* and *in vivo* conditions, the high cost and complexity of experiments, and difficulties in interpreting observed phenomena.

Acknowledgement: The work was supported by the Russian Science Foundation, Agreement 24-11-00073, https://rscf.ru/project/24-11-00073/.

Anastasiia Mozokhina, S. M. Nikol'skii Mathematical Institute, RUDN University, 6 Miklukho-Maklaya St, 117198 Moscow, Russia, e-mail: anastasia_mozohina@mail.ru
Vitaly Volpert, University Lyon 1, Institut Camille Jordan, UMR 5208 CNRS, 43, bd du 11 novembre 1918 42, 69622 Villeurbanne, France, e-mail: volpert@math.univ-lyon1.fr

https://doi.org/10.1515/9783112216620-014

Mathematical modeling has emerged as an indispensable tool for overcoming these limitations, offering a means to systematically analyze infection dynamics under varied conditions. Viral spread in tissues or cell cultures is often conceptualized as a reaction–diffusion process, described by systems of differential or integrodifferential equations. Key characteristics of this process include the virus replication number (analogous to the basic reproduction number in epidemiology), the total viral load, and the speed of wave-like infection fronts.

The virus replication number defines the conditions for infection establishment and spread in tissue or cell culture and mathematically corresponds to the stability of the "healthy" steady state. The total viral load—quantified as the integral of viral concentration over space—is a critical measure of infectivity, informing estimates of transmission risk. Meanwhile, the wave propagation speed reflects the advance of the infection front and is associated with viral virulence and disease severity.

Early mathematical models in immunology primarily used systems of ordinary differential equations (ODEs), assuming a well-mixed environment. Notable early models incorporated cellular compartments such as target, plasma, and memory cells, sometimes with delays to represent immune responses (see, e. g. [4, 13]). This approach was further developed in numerous works such as [18, 11, 10, 20, 21, 19, 12] (see also a review in [7]).

However, many modern immunological problems require explicit spatial modeling to capture local interactions and diffusion-limited spread [14, 25, 9]. There are different ways to account for spatial aspects in models [11]. Reaction–diffusion systems provide a natural framework to model these spatial aspects, allowing the study of traveling wave solutions that describe the progression of infection fronts in tissue. Prior research has applied such approaches to model interactions between viruses and immune responses—including interferon dynamics, cytotoxic T lymphocyte activity, and antibody neutralization [3, 2, 16, 17, 23, 5, 1].

In this work, we propose a comprehensive reaction–diffusion model that integrates innate and adaptive immune responses, unifying and generalizing several earlier models. We then focus on analyzing a minimal representative version of this class of models, deriving explicit conditions for infection clearance or spread, and characterizing traveling wave solutions. This approach aims to clarify the roles of immune regulation and spatial heterogeneity in shaping infection dynamics, providing both theoretical insight and potential guidance for experimental and clinical research.

2 General model of infection progression

We consider a general model of immune response to viral infection illustrated by the scheme in Figure 10.1. The left part of the scheme contains local factors in the infected

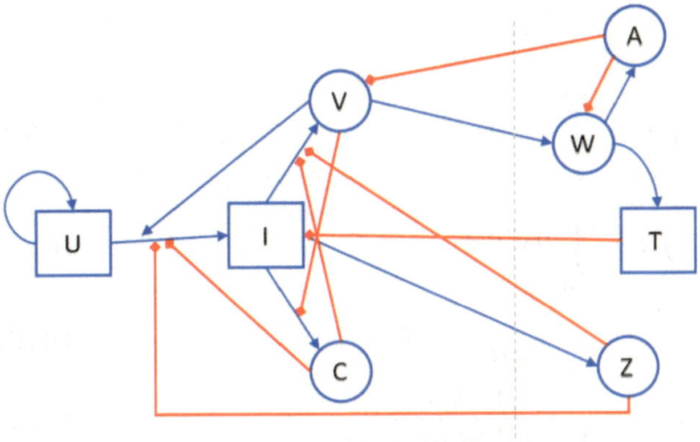

Figure 10.1: Schematic representation of the model, described by system (1)–(12). Here, U is uninfected cells, I is infected cells, V is extracellular virus, W is circulating virus, C is locally produced interferon (IFN), Z is globally produced IFN, A is antibodies, T is cytotoxic T-lymphocytes (CTL); details are in text. The left part shows the players acting in the local area, and the right part shows players produced and coming from the systemic blood circulation. Blue arrows denote production or up-regulation, while red arrows correspond to inhibition.

tissue, they are uninfected cells U, infected cells I, virus V, and locally produced interferon C. The right part shows systemic factors produced in the organism and distributed by blood circulation, they are circulating viral particles W, cytotoxic T-lymphocytes T, antibodies A, systemically produced interferon Z. Innate immune response is represented by interferon (IFN) locally produced by infected cells and systemically by circulating immune cells. Adaptive immune response is represented by cytotoxic T-lymphocytes (CTL) and antibodies. Overall, the model is described by the following system of equations:

$$\frac{\partial U}{\partial t} = \lambda f(U,I) - g(V,C,Z)UV, \tag{10.1}$$

$$\frac{\partial I}{\partial t} = g(V,C,Z)UV - (\beta_0 + \beta_{11}T)I, \tag{10.2}$$

$$\frac{\partial V}{\partial t} = D\frac{\partial^2 V}{\partial x^2} + \frac{b_1}{1 + k_1 Z + k_6 C} I(t-\tau_1) - (\sigma_{10} + \sigma_{11}A)V - \gamma g(V,C,Z)UV, \tag{10.3}$$

$$\frac{\partial C}{\partial t} = D_2\frac{\partial^2 C}{\partial x^2} + \frac{b_2}{1 + k_2 V} I(t-\tau_2) - \sigma_c C, \tag{10.4}$$

$$\frac{dW}{dt} = b_2 J(V) - (\sigma_{20} + \sigma_{21}A)W, \tag{10.5}$$

$$\frac{dZ}{dt} = b_3 J(I) e^{-k_3 J(V)} - \sigma_3 Z, \tag{10.6}$$

$$\frac{dT}{dt} = b_4 W(t-\tau_3) - \sigma_4 T J(I), \tag{10.7}$$

$$\frac{dA}{dt} = b_5 W(t-\tau_4) - \sigma_5 WA - \sigma_6 J(V)A, \tag{10.8}$$

where

$$J(V) = \int_{-\infty}^{+\infty} V(x,t)\,\mathrm{d}x, \tag{10.9}$$

$$J(I) = \int_{-\infty}^{+\infty} I(x,t)\,\mathrm{d}x, \tag{10.10}$$

$$f(U,I) = \frac{U}{\varkappa_1 + \varkappa_2 U}(U_0 - U - \chi I), \tag{10.11}$$

$$g(V,C,Z) = \frac{a + k_5 V}{1 + k_{31}C + k_{32}Z + k_4 V}. \tag{10.12}$$

Various particular cases of this model were considered in [3, 2, 16, 17] (models 1, 6–10, 13 in Figure 10.2). Here, $U(x,t)$ is the concentration of uninfected cells, $I(x,t)$ is the concentration of infected cells, $V(x,t)$ is the concentration of virus particles in the infection area, $C(x,t)$ is the concentration of IFN produced by infected cells, $W(t)$ is the concentration of virus particles in general circulation (in blood and lymph), $Z(t)$ is the concentration of IFN produced by immune cells in the circulation, $T(t)$ is cytotoxic T-lymphocytes (CTL) concentration, $A(t)$ is the concentration of antibodies.

The first term in the right-hand side of the first equation describes the production of uninfected cells; possible regimes will be discussed below. The second term represents the infection of uninfected cells. The infection rate $g(V,C,Z)$ is downregulated by interferon (IFN) concentration, which itself is inhibited by the virus concentration [22, 16]. As a result, uninfected cells become infected and, therefore, the same term appears on the right-hand side of the second equation.

The second term on the right-hand side of the second equation describes the death of infected cells, caused by cytotoxic T lymphocytes (CTLs) and other factors.

The third equation governs virus dynamics. It includes random motion of virus particles in the extracellular space (modeled by a diffusion term), production of new viral particles by infected cells with a delay τ_1 and a rate inhibited by IFN, degradation due to antibodies and other mechanisms, and consumption of virus particles during infection. Note that more than one viral particle may be required to infect a single cell; therefore, an additional coefficient $\gamma > 1$ is introduced in front of the infection rate $g(V,C,Z)$.

The fourth equation describes the dynamics of IFN produced by infected cells within the infection site. This IFN diffuses through the extracellular space and is produced by infected cells with a delay τ_2, at a rate inhibited by virus concentration. The final term accounts for IFN degradation and its transport into the bloodstream, away from the infection area.

Equation (10.5) describes the dynamics of antigen W in circulation. This variable, which we may refer to as the “global virus,” includes intact viral particles as well as viral components and degradation products that activate the adaptive immune response. It

1

$$\frac{\partial U}{\partial t} = -aUV, \quad \frac{\partial I}{\partial t} = aUV - \beta I, \quad \frac{\partial V}{\partial t} = D\frac{\partial^2 V}{\partial x^2} + bI(t-\tau) - \sigma V$$

6

$$\frac{\partial U}{\partial t} = -aUV, \quad \frac{\partial I}{\partial t} = aUV - \beta I,$$
$$\frac{\partial V}{\partial t} = D_1\frac{\partial^2 V}{\partial x^2} + \frac{b_1}{1+k_1C}I(t-\tau_1) - \sigma_1 V,$$
$$\frac{\partial C}{\partial t} = D_2\frac{\partial^2 C}{\partial x^2} + \frac{b_2}{1+k_2V}I(t-\tau_2) - \sigma_2 C$$

7

$$\frac{\partial U}{\partial t} = -\frac{a}{1+k_3C}UV,$$
$$\frac{\partial I}{\partial t} = \frac{a}{1+k_3C}UV - \beta I,$$
$$\frac{\partial V}{\partial t} = D_1\frac{\partial^2 V}{\partial x^2} + \frac{b_1}{1+k_1C}I(t-\tau_1) - \sigma_1 V,$$
$$\frac{\partial C}{\partial t} = D_2\frac{\partial^2 C}{\partial x^2} + \frac{b_2}{1+k_2V}I(t-\tau_2) - \sigma_2 C$$

13

$$\frac{\partial U}{\partial t} = -\frac{a+k_5V}{1+k_3C+k_4V}UV,$$
$$\frac{\partial I}{\partial t} = \frac{a+k_5V}{1+k_3C+k_4V}UV - \beta I,$$
$$\frac{\partial V}{\partial t} = D_1\frac{\partial^2 V}{\partial x^2} + \frac{b_1}{1+k_1C}I(t-\tau_1) - \sigma_1 V,$$
$$\frac{\partial C}{\partial t} = D_2\frac{\partial^2 C}{\partial x^2} + \frac{b_2}{1+k_2V}I(t-\tau_2) - \sigma_2 C$$

8

$$\frac{\partial U}{\partial t} = -aUV,$$
$$\frac{\partial I}{\partial t} = aUV - \beta I,$$
$$\frac{\partial V}{\partial t} = D_1\frac{\partial^2 V}{\partial x^2} + \frac{b_1}{1+k_1Z}I(t-\tau_1) - \sigma_1 V,$$
$$\frac{dZ}{dt} = b_2 J(I)e^{-k_2 J(V)} - \sigma_2 Z,$$
$$J(f) = \int_{-\infty}^{+\infty} f(x,t)\,dx$$

9

$$\frac{\partial U}{\partial t} = -aUV,$$
$$\frac{\partial I}{\partial t} = aUV - (\beta_0 + \beta_1 T)I,$$
$$\frac{\partial V}{\partial t} = D_1\frac{\partial^2 V}{\partial x^2} + \frac{b_1}{1+k_1Z}I(t-\tau_1) - (\sigma_{10} + \sigma_{11}A)V,$$
$$\frac{dZ}{dt} = b_2 J(I)e^{-k_2 J(V)} - \sigma_2 Z,$$
$$\frac{dT}{dt} = b_3 J(V) - \sigma_3 T J(I),$$
$$\frac{dA}{dt} = b_4 J(V) - \sigma_4 A J(V),$$
$$J(f) = \int_{-\infty}^{+\infty} f(x,t)\,dx$$

10

$$\frac{\partial U}{\partial t} = -aUV,$$
$$\frac{\partial I}{\partial t} = aUV - (\beta_0 + \beta_1 T)I,$$
$$\frac{\partial V}{\partial t} = D_1\frac{\partial^2 V}{\partial x^2} + \frac{b_1}{1+k_1Z}I(t-\tau_1) - (\sigma_{10} + \sigma_{11}A)V,$$
$$\frac{dW}{dt} = b_2 J(V) - (\sigma_{20} + \sigma_{21}A)W$$
$$\frac{dZ}{dt} = b_3 J(I)e^{-k_3 J(V)} - \sigma_3 Z,$$
$$\frac{dT}{dt} = b_4 W(t-\tau_2) - \sigma_4 T J(I),$$
$$\frac{dA}{dt} = b_5 W(t-\tau_3) - \sigma_5 AW,$$
$$J(f) = \int_{-\infty}^{+\infty} f(x,t)\,dx$$

Figure 10.2: Different models of infection progression in tissue or cell culture, as particular cases of models (10.1)–(10.12). Each bottom model includes the physiological mechanisms in more detail. The models of the right branch are multiscale—they include factors, produced outside the infected region. Models 2–5, 11, 12 are not the special cases of (10.1)–(10.12); they are shown in A.

also includes antigens presented on the surface of antigen-presenting cells (APCs). Production of W is proportional to the amount of virus in the infection site, while the final term describes its degradation by antibodies and other processes.

Equation (10.6) models the dynamics of IFN produced by immune cells in circulation. Its production is proportional to the total number of infected cells and is downregulated by the total viral load in the infection area. The last term accounts for its degradation.

Equations (10.7) and (10.8) describe the dynamics of CTLs and antibodies, respectively. Their production is proportional to the "global virus" in circulation. The final term in (10.7) accounts for CTL consumption during the elimination of infected cells, as CTLs lose their cytotoxic capacity. The last two terms in (10.8) represent the consumption of antibodies in eliminating both the global virus and virus particles within the infection area.

Summarizing, the following processes are taken into account in model (10.1)–(10.8):
- mutual inhibition of virus and IFN production by each other;
- "antiviral state" in uninfected cells—reduced rate of virus invasion as a result of IFN products action;
- virus interference of "antiviral state";
- production of IFN by infected cells;
- production of CTL and antibodies as a result of virus presentation of antigen-presenting cells somewhere in the body (e. g., in lymph nodes);
- neutralization of virus by antibodies;
- destruction of infected cells by CTL.

For the rate of production of uninfected cells $f(U,I)$, the following particular cases can be considered:
- If $\lambda = 0$, then $f(U,I) \equiv 0$. It corresponds to the absence of the production of uninfected cells. This case is appropriate if the time of infection progression is much less than the time of cell reproduction.
- If $\varkappa_1 = 0, \chi = 0$, then $f(U,I) = (U_0 - U)$. It describes the influx of uninfected cells into the considered area, and the total number of uninfected cells tends to U_0 in the absence of infection.
- If $\varkappa_2 = 0, \chi = 0$, then $f(U,I) = U(U_0 - U)$. In this case, the production of uninfected cells without infection is governed by a logistic equation. The total number of uninfected cells without infection tends to U_0.
- If $\varkappa_1 = 0$ and $\chi = 1$, $f(U,I) = (U_0 - U - I)$. In this case, the reproduction term takes into account not only the number of uninfected cells, but also the number of infected cells.
- If $\varkappa_2 = 0$ and $\chi = 1$, $f(U,I) = U(U_0 - U - I)$. It is the logistic term with respect to the number of infected cells as well.

Note that the density-dependent proliferation can take into account infected cells, $\chi = 1$. If this is not the case, the total cell concentration can exceed the maximal physiological value U_0.

Various particular cases of this model can be obtained under the following conditions (Figure 10.2):

- If $\lambda = \gamma = 0$, $k_5 = k_{31} = k_{32} = k_4 = 0$, $k_1 = k_6 = 0$, $\beta_{11} = \sigma_{11} = 0$, we obtain Model 1.
- If $\lambda = \gamma = 0$, $k_5 = k_{31} = k_{32} = k_4 = 0$, $\beta_{11} = \sigma_{11} = 0$, $k_1 = 0$, we obtain Model 6 [2].
- If $\lambda = \gamma = 0$, $k_5 = k_{32} = k_4 = 0$, $\beta_{11} = \sigma_{11} = 0$, $k_1 = 0$, we obtain Model 7 [16].
- If $\lambda = \gamma = 0$, $k_{32} = 0$, $\beta_{11} = \sigma_{11} = 0$, $k_1 = 0$, we obtain Model 13 [16].
- If $\lambda = \gamma = 0$, $k_5 = k_{31} = k_{32} = k_4 = 0$, $\beta_{11} = \sigma_{11} = 0$, $k_6 = 0$, we obtain Model 8 [2].
- If $\lambda = \gamma = 0$, $k_5 = k_{31} = k_{32} = k_4 = 0$, $k_6 = 0$, $\sigma_{21} = 0$, $\tau_2 = \tau_3 = 0$, and under quasistationary approximation for W (in this case, $W \sim J(V)$), we obtain Model 9 [2].
- If $\lambda = \gamma = 0$, $k_5 = k_{31} = k_{32} = k_4 = 0$, $k_6 = 0$, $\sigma_6 = 0$, we obtain Model 10 [17].

In the next section, we discuss Model 1 in more detail. It admits an analytical assessment of infection progression and clear biological interpretations of the results.

3 Basic model

Let the coefficients $k_1, k_4, k_5, k_6, k_{31}, k_{32}, \lambda, \gamma, \beta_{11}, \sigma_{11}$ in model (10.1)–(10.12) equal 0. Then, we obtain Model 1 as

$$\frac{\partial U}{\partial t} = -aUV, \tag{10.13}$$

$$\frac{\partial I}{\partial t} = aUV - \beta I, \tag{10.14}$$

$$\frac{\partial V}{\partial t} = D\frac{\partial^2 V}{\partial x^2} + bI(t-\tau) - \sigma V. \tag{10.15}$$

3.1 Basic model without diffusion

First, consider system (10.13)–(10.15) with $D = 0$:

$$\frac{dU}{dt} = -aUV, \tag{10.16}$$

$$\frac{dI}{dt} = aUV - \beta I, \tag{10.17}$$

$$\frac{dV}{dt} = bI(t-\tau) - \sigma V, \tag{10.18}$$

with initial conditions $U(0) = U_0$, $I(0) = I_0$, $I(t) = \phi(t) \geq 0$, for $t \in (-\tau; 0)$, and $V(0) = V_0$. Here, $\phi(t)$ is some given nonnegative continuous function, bounded on $[-\tau; 0]$. This system satisfies the conditions of the existence and uniqueness theorem for any initial condition. Also, it can be shown that solutions of the system are nonnegative and bounded for any nonnegative initial conditions.

3.1.1 Linear stability analysis and virus replication number

The stationary points of the system (10.16)–(10.18) have the form $(U = U_0, 0, 0)$ for any positive U_0. Suppose that we have an initial condition $U(0) = U_0$ and consider the stationary point $(U_0, 0, 0)$. This stationary point corresponds to the healthy state, where there is no virus and no infected cells. Linearize system (10.16)–(10.18) around this point and obtain the following linear system:

$$\frac{dI}{dt} = -\beta I + aU_0 V, \tag{10.19}$$

$$\frac{dV}{dt} = bI(t-\tau) - \sigma V. \tag{10.20}$$

Here, U and V are the perturbations of the stationary solution. Let us find the solution of this linearized system in the form

$$I = pe^{\lambda t}, \quad V = qe^{\lambda t}.$$

We substitute this solution into the system (10.19)–(10.20) and obtain the following characteristic equation:

$$(\beta + \lambda)(\sigma + \lambda) = abU_0 e^{-\tau\lambda}. \tag{10.21}$$

In (10.21), the left side is the parabola with roots $\lambda = -\beta < 0$ and $\lambda = -\sigma < 0$, which intersects the vertical axis at point $y = \beta\sigma$. In the right side, there is a decaying exponent, which intersects the vertical axis at the point $y = abU_0$. Both roots of the polynomial will be negative if $\beta\sigma > abU_0$, i. e.,

$$R_v = \frac{abU_0}{\beta\sigma} < 1. \tag{10.22}$$

Otherwise, the polynomial has one positive root. Thus, we have proved the following statement.

Proposition 3.1. *The stationary point* $(U_0, 0, 0)$ *of the system* (10.16)–(10.18) *is stable if and only if condition* (10.22) *is satisfied.*

If this condition is satisfied, the stationary point $(U_0, 0, 0)$ corresponding to the healthy state is stable, which means that infection is eliminated. The virus replication number R_v has the same meaning as the basic reproduction number in epidemiology, i. e., if $R_v > 1$ then infection develops, if the inequality is opposite, it is eliminated.

Note that characteristic equation (10.21) has two real solutions. Hence, oscillatory regimes are not possible. One of these solutions is negative, another one is also negative if condition (10.22) is satisfied and positive otherwise. In the first case, the stationary point is a stable node, and in the second case, it is a saddle point.

3.1.2 Stability with the Lyapunov's function

Linear analysis of stationary points $(U_0, 0, 0)$ in system (10.16)–(10.18) is formal because they are not isolated. Let us justify this result by constructing a Lyapunov function [8].

Theorem 3.1. *For any nonnegative initial condition* $E_0 = (U_0, 0, 0)$, *the solution of system* (10.16)–(10.18) *converges to a stationary point* $(U_f, 0, 0)$. *Here,* U_f *satisfies the equation*

$$\ln\left(\frac{U_f}{U_0}\right) = R_v\left(\frac{U_f}{U_0} - 1\right) - J_0, \tag{10.23}$$

where

$$R_v = \frac{abU_0}{\beta\sigma}, \quad J_0 = \frac{abI_0}{\beta\sigma} + \frac{aV_0}{\sigma} + \frac{ab}{\sigma}\int_{-\tau}^{0} \phi(t)\,\mathrm{d}t.$$

Proof. We construct the Lyapunov function in the form $\Phi = p_1U + p_2I + p_3V + p_4(U(t-\tau) + I(t-\tau))$ with $p_i > 0$, $i = 1, 2, 3, 4$. This function is defined for $t \geq \tau$. Then

$$\begin{aligned}\frac{\mathrm{d}\Phi}{\mathrm{d}t} &= p_1\frac{\mathrm{d}U}{\mathrm{d}t} + p_2\frac{\mathrm{d}I}{\mathrm{d}t} + p_3\frac{\mathrm{d}V}{\mathrm{d}t} + p_4\left(\frac{\mathrm{d}U(t-\tau)}{\mathrm{d}t} + \frac{\mathrm{d}I(t-\tau)}{\mathrm{d}t}\right)\\ &= aUV(p_2 - p_1) - p_2\beta I + I(t-\tau)(bp_3 - \beta p_4) - \sigma V p_3\\ &< aUV(p_2 - p_1) + I(t-\tau)(bp_3 - \beta p_4).\end{aligned}$$

In order to get the inequality $\mathrm{d}\Phi/\mathrm{d}t < 0$, the following two conditions should be satisfied:

$$p_2 < p_1, \quad bp_3 < \beta p_4.$$

Therefore, we can choose p_i, $i = 1, 2, 3$ such that $\mathrm{d}\Phi/\mathrm{d}t < 0$. Function $\Phi(t)$ is nonnegative as a linear combination of nonnegative functions, and as $\mathrm{d}\Phi/\mathrm{d}t < 0$ it decreases for $t \to +\infty$. Thus, it has a limit. Hence, $\Phi' \to 0$ when $t \to +\infty$. Therefore, I and V tend to zero when $t \to +\infty$, and U tends to some value U_f. Let us find this value. Consider system (10.16)–(10.18) assuming

$$U(+\infty) = U_f, \quad I(+\infty) = V(+\infty) = 0.$$

From the system (10.16)–(10.18), we derive the following equations:

$$\ln\left(\frac{U_f}{U_0}\right) = -aJ(V),$$
$$U_f - U_0 - I_0 = -\beta J(I),$$
$$bJ(I) = \sigma J(V) - V_0 - b\int_{-\tau}^{0} \phi(t)\,dt,$$

where $J(f) = \int_0^{+\infty} f(t)\,dt$. From these equations, we exclude integrals and get

$$\ln\left(\frac{U_f}{U_0}\right) = R_v\left(\frac{U_f}{U_0} - 1\right) - J_0, \tag{10.24}$$

where

$$J_0 = \frac{abI_0}{\beta\sigma} + \frac{aV_0}{\sigma} + \frac{ab}{\sigma}\int_{-\tau}^{0} \phi(t)\,dt.$$

Thus, we conclude that for any initial condition $E_0 = (U_0, I_0, V_0)$ the solution converges to the point $E_f = (U_f, 0, 0)$ where E_f satisfies equation (10.24). □

Using the result of this theorem, we determine the conditions under which the stationary point $E_s = (U_s, 0, 0)$ is stable in the sense of the following definition.

Definition 3.1. A stationary point $E_s = (U_0, 0, 0)$ with $U_0 > 0$ is stable if for any $\varepsilon > 0$ there exists $\delta > 0$ such that for any initial condition from the δ-neighborhood of E_s, the solution of system (10.16)–(10.18) converges to a stationary point in the ε-neighborhood of E_s [8].

Corollary 3.1. *Let $R_v = \frac{abU_0}{\beta\sigma} (> 0)$. The stationary point $E_s = (U_0, 0, 0)$ is stable if $R_v < 1$ and unstable otherwise.*

Proof. There are two possible cases:
- $R_v < 1$: the stationary point $E_s = (U_0, 0, 0)$ is stable. Indeed, if the initial condition $E(0)$ is taken such that the norm $|E(0) - E_s|$ is sufficiently small, then the solution converges to the stationary point $E_f = (U_f, 0, 0)$ where E_f converges to E_s if J_0 converges to zero.
- $R_v > 1$: the stationary point $E_s = (U_0, 0, 0)$ is unstable. Indeed, if we consider any initial condition $E(0)$ such that $I(0) + V(0) > 0$, then the solution converges to the stationary point $E_f = (U_f, 0, 0)$, where E_f is not close to E_s in the sense that the stationary point E_f does not converge to E_s if J_0 converges to zero. □

Stability analysis with the Lyapunov's function gives the same result and the same value of the virus replication number R_v as the linear stability analysis provided in the previous section.

3.1.3 Analytical solution

Consider system (10.16)–(10.18) in the interval $0 \le t < \tau$ with the following initial conditions:

$$U(\theta) = U(0), \quad I(\theta) = 0, \quad V(\theta) = 0, \quad -\tau \le \theta < 0, \tag{10.25}$$

$$U(0) = U_i, \quad I(0) = I_i, \quad V(0) = V_i. \tag{10.26}$$

In this interval, the last equation of the system turns into

$$\frac{dV}{dt} = -\sigma V,$$

and the system admits the following solution for $t \in [0, \tau)$:

$$V_0(t) = V_i e^{-\sigma t}, \tag{10.27}$$

$$U_0(t) = U_i \exp\left(\frac{aV_i}{\sigma}(e^{-\sigma t} - 1)\right), \tag{10.28}$$

$$I_0(t) = I_i e^{-\beta t} + aU_i V_i e^{-\beta t} \int_0^t \exp\left(\theta(\beta - \sigma) + \frac{aV_i}{\sigma}(e^{-\sigma\theta} - 1)\right) d\theta. \tag{10.29}$$

Consider next the intervals $n\tau \le t < (n+1)\tau$, $n = 1, 2, 3, \ldots$. Note that for $n = 0$ it is the previous interval, and the solution is given by (10.27)–(10.29). For $n = 1$, the last equation of the system gives

$$\frac{dV}{dt} = bI(t - \tau) - \sigma V,$$

where $0 \le t - \tau < \tau$ and $I(t - \tau) = I_0(t - \tau)$, which is already found. Solving this equation as a linear inhomogeneous equation with the known function $I_0(t - \tau)$, we obtain for $V(t)$,

$$V_1(t) = V_0(\tau)e^{\sigma(\tau - t)} + b \int_\tau^t e^{\sigma(\theta - t)} I_0(\theta - \tau)\, d\theta.$$

Subscripts here denote the respective time interval, $n = 1$ or $n = 0$. We also use here the continuity condition $V_0(\tau) = V_1(\tau)$. Substituting $V_1(t)$ into the first equation, we obtain $U_1(t)$. Substituting $V_1(t)$ and $U_1(t)$ into the second equation, we obtain $I_1(t)$. In both cases, they are linear homogeneous or inhomogeneous equations. We follow this procedure for all time intervals $n\tau \le t < (n+1)\tau$, $n = 1, 2, 3, \ldots$ Hence, solution of system (10.16)–(10.18) with initial conditions (10.25), (10.26) is composed by the following functions:

$$V_n(t) = V_{n-1}(n\tau)e^{\sigma(n\tau - t)} + b\int_{n\tau}^{t} e^{\sigma(\theta - t)} I_{n-1}(\theta - \tau)\,d\theta, \tag{10.30}$$

$$U_n(t) = U_{n-1}(n\tau)e^{-a\int_{n\tau}^{t} V_n(\theta)\,d\theta}, \tag{10.31}$$

$$I_n(t) = I_{n-1}(n\tau)e^{\beta(n\tau - t)} + a\int_{n\tau}^{t} e^{\beta(\theta - t)} U_n(\theta)V_n(\theta)\,d\theta, \tag{10.32}$$

$$V_n(n\tau) = V_{n-1}(n\tau), \quad U_n(n\tau) = U_{n-1}(n\tau), \quad I_n(n\tau) = I_{n-1}(n\tau), \tag{10.33}$$

$$n\tau \le t < (n+1)\tau, \quad n = 1, 2, 3, \ldots, \tag{10.34}$$

where the functions $V_0(t)$, $U_0(t)$, and $I_0(t)$ are defined on the interval $0 \le t < \tau$ and given by (10.27)–(10.29).

The plot of the solution according to formulas (10.27)–(10.29), (10.30)–(10.34) is presented in Figure 10.3. We note that for sufficiently large delay τ, there are two peaks (Figure 10.3b). In the case with reproduction of uninfected cells with $\lambda \ne 0$, time delay can lead to the emergence of oscillatory solution.

3.2 Basic model with diffusion

Consider system (10.13)–(10.15):

$$\frac{\partial U}{\partial t} = -aUV, \quad \frac{\partial I}{\partial t} = aUV - \beta I, \quad \frac{\partial V}{\partial t} = D\frac{\partial^2 V}{\partial x^2} + bI(t-\tau) - \sigma V, \quad t > 0$$

on the real axis with the following boundary conditions:

$$U(+\infty) = U_0, \quad U(-\infty) = U_f, \quad I(\pm\infty) = V(\pm\infty) = 0, \tag{10.35}$$

where U_f is the unknown concentration of the uninfected cells at the end of infection.

3.2.1 Final concentration of uninfected cells and total viral load

We consider the wave solution of the system (10.13)–(10.15), i. e., a bounded solution, which has a constant profile in space, constant limits at $\pm\infty$ and which moves from left to right with some constant speed $c > 0$.

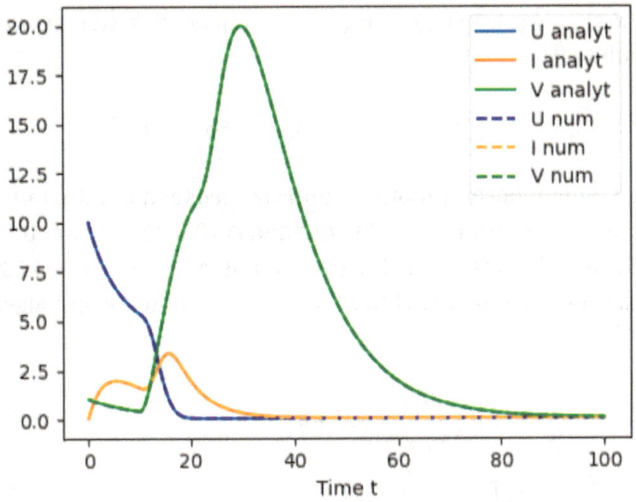

A

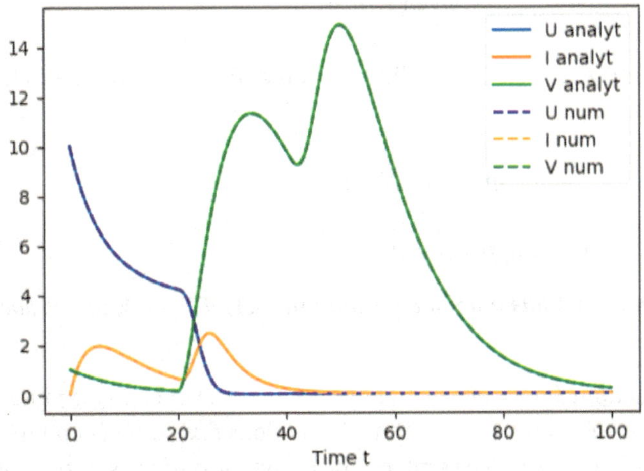

B

Figure 10.3: The analytical solution of system (10.16)–(10.18) with formulas (10.28), (10.29), (10.30)–(10.34) (solid) and its numerical solution (dashed). It can be seen that they well correspond to each other. The parameters are: $a = 0.1, \beta = 0.2, \sigma = 0.1, b = 1$, and $\tau = 10$ in (A), $\tau = 20$ in (B). Initial conditions are $U(0) = 10, I(0) = 0, V(0) = 1$. For these parameters, $R_v = 50$.

Set $\xi = x - ct$, where $c > 0$ is the speed of wave propagation. We obtain from system (10.13)–(10.15):

$$cu' = auv, \tag{10.36}$$

$$cw' + auv - \beta w = 0, \tag{10.37}$$

$$Dv'' + cv' + bw(\xi + c\tau) - \sigma v = 0. \tag{10.38}$$

Here, $U(x,t) = u(\xi)$, $I(x,t) = w(\xi)$, $V(x,t) = v(\xi)$. We complete this system with the following boundary conditions:

$$u(+\infty) = u_0, \quad u(-\infty) = u_f, \quad w(\pm\infty) = v(\pm\infty) = 0. \tag{10.39}$$

Here, constant u_f is an unknown concentration of uninfected cells at the end of infection.

In order to find u_f, we separate variables in equation (10.36), take a sum of equations (10.36) and (10.37), and integrate resulting equations and equation (10.38) in $(-\infty, +\infty)$ taking into account conditions (10.39). We obtain the following algebraic system:

$$c \ln\left(\frac{u_f}{u_0}\right) = -aJ(v), \tag{10.40}$$

$$cu_0\left(1 - \frac{u_f}{u_0}\right) = \beta J(w), \tag{10.41}$$

$$bJ(w) = \sigma J(v), \tag{10.42}$$

where $J(f) = \int_{-\infty}^{+\infty} f(\xi)\,\mathrm{d}\xi, f \in \{v, w\}$. Excluding the integrals, we obtain the following equation:

$$\ln X = R_v(X - 1), \tag{10.43}$$

where $R_v = (abU_0)/(\beta\sigma)$, and $X = u_f/u_0, X \in (0,1)$.

Theorem 3.2. *Equation* (10.43) *has a unique solution in* $(0,1)$ *if* $R_v > 1$, *and it does not have a solution in* $(0,1)$ *if* $R_v < 1$.

Proof. We consider functions $f(x) = \ln(x)$ and $g(x) = R_v(x-1)$. Function $g(x)$ intersects the vertical axis at point $-R_v$ and $f'(1) = 1$. Hence, functions $f(x)$ and $g(x)$ intersect in $x \in (0,1)$ only if $-R_v < -1$ or $R_v > 1$. The result on the uniqueness of the solution follows from the monotony of both functions on $(0,1)$. □

Thus, if the "healthy" stationary point $(U_0, 0, 0)$ loses its stability in the homogeneous case, there can exist a wave solution in the space-dependent case.

For $R_v > 1$, we can find the total viral load from (10.40):

$$J(v) = -\frac{c}{a} \ln X, \tag{10.44}$$

where X is a solution of (10.43). For $X \ll 1$, we also can use the approximate formula for $J(v)$:

$$J(v) \approx \frac{cR_v}{a} = \frac{cbU_0}{\beta\sigma}. \tag{10.45}$$

3.2.2 The speed of wave propagation

Here, we will estimate the speed of the wave propagation in system (10.13)–(10.15) using the linearization method [24]. We consider the variable $\xi = x - ct$, where $c > 0$ is the wave speed, and obtain system (10.36)–(10.38). Then we linearize this system about $(u_0, 0, 0)$, and obtain the following linear system:

$$cw' + au_0 v - \beta w = 0, \tag{10.46}$$

$$Dv'' + cv' + bw(\xi + c\tau) - \sigma v = 0. \tag{10.47}$$

We would like to find the minimal $c > 0$ such that the system (10.46)–(10.47) has a monotonically decreasing solution. We seek this solution in the form:

$$w = pe^{-\lambda\xi}, \quad v = qe^{-\lambda\xi}, \quad \lambda > 0. \tag{10.48}$$

Substitute it into (10.46)–(10.47) and obtain

$$-(c\lambda + \beta)p + au_0 q = 0,$$

$$be^{-c\lambda\tau}p + (D\lambda^2 - c\lambda - \sigma)q = 0.$$

This system has nontrivial solutions (p, q) only if the determinant of its matrix is zero:

$$\begin{vmatrix} -c\lambda - \beta & au_0 \\ be^{-c\lambda\tau} & D\lambda^2 - c\lambda - \sigma \end{vmatrix} = -(D\lambda^2 - c\lambda - \sigma)(c\lambda + \beta) - abu_0 e^{-c\lambda\tau} = 0.$$

We look for the minimal $c > 0$ for which this characteristic equation has real solutions. Set $\mu = \lambda c$. Then the characteristic equation takes the form:

$$D\frac{\mu^2}{c^2} - \mu - \sigma = -\frac{abu_0 e^{-\tau\mu}}{\mu + \beta},$$

and we can find c as some minimum of the function depending on μ:

$$c^2 = \min_{\mu > \mu_0} F(\mu), \quad F(\mu) = \frac{D\mu^2(\mu + \beta)}{(\mu + \beta)(\mu + \sigma) - abu_0 e^{-\tau\mu}}. \tag{10.49}$$

Function $F(\mu)$ has a positive root μ_0 of the denominator if $\beta\sigma < abu_0$, i. e., for $R_v > 1$, and $F(\mu) \to +\infty$ for $\mu \to \mu_0 + 0$. Thus, it has the minimum (Figure 10.4) which gives the minimal value of speed $c > 0$.

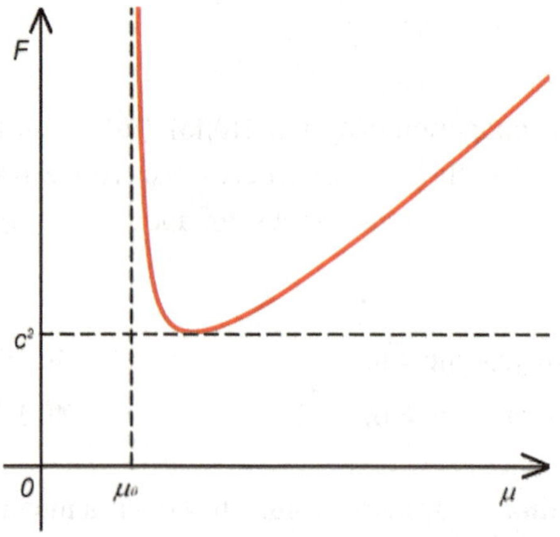

Figure 10.4: Function $F(\mu)$ in the right-hand side of equation for the wave speed (10.49), its minimum provides the square of the minimal wave speed. This function has a minimum for the argument $\mu > \mu_0$, where μ_0 is zero of the denominator of $F(\mu)$. For the basic replication number R_v greater than 1, μ_0 is positive.

3.2.3 Approximate wave solution for $\tau = 0$

We consider system (10.36)–(10.38) with boundary conditions (10.39). Suppose that $\tau = 0$, $c > 0$ is a known wave speed, and

$$u(\xi) = \begin{cases} u_0, & \xi > 0, \\ 0, & \xi < 0. \end{cases}$$

We substitute it into the system (10.36)–(10.38), and obtain the systems of equations:

$$cw' + a\bar{u}v - \beta w = 0, \tag{10.50}$$

$$Dv'' + cv' + bw - \sigma v = 0, \tag{10.51}$$

where $\bar{u} = 0$ for $\xi < 0$ and $\bar{u} = u_0$ for $\xi > 0$.

First, we consider $\xi < 0$. In this case, we have

$$w(t) = Ae^{\frac{\beta}{c}\xi}, \tag{10.52}$$

$$v(t) = Be^{\lambda_1 \xi} - \frac{bA}{D\beta^2/c^2 + \beta - \sigma} e^{\frac{\beta}{c}\xi}, \tag{10.53}$$

where

$$\lambda_1 = \frac{1}{2}\left(-\frac{c}{D} + \sqrt{\left(\frac{c}{D}\right)^2 + \frac{4\sigma}{D}} \right) > 0. \tag{10.54}$$

Next, we consider $\xi > 0$. In this case, we have a system of equations

$$cw' - \beta w + au_0 v = 0,$$
$$bw + Dv'' + cv' - \sigma v = 0.$$

This is a system of linear equations. We will find its solution in the form $w(\xi) = pe^{-\lambda\xi}$, $v(\xi) = qe^{-\lambda\xi}$, $\lambda > 0$. Thus, the characteristic equation takes the form

$$\left.\begin{aligned} \begin{vmatrix} -c\lambda - \beta & au_0 \\ b & D\lambda^2 - c\lambda - \sigma \end{vmatrix} &= -(c\lambda + \beta)(D\lambda^2 - c\lambda - \sigma) - abu_0 \\ &= 0. \end{aligned}\right\} \tag{10.55}$$

Taking into account the definition of the minimal wave speed c, i. e., it is the minimal c for which there exists a monotonously decreasing solution of the system (10.46), (10.47), we construct the solution of the considered system in the form:

$$\left.\begin{aligned} w(\xi) &= (P + Q\xi)e^{-\lambda\xi}, \\ v(\xi) &= \left(\frac{c\lambda + \beta}{au_0}P - \frac{c}{au_0}Q + \frac{c\lambda + \beta}{au_0}Q\xi\right)e^{-\lambda\xi}, \end{aligned}\right\} \tag{10.56}$$

where λ is a positive double root of the characteristic equation (10.55).

Next, we define the constants A, B, P, Q in (10.52), (10.53), and (10.56) from the continuity conditions:

$$w(-0) = w(+0), \quad v(-0) = v(+0), \quad v'(-0) = v'(+0), \tag{10.57}$$

and from the expression for the total viral load:

$$\left.\begin{aligned} J(v) &= \int_{-\infty}^{+\infty} v(\xi)\,\mathrm{d}\xi \\ &= -\frac{c}{a}\ln X, \end{aligned}\right\} \tag{10.58}$$

where X is the solution of (10.43).

Substituting the expressions (10.52), (10.53), (10.56) in these conditions we finally obtain the approximate solution of the system (10.36)–(10.38) with boundary conditions (10.39):

$$\left.\begin{aligned} w(\xi) &= \begin{cases} Ae^{\frac{\beta}{c}\xi}, & \xi < 0 \\ (P + Q\xi)e^{-\lambda\xi}, & \xi > 0 \end{cases} \\ v(\xi) &= \begin{cases} Be^{\lambda_1\xi} - \dfrac{bA}{D\beta^2/c^2 + \beta - \sigma}e^{\frac{\beta}{c}\xi}, & \xi < 0 \\ \left(\dfrac{c\lambda + \beta}{au_0}P - \dfrac{c}{au_0}Q + \dfrac{c\lambda + \beta}{au_0}Q\xi\right)e^{-\lambda\xi}, & \xi > 0 \end{cases} \end{aligned}\right\} \tag{10.59}$$

where λ_1 is defined by (10.54), λ is a positive double-root of (10.55), and constants A, B, P, and Q are as follows:

$$
\left.
\begin{aligned}
& A = P, \quad Q = rP, \quad B = r_2 P, \\
& r = \frac{\frac{au_0 b(c\lambda_1 - \beta)}{c(D\frac{\beta^2}{c^2} + \beta - \sigma)} + (c\lambda + \beta)(\lambda_1 + \lambda)}{c\lambda_1 + 2c\lambda + \beta}, \\
& r_2 = \frac{b}{D\frac{\beta^2}{c^2} + \beta - \sigma} + \frac{c\lambda + \beta - cr}{au_0}, \\
& P = \frac{-c/a \ln X}{\frac{r_2}{\lambda_1} - \frac{b}{D\frac{\beta^2}{c^2} + \beta - \sigma}\frac{c}{\beta} + \frac{c\lambda + \beta}{au_0\lambda} - \frac{cr}{au_0\lambda} + \frac{c\lambda + \beta}{au_0\lambda^2} r}.
\end{aligned}
\right\}
\tag{10.60}
$$

The results of the comparison of the approximate solution given by the formulas above and the stabilized numerical solution are presented in Figure 10.5. They are similar to each other. The difference in the left-hand part of the curve is explained by the approximation of $u(\xi) = 0$ for $\xi < 0$ while it has some small value u_f (see beginning of Section 3.2.1).

4 Discussion

In this work, we have considered models of virus infection in tissue and cell culture. Tissue or cell culture in the model can be considered as a layer of cells infected by virus. For example, it can be epithelial cells from the respiratory tract. The dynamics of the model are considered only in one direction in the tissue; however, we considered two-dimensional case earlier, and shown that results for viral load in the one-dimensional case are the lower estimate for the results in the two-dimensional case [1].

Also, in the considered models, we did not specify the units of the included parameters and variables, as they depend on the precise experiments, which we are not considering in this work. We can only show the dependencies between these units. For the model (10.13)–(10.15), the basic units are the units for the concentration of viral particles in the interstitial space $[V]$, units for concentration of uninfected cells $[U]$, units for time $[t]$, and units for space $[x]$. All others units are expressed through them as follows:

$$
\begin{aligned}
& [I] = [U], \quad [a] = \frac{1}{[t][V]}, \quad [\beta] = [\sigma] = \frac{1}{[t]}, \quad [D] = \frac{[x]^2}{[t]}, \\
& [b] = \frac{[V]}{[t][U]}, \quad [\tau] = [t].
\end{aligned}
\tag{10.61}
$$

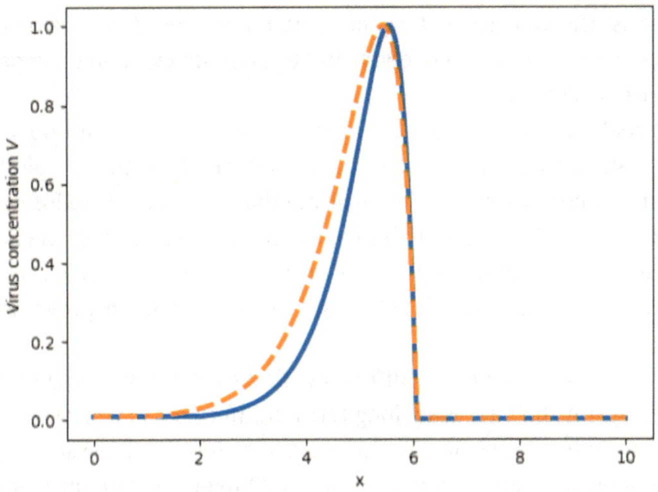

A

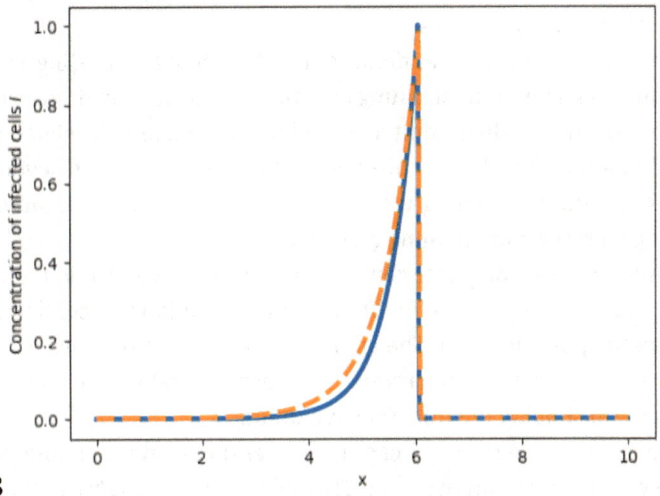

B

Figure 10.5: The spatial distributions of virus $V(x, t)$ (A) and infected cells $I(x, t)$ (B) at time moment $t = 80$. Values are normalized on their maximums: 3671 for V and 0.98 for I. The approximate analytical solution of the problem (10.36)–(10.38), (10.39) constructed in Section 3.2.3 is shown by blue solid lines, and the numerical solution is shown by orange dashed lines. Values of parameters are $a = 0.1$, $b = 1000$, $\beta = 0.1$, $\sigma = 0.1$, $D = 10^{-4}$, $u_f = 0$, $u_0 = 1$, $L = 10$. For these parameters, $c = 0.054$ and $R_v = 10000$.

For each particular case, the parameters of the model are adapted using experimental data. We have done it earlier for the dynamics of Omicron and Delta strains of SARS-CoV-2 virus in different cell cultures in works [3, 23].

This study presents a comprehensive framework for modeling viral infection progression in tissue, integrating both innate and adaptive immune responses through reaction–diffusion equations. The analysis of the general and the reduced models reveals several important insights into the interplay between viral dynamics, spatial propagation, and immune defense mechanisms.

First, our findings reaffirm the critical role of the virus replication number R_v, which is a threshold parameter determining infection outcome. In both the spatially homogeneous and heterogeneous cases, we demonstrate that $R_v < 1$ leads to infection elimination, whereas $R_v > 1$ enables sustained infection spread. This parallels the basic reproduction number in classical epidemiology, yet in our tissue-level model, it also controls the emergence of traveling wavefronts that describe infection progression through tissue volume.

The simplified model with delay and no diffusion allows for analytical solutions, providing clear expressions for the viral load, long-term uninfected cell concentration, and conditions for steady states. In particular, the exact solution for the system with delay (but without spatial terms) clarifies the transient and long-term dynamics under varying parameter regimes. The derivation of a Lyapunov function and the use of delay differential equations reinforce the robustness of the infection-clearing condition based on R_v, extending beyond local linear analysis.

In the spatially heterogeneous case, we identify and characterize traveling wave solutions, a phenomenon crucial for understanding infection propagation in tissues. The wave speed, total viral load, and residual uninfected cell population are all related via explicit integrals, providing biologically interpretable indicators of infection severity. Notably, we establish that a stable traveling wave only exists when the infection can overcome the tissue's regenerative and immune responses, i. e., when $R_v > 1$.

Additionally, our approximate analytical construction of the wave profile for $\tau = 0$ and small diffusion coefficients agrees well with numerical simulations, confirming the validity of the proposed approximation. The analytical formula for the wave speed further highlights the role of key biological parameters—such as viral production rate, clearance rate, and delay—in shaping the infection dynamics.

From a biological perspective, our model captures several essential immune processes, including interferon-mediated suppression, CTL and antibody responses, and the mutual inhibition between virus and interferon production. The modular structure of the model (Figure 10.2) allows for multiple simplifications, enabling focused analysis of distinct aspects of the immune response or viral strategy. These features make the model adaptable for a variety of experimental systems or infection scenarios.

Importantly, our analysis lays the groundwork for future investigations. Extensions of the model could include: heterogeneous tissue structures (e. g., stratified epithelia), intracellular dynamics of virus replication, coinfections or viral competition (as considered in other models in the A), or immune evasion strategies used by different viruses.

From a clinical or therapeutic viewpoint, the derived conditions for infection elimination (especially in terms of delays and immune efficiency) may inform strategies aimed at reducing viral spread in tissues, e. g., through enhancement of IFN signaling or CTL response timing. The results may also be useful for interpreting *in vitro* experiments where infection fronts are observed in tissue cultures or organoids.

In summary, the combination of analytical and numerical approaches employed in this work provides a flexible and biologically grounded toolkit for studying infection progression at the tissue level. The insights into spatial effects and immune regulation deepen our understanding of the fundamental mechanisms that govern viral spread and control, and offer quantitative criteria for distinguishing between containment and progression of infection.

5 Conclusion

In this work, we have presented a general model of viral infection progression in cell culture or tissue with immune response. This model includes a set of previously studied models as particular cases. In the basic simplest case, the basic reproduction number along with the total viral load and the wave speed are determined. Also, for this case, the approximate solution for the presence of diffusion and zero time delay, and the exact solution for zero diffusion and nonzero time delay are constructed.

Bibliography

[1] L. Ait Mahiout, N. Bessonov, B. Kazmierczak, G. Sadaka, and V. Volpert. Infection spreading in cell culture as a reaction-diffusion wave. *ESAIM: Mathematical Modelling and Numerical Analysis*, 56(3):791–814, 2022.

[2] L. Ait Mahiout, A. Mozokhina, A. Tokarev, and V. Volpert. The influence of immune response on spreading of viral infection. *Lobachevskii Journal of Mathematics*, 43(10):2699–2713, 2022.

[3] L. Ait Mahiout, A. Mozokhina, A. Tokarev, and V. Volpert. Virus replication and competition in a cell culture: Application to the SARS-CoV-2 variants. *Applied Mathematics Letters*, 133:108217, 2022.

[4] G. I. Bell. Mathematical model of clonal selection and antibody production. *Nature*, 228(5273):739–744, 1970.

[5] N. Bessonov, G. Bocharov, A. Mozokhina, and V. Volpert. Viral infection spreading in cell culture with intracellular regulation. *Mathematics*, 11(6):1526, 2023.

[6] M. Bouzari, L. Ait Mahiout, A. Mozokhina, and V. Volpert. Infection propagation in a tissue with resident macrophages. *Mathematical Biosciences*, 381:109399, 2025.

[7] R. Eftimie, J. J. Gillard, and D. A. Cantrell. Mathematical models for immunology: Current state of the art and future research directions. *Bulletin of Mathematical Biology*, 78(10):2091–2134, 2016.

[8] W. El Hajj, N. El Khatib, and V. Volpert. Inflammation propagation modeled as a reaction–diffusion wave. *Mathematical Biosciences*, 365:109074, 2023.

[9] T. F. A. Freire, Z. Hu, K. B. Wood, and E. Gjini. Modeling spatial evolution of multi-drug resistance under drug environmental gradients. *PLoS Computational Biology*, 20(5):e1012098, 2024.

[10] M. R. Fumagalli, S. Zapperi, and C. A. M. La Porta. Role of body temperature variations in bat immune response to viral infections. *Journal of the Royal Society Interface*, 18(180):20210211, 2021.

[11] F. Graw and A. S. Perelson. *Spatial Aspects of HIV Infection*, pages 3–31. Springer New York, New York, USA, 2012.

[12] M. Locke, D. Grebennikov, I. Sazonov, M. López-García, M. Loguinova, A. Meyerhans, G. Bocharov, and C. Molina-París. Exploring the therapeutic potential of defective interfering particles in reducing the replication of SARS-CoV-2. *Mathematics*, 12(12):1904, 2024.

[13] G. I. Marchuk. *Mathematical Models in Immunology*. Nauka, Moscow, Russia, 1980.

[14] S. Means, M. A. Ali, H. Ho, and J. Heffernan. Mathematical modeling for hepatitis b virus: Would spatial effects play a role and how to model it? *Frontiers in Physiology*, 11, 2020.

[15] A. Mozokhina, L. Ait Mahiout, and V. Volpert. Modeling of viral infection with inflammation. *Mathematics*, 11(19):4095, 2023.

[16] A. Mozokhina and V. Volpert. *The Interferon Influence on the Infection Wave Propagation* pages 255–261. Springer Nature Switzerland, Birkhäuser, Cham, 2024.

[17] A. S. Mozokhina and K. A. Ryumina. Multiscale mathematical model of the spread of respiratory infection considering the immune response. *Journal of Mathematical Sciences*, 292(1):125–137, 2025.

[18] A. S. Perelson, A. U. Neumann, M. Markowitz, J. M. Leonard, and D. D. Ho. Hiv-1 dynamics in vivo: Virion clearance rate, infected cell life-span, and viral generation time. *Science*, 271(5255):1582–1586, 1996.

[19] N. V. Pertsev. Stability of solutions of linear systems of differential equations of population dynamics with variable delay. *Siberian Advances in Mathematics*, 35(1):50–62, 2025.

[20] H. Ramaswamy, A. A. Oberai, and Y. C. Yortsos. A comprehensive spatial-temporal infection model. *Chemical Engineering Science*, 233:116347, 2021.

[21] N. Ringa, M. L. Diagne, H. Rwezaura, A. Omame, S. Y. Tchoumi, and J. M. Tchuenche. HIV and COVID-19 co-infection: A mathematical model and optimal control. *Informatics in Medicine Unlocked*, 31:100978, 2022.

[22] J. Shim, J. Kim, T. Tenson, J.-Y. Min, and D. Kainov. Influenza virus infection, interferon response, viral counter-response, and apoptosis. *Viruses*, 9(8):223, 2017.

[23] A. Tokarev, A. Mozokhina, and V. Volpert. Competition of SARS-CoV-2 variants in cell culture and tissue: Wins the fastest viral autowave. *Vaccines*, 10(7):995, 2022.

[24] V. A. Volpert, A. I. Volpert, and V. A. Volpert. *Traveling Wave Solutions of Parabolic Systems*. Translations of Mathematical Monographs. American Mathematical Society, Providence, Rhode Island USA, 2000.

[25] T. Williams, J. M. McCaw, and J. M. Osborne. Spatial information allows inference of the prevalence of direct cell–to–cell viral infection. *PLoS Computational Biology*, 20(7):e1012264, 2024.

Appendix A. Additional models

Models 2–5, 11, 12 are presented at Figure A.1. Model 2 describes the infection with two viruses, which are concurrent on the uninfected cells, but cannot infect one cell simultaneously. It is considered in [3]. Model 3 has additional concentrations of cells which are infected by not yet infectious, i. e. they are infected but cannot infect others. Model 4 is the concurrency of two viruses, whose dynamic is described by model 3 in the absence of the second virus. These models are considered in [23]. Model 5 describes the virus infection with respect to extracellular virus, products of several stages of its replication inside the infected cell, and intracellular resources for virus production. This model incorporates the intracellular level into the spatially distributed description. It is considered in [5]. Models 11 and 12 describe the virus infection with inflammation. In model 11, inflammation is represented by pro-inflammatory cytokines only, and in model 12, the influence of tissue resident macrophages is taken into account. These models are considered in [15] and [6], respectively.

1
$$\frac{\partial U}{\partial t} = -aUV, \quad \frac{\partial I}{\partial t} = aUV - \beta I, \quad \frac{\partial V}{\partial t} = D\frac{\partial^2 V}{\partial x^2} + bI(t-\tau) - \sigma V$$

2
$$\frac{\partial U}{\partial t} = -a_1UV_1 - a_2UV_2$$
$$\frac{\partial I_i}{\partial t} = a_iUV_i - \beta_iI_i$$
$$\frac{\partial V_i}{\partial t} = D_i\frac{\partial^2 V_i}{\partial x^2} + b_iI_i(t-\tau_i) - \sigma_iV_i \ ,$$
$$i = 1,2$$

3
$$\frac{\partial U}{\partial t} = -aUV,$$
$$\frac{\partial E}{\partial t} = aUV - \gamma E,$$
$$\frac{\partial I}{\partial t} = \gamma E - \beta I,$$
$$\frac{\partial V}{\partial t} = D\frac{\partial^2 V}{\partial x^2} - paUV + N\beta I - \sigma V$$

4
$$\frac{\partial U}{\partial t} = -a_1UV_1 - a_2UV_2,$$
$$\frac{\partial E_i}{\partial t} = a_iUV_i - \gamma_iE_i,$$
$$\frac{\partial I_i}{\partial t} = \gamma_iE_i - \beta_iI_i,$$
$$\frac{\partial V_i}{\partial t} = D_i\frac{\partial^2 V_i}{\partial x^2} - p_ia_iUV_i + N_i\beta_iI_i - \sigma_iV_i,$$
$$i = 1,2$$

5
$$\frac{\partial V_e}{\partial t} = D\frac{\partial^2 V_e}{\partial x^2} - q_1V_e + q_2V_r - \sigma_1V_e,$$
$$\frac{\partial V_i}{\partial t} = q_1V_e - k_1V_i,$$
$$\frac{\partial R}{\partial t} = k_1V_i - \sigma_2R,$$
$$\frac{\partial V_r}{\partial t} = k_3AR - q_2V_r \ ,$$
$$\frac{\partial A}{\partial t} = -k_4AR$$

11
$$\frac{\partial U}{\partial t} = k_1(U_0 - U) - k_2UV,$$
$$\frac{\partial W}{\partial t} = k_2UV - k_3SW - \sigma_1W,$$
$$\frac{\partial V}{\partial t} = D_1\frac{\partial^2 V}{\partial x^2} + bW - \sigma_3V,$$
$$\frac{dS}{dt} = k_6J(W) - k_7J(W)S - \sigma_4S$$

12
$$\frac{\partial U}{\partial t} = a_1UV,$$
$$\frac{\partial F}{\partial t} = a_1UV - k_1FS - \sigma_1F,$$
$$\frac{\partial M}{\partial t} = -a_2MV,$$
$$\frac{\partial N}{\partial t} = a_2MV - \sigma_2N,$$
$$\frac{\partial V}{\partial t} = D\frac{\partial^2 V}{\partial x^2} + bF - \sigma_3V,$$
$$\frac{dS}{dt} = a_3J(F) + dJ(N) - k_2J(F)S - \sigma_4S$$

Figure A.1: Models 2–5, 11, 12. Model 2 describes competition of two viruses, models 3 and 4 have intermediate concentration of infected but not infectious (exposed) cells, model 5 includes intracellular regulation, and models 11 and 12 include inflammation. Details are in text. Arrows go from less detailed to more detailed model, and the previous one can be obtained from the next one if some parameters are zero.

Part II-B: Person’s ecosphere exposures

Rationale: 5

Part II-B: **Person's ecosphere exposures**

Conceptually, health impacts of exposure in person's habitual natural or man-made environment, preventive measures, protective interventions, and epidemiology of adverse effects.

Three classes are considered:
- living-nature cohabitation
- civilization
- geography

Part II-B.1: **Living nature cohabitation**

This class includes
- highly poisonous exposures
- transmission of pathogens among plants, animals, and humans
- exposures to household contaminants: bacteria, archaea, protists, fungi
- hazardous encounters with wildlife

In this volume:
- Recent developments in epidemiological delay models
- Shanghai 2022 outbreak of SARS-CoV-2 omicron revisited

In former volume BIOKYBERNETIKA:
- Spread of SARS-CoV-2 and the evolution of properties of the pathogen
- Agent-based modeling of epidemic spread
- Control of outbreaks in China's provinces 2020

https://doi.org/10.1515/9783112216620-015

Masoud Saade, Samiran Ghosh, Malay Banerjee, and Vitaly Volpert

Recent developments in epidemiological delay models

Abstract: Mathematical modeling is essential for understanding and predicting infectious disease dynamics. While traditional compartmental models like the SIR framework assume instantaneous transitions, this simplification often overlooks delays in infection onset, symptom development, or recovery. Delay differential equation (DDE) models address these limitations by explicitly incorporating temporal lags, offering more realistic descriptions of epidemic progression.

This review examines recent advances in DDE-based epidemiological models, including their theoretical foundations, types of delays (e. g., incubation, latency, reporting), and practical applications in studying diseases such as influenza and COVID-19. Evidence from literature highlights the ability of DDE models to capture nonlinear epidemic dynamics, improve peak predictions, and support more effective intervention strategies.

Despite these advantages, challenges remain in estimating delay parameters and integrating complex delays into real-time forecasting. Future directions include adaptive parameter estimation, enhanced computational methods, and applications to multipathogen and multihost systems. Overall, DDE models provide a powerful framework to bridge theoretical modeling with public health decision-making.

Keywords: Epidemiological modeling, infectious disease dynamics, immunity waning, reinfection, two strains, nonlinear dynamics, delay differential equations (DDEs), public health, intervention strategies

MSC 2020: 34K18, 92B05

1 Introduction

Mathematical modeling has emerged as a critical framework for analyzing and addressing the public health challenges posed by epidemics. Historically, its development in epi-

Masoud Saade, Peoples Friendship University of Russia (RUDN University), 6 Miklukho-Maklaya St, 117198 Moscow, Russia, e-mail: 1042225269@pfur.ru
Samiran Ghosh, Bigdata Research and Supercomputing Devision, CSIR Fourth Paradigm Institute, Bengaluru-560037, India, e-mail: samiran.4pi@csir.res.in
Malay Banerjee, Department of Mathematics & Statistics, Indian Institute of Technology Kanpur, Kanpur 208016, India, e-mail: malayb@iitk.ac.in
Vitaly Volpert, Institut Camille Jordan, UMR 5208 CNRS, University of Lyon 1, 69622 Villeurbanne, France, e-mail: volpert@math.univ-lyon1.fr

https://doi.org/10.1515/9783112216620-016

demiology has been driven by the periodic emergence of large-scale outbreaks, including HIV (1980s–present) [23, 14], SARS (SARS-CoV-1) (2002–2003) [3, 40], H5N1 influenza (2005) [12, 38], H1N1 (2009) [34, 26], and Ebola (2014) [8, 39]. The unprecedented global disruption caused by the COVID-19 pandemic further underscored the profound societal, economic, and public health consequences of infectious diseases, solidifying the role of modeling in crisis response.

The foundations of modern epidemic modeling trace back to the 1918–1919 Spanish influenza pandemic and the pioneering compartmental frameworks of Kermack and McKendrick [35, 2]. Contemporary extensions include multi-compartment models, which remain central to epidemiological research. These models have diversified to address complex scenarios: analyzing historical outbreaks, projecting disease trajectories [54, 7, 36], incorporating nonlinear transmission dynamics [19, 55], spatial heterogeneity via multi-patch systems [5, 44], population stratification in multi-group models [20], and control strategies such as vaccination [4, 21]. Spatiotemporal frameworks further account for human mobility patterns to map susceptibility and infection spread [1, 22]. Comprehensive reviews of these advancements are detailed in monographs [43, 6] and surveys [29, 32]. In epidemic modeling, time delays are incorporated to reflect the latent period between initial infection and the onset of infectiousness in individuals. For vector-borne diseases, this delay also represents the duration required for pathogens to accumulate within a vector to a critical threshold necessary for successful transmission to a host.

Under the assumption that the infection transmission rate at time t is influenced by both the population of infected individuals at a prior time $t - \tau$ and the susceptible population at the current time t, the delayed SIR (Susceptible-Infected-Recovered) model for vector-borne diseases can be described by the following system of equations:

$$
\begin{aligned}
\frac{\mathrm{d}S(t)}{\mathrm{d}t} &= -f(S(t), I(t-\tau)), \\
\frac{\mathrm{d}I(t)}{\mathrm{d}t} &= f(S(t), I(t-\tau)) - \delta I(t),
\end{aligned}
$$

where $S(t)$ represents susceptible and I(t) infectious compartments, the function $f(S(t), I(t))$ characterizes the disease transmission rate, δ is the clearance rate, and is the time delay from the moment of infection to disease transmission. This formulation, explored in [31] and related studies, captures the interplay between temporal delays and disease dynamics.

A distinct class of delay models captures temporary immunity in epidemiological systems [6]. These models describe scenarios where recovered individuals gradually lose protective immunity and rejoin the susceptible population after a finite period. The dynamics are governed by the following system:

$$
\frac{\mathrm{d}S(t)}{\mathrm{d}t} = -f(S(t), I(t)) + \delta I(t-\omega),
$$

$$\frac{\mathrm{d}I(t)}{\mathrm{d}t} = f(S(t), I(t)) - \delta I(t),$$

where δ is the clearance rate and ω is the immunity duration, i. e., the delay between recovery and the loss of immunity, after which individuals transition back to the susceptible class.

In this framework, the term $\delta I(t-\omega)$ represents an individual's regaining susceptibility ω time units after recovery. The term $-\delta I(t)$ accounts for the outflow from the infected compartment due to recovery.

In long-lasting epidemics such as HIV, the influx of new susceptible individuals through demographic processes (e. g., birth or maturation) plays a crucial role in sustaining transmission dynamics. For such systems, time delays are introduced to model the maturation period τ during which individuals transition into the susceptible population. This framework, studied in [45, 15, 42] and related works, is captured by the following delay-differential equations:

$$\frac{\mathrm{d}S(t)}{\mathrm{d}t} = g(N(t-\tau)) - h(S(t), I(t)),$$
$$\frac{\mathrm{d}I(t)}{\mathrm{d}t} = h(S(t), I(t)) - \delta I(t),$$

where $N(t) = S(t)+I(t)$ denotes the total population size, the recruitment term $g(N(t-\tau))$ represents the entry of new susceptibles from the healthy population, accounting for the maturation delay τ, the transmission term $h(S, I)$ governs the rate of new infections, dependent on interactions between susceptible S and infected I individuals, and the recovery rate δ reflects the removal of individuals from the infected compartment due to recovery, immunity, or other mechanisms. The integration of time delays into infection dynamics within epidemiological models can be approached in multiple ways. One such method, termed the SIRS model [16, 37, 33], is defined by the following delay differential equations:

$$\frac{\mathrm{d}S(t)}{\mathrm{d}t} = -\frac{\beta}{N} S(t-\tau) I(t-\tau) + \mu(N - S(t) - I(t)),$$
$$\frac{\mathrm{d}I(t)}{\mathrm{d}t} = \frac{\beta}{N} S(t-\tau) I(t-\tau) - \rho I(t).$$

Here, the model posits a delay τ between the contact of a susceptible individual S and an infectious individual I, and the subsequent transition of the susceptible individual to the infectious state. During this latency period τ, the exposed individual remains classified as part of the susceptible population. Infectious individuals recover at a rate ρ, entering the removed/recovered class. Recovered individuals, in turn, lose immunity at a rate μ, rejoining the susceptible pool. The total population $N = S(t) + I(t) + R(t)$ is held constant over time, reflecting a closed system with no births, deaths, or migration.

This formulation explicitly accounts for the delayed progression from susceptibility to infectiousness, while maintaining mass-action transmission dynamics and homogeneous mixing within the population. The integration of transient immunity into SIRS frameworks can be achieved through the strategic incorporation of time delays, as demonstrated in [29]. While earlier approaches employed distributed delays, this formulation introduces three discrete, stage-specific latency periods: τ_1 represents delay between susceptible-exposed contact and onset of infectiousness, τ_2 is the minimum infectious period before recovery, and τ_3 is the minimum duration of immunity before waning. The resulting model is governed by the equations:

$$\begin{aligned}
\frac{\mathrm{d}S(t)}{\mathrm{d}t} &= -\frac{\beta}{N}S(t-\tau_1)I(t-\tau_1) + \mu R(t-\tau_3),\\
\frac{\mathrm{d}I(t)}{\mathrm{d}t} &= \frac{\beta}{N}S(t-\tau_1)I(t-\tau_1) - \rho I(t-\tau_2),\\
\frac{\mathrm{d}R(t)}{\mathrm{d}t} &= \rho I(t-\tau_2) - \mu R(t-\tau_3).
\end{aligned}$$

To analyze how delays influence system dynamics, distinct delay formulations are implemented and their outcomes compared. The proposed model involves two plant populations: susceptible (healthy) plants, $S(t)$, and infected plants, $I(t)$. As plants generally do not recover from infection, a recovered class is omitted. Similarly, vectors are categorized into susceptible vectors, $C(t)$, and infected vectors, $F(t)$. Vectors solely transmit the disease without experiencing recovery.

The delay-infection model assumes that susceptible plants transition to a latent state immediately after exposure but require a time lag θ to become infectious. The term $e^{-\alpha\theta}$ quantifies the probability of newly infected plants surviving the latency period θ before becoming infectious [13]. The governing equations are

$$\begin{aligned}
\frac{\mathrm{d}S(t)}{\mathrm{d}t} &= \mu(P - S(t)) + \sigma I(t) - \beta F(t)S(t),\\
\frac{\mathrm{d}I(t)}{\mathrm{d}t} &= e^{-\alpha\theta}\beta F(t-\theta)S(t-\theta) - (\alpha+\sigma)I(t),\\
\frac{\mathrm{d}C(t)}{\mathrm{d}t} &= \Lambda - \beta_1 I(t)C(t) - \eta C(t),\\
\frac{\mathrm{d}F(t)}{\mathrm{d}t} &= \beta_1 I(t)C(t) - \eta F(t).
\end{aligned}$$

The study presented in [10] investigates a vector-borne disease model without mortality or immunity. The constant population $S+E+I=1$ assumes homogeneous mixing, infectious vectors proportional to I, an exposure delay τ, and the infective individuals are cured and return to the susceptible group at a rate c proportional to their number. The resulting dynamics are described by the following delay differential equations for S, E, and I,

$$\begin{aligned}
\frac{\mathrm{d}S(t)}{\mathrm{d}t} &= -\beta S(t)I(t) + cI(t),\\
\frac{\mathrm{d}E(t)}{\mathrm{d}t} &= \beta S(t)I(t) - \beta S(t-\tau)I(t-\tau),
\end{aligned}$$

$$\frac{dI(t)}{dt} = \beta S(t-\tau)I(t-\tau) - cI(t).$$

Classical SIR (Susceptible–Infected–Recovered) models have been extended to single and multistrain systems, often assuming instantaneous recovery or mortality rates. To address this limitation, delay differential equation (DDE) models have gained prominence for their ability to integrate temporal delays inherent in disease progression and transmission. DDE frameworks have been applied to influenza (single and multistrain) and COVID-19 [25, 51, 24], including cross-immunity dynamics in two-strain systems [52]. Recent work also explores strain competition without cross-immunity [46] and validates DDEs as approximations for distributed recovery/death rates [24]. Additional extensions incorporate vaccination [50] and periodic transmission rates in single and dual strain contexts [53].

The COVID-19 pandemic catalyzed widespread societal and economic upheaval, prompting governments to adopt partial lockdowns—a strategy balancing infection control with economic functionality [56]. Studies demonstrate that such measures reduce viral reproduction numbers [30] but incur significant economic costs, including GDP contraction, unemployment, and heightened poverty [41]. These trade-offs highlight the necessity of integrated policy frameworks. For instance, economic-demographic models [59] reveal scenarios where transient population growth may precede decline, emphasizing the delicate interplay between public health interventions and socioeconomic stability.

Prior research has examined the interconnected dynamics of partial containment policies, epidemic evolution, and socioeconomic effects. For example, the epidemic-economic framework in [47] applies delay differential equations (DDEs) to simulate time-lagged disease transmission processes, explicitly connecting public health outcomes to economic performance. The model evaluates both short-term impacts such as abrupt shifts in infection rates and long-term consequences such as workforce disruptions and productivity trends.

In Section 2, we discuss the distributed model and derive the delay model where the discrete time delay estimates average disease duration. We also obtain epidemic characterization of the delay model. Then we perform a numerical comparison among the distributed model, delay model, and conventional SIR model with the equivalent parameter values. Next, we discuss a method to estimate the value of disease duration using the real data of disease incidence and we validate our delay model with epidemiological data collected during the COVID-19 epidemic. In Section 3, we introduce an epidemic delay model in the case of a single strain. We show how this model can be reduced to a single integral equation and use it to study the existence and stability of stationary solutions. We also study epidemic dynamics for the two-strain model. We find stationary solutions of the model and study their stability. After that, we present numerical simulations to study the dynamics of solutions. Finally, we end this work with conclusions and further perspectives.

2 An epidemic model with time-distributed recovery and death rates

Ongoing pandemic stimulated important modelling efforts directed to the application of the existing models and to their critical rethinking. Compartmental epidemiological models, like the classical SIR model, are based on the assumptions that newly infected individuals at time t appear with the rate proportional to the product of the numbers of susceptible individuals $S(t)$ and infected individuals $I(t)$, and that the recovery and death rates are proportional to the number of infected individuals. The first assumption is justified for homogeneous populations, but the second assumption has a limited applicability. Indeed, assuming that an average disease duration is τ, we conclude that the recovery and death rates at time t are determined by the number of infected individuals at time $I(t-\tau)$ (disease onset), which can be very different from $I(t)$, unless the epidemic progression is slow (basic reproduction number is close to 1). In a more detailed description, we do not consider a fixed disease duration but take into account that the recovery and death rates depend on the disease status of infected individuals, that is on time passed after the disease onset. We propose a compartmental model based on integrodifferential equations where we take into account that recovery and death rates at time t depend on the time interval $t-\eta$ from the disease onset for the individuals infected at time η [32].

2.1 Model formulation

We propose an integrodifferential equation model where the recovery and death rates depend on the time-since-infection of the infected individuals. Let $J(t)$ be the number of newly infected individuals at time t, while $S(t)$, $I(t)$, $R(t)$, and $D(t)$ denote the total numbers of susceptible, infected, recovered, and dead individuals at time t. The total number of infected at time t is given by the following expression:

$$I(t) = \int_0^t J(\eta)\mathrm{d}\eta - R(t) - D(t). \tag{11.1}$$

We assume that the total population size remains constant, $S(t)+I(t)+R(t)+D(t)=N$, i. e., natural natality and mortality rates are assumed to be equal to each other. Using the equality $I(t)+R(t)+D(t) = N - S(t)$ and differentiating equality (11.1), we obtain $\frac{\mathrm{d}S}{\mathrm{d}t} = -J(t)$. On the other hand, the rate of change of the susceptible population is given by the equation:

$$\frac{\mathrm{d}S}{\mathrm{d}t} = -\beta\frac{S}{N}I \quad (= -J(t)),$$

where β is the disease transmission rate.

Let $r(t-\eta)$ and $d(t-\eta)$ be the recovery and death rates depending on the time since infection $t-\eta$. Then the number of infected individuals who will recover at time t is given by the expression:

$$\int_0^t r(t-\eta)J(\eta)\mathrm{d}\eta.$$

Similarly, we determine the number of infected individuals who will die at time t:

$$\int_0^t d(t-\eta)J(\eta)\mathrm{d}\eta.$$

Thus, the rate of change of the infected compartment $I(t)$ is given by the following equation:

$$\frac{\mathrm{d}I}{\mathrm{d}t} = \beta\frac{S}{N}I - \int_0^t r(t-\eta)J(\eta)\mathrm{d}\eta - \int_0^t d(t-\eta)J(\eta)\mathrm{d}\eta.$$

The rates of change of the recovered $R(t)$ and the death compartment $D(t)$ are given, respectively, by the equations:

$$\frac{\mathrm{d}R}{\mathrm{d}t} = \int_0^t r(t-\eta)J(\eta)\mathrm{d}\eta,$$

$$\frac{\mathrm{d}D}{\mathrm{d}t} = \int_0^t d(t-\eta)J(\eta)\mathrm{d}\eta.$$

Hence, we obtain the following model:

$$\frac{\mathrm{d}S}{\mathrm{d}t} = -\beta\frac{S}{N}I, \tag{11.2a}$$

$$\frac{\mathrm{d}I}{\mathrm{d}t} = \beta\frac{S}{N}I - \int_0^t r(t-\eta)J(\eta)\mathrm{d}\eta - \int_0^t d(t-\eta)J(\eta)\mathrm{d}\eta, \tag{11.2b}$$

$$\frac{\mathrm{d}R}{\mathrm{d}t} = \int_0^t r(t-\eta)J(\eta)\mathrm{d}\eta, \tag{11.2c}$$

$$\frac{\mathrm{d}D}{\mathrm{d}t} = \int_0^t d(t-\eta)J(\eta)\mathrm{d}\eta. \tag{11.2d}$$

This system of equations should be completed by the initial condition $S(0) = N, I(0) = I_0 > 0, R(0) = 0, D(0) = 0$, and $I_D(t) = 0$ for $t \le 0$. We will study below some properties of this model and will apply it to assess the epidemic progression.

2.2 Reduction to the SIR model

In this section, we show that model (11.2a)–(11.2d) can be reduced to the conventional SIR model under some assumptions. Consider the recovery and death rates in the form:

$$\left.\begin{aligned} r(t-\eta) &= \begin{cases} r_0, & t-\tau < \eta \le t \\ 0, & \eta < t-\tau \end{cases} \\ d(t-\eta) &= \begin{cases} d_0, & t-\tau < \eta \le t \\ 0, & \eta < t-\tau \end{cases} \end{aligned}\right\} \tag{11.3}$$

where $\tau > 0$ is disease duration, r_0 and d_0 are some constants. Substituting these functions into (11.2c) and (11.2d), we get

$$\left.\begin{aligned} \frac{dR}{dt} &= r_0 \int_{t-\tau}^{t} J(\eta)d\eta, \\ \frac{dD}{dt} &= d_0 \int_{t-\tau}^{t} J(\eta)d\eta. \end{aligned}\right\} \tag{11.4}$$

Integrating these equalities from $t - \tau$ to t, we obtain

$$R(t) - R(t-\tau) = r_0 \int_{t-\tau}^{t} \left(\int_{s-\tau}^{s} J(\eta)d\eta \right) ds,$$

$$D(t) - D(t-\tau) = d_0 \int_{t-\tau}^{t} \left(\int_{s-\tau}^{s} J(\eta)d\eta \right) ds.$$

Since we assume that the disease duration is τ, then (11.1) can be written as follows:

$$I(t) = \int_{t-\tau}^{t} J(\eta)d\eta - (R(t) - R(t-\tau)) - (D(t) - D(t-\tau)), \tag{11.5}$$

where $(R(t)-R(t-\tau))$ and $(D(t)-D(t-\tau))$ represent the numbers of recovered and dead during the time interval $(t - \tau,\ t)$, respectively. Hence, from (11.5), we have

$$I(t) = \int_{t-\tau}^{t} J(\eta)\mathrm{d}\eta - (r_0 + d_0) \int_{t-\tau}^{t} \left(\int_{s-\tau}^{s} J(\eta)\mathrm{d}\eta \right) \mathrm{d}s. \tag{11.6}$$

Now, from (11.2b) and (11.6),

$$\begin{aligned} \frac{\mathrm{d}I}{\mathrm{d}t} &= \beta \frac{S}{N} I - (r_0 + d_0) \int_{t-\tau}^{t} J(\eta)\mathrm{d}\eta, \\ &= \beta \frac{S}{N} I - (r_0 + d_0) \left[I(t) + (r_0 + d_0) \int_{t-\tau}^{t} \left(\int_{s-\tau}^{s} J(\eta)\mathrm{d}\eta \right) \mathrm{d}s \right]. \end{aligned}$$

Assuming that $(r_0 + d_0)$ is small enough, we neglect the term involving $(r_0 + d_0)^2$. Hence, we obtain

$$\frac{\mathrm{d}I}{\mathrm{d}t} \approx \beta \frac{S}{N} I - (r_0 + d_0)I.$$

In this case, system (11.2a)–(11.2d) is reduced to the conventional SIR model:

$$\frac{\mathrm{d}S}{\mathrm{d}t} = -\beta \frac{S}{N} I, \tag{11.8a}$$

$$\frac{\mathrm{d}I}{\mathrm{d}t} = \beta \frac{S}{N} I - (r_0 + d_0)I, \tag{11.8b}$$

$$\frac{\mathrm{d}R}{\mathrm{d}t} = r_0 I, \tag{11.8c}$$

$$\frac{\mathrm{d}D}{\mathrm{d}t} = d_0 I. \tag{11.8d}$$

Thus, assuming uniform distribution of recovery and death rates (11.3) and that they are small enough, we can reduce model (11.2a)–(11.2d) to the classical SIR model. However, in general, these assumptions do not hold, and we need to take into account more realistic recovery and death rate distributions.

2.3 Estimation of recovery and death rate functions

2.3.1 Gamma distribution

We estimate the recovery $r(t - \eta)$ and death $d(t - \eta)$ distributions in the case of COVID-19 epidemic. The data for 120 recovered patients and 31 dead individuals from [27] were used to fit a gamma distribution. Note that there are no recovery or death during the first 2 days after infection, and the maximums of these distributions are reached between 10 and 15 days (Figure 11.1).

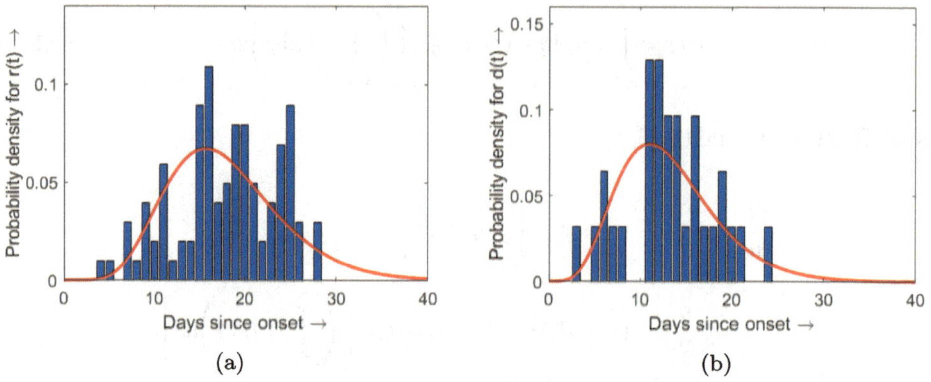

Figure 11.1: Probability distribution of recovery (a) and death (b) as a function of time (in days) after the onset of infection. The red curves show the best fit gamma distributions (the values of parameters are given in the text). This figure is taken from [25] with permission from SNCSC.

We estimate the mean time from the disease onset to recovery as 17.85 days and the mean time to death as 13.19 days. The best fitted gamma distribution corresponding to recovery, which is shown by the red curve in Figure 11.1a, is given by the expression

$$f_1(t) = \frac{1}{b_1^{a_1}\Gamma(a_1)} t^{a_1-1} e^{-\frac{t}{b_1}}$$

with the estimated parameter values $a_1 = 8.06275$ and $b_1 = 2.21407$. Similarly, the best fitted gamma distribution corresponding to death shown by the red curve in Figure 11.1b, is given by

$$f_2(t) = \frac{1}{b_2^{a_2}\Gamma(a_2)} t^{a_2-1} e^{-\frac{t}{b_2}},$$

where $a_2 = 6.00014$ and $b_2 = 2.19887$.

These functions are normalized in such a way that the total probability of recovery and death equals 1. We set $r(t) = p_0 f_1(t)$, $d(t) = (1 - p_0) f_2(t)$, where p_0 is the survival probability. Its value is estimated from the data as $p_0 = 0.97$ [49].

2.3.2 Model validation with epidemiological data

In order to validate the model with distributed recovery and death rates, we compare the results of modeling with the epidemiological data. Distributed recovery $r(t - \eta)$ and death $d(t - \eta)$ rates are estimated in Section 2.3.1 from the data in China in February 2020 [27, 57]. Once these functions are determined, we take the number $J(t)$ of daily infected individuals from the epidemiological data and find the sum of daily recoveries and deaths from the expression:

$$\Sigma(t) = \int_0^t r(t-\eta)J(\eta)\mathrm{d}\eta + \int_0^t d(t-\eta)J(\eta)\mathrm{d}\eta. \tag{11.9}$$

These results are compared with the sum of recoveries and deaths in the data. Figure 11.2 shows the result of such comparison for China from January 23, 2020, to April 15, 2020, with the data from [58] (7-day moving average).

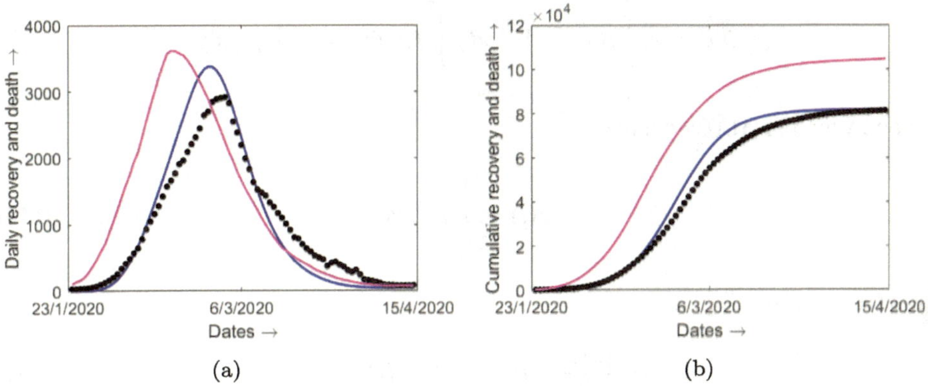

Figure 11.2: In the left panel, the blue curve shows the number $\Sigma(t)$ of recovered and dead in the distributed model, the magenta curve corresponds to $\sigma(t)$ in the SIR model, and the black dots correspond to the 7-day moving average of daily recoveries and deaths in China. The right panel shows the corresponding cumulative recovery and death. This figure is taken from [25] with permission from SNCSC.

2.4 Delay model

Let us assume that disease duration is τ, and the individuals $J(t-\tau)$ infected at time $t-\tau$ recover or die at time t with certain probabilities. This assumption corresponds to the following choice of the functions $r(t-\eta)$ and $d(t-\eta)$:

$$r(t-\eta) = r_0\delta(t-\eta-\tau),$$
$$d(t-\eta) = d_0\delta(t-\eta-\tau),$$

where r_0, d_0 are constants, $r_0 + d_0 = 1$, and δ is the Dirac delta-function. Then

$$\frac{\mathrm{d}R(t)}{\mathrm{d}t} = \int_0^t r(t-\eta)J(\eta)\mathrm{d}\eta = r_0 J(t-\tau),$$

$$\frac{\mathrm{d}D(t)}{\mathrm{d}t} = \int_0^t d(t-\eta)J(\eta)\mathrm{d}\eta = d_0 J(t-\tau).$$

Note that $J(t)$ is the number of newly infected individuals that appeared at time t. If we assume that the first infected case was reported at time $t = 0$, then we can set $J(t) = 0$ for all $t < 0$. Now integrating the above two equations from 0 to t and assuming that $R(0) = D(0) = 0$, we get

$$R(t) = r_0 \int_0^t J(s-\tau)\mathrm{d}s = r_0 \int_{-\tau}^{t-\tau} J(y)\mathrm{d}y = r_0 \int_0^{t-\tau} J(y)\mathrm{d}y,$$

$$D(t) = d_0 \int_0^t J(s-\tau)\mathrm{d}s = d_0 \int_{-\tau}^{t-\tau} J(y)\mathrm{d}y = d_0 \int_0^{t-\tau} J(y)\mathrm{d}y.$$

Then instead of (11.1), we have

$$I(t) = \int_{t-\tau}^{t} J(s)\mathrm{d}s, \tag{11.10}$$

and from (11.2b)

$$\frac{\mathrm{d}I(t)}{\mathrm{d}t} = \beta\frac{S(t)}{N} I(t) - J(t-\tau). \tag{11.11}$$

From (11.10), we obtain

$$J(t) = \beta\frac{S(t)}{N}\int_{t-\tau}^{t} J(s)\mathrm{d}s \quad \text{and} \quad \frac{\mathrm{d}S(t)}{\mathrm{d}t} = -J(t). \tag{11.12}$$

Hence,

$$\left.\begin{aligned} \frac{\mathrm{d}S(t)}{\mathrm{d}t} &= -\beta\frac{S(t)}{N}\int_{t-\tau}^{t} J(s)\mathrm{d}s \\ &= \beta\frac{S(t)}{N}\int_{t-\tau}^{t}\frac{\mathrm{d}S(s)}{\mathrm{d}s}\mathrm{d}s \\ &= -\beta\frac{S(t)}{N}\left(S(t-\tau) - S(t)\right). \end{aligned}\right\} \tag{11.13}$$

Once we get the solution $S(t)$ from (11.13), we can then find $I(t)$ using the following relation:

$$I(t) = \int_{t-\tau}^{t} J(s)\mathrm{d}s = -\int_{t-\tau}^{t}\frac{\mathrm{d}S(s)}{\mathrm{d}s}\mathrm{d}s = S(t-\tau) - S(t).$$

Hence, system (11.2a)–(11.2d) is reduced to the following delay model:

$$\frac{dS(t)}{dt} = -J(t), \tag{11.14a}$$
$$\frac{dI(t)}{dt} = J(t) - J(t-\tau), \tag{11.14b}$$
$$\frac{dR(t)}{dt} = r_0 J(t-\tau), \tag{11.14c}$$
$$\frac{dD(t)}{dt} = d_0 J(t-\tau), \tag{11.14d}$$
$$J(t) = \beta \frac{S(t)}{N} I(t), \tag{11.14e}$$

with $J(t) = 0$ for all $t < 0$.

2.5 Epidemic characteristics

In this section, we determine the basic reproduction number, the final size of epidemic, and the maximum number of infected individuals for the delay model (11.14).

2.5.1 Basic reproduction number

To determine the initial exponential growth rate and the basic reproduction number $\mathcal{R}_0$, we will use the equations for the infected compartments $I(t)$ considered in the form:

$$\frac{dI(t)}{dt} = \beta \frac{S(t)}{N} I(t) - \beta \frac{S(t-\tau)}{N} I(t-\tau). \tag{11.15}$$

Suppose that, in the beginning of the epidemic, $S(t) \approx S_0$ and $S(t-\tau) = S_0$. Then from (11.15), we have

$$\frac{dI(t)}{dt} = \beta \frac{S_0}{N} (I(t) - I(t-\tau)).$$

Substituting $I(t) = I_0 e^{\lambda t}$ we have

$$\lambda = \beta \frac{S_0}{N} (1 - e^{-\lambda\tau}). \tag{11.16}$$

Let $\mathcal{G}(\lambda) = \beta \frac{S_0}{N} (1 - e^{-\lambda\tau})$. Clearly, (11.16) has a solution $\lambda = 0$ and a nonzero solution with sign determined by $\mathcal{G}'(0)$. Denote

$$\mathcal{R}_0 = \mathcal{G}'(0) = \beta\tau \frac{S_0}{N}.$$

Suppose that $\mathcal{R}_0 > 1$. This implies that $\mathcal{G}'(0) > 1$. We observe that $\mathcal{G}(\lambda)$ is an increasing function of λ and $\mathcal{G}(\lambda) \to \beta \frac{S_0}{N}$ as $\lambda \to \infty$. This implies that Equation (11.16) has a positive solution, i. e., there exists $\lambda_* > 0$ such that $\mathcal{G}(\lambda_*) = \lambda_*$. If $\mathcal{R}_0 < 1$, i. e., $\mathcal{G}'(0) < 1$, then

$$\mathcal{G}'(\lambda) = \beta\tau\frac{S_0}{N}e^{-\lambda\tau} < 1$$

for all $\lambda \geq 0$. This implies that the equation $\mathcal{G}(\lambda) = \lambda$ has no positive solution. Hence, the basic reproduction number is given by the following expression:

$$\mathcal{R}_0 = \beta\tau\frac{S_0}{N}. \tag{11.17}$$

Letting $\hat{\lambda} = \lambda/(\beta\frac{S_0}{N})$, we can write Equation (11.16) in the following form:

$$\hat{\lambda} = 1 - e^{-\mathcal{R}_0\hat{\lambda}}.$$

The solution of this equation determines the initial exponential growth rate, which depends on the single parameter $\mathcal{R}_0$.

2.5.2 Final size of epidemic

Next, we determine the final size of the susceptible compartment $S_f = \lim_{t\to\infty} S(t)$. From (11.2a), we have

$$\frac{\mathrm{d}S(t)}{\mathrm{d}t} = -\beta\frac{S(t)}{N}\int_{t-\tau}^{t} J(\eta)\mathrm{d}\eta.$$

Then, integrating from 0 to ∞, we obtain

$$\int_0^\infty \frac{\mathrm{d}S}{S} = -\frac{\beta}{N}\int_0^\infty\left(\int_{t-\tau}^{t} J(\eta)\mathrm{d}\eta\right)\mathrm{d}t.$$

Changing the order of integration, we get

$$\begin{aligned}\ln\frac{S_0}{S_f} &= \frac{\beta}{N}\left[\int_{-\tau}^{0}\left(\int_0^{\eta+\tau} J(\eta)\mathrm{d}t\right)\mathrm{d}\eta + \int_0^\infty\left(\int_\eta^{\eta+\tau} J(\eta)\mathrm{d}t\right)\mathrm{d}\eta\right] \\ &= \frac{\beta}{N}\left[\int_{-\tau}^{0}(\eta+\tau)J(\eta)\mathrm{d}\eta + \int_0^\infty \tau J(\eta)\mathrm{d}\eta\right].\end{aligned}$$

Now, integrating (11.14a) from 0 to ∞, we obtain

$$\int_0^\infty J(\eta)\mathrm{d}\eta = S_0 - S_f. \tag{11.18}$$

Since $J(t) = 0$ for all $t \in [-\tau, 0]$, then

$$\ln\frac{S_0}{S_f} = \frac{\beta}{N}\left[\int_0^\infty \tau J(\eta)\mathrm{d}\eta\right] = \frac{\beta}{N}\tau(S_0 - S_f). \tag{11.19a}$$

Thus, the final size can be obtained from the equation:

$$\ln\frac{S_0}{S_f} = \frac{\beta}{N}\tau(S_0 - S_f). \tag{11.20}$$

Integrating (11.14c) and (11.14d) from 0 to ∞ and using (11.18), we have the following equalities:

$$R_f \equiv \lim_{t\to\infty} R(t) = r_0\int_0^\infty J(s-\tau)\mathrm{d}s = r_0\int_0^\infty J(\eta)\mathrm{d}\eta = r_0(S_0 - S_f),$$

$$D_f \equiv \lim_{t\to\infty} D(t) = d_0\int_0^\infty J(s-\tau)\mathrm{d}s = d_0\int_0^\infty J(\eta)\mathrm{d}\eta = d_0(S_0 - S_f).$$

Final size S_f obtained from the numerical simulation and using the formula (11.20) is verified for three different values of τ and a range of values of β (Figure 11.3). Observe that Equation (11.20) can be written as follows:

$$\ln\frac{S_0}{S_f} = \mathcal{R}_0\left(1 - \frac{S_f}{S_0}\right). \tag{11.21}$$

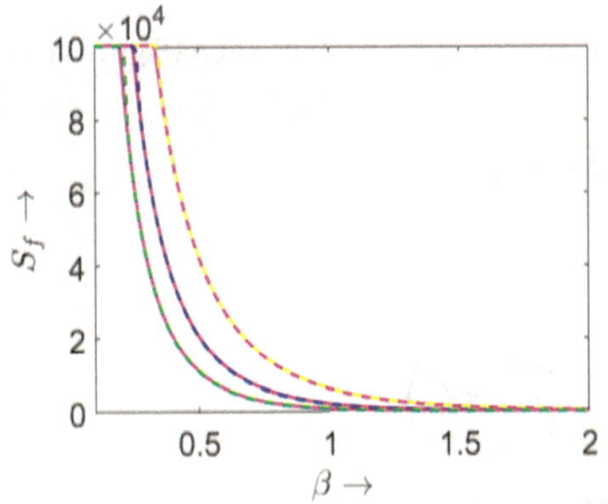

Figure 11.3: Dependence of S_f on β found analytically by formula (11.20) and in numerical simulations of (11.13) for $\tau = 3$ (upper curves), $\tau = 4$ (middle curves), $\tau = 5$ (lower curves). Analytical and numerical solutions coincide.

Note that the SIR model (11.8a)–(11.8d) gives the same equation for the final size if we assume that $r_0 + d_0 = 1/\tau$.

2.5.3 Maximum number of infected individuals

We now derive an approximate formula for the maximal number of infected individuals for the delay model (11.14). We have $I(t) = S(t-\tau) - S(t)$. Suppose that $I(t)$ attains its maximum at $t = t_m$. Set $I_m = I(t_m)$, $S_m = S(t_m)$. From the equality $I'(t_m) = 0$, we get $S'(t_m - \tau) = S'(t_m)$. This implies

$$\frac{\beta S(t_m-\tau)I(t_m-\tau)}{N} = \frac{\beta S(t_m)I(t_m)}{N}. \tag{11.22}$$

Substituting the relation $I(t_m) = S(t_m - \tau) - S(t_m)$ in (11.22), we obtain

$$I(t_m - \tau) = \frac{S_m I_m}{S_m + I_m}. \tag{11.23}$$

From (11.13), we have

$$\frac{\mathrm{d}S}{\mathrm{d}t} = -\frac{\beta S(t)}{N}(S(t-\tau) - S(t)). \tag{11.24}$$

Integrating (11.24) from 0 to t_m and changing the variable inside the first integral of the right-hand side, we get

$$\int_0^{t_m} \frac{\mathrm{d}S}{S} = -\frac{\beta}{N}\left(\int_{-\tau}^{0} S(t)\mathrm{d}t - \int_{t_m-\tau}^{t_m} S(t)\mathrm{d}t\right). \tag{11.25a}$$

We assume that $S(t) = S_0$ for all $t \in [-\tau, 0]$ and use the approximation (Figure 11.4):

$$\int_{t_m-\tau}^{t_m} S(t)\mathrm{d}t \approx \frac{\tau}{2} I_m + \tau S_m.$$

Then from (11.25a), we have

$$\ln\frac{S_m}{S_0} = -\frac{\beta}{N}\left(\tau S_0 - \frac{\tau}{2} I_m - \tau S_m\right).$$

Let $x = \frac{I_m}{S_0}, y = \frac{S_m}{S_0}$ and $\frac{S_0}{N} \approx 1$. Then we have

$$\ln y = -\beta\tau\left(1 - \frac{1}{2}x - y\right). \tag{11.26}$$

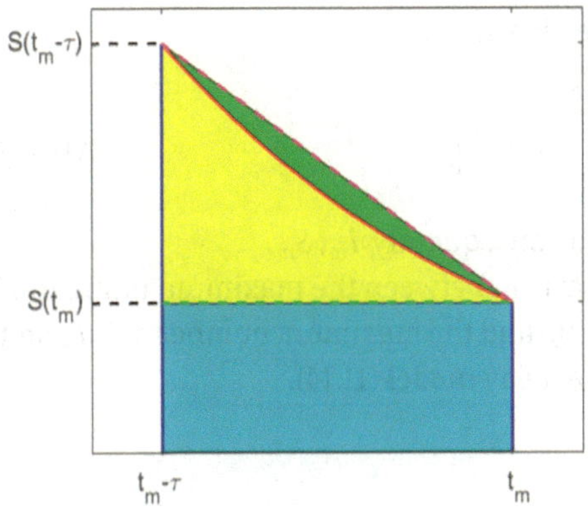

Figure 11.4: The red curve represents $S(t)$. The integral $\int_{t_m-\tau}^{t_m} S(t)dt$, i. e., the area under the red curve (yellow color region + cyan color region) is approximated to the sum of the areas of the cyan color region, the yellow color region, and the green color region.

Integrating (11.24), from $t_m - \tau$ to t_m we get

$$\int_{t_m-\tau}^{t_m} \frac{dS}{S} = -\frac{\beta}{N}\left(\int_{t_m-\tau}^{t_m} S(t-\tau)dt - \int_{t_m-\tau}^{t_m} S(t)dt\right).$$

Changing the variable inside the first integral of the right-hand side, we obtain

$$\ln \frac{S_m}{S(t_m-\tau)} = -\frac{\beta}{N}\left(\int_{t_m-2\tau}^{t_m-\tau} S(t)dt - \int_{t_m-\tau}^{t_m} S(t)dt\right).$$

Now, using the approximation described in Figure 11.4, we have the next equation:

$$\ln \frac{S_m}{S(t_m-\tau)} = -\frac{\beta}{N}\left(\tau S(t_m-\tau) + \frac{\tau}{2}I(t_m-\tau) - \tau S_m - \frac{\tau}{2}I_m\right). \tag{11.27}$$

Using the relation $I(t_m) = S(t_m - \tau) - S(t_m)$, after some manipulation, (11.27) can be written as follows:

$$\ln \frac{S_m}{S_m + I_m} = -\frac{\beta}{N}\frac{\tau}{2}(I_m + I(t_m - \tau)).$$

Using (11.23), we can write the previous equality as

$$\ln \frac{S_m}{S_m + I_m} = -\frac{\beta}{N}\frac{\tau}{2}\left(I_m + \frac{S_m I_m}{S_m + I_m}\right).$$

Substituting $x = \frac{I_m}{S_0}$, $y = \frac{S_m}{S_0}$, and $\frac{S_0}{N} \approx 1$, we obtain

$$\ln \frac{y}{x+y} = -\frac{\beta\tau}{2}\left(x + \frac{xy}{x+y}\right). \tag{11.28}$$

Solving (11.26) and (11.28), we can find x, y, and consequently, I_m, S_m.

In the Figure 11.5, we have shown a comparison between the maximum number of infected obtained by the Equations (11.26), (11.28), and the maximum number of infected obtained by direct numerical simulation of the delay model (11.14).

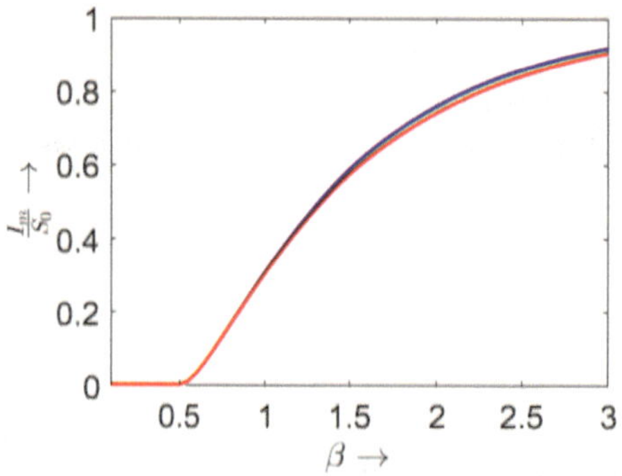

Figure 11.5: The red curve and the blue curve show the maximum number of infected using the direct numerical simulation of the delay model and using the Equations (11.26), (11.28), respectively. Parameter values: $N = 10^5$, $\tau = 4$, $S_0 = N - 1$, $I_0 = 1$.

2.6 Comparison of models (11.2), (11.14), and SIR (11.8)

We compare the results obtained from the distributed model (11.2), the delay model (11.14) and the conventional SIR model (11.8). It was shown above that model (11.2) can describe the epidemic spread more precisely, and it can exactly capture the recovery and death dynamics by using suitable distributed recovery and death rates. However, the main constraint in using the distributed model (11.2) is the availability of distributed recovery and death rates. Instead, the average duration of the disease is easier to determine. We show that, in the delay model (11.14), if we take τ as the mean of the recovery and death distribution functions involved in the distributed model (11.2), then the delay model (11.14) gives a close solution to the distributed model (11.2).

To estimate the recovery and death distributions in the model (11.2), we use the function *fitdist(:,gamma)* in MATLAB. This function is used to fit a vector of data $X = (x_1, x_2, \ldots, x_n)$ by a gamma distribution of the form $\frac{1}{b^a\Gamma(a)} x^{a-1} e^{-x/b}$, where a and b are the shape and scale parameters. This function gives the maximum likelihood estimators of a and b for the gamma distribution, which are the solutions of the simultaneous equations

$$\log \hat{a} - \Psi(\hat{a}) = \log\left(\bar{X}/\left(\prod_{i=1}^{n} x_i\right)^{1/n}\right),$$

$$\hat{b} = \bar{X}/\hat{a},$$

where $\bar{X}$ is the sample mean of the data X and Ψ is the digamma function given by

$$\Psi(x) = \Gamma'(x)/\Gamma(x).$$

The function *fitdist(:,gamma)* estimates the shape and scale parameters with 95 % confidence interval.

Using this statistical technique, we have estimated the recovery and death distributions for the model (11.2) for the data used in [25] and given by the formulas:

$$r(t) = \frac{p_0}{b_1^{a_1}\Gamma(a_1)} t^{a_1-1} e^{-\frac{t}{b_1}}, \quad d(t) = \frac{(1-p_0)}{b_2^{a_2}\Gamma(a_2)} t^{a_2-1} e^{-\frac{t}{b_2}},$$

where the estimated parameter values are given in Table 11.1. Here, p_0 is the survival probability and its estimated value, from the data, is $p_0 = 0.97$ [49].

Table 11.1: Estimated parameter values (gamma distributions).

Parameters	Estimated value	95 % confidence interval
a_1	8.06275	[6.15314, 10.565]
b_1	2.2140	[1.67523, 2.92623]
a_2	6.00014	[3.69566, 9.74161]
b_2	2.19887	[1.32639, 3.64526]

The estimated average time to recovery is 17.85 days and to death is 13.19 days. The value of survival probability p_0 is 0.97 [49], i. e., out of 100 infected individuals, 97 infected will recover. This estimate matches with most of the COVID-19 epidemic data from various countries [49, 58]. Thus, we can take average disease duration τ as 17.7 days. The corresponding value in SIR model is $r_0 + d_0 \approx 1/17.7\ \text{days}^{-1}$.

It is essential to mention here that, instead of gamma distribution we can use the Erlang distribution $E_k = \frac{\lambda^k x^{k-1} e^{-\lambda x}}{(k-1)!}$, $x, \lambda \geq 0$, and $k \in \mathbb{N}$. Since, for lower values of k, e. g., E_1, E_2 distributions, with reasonable choice of λ, give significant probability of recovery or death in the beginning of infection, we should look for Erlang distribution with higher values of k.

We have found that Erlang-8 distribution:

$$E_8 = \frac{\lambda^8 x^7 e^{-\lambda x}}{7!},$$

with $\lambda = 0.4545$, and the Erlang-6 distribution

$$E_6 = \frac{\lambda^6 x^5 e^{-\lambda x}}{5!},$$

with $\lambda = 0.4548$, are closely matching with the gamma distributions for recovery and death rates, respectively, as estimated above. The estimated values of the shape parameters a_1 and a_2 associated with the gamma distributions can be approximated as $a_1 \approx 8$ and $a_2 \approx 6$ for the recovery and death rate distributions respectively. Thus, one can use E_8 and E_6 distributions instead of gamma distributions for recovery and death rates. However, the result will be similar in both cases and we continue our numerical simulation with the gamma distributions.

Though the parameters of the three models correspond to each other, the distributed model (11.2) and the delay model (11.14) both give far different dynamics as compared to the SIR model (11.8), whereas the distributed model (11.2) and the delay model (11.14) give reasonably close dynamics to each other (Figure 11.6).

3 Delay epidemic models determined by latency, infection, and immunity duration

3.1 A single strain delay model

While the analysis in the previous section assumed permanent immunity after infection, we now consider another scenario in which acquired immunity wanes over time. In this model, individuals who recover from infection lose immunity after a finite duration and become susceptible again.

3.1.1 Model formulation

The SEIRS model tracks susceptible S, exposed E, infectious I, and recovered R populations, with constant total $N = S + E + I + R$. New exposures $J(t) = \frac{\beta}{N} S(t)I(t)$ occur at transmission rate β. Fixed periods govern transitions: latency τ_0, infectious τ_1, and immunity τ_2. Exposed individuals at $t - \tau_0$ become infectious at t, recover at $t + \tau_1$, and rejoin S at $t + \tau_0 + \tau_1 + \tau_2$. The nonsusceptible population satisfies

$$I(t) + E(t) + R(t) = \int_{t-\tau_0-\tau_1-\tau_2}^{t} J(\eta)\mathrm{d}\eta.$$

Differentiating this equality and using $N = S + E + I + R$, we derive

$$\frac{\mathrm{d}S}{\mathrm{d}t} = -J(t) + J(t - \tau_0 - \tau_1 - \tau_2).$$

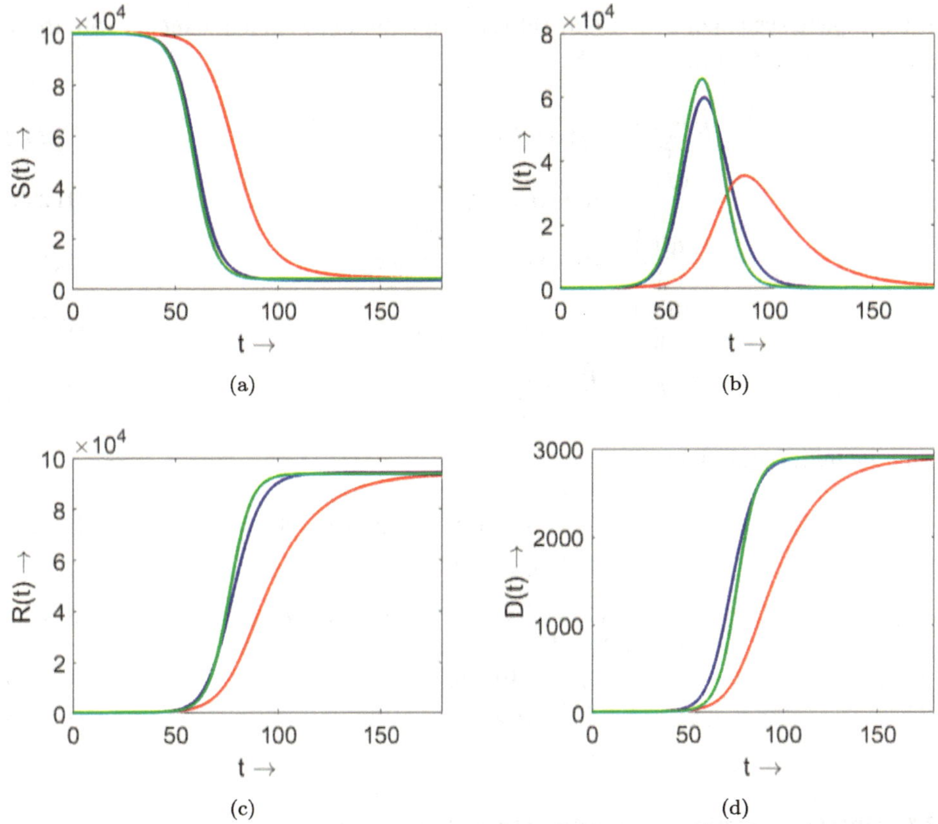

Figure 11.6: Comparison of solutions of the distributed model (11.2) (blue curve), the delay model (11.14) (green curve), and the conventional SIR model (11.8) (red curve). (a) susceptible ($S(t)$), (b) infected ($I(t)$), (c) recovered ($R(t)$), and (d) dead ($D(t)$). The values of parameters: $N = 10^5, I_0 = 1, \beta = 0.2$, for the SIR model $r_0 + d_0 = 1/17.7$; for delay model $\tau = 17.7$; and for the distributed model (11.2): $a_1 = 8.06275, b_1 = 2.21407, a_2 = 6.00014, b_2 = 2.19887, p_0 = 0.97$.

Since $J(t)$ is the number of newly exposed individuals at time t, the number of newly infectious, newly recovered and newly susceptible individuals at time t are determined by $J(t-\tau_0), J(t-\tau_0-\tau_1)$, and $J(t-\tau_0-\tau_1-\tau_2)$, respectively. Therefore, the governing equations for the exposed, infectious, and recovered compartments in terms of the daily number of new exposed individuals are

$$\frac{dE}{dt} = J(t) - J(t-\tau_0),$$
$$\frac{dI}{dt} = J(t-\tau_0) - J(t-\tau_0-\tau_1),$$
$$\frac{dR}{dt} = J(t-\tau_0-\tau_1) - J(t-\tau_0-\tau_1-\tau_2).$$

The complete system of DDE (Delay Differential Equations) epidemic model with three time delays is as follows:

$$\frac{dS(t)}{dt} = -J(t) + J(t - \tau_0 - \tau_1 - \tau_2), \tag{11.29a}$$

$$\frac{dE(t)}{dt} = J(t) - J(t - \tau_0), \tag{11.29b}$$

$$\frac{dI(t)}{dt} = J(t - \tau_0) - J(t - \tau_0 - \tau_1), \tag{11.29c}$$

$$\frac{dR(t)}{dt} = J(t - \tau_0 - \tau_1) - J(t - \tau_0 - \tau_1 - \tau_2), \tag{11.29d}$$

$$J(t) = \frac{\beta}{N} S(t)I(t), \tag{11.29e}$$

with the initial conditions:

$$S(\xi) = N, \quad S(0) = S_0 > 0, \tag{11.30a}$$

$$E(0) = 0, \tag{11.30b}$$

$$\bigwedge_{\xi \in [-(\tau_0+\tau_1+\tau_2),0)} \quad I(\xi) = 0, \quad I(0) = I_0 > 0, \tag{11.30c}$$

$$R(0) = 0, \tag{11.30d}$$

$$S_0 + I_0 = N. \tag{11.30e}$$

3.1.2 Integral equations and stationary points

Any constant values S and I provide a stationary solution of the system (11.29). However, this stationary solution does not take into account the initial condition. In order to determine the stationary point specific for a given initial condition, we integrate Equation (11.29a) from 0 to t:

$$\begin{aligned} S(t) - S_0 &= -\int_0^t J(s)\mathrm{d}s + \int_0^t J(s - \tau_0 - \tau_1 - \tau_2)\mathrm{d}s \\ &= -\int_0^t J(s)\mathrm{d}s + \int_0^{t-\tau_0-\tau_1-\tau_2} J(s)\mathrm{d}s = -\int_{t-\tau_0-\tau_1-\tau_2}^{t} J(s)\mathrm{d}s. \end{aligned}$$

We take into account here that $J(t) = 0$ for $t < 0$. Similarly, we obtain from Equation (11.29c):

$$I(t) = I_0 + \int_{t-\tau_0-\tau_1}^{t-\tau_0} J(s)\mathrm{d}s.$$

From Equation (11.29b), we obtain

$$E(t) = E_0 + \int_{t-\tau_0}^{t} J(s)\mathrm{d}s.$$

Next, we express $S(t)$ and $I(t)$ from the last two equalities and take their product:

$$J(t) = \frac{\beta}{N}\left(S_0 - \int_{t-\tau_0-\tau_1-\tau_2}^{t} J(s)\mathrm{d}s \right)\left(I_0 + \int_{t-\tau_0-\tau_1}^{t-\tau_0} J(s)\mathrm{d}s \right). \tag{11.31}$$

Thus, we have reduced system (11.29) to a single integral equation (11.31).

Stationary points of this equation can be found from the following algebraic equation:

$$J_s = \frac{\beta}{N}(S_0 - (\tau_0 + \tau_1 + \tau_2)J_s)(I_0 + \tau_1 J_s), \tag{11.32}$$

where J_s is a stationary solution of the function J. From the previous equation, we obtain the following quadratic equation with respect to J_s:

$$\tau_1(\tau_0 + \tau_1 + \tau_2)J_s^2 + \left(\frac{N}{\beta} - S_0\tau_1 + I_0(\tau_0 + \tau_1 + \tau_2)\right)J_s - S_0 I_0 = 0.$$

It has a positive solution

$$J_s = \frac{-(\frac{N}{\beta} - S_0\tau_1 + I_0(\tau_0 + \tau_1 + \tau_2)) + \sqrt{\Delta}}{2\tau_1(\tau_0 + \tau_1 + \tau_2)} > 0,$$

where

$$\Delta = \left(\frac{N}{\beta} - S_0\tau_1 + I_0(\tau_0 + \tau_1 + \tau_2)\right)^2 + 4S_0 I_0 \tau_1(\tau_0 + \tau_1 + \tau_2).$$

If $I_0 = 0$ and $S_0 = N$, then we find two approximate solutions of Equation (11.32):

$$J_s = 0, \quad J_s = \frac{N(\beta\tau_1 - 1)}{\beta\tau_1(\tau_0 + \tau_1 + \tau_2)}.$$

Hence, there exists a positive stationary solution if the basic reproduction number $\mathbb{R}_0 = \beta\tau_1 > 1$.

We can now determine the stationary values of numbers of susceptible, exposed, infected, and recovered in case of $J_s > 0$ as follows:

$$S_s = \frac{N}{\beta\tau_1},$$

$$E_s = \frac{N\tau_0(\beta\tau_1 - 1)}{\beta\tau_1(\tau_0 + \tau_1 + \tau_2)},$$
$$I_s = \frac{N(\beta\tau_1 - 1)}{\beta(\tau_0 + \tau_1 + \tau_2)},$$
$$R_s = \frac{N\tau_2(\beta\tau_1 - 1)}{\beta\tau_1(\tau_0 + \tau_1 + \tau_2)}.$$

Moreover, we have

$$\frac{dR_s(\tau_2)}{d\tau_2} = \frac{N(\beta\tau_1 - 1)(\tau_0 + \tau_1)}{\beta\tau_1(\tau_0 + \tau_1 + \tau_2)^2} > 0$$

as $\mathcal{R}_0 = \beta\tau_1 > 1$.

Let us note that S_s is a decreasing function of β and τ_1, and independent of τ_0 and τ_2. Also, if $\beta\tau_1 > 1$, then E_s increases as τ_0 increases, and decreases as τ_2 increases. I_s decreases as τ_0 and τ_2 increase. R_s decreases as a function of τ_0 and increases as a function of τ_2.

3.1.3 Stability of the stationary solution

In order to linearize Equation (11.31) about the stationary solution, we set $J(t) = J_s + v(t)$ and keep the first-order terms with respect to small perturbation v:

$$v(t) = -a_1 \int_{t-\tau_0-\tau_1-\tau_2}^{t} v(s)ds + a_2 \int_{t-\tau_0-\tau_1}^{t-\tau_0} v(s)ds. \tag{11.33}$$

Here,

$$a_1 = \frac{\beta}{N}(I_0 + \tau_1 J_s), \quad a_2 = \frac{\beta}{N}(S_0 - (\tau_0 + \tau_1 + \tau_2)J_s). \tag{11.34}$$

Let $v(t) = e^{\lambda t}$. Then from Equation (11.33), we obtain

$$\lambda = -a_1(1 - e^{-(\tau_0+\tau_1+\tau_2)\lambda}) + a_2 e^{-\tau_0\lambda}(1 - e^{-\tau_1\lambda}). \tag{11.35}$$

Clearly, $\lambda = 0$ is a solution of Equation (11.35). We will study this equation from the point of view of existence of solutions with a positive real part, which determines the loss of stability of the stationary solution. In order to simplify this analysis, we set $I_0 = 0$, $S_0 = N$ in (11.34). Then, for $J_s > 0$, we have

$$a_1 = \frac{\beta\tau_1 - 1}{\tau_0 + \tau_1 + \tau_2} > 0, \quad a_2 = \frac{1}{\tau_1} > 0,$$

and for $J_s = 0$, we have $a_1 = 0$ and $a_2 = \beta > 0$.

Based on these assumptions, we proceed to the first theorem.

Theorem 3.1. *The following statements hold:*
- *If* $\mathcal{R}_0 > 1$ *and* $J_s > 0$, *then Equation* (11.35) *does not have nontrivial positive real solutions.*
- *If* $\mathcal{R}_0 > 1$ *and* $J_s = 0$, *then Equation* (11.35) *has exactly one positive real solution. If* $\mathcal{R}_0 < 1$ *and* $J_s = 0$, *then Equation* (11.35) *has exactly one negative real solution.*
- *There is a value* $\mathcal{R}_0 = \mathcal{R}_c > 1$, *for which Equation* (11.35) *has a pure imaginary solution for any positive* τ_0, τ_1, *and* τ_2 *such that* $\tau_0 < \tau_2$.

Proof. The proof is given in [52]. □

Dependence of the stability boundary on τ_2 is shown in Figure 11.7. Let us recall that this parameter characterizes duration of acquired immunity in recovered individuals. Unexpectedly, stability region on the (τ_1, β)-plane (under the curves) decreases with the increase of this parameter, i. e., the stationary solution becomes less stable. On the other hand, the time period between the outbreaks increases.

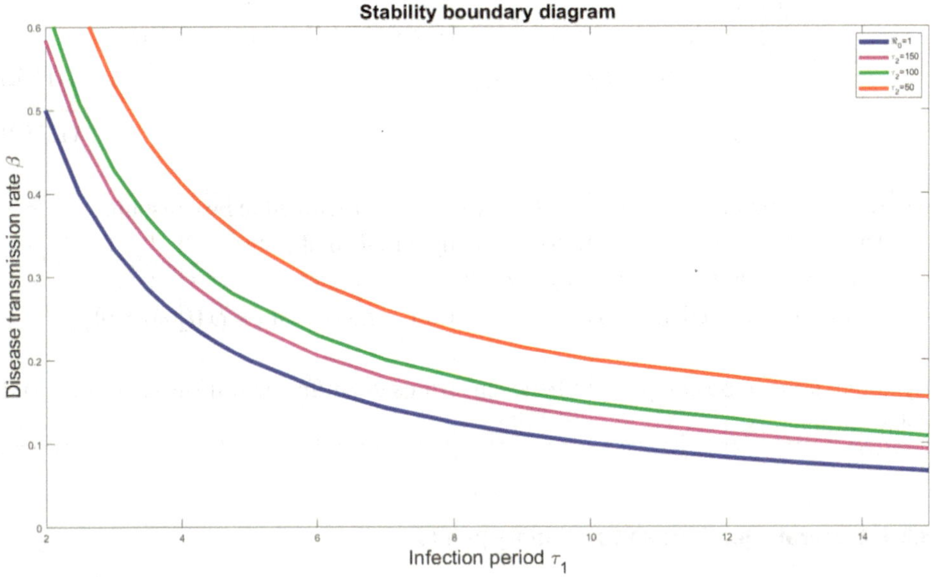

Figure 11.7: Stability boundary in the (τ_1, β)-plane for different values of τ_2 and $\tau_0 = 4$. Stability region is below the curves. The blue curve corresponds to $\mathcal{R}_0 = \beta\tau_1 = 1$. Positive endemic equilibrium exists in the region above the curve.

Summarizing analytical and numerical results, we conclude that the stationary solution $J_s = 0$ loses its stability for $\mathcal{R}_0 > 1$, and another stationary solution $J_s > 0$ appears. Next, there exists a critical value $\mathcal{R}_0 = \mathcal{R}_c > 1$ for which the oscillatory instability of the positive stationary solution occurs.

3.2 Two-strain delay model with cross-immunity

A model for two competing strains ($j = 1, 2$) incorporates distinct transmission rates (β_j), latency (τ_{0j}), infectious (τ_{1j}), and immunity periods (τ_{2j}). Cross-immunity assumes recovery from one strain prevents infection by the other until susceptibility resumes. The dynamics are governed by delay differential equations:

$$\frac{dS(t)}{dt} = -J_1(t) - J_2(t) + J_1(t - \tau_{01} - \tau_{11} - \tau_{21}) + J_2(t - \tau_{02} - \tau_{12} - \tau_{22}), \tag{11.36a}$$

$$\frac{dE_1(t)}{dt} = J_1(t) - J_1(t - \tau_{01}), \tag{11.36b}$$

$$\frac{dE_2(t)}{dt} = J_2(t) - J_2(t - \tau_{02}), \tag{11.36c}$$

$$\frac{dI_1(t)}{dt} = J_1(t - \tau_{01}) - J_1(t - \tau_{01} - \tau_{11}), \tag{11.36d}$$

$$\frac{dI_2(t)}{dt} = J_2(t - \tau_{02}) - J_2(t - \tau_{02} - \tau_{12}), \tag{11.36e}$$

$$\frac{dR(t)}{dt} = J_1(t - \tau_{01} - \tau_{11}) + J_2(t - \tau_{02} - \tau_{12}) - J_1(t - \tau_{01} - \tau_{11} - \tau_{21}) - J_2(t - \tau_{02} - \tau_{12} - \tau_{22}), \tag{11.36f}$$

$$J_j(t) = \frac{\beta_j}{N} S(t) I_j(t), \tag{11.36g}$$

where total population $N = S + E_1 + E_2 + I_1 + I_2 + R$ is assumed to be constant.

The model is completed with the following initial conditions:

- At $t < 0$: $S = N, I_1 = I_2 = E_1 = E_2 = R = 0$.
- Strain 1 introduced at $t = 0$ ($I_1(0) > 0$); strain 2 emerges at $t = \omega$ ($I_2(\omega) > 0$).

Dynamics depend on competition between strains via differing transmission rates and delay parameters.

3.2.1 Integral equations and stationary points

Let us note that any constant values of J_1 and J_2, i. e., of S, I_1 and I_2 provide a stationary solution of system (11.36). However, this stationary solution does not take into account the initial condition. To determine the stationary point specific for a given initial condition, we integrate Equation (11.36a) from 0 to t:

$$S(t) - S_0 = -\int_0^t J_1(s)ds + \int_0^t J_1(s - \tau_{01} - \tau_{11} - \tau_{21})ds$$

$$
\begin{aligned}
&- \int_0^t J_2(s)\mathrm{d}s + \int_0^t J_2(s - \tau_{02} - \tau_{12} - \tau_{22})\mathrm{d}s \\
&= - \int_0^t J_1(s)\mathrm{d}s + \int_0^{t-\tau_{01}-\tau_{11}-\tau_{21}} J_1(s)\mathrm{d}s - \int_0^t J_2(s)\mathrm{d}s + \int_0^{t-\tau_{02}-\tau_{12}-\tau_{22}} J_2(s)\mathrm{d}s \\
&= - \int_{t-\tau_{01}-\tau_{11}-\tau_{21}}^{t} J_1(s)\mathrm{d}s - \int_{t-\tau_{02}-\tau_{12}-\tau_{22}}^{t} J_2(s)\mathrm{d}s.
\end{aligned}
$$

We take into account here that $J_1(t) = 0, J_2(t) = 0$ for $t < 0$. Similarly, we obtain from Equations (11.36d) and (11.36e):

$$
I_1(t) = I_{01} + \int_{t-\tau_{01}-\tau_{11}}^{t-\tau_{01}} J_1(s)\mathrm{d}s,
$$

$$
I_2(t) = I_{02} + \int_{t-\tau_{02}-\tau_{12}}^{t-\tau_{02}} J_2(s)\mathrm{d}s.
$$

Next, we express $S(t), I_1(t)$, and $I_2(t)$ from the last three equalities and take their product:

$$
J_1(t) = \frac{\beta_1}{N}\left(S_0 - \int_{t-\tau_{01}-\tau_{11}-\tau_{21}}^{t} J_1(s)\mathrm{d}s - \int_{t-\tau_{02}-\tau_{12}-\tau_{22}}^{t} J_2(s)\mathrm{d}s \right) \left(I_{01} + \int_{t-\tau_{01}-\tau_{11}}^{t-\tau_{01}} J_1(s)\mathrm{d}s \right) \tag{11.37a}
$$

$$
J_2(t) = \frac{\beta_2}{N}\left(S_0 - \int_{t-\tau_{01}-\tau_{11}-\tau_{21}}^{t} J_1(s)\mathrm{d}s - \int_{t-\tau_{02}-\tau_{12}-\tau_{22}}^{t} J_2(s)\mathrm{d}s \right) \left(I_{02} + \int_{t-\tau_{02}-\tau_{12}}^{t-\tau_{02}} J_2(s)\mathrm{d}s \right). \tag{11.37b}
$$

Thus, we have reduced system (11.36) to the two integral equations (11.37a) and (11.37b). Stationary points of these equations can be found by solving the following two algebraic equations:

$$
J_{1s} = \frac{\beta_1}{N}(S_0 - (\tau_{01}+\tau_{11}+\tau_{21})J_{1s} - (\tau_{02}+\tau_{12}+\tau_{22})J_{2s})(I_{01}+\tau_{11}J_{1s}), \tag{11.38a}
$$

$$
J_{2s} = \frac{\beta_2}{N}(S_0 - (\tau_{01}+\tau_{11}+\tau_{21})J_{1s} - (\tau_{02}+\tau_{12}+\tau_{22})J_{2s})(I_{02}+\tau_{12}J_{2s}). \tag{11.38b}
$$

If $I_{01} = 0, I_{02} = 0$, and $S_0 = N$, then Equations (11.38a) and (11.38b) can be written as

$$J_{1s}\left(\frac{\beta_1\tau_{11} - 1}{\beta_1\tau_{11}} - \frac{\tau_{01}+\tau_{11}+\tau_{21}}{N}J_{1s} - \frac{\tau_{02}+\tau_{12}+\tau_{22}}{N}J_{2s}\right) = 0, \tag{11.39a}$$

$$J_{2s}\left(\frac{\beta_2\tau_{12} - 1}{\beta_2\tau_{12}} - \frac{\tau_{01}+\tau_{11}+\tau_{21}}{N}J_{1s} - \frac{\tau_{02}+\tau_{12}+\tau_{22}}{N}J_{2s}\right) = 0. \tag{11.39b}$$

Recalling that $\mathcal{R}_{01} = \beta_1\tau_{11}$ and $\mathcal{R}_{02} = \beta_2\tau_{12}$, we analyze the stationary solutions depending on the basic reproduction numbers.

Stationary solutions

There are up to four nonnegative solutions (J_{1s}, J_{2s}) of system (11.39a), (11.39b).

– zero solution exists for all values of parameters:

$$J_{1s} = 0, J_{2s} = 0. \tag{11.40}$$

– if $\mathcal{R}_{01} > 1$, then there is a solution with the first positive component:

$$J_{1s} = \frac{N}{\beta_1\tau_{11}}\frac{\beta_1\tau_{11} - 1}{\tau_{01} + \tau_{11} + \tau_{21}}, J_{2s} = 0. \tag{11.41}$$

– if $\mathcal{R}_{02} > 1$, then there is a solution with the second positive component:

$$J_{1s} = 0, J_{2s} = \frac{N}{\beta_2\tau_{12}}\frac{\beta_2\tau_{12} - 1}{\tau_{02} + \tau_{12} + \tau_{22}}. \tag{11.42}$$

– if $\mathcal{R}_{01}, \mathcal{R}_{02} > 1$, then there is a solution for which both J_{1s} and J_{2s} are positive. Then

$$(\tau_{01} + \tau_{11} + \tau_{21})J_{1s} + (\tau_{02} + \tau_{12} + \tau_{22})J_{2s} = \frac{\beta_1\tau_{11} - 1}{\beta_1\tau_{11}}N = \frac{\beta_2\tau_{12} - 1}{\beta_2\tau_{12}}N. \tag{11.43}$$

Hence, $\mathcal{R}_{01} = \mathcal{R}_{02} > 1$. In this case, J_{1s} and J_{2s} are obtained from (11.39a), (11.39b) and (11.43):

$$J_{1s} = \frac{\beta_1(\mathcal{R}_{01} - 1)N}{\mathcal{R}_{01}(\beta_1(\tau_{01} + \tau_{11} + \tau_{21}) + \beta_2(\tau_{02} + \tau_{12} + \tau_{22}))}, \tag{11.44a}$$

$$J_{2s} = \frac{\beta_2(\mathcal{R}_{02} - 1)N}{\mathcal{R}_{02}(\beta_1(\tau_{01} + \tau_{11} + \tau_{21}) + \beta_2(\tau_{02} + \tau_{12} + \tau_{22}))}. \tag{11.44b}$$

We can observe from (11.44a) and (11.44b) that $\frac{J_{1s}}{J_{2s}} = \frac{\beta_1}{\beta_2}$.

3.2.2 Stability of stationary solutions

In order to simplify the analysis, we set $I_{01}, I_{02} = 0, S_0 = N$. Linearizing Equations (11.37a) and (11.37b), about the stationary solution, we substitute $J_1(t) = J_{1s} + \epsilon_1 e^{\lambda t}, J_2(t) = J_{2s} + \epsilon_2 e^{\lambda t}$ and keep the first-order terms with respect to ϵ_1, ϵ_2. We obtain the following linear system of two equations with respect to ϵ_1, ϵ_2:

$$a_{11}\epsilon_1 + a_{12}\epsilon_2 = 0, \tag{11.45}$$

$$a_{21}\epsilon_1 + a_{22}\epsilon_2 = 0, \tag{11.46}$$

where

$$\begin{aligned} a_{11} &= \lambda - \frac{\beta_1}{N}\big[\,(N-(\tau_{01}+\tau_{11}+\tau_{21})J_{1s}-(\tau_{02}+\tau_{12}+\tau_{22})J_{2s})e^{-\tau_{01}\lambda}(1-e^{-\tau_{11}\lambda}) \\ &\quad - \tau_{11}J_{1s}(1-e^{-(\tau_{01}+\tau_{11}+\tau_{21})\lambda})\,\big], \end{aligned} \tag{11.47}$$

$$a_{12} = \frac{\beta_1}{N}\tau_{11}J_{1s}(1-e^{-(\tau_{02}+\tau_{12}+\tau_{22})\lambda}), \tag{11.48}$$

$$a_{21} = \frac{\beta_2}{N}\tau_{12}J_{2s}(1-e^{-(\tau_{01}+\tau_{11}+\tau_{21})\lambda}), \tag{11.49}$$

$$\begin{aligned} a_{22} &= \lambda - \frac{\beta_2}{N}\big[\,(N-(\tau_{01}+\tau_{11}+\tau_{21})J_{1s}-(\tau_{02}+\tau_{12}+\tau_{22})J_{2s})e^{-\tau_{02}\lambda}(1-e^{-\tau_{12}\lambda}) \\ &\quad - \tau_{12}J_{2s}(1-e^{-(\tau_{02}+\tau_{12}+\tau_{22})\lambda})\,\big], \end{aligned} \tag{11.50}$$

System (11.45), (11.46) has a nonzero solution, if

$$\det\begin{bmatrix} a_{11} & a_{12} \\ a_{21} & a_{22} \end{bmatrix} = 0. \tag{11.51}$$

Theorem 3.2. *The following statements hold:*
1. *Let* $(J_{1s}, J_{2s}) = (0,0)$ *(see* (11.40)*). If* $\mathcal{R}_{01} > 1$ *and* $\mathcal{R}_{02} > 1$*, then* (11.51) *has two positive real solutions. If* $\mathcal{R}_{01} < 1$ *and* $\mathcal{R}_{02} < 1$*, then* (11.51) *has two negative real solutions.*
2. *Consider stationary solution* (11.41)*. If* $1 < \mathcal{R}_{02} < \mathcal{R}_{01}$*, then* (11.51) *has only negative real solution. If* $1 < \mathcal{R}_{01} < \mathcal{R}_{02}$*, then* (11.51) *has a positive real solution.*

Proof. The proof is given in [52]. □

Summarizing this stability analysis, we conclude that for $\mathcal{R}_{01}, \mathcal{R}_{02} > 1$, $\mathcal{R}_{01} \neq \mathcal{R}_{02}$, the stationary point $(0,0)$ is unstable. Among the two other nonnegative stationary points, one is stable, another one is unstable. Stable point corresponds to the larger basic reproduction number.

3.3 Numerical simulations

In this section, we present some numerical simulations to study the dynamics of the two-strain model (11.36). The behavior of the solution depends on the basic reproduction numbers $\mathcal{R}_{01}$, $\mathcal{R}_{02}$. If both of them are less than 1, then the only nonnegative stationary solution is $J_{1s} = J_{2s} = 0$, and solution of system (11.36) converges to this stationary solution.

Oscillatory stability boundary is shown in Figure 11.8(a). It separates stability region (below the curves) where the solution converges to the stationary solution (Fig-

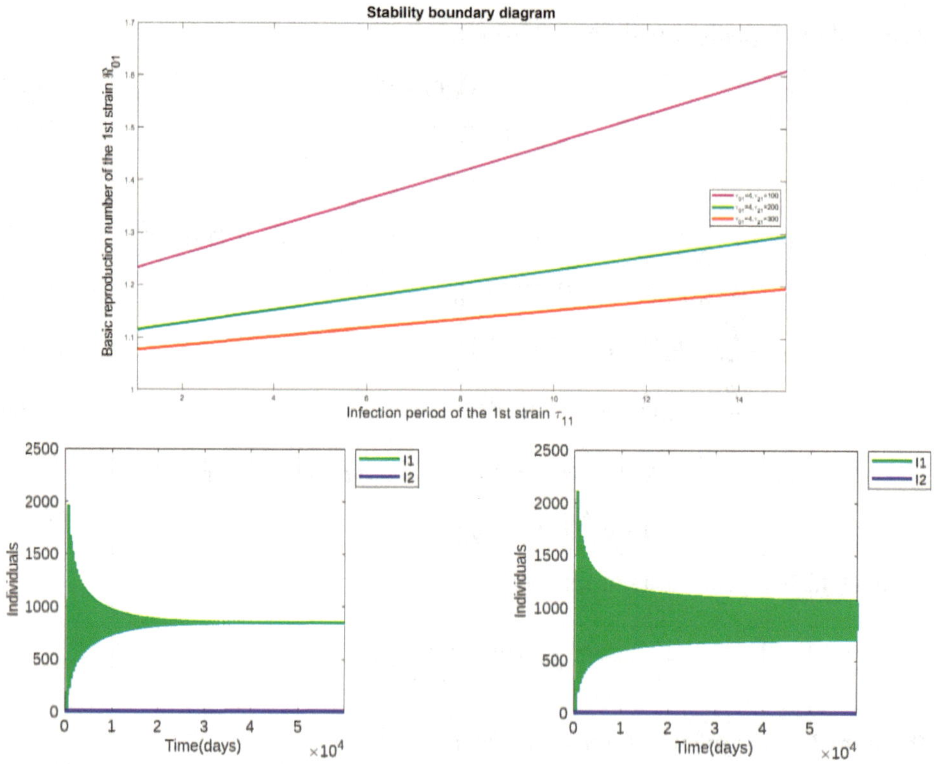

Figure 11.8: (a) Stability boundaries in the $(\tau_{11}, \mathcal{R}_{01})$-plane for $\tau_{01} = 4$. The point $(\tau_{11} = 10, \mathcal{R}_c = 1.22859)$ on the green line is related to panels (b) and (c) showing the emergence of oscillations when crossing the stability boundary. (b) Numerical simulations of two-strain system (11.36) with cross-immunity for the values of the initial conditions $N = 10^5, S_0 = N - 0.0001, I_1(0) = 0.0001, I_2(\omega) = 0.0001, \omega = 360$, and the following values of parameters: $\beta_1 = 0.122, \beta_2 = 0.12, \tau_{01} = \tau_{02} = 4, \tau_{11} = \tau_{12} = 10, \tau_{21} = \tau_{22} = 200 \Rightarrow \mathcal{R}_{01} = 1.22 < \mathcal{R}_c$, which correspond to stability of the endemic equilibrium. (c) Numerical simulations of system (11.36) for the values of the initial conditions $N = 10^5, S_0 = N - 0.0001, I_1(0) = 0.0001, I_2(\omega) = 0.0001, \omega = 360$, and the following values of parameters: $\beta_1 = 0.123, \beta_2 = 0.12, \tau_{01} = \tau_{02} = 4, \tau_{11} = \tau_{12} = 10, \tau_{21} = \tau_{22} = 200 \Rightarrow \mathcal{R}_{01} = 1.23 > \mathcal{R}_c$, which correspond to instability of the endemic equilibrium.

ure 11.8(b)) and instability region (above the curve) with periodic oscillations (Figure 11.8(c)). Similar to the model with a single strain, the instability region increases with the increase of the parameter τ_{21}.

Dynamics of oscillating strains is quite complex with a number of different cases. Figure 11.9(a, b) shows that for equal basic reproduction numbers, the disease transmission rate β determines which strain dominates. However, the second one is not completely eliminated. Panels (c, d) indicate that larger basic reproduction number gives larger peaks. However, this description remains true near the stability boundary ($\mathcal{R}_{01}$ close to $\mathcal{R}_c$). Far from the stability boundary (large $\mathcal{R}_{01}$), these dynamics can be different (see Figure 11.11 below and comments). Panels Figure 11.10(e, f) show that the strain with

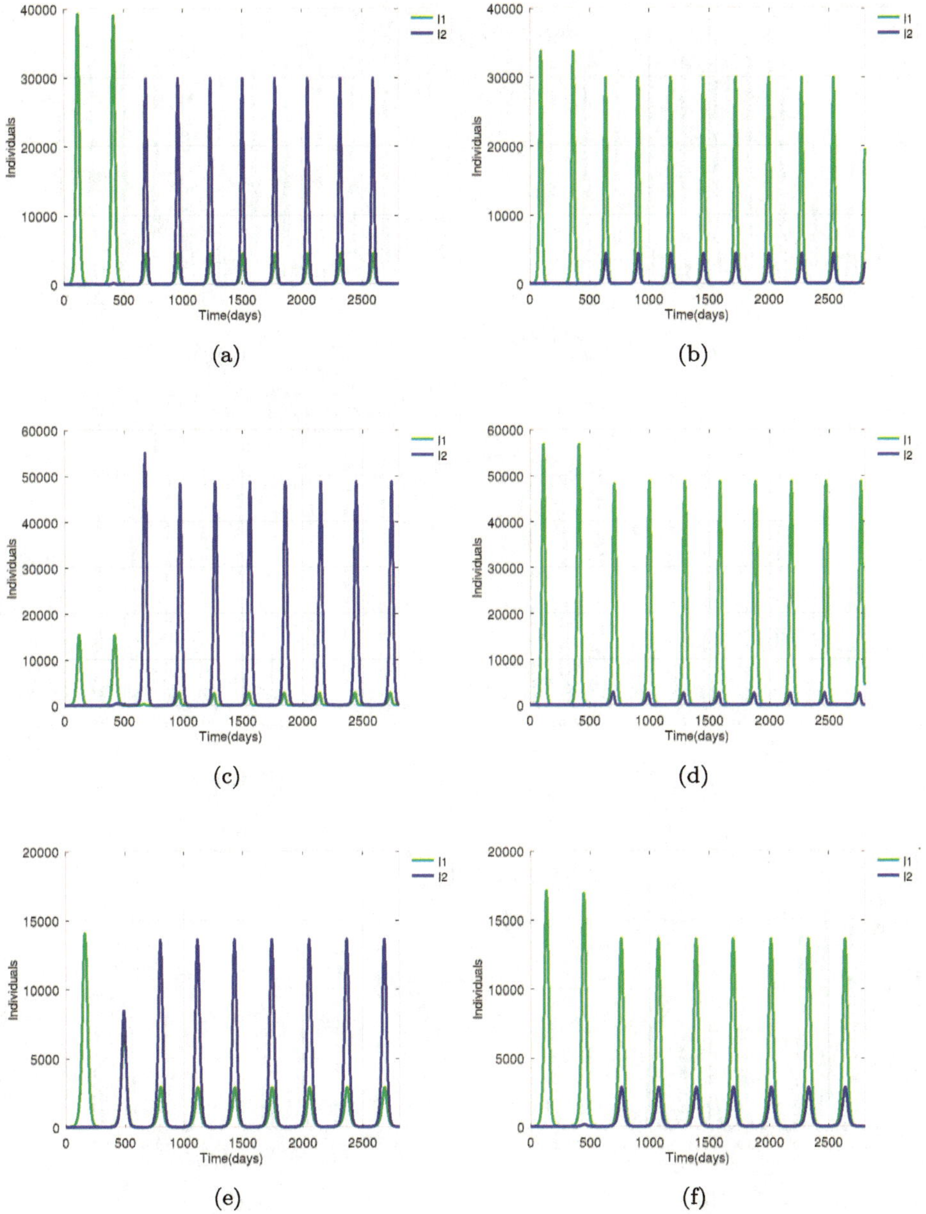

Figure 11.9: Numerical simulations of the two-strain delay model with cross immunity (11.36) for the values of the initial conditions $N = 10^5$, $S_0 = N - 1$, $I_1(0) = 1$, $E_1(0) = R(0) = 0$, $\omega = 360$, $I_2(\omega) = 1$, $E_2(\omega) = 0$, and the following values of parameters given in Table 11.2.

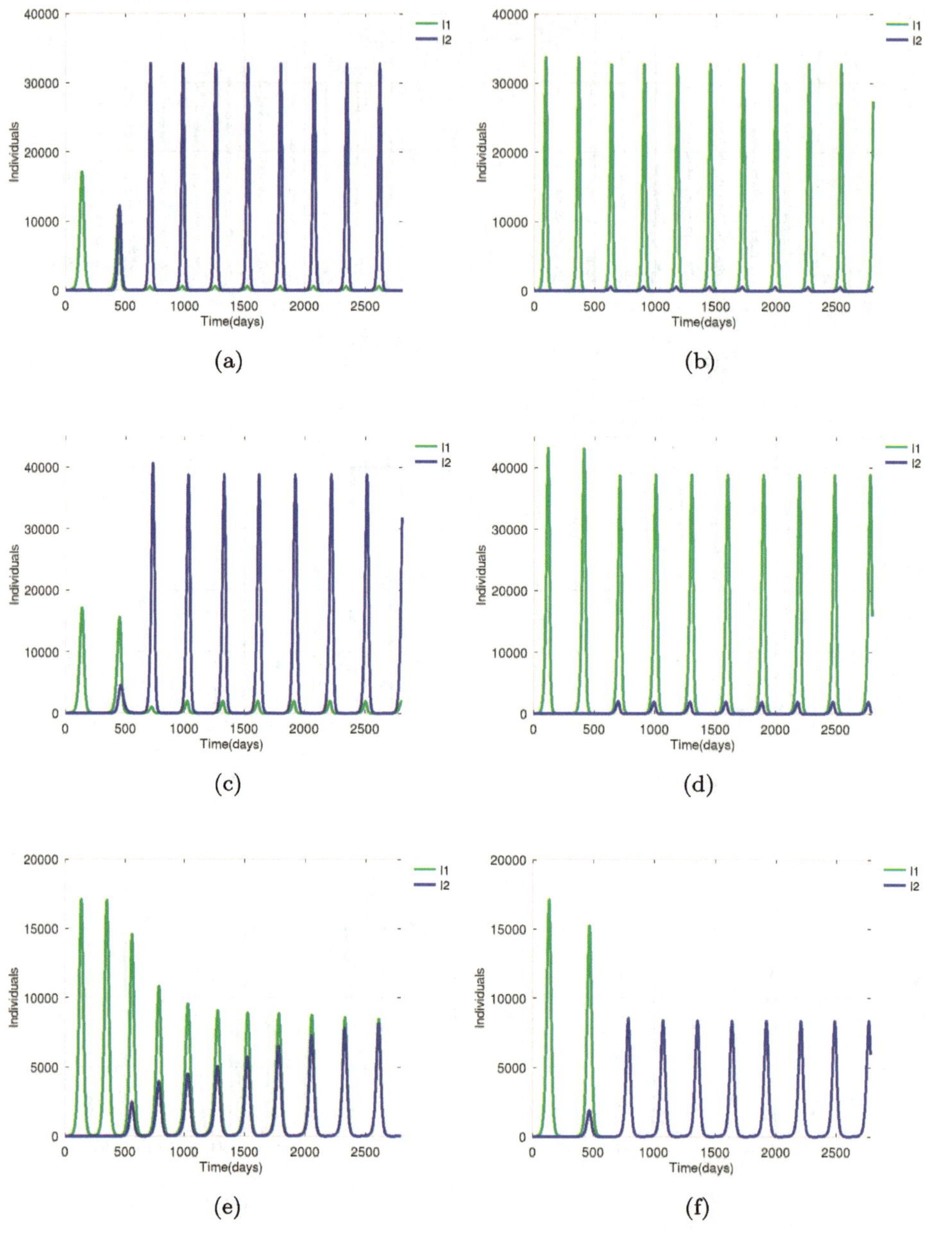

Figure 11.10: Figure 11.9 (continued): Numerical simulations of the two-strain delay model with cross immunity (11.36) for the values of the initial conditions $N = 10^5$, $S_0 = N - 1$, $I_1(0) = 1$, $E_1(0) = R(0) = 0$, $\omega = 360$, $I_2(\omega) = 1$, $E_2(\omega) = 0$, and the following values of parameters given in Table 11.2.

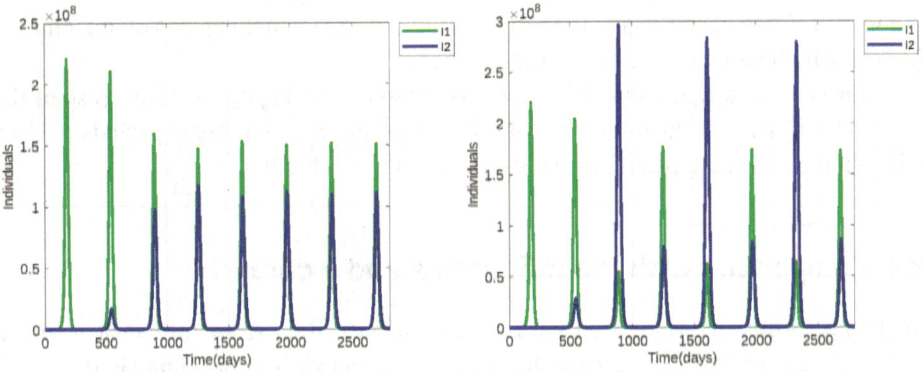

(a) (b)

Figure 11.11: Numerical simulations of the two-strain delay model with cross immunity (11.36) for the values of the initial conditions $N = 10^9$, $S_0 = N - 1$, $I_1(0) = 1$, $E_1(0) = R(0) = 0$, $\omega = 360$, $I_2(\omega) = 1$, $E_2(\omega) = 0$, and the following values of parameters $\beta_1 = 0.3$, $\beta_2 = 0.2$, $\tau_{01} = \tau_{02} = 4$, $\tau_{21} = \tau_{22} = 180$. (a) $\tau_{11} = 8$, $\tau_{12} = 14 \Rightarrow \mathcal{R}_{01} = 2.4$, $\mathcal{R}_{02} = 2.8$; (b) $\tau_{11} = 8$, $\tau_{12} = 15 \Rightarrow \mathcal{R}_{01} = 2.4$, $\mathcal{R}_{02} = 3$.

Table 11.2: Parameters of model (11.36) used in numerical simulations in Figures 11.9, 11.10.

Parameters	9a	9b	9c	9d	9e	9f	10a	10b	10c	10d	10e	10f
β_1	0.2	0.3	0.25	0.2	0.2	0.2	0.2	0.3	0.2	0.2	0.2	0.2
β_2	0.3	0.2	0.2	0.25	0.2	0.2	0.3	0.2	0.2	0.2	0.2	0.2
τ_{01}	4	4	4	4	6	4	4	4	4	4	4	4
τ_{02}	4	4	4	4	4	6	4	4	4	4	4	4
τ_{11}	15	10	8	20	10	10	10	10	10	16	10	10
τ_{12}	10	15	20	8	10	10	10	10	16	10	10	10
τ_{21}	180	180	180	180	180	180	180	180	180	180	100	200
τ_{22}	180	180	180	180	180	180	180	180	180	180	200	100
$\mathcal{R}_{01}$	3	3	2	4	2	2	2	3	2	3.2	2	2
$\mathcal{R}_{02}$	3	3	4	2	2	2	3	2	3.2	2	2	2

the smaller latency period prevails, while panels Figure 11.10(g, h) reveal that the strain with the higher disease transmission rate dominates in the population. For larger $I_2(\omega)$, the stationary solution is not exactly 0. This can explain the small spikes for I_2 in Figure 11.9(b). While for a value of $I_2(\omega)$ close to 0, the small spikes for I_2 completely disappear in Figure 11.10(b). Similar conclusions are valid for Figure 11.10(a). Figure 11.10(c, d) illustrate that the strain with the higher infectious period will dominate. Figure 11.10(e) reveals that if the duration of immunity of the second strain is greater than that of the first strain, then infection in the second strain starts increasing and the distance between the last consequent peaks increases. On the other hand, Figure 11.10(f) shows that if the duration of immunity of the first strain is greater than that of the second one, then the number of peaks becomes less than in the previous case panel Figure 11.10(e), and the peaks for both strains coincide with each other.

More complex dynamics of solutions is presented in Figure 11.11. Increase of the parameter τ_{12} leads to the transition from simple periodic oscillations to period doubling where large and small peaks of each strain alternate each other.

3.4 Comparison with the Influenza A and B data

Modeling results are compared with epidemiological data from Influenza A and B outbreaks in Canada. The disease duration for influenza is approximately one week [11], while immunity against influenza typically lasts 6 months [18, 28]. The two-strain model (11.36) incorporates cross-immunity, wherein infection with one strain confers temporary immunity against the other. Although epidemiological records indicate rare instances of consecutive A and B influenza infections [48], such cases remained below 0.2 % across all observed years, justifying their exclusion from the model. Surveillance data for Canada were sourced from [17].

Transmission rates β_1 and β_2 were estimated through model calibration against empirical data, utilizing nonlinear least-squares optimization (SSE minimization) implemented via MATLAB's "fmincon" and "patternsearch" solvers applied to weekly incidence data.

Figure 11.12 compares modeled incidence trajectories with empirical trends. Both strains exhibit periodic outbreaks in surveillance records, with Influenza B demonstrat-

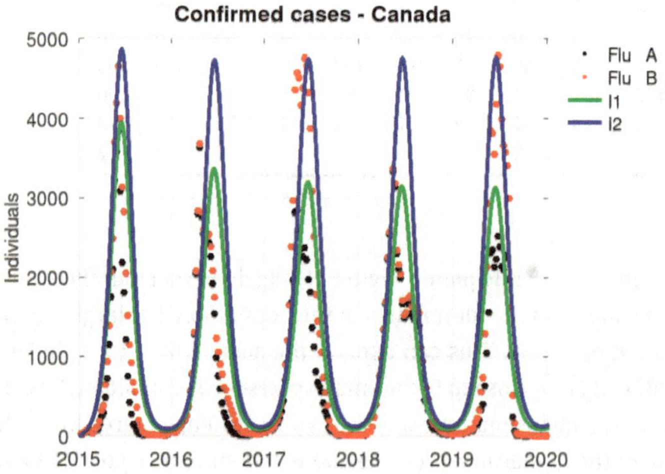

Figure 11.12: Simulation of the two-strain delay model with cross-immunity (11.36) showing the active cases of Flu A and Flu B in Canada with parameters $\beta_1 = 1.7657$, $\beta_2 = 1.777$, $\tau_{01} = \tau_{02} = 0.2$, $\tau_{11} = \tau_{12} = 0.65$, $\tau_{21} = 28$, $\tau_{22} = 27$, $\omega = 0.2$, initial conditions $I_1(t < 0) = 0$, $I_2(t < \omega) = 0$, $I_1(0) = 5$, $I_2(\omega) = 5$, $N = 751900$ [9], $S_0 = N - 5$, $E_1(0) = E_2(\omega) = 0$. The simulation period is from 19.10.2015 to 27.07.2020, with the time unit corresponding approximately to one week.

ing larger peak magnitudes. Notably, while empirical outbreak amplitudes exhibit periodic oscillations, simulated peaks for each strain maintain constant magnitudes under the current model assumptions. A similar discrepancy arises in the periodic amplitude modulation observed in Figure 11.11(b), which is absent in the model outputs. This inconsistency suggests that additional mechanisms, such as time-varying transmission rates or waning immunity, may be required to fully capture the oscillatory dynamics observed in surveillance data.

4 Discussion

The classical SIR model, while a cornerstone of epidemiological analysis, faces limitations due to its assumption of abrupt transitions from infection to recovery. Such idealized assumptions can lead to inaccuracies, including underestimation of the basic reproduction number ($\mathcal{R}_0$), particularly during fast-evolving outbreaks where infection and recovery rates fluctuate dynamically.

Delay-enhanced models address these shortcomings by integrating disease duration and variable transmission rates into their framework. By incorporating time-delayed recovery and mortality processes, these models align more closely with empirical data than the classical SIR framework. For instance, point-specific delay models improve precision in estimating $\mathcal{R}_0$ and infection timelines, while also revealing nuanced dynamics—such as oscillatory infection patterns—that traditional approaches overlook.

Expanding the SIR model to account for latency periods and declining immunity demonstrates how recurrent outbreaks can emerge independently of viral mutations. Delay-based analyses illustrate that waning immunity triggers periodic infection waves, even in the absence of new variants. Furthermore, embedding continuous genotype variables to model viral evolution offers insights into variant emergence, a critical factor for understanding pathogen adaptation and epidemic trajectories, as evidenced by COVID-19.

The delay differential equation model generates increasingly intricate, aperiodic oscillations when incorporating vaccination dynamics [50] or a periodic transmission rate [53] designed to reflect seasonal variations in epidemic spread. These patterns align with empirical observations of Influenza A progression documented in Norway.

Nonpharmaceutical interventions (e. g., lockdowns) are emphasized as pivotal tools for curbing transmission [47]. Optimizing isolation protocols through delay models not only reduces infection rates but also balances socioeconomic costs. Such frameworks enable policymakers to quantify the ideal scope of population-level restrictions.

5 Conclusion

In summary, augmenting classical epidemiological models with delayed recovery phases, immunity decay, and mutation dynamics significantly enhances their realism and utility. These refinements improve forecasting accuracy and uncover intricate outbreak patterns, equipping public health systems with actionable insights during crises.

Bibliography

[1] N. Ahmed, Z. Wei, D. Baleanu, M. Rafiq, and M. A. Rehman. Spatio-temporal numerical modeling of reaction-diffusion measles epidemic system. *Chaos: An Interdisciplinary Journal of Nonlinear Science*, 29(10), 2019.

[2] R. Almeida and S. Qureshi. A fractional measles model having monotonic real statistical data for constant transmission rate of the disease. *Fractal and Fractional*, 3(4):53, 2019.

[3] R. M. Anderson, C. Fraser, A. C. Ghani, C. A. Donnelly, S. Riley, N. M. Ferguson, G. M. Leung, T. H. Lam, and A. J. Hedley. Epidemiology, transmission dynamics and control of sars: The 2002–2003 epidemic. *Philosophical Transactions of the Royal Society of London. Series B, Biological Sciences*, 359(1447):1091–1105, 2004.

[4] S. Aniţa, M. Banerjee, S. Ghosh, and V. Volpert. Vaccination in a two-group epidemic model. *Applied Mathematics Letters*, 119:107197, 2021.

[5] D. Bichara and A. Iggidr. Multi-patch and multi-group epidemic models: A new framework. *Journal of Mathematical Biology*, 77(1):107–134, 2018.

[6] F. Brauer, C. Castillo-Chavez, Z. Feng, et al. *Mathematical Models in Epidemiology*, volume 32. Springer, 2019.

[7] F. Brauer, P. Van den Driessche, J. Wu, and L. J. Allen. *Mathematical Epidemiology*, volume 1945. Springer, 2008.

[8] S. Briand, E. Bertherat, P. Cox, P. Formenty, M.-P. Kieny, J. K. Myhre, C. Roth, N. Shindo, and C. Dye. The international Ebola emergency. *The New England Journal of Medicine*, 371(13):1180–1183, 2014.

[9] S. Buckrell, M. B. Moussa, T. Bui, A. Rahal, K. Schmidt, L. Lee, N. Bastien, and C. Bancej. National influenza annual report, Canada, 2021–2022. *Canada Communicable Disease Report*, 48(10):473–483, 2022.

[10] S. Busenberg and K. L. Cooke. The effect of integral conditions in certain equations modelling epidemics and population growth. *Journal of Mathematical Biology*, 10:13–32, 1980.

[11] CDC US Government. Clinical signs and symptoms of influenza. https://www.cdc.gov/flu/. Accessed: October 2023.

[12] S.-H. Chen, F. Mallamace, C.-Y. Mou, M. Broccio, C. Corsaro, A. Faraone, and L. Liu. The violation of the Stokes–Einstein relation in supercooled water. *Proceedings of the National Academy of Sciences*, 103(35):12974–12978, 2006.

[13] B. Chen-Charpentier. On population models with delays and dependence on past values. *Axioms*, 13(3):206, 2024.

[14] C. Chintu, U. H. Athale, and P. S. Patil. Childhood cancers in Zambia before and after the hiv epidemic. *Archives of Disease in Childhood*, 73(2):100–105, 1995.

[15] K. L. Cooke, P. Van den Driessche, and X. Zou. Interaction of maturation delay and nonlinear birth in population and epidemic models. *Journal of Mathematical Biology*, 39:332–352, 1999.

[16] K. L. Cooke and J. A. Yorke. Some equations modelling growth processes and gonorrhea epidemics. *Mathematical Biosciences*, 16(1–2):75–101, 1973.

[17] S. Dattani, F. Spooner, E. Mathieu, H. Ritchie, and M. Roser. Influenza. https://ourworldindata.org/influenza. Accessed: October 2023.

[18] K. Duda. How long does a flu shot last? https://www.verywellhealth.com/. Accessed: October 2023.

[19] A. d'Onofrio, M. Banerjee, and P. Manfredi. Spatial behavioural responses to the spread of an infectious disease can suppress Turing and Turing–Hopf patterning of the disease. *Physica A: Statistical Mechanics and its Applications*, 545:123773, 2020.

[20] E. H. Elbasha and A. B. Gumel. Vaccination and herd immunity thresholds in heterogeneous populations. *Journal of Mathematical Biology*, 83(6–7):73, 2021.

[21] T. S. Faniran, A. Ali, N. E. Al-Hazmi, J. Kiddy, K. Asamoah, T. A. Nofal, and M. O. Adewole. New variant of sars-cov-2 dynamics with imperfect vaccine. *Complexity*, 2022, 2022.

[22] J. A. N. Filipe and M. M. Maule. Effects of dispersal mechanisms on spatio-temporal development of epidemics. *Journal of Theoretical Biology*, 226(2):125–141, 2004.

[23] S. P. Fisher-Hoch and L. Hutwagner. Opportunistic candidiasis: An epidemic of the 1980s. *Clinical Infectious Diseases*, 21(4):897–904, 1995.

[24] S. Ghosh, V. Volpert, and M. Banerjee. An epidemic model with time delay determined by the disease duration. *Mathematics*, 10(15):2561, 2022.

[25] S. Ghosh, V. Volpert, and M. Banerjee. An epidemic model with time-distributed recovery and death rates. *Bulletin of Mathematical Biology*, 84(8):78, 2022.

[26] M. P. Girard, J. S. Tam, O. M. Assossou, and M. P. Kieny. The 2009 a (h1n1) influenza virus pandemic: A review. *Vaccine*, 28(31):4895–4902, 2010.

[27] GitHub, Inc. Covid19_cfr_submission. https://github.com/mrc-ide/COVID19_CFR_submission.

[28] T. Haelle. What offers better immunity: The flu vaccine or the flu itself? https://www.kqed.org/stateofhealth/152601/better-immunity-from. Accessed: October 2023.

[29] H. W. Hethcote. The mathematics of infectious diseases. *SIAM Review*, 42(4):599–653, 2000.

[30] S. Hsiang, D. Allen, S. Annan-Phan, K. Bell, I. Bolliger, T. Chong, H. Druckenmiller, L. Y. Huang, A. Hultgren, E. Krasovich, et al. The effect of large-scale anti-contagion policies on the covid-19 pandemic. *Nature*, 584(7820):262–267, 2020.

[31] G. Huang, Y. Takeuchi, W. Ma, and D. Wei. Global stability for delay sir and seir epidemic models with nonlinear incidence rate. *Bulletin of Mathematical Biology*, 72:1192–1207, 2010.

[32] H. S. Hurd and J. B. Kaneene. The application of simulation models and systems analysis in epidemiology: A review. *Preventive Veterinary Medicine*, 15(2–3):81–99, 1993.

[33] M. Jackson and B. M. Chen-Charpentier. Modeling plant virus propagation with delays. *Journal of Computational and Applied Mathematics*, 309:611–621, 2017.

[34] S. Jain, L. Kamimoto, A. M. Bramley, A. M. Schmitz, S. R. Benoit, J. Louie, D. E. Sugerman, J. K. Druckenmiller, K. A. Ritger, R. Chugh, et al. Hospitalized patients with 2009 h1n1 influenza in the United States, April–June 2009. *The New England Journal of Medicine*, 361(20):1935–1944, 2009.

[35] M. Kapralov, S. Khanna, and M. Sudan. Approximating matching size from random streams. In *Proceedings of the Twenty-Fifth Annual ACM-SIAM Symposium on Discrete Algorithms*, pages 734–751. SIAM, 2014.

[36] M. J. Keeling and P. Rohani. *Modeling Infectious Diseases in Humans and Animals*. Princeton university press, 2011.

[37] Q. J. A. Khan and E. V. Krishnan. An epidemic model with a time delay in transmission. *Applications of Mathematics*, 48:193–203, 2003.

[38] A. M. Kilpatrick, A. A. Chmura, D. W. Gibbons, R. C. Fleischer, P. P. Marra, and P. Daszak. Predicting the global spread of h5n1 avian influenza. *Proceedings of the National Academy of Sciences*, 103(51):19368–19373, 2006.

[39] B. Kreuels, D. Wichmann, P. Emmerich, J. Schmidt-Chanasit, G. de Heer, S. Kluge, A. Sow, T. Renné, S. Günther, A. W. Lohse, et al. A case of severe Ebola virus infection complicated by gram-negative septicemia. *The New England Journal of Medicine*, 371(25):2394–2401, 2014.

[40] W. K. Lam, N. S. Zhong, and W. C. Tan. Overview on sars in Asia and the world. *Respirology*, 8:S2–S5, 2003.
[41] N. Loayza. Costs and trade-offs in the fight against the covid-19 pandemic: A developing country perspective. In *Research and Policy Briefs* 148535. The World Bank, 2020.
[42] L. Lou and X.-Q. Zhao. Threshold dynamics in a time-delayed periodic sis epidemic model. *Discrete and Continuous Dynamical Systems. Series B*, 126(1):169–186, 2009.
[43] M. Martcheva. *An Introduction to Mathematical Epidemiology* volume 61. Springer, 2015.
[44] R. K. McCormack and L. J. Allen. Multi-patch deterministic and stochastic models for wildlife diseases. *Journal of Biological Dynamics*, 1(1):63–85, 2007.
[45] M. Mehdaoui. A review of commonly used compartmental models in epidemiology, arXiv preprint arXiv:2110.09642, 2021.
[46] M. Meziane, A. Moussaoui, and V. Volpert. On a two-strain epidemic model involving delay equations. *Mathematical Biosciences and Engineering*, 20(12):20683–20711, 2023.
[47] A. Mozokhina, I. Popravka, M. Saade, and V. Volpert. Modeling the influence of lockdown on epidemic progression and economy. *Mathematics*, 12(19):3106, 2024.
[48] M. Nairz, T. Todorovic, C. M. Gehrer, P. Grubwieser, F. Burkert, M. Zimmermann, K. Trattnig, W. Klotz, I. Theurl, R. Bellmann-Weiler, and G. Weiss. Single-center experience in detecting influenza virus, rsv and sars-cov-2 at the emergency department. *Viruses*, 15(2), 2023.
[49] S. Paul and E. Lorin. Estimation of covid-19 recovery and decease periods in Canada using delay model. *Scientific Reports*, 11(1):23763, 2021.
[50] M. Saade, S. Aniţa, and V. Volpert. Dynamics of persistent epidemic and optimal control of vaccination. *Mathematics*, 11(17):3770, 2023.
[51] M. Saade, S. Ghosh, M. Banerjee, and V. Volpert. An epidemic model with time delays determined by the infectivity and disease durations. *Mathematical Biosciences and Engineering*, 20(7):12864–12888, 2023.
[52] M. Saade, S. Ghosh, M. Banerjee, and V. Volpert. Delay epidemic models determined by latency, infection, and immunity duration. *Mathematical Biosciences*, 109155, 2024.
[53] M. Saade, S. Ghosh, M. Banerjee, and V. Volpert. Dynamics of delay epidemic model with periodic transmission rate. *Applied Mathematical Modelling*, 115802, 2024.
[54] S. Sharma, V. Volpert, and M. Banerjee. Extended seiqr type model for covid-19 epidemic and data analysis. *MedRxiv*, pages 2020–08, 2020.
[55] G.-Q. Sun, Z. Jin, Q.-X. Liu, and L. Li. Chaos induced by breakup of waves in a spatial epidemic model with nonlinear incidence rate. *Journal of Statistical Mechanics: Theory and Experiment*, 2008(08):P08011, 2008.
[56] L. Varona and J. R. Gonzales. Dynamics of the impact of covid-19 on the economic activity of Peru. *PLoS ONE*, 16(1):e0244920, 2021.
[57] R. Verity, L. C. Okell, I. Dorigatti, P. Winskill, C. Whittaker, N. Imai, G. Cuomo-Dannenburg, H. Thompson, P. G. T. Walker, H. Fu, et al. Estimates of the severity of coronavirus disease 2019: A model-based analysis. *Lancet. Infectious Diseases*, 20:669–677, 2020.
[58] Worldometer. Coronavirus tracker. https://www.worldometers.info/coronavirus/, April 13th, 2024.
[59] A. Zincenko, S. Petrovskii, and V. Volpert. An economic-demographic dynamical system. *Mathematical Modelling of Natural Phenomena*, 13(3):27, 2018.

Jochen Mau and Wenhua Tian

Shanghai 2022 outbreak of SARS-CoV-2 omicron revisited

Analysis of Daily Briefings of China National Health Commission

Abstract: Another outbreak of SARS-CoV-2 hit China in Spring 2022 in Shanghai with the same order of magnitude as in Hubei at the Spring Festival 2020. Again, the Administration ordered a strict lockdown, which brought this outbreak under control within 6 weeks, as in 2020. At that time, logarithms of daily numbers of new infected cases exhibited a sugarloaf shape that matched a Bateman function up to peak-case numbers; also, the rapid descent to level zero matched the Bateman's power-series (up to power ten) approximation; effectiveness of counterepidemic management in 2020 was calculated as 78 %, using relative reduction in area under the curves (AUC) of the two functions. But in 2022, hidden spreaders of the predominant omicron variant were of primary concern and compulsory repeated mass screening revealed the asymptomatic infected in large numbers, appearing on top of the concurrently emerging clinically symptomatic COVID-19 cases. The first issue was day-to-day monitoring of daily new cases of asymptomatic infected (nAIC), of clinically symptomatic (nCD-19), either without AIC history (nCD-19*), or with conversion from previous AIC (nCONV), for an estimate of the relative AIC population and early signs of slow down. The district-by-district screening design let AIC appear as bulk infections, appearing across the megacity as escalating waves, far from the monolithic shapes seen in 2020. Using daily reported numbers of new cases from the China National Health Commission's Daily Briefings, we found that new AIC emerged daily in 36-fold of CD-19* in Shanghai, compared to only four-fold in mainland China without Shanghai. Two percent of AIC turned clinically symptomatic 9 to 10 days later, in Shanghai and in Mainland China without Shanghai. Separate Bateman functions were fitted individually to the escalating waves of outbreaks, and sharp descent was captured without extra cost. Bateman effectiveness of counterepidemic management reached 85 % in asymptomatically infected. No mechanistic epidemic models were used, as the focus was only on intensity (rate of incidence), not on occupancy.

Acknowledgement: Thanks to the National Health Commission (NHC) of the People's Republic of China, detailed daily briefings [50], all data could be obtained manually.

Jochen Mau, Medical Faculty and University Hospital Düsseldorf, Heinrich Heine University Düsseldorf, Moorenstr. 5, 40225 Düsseldorf, Germany, e-mail: j.mau@hhu.de
Wenhua Tian, School of Social Development and Public Policy, Fudan University, Handan Road 220, Shanghai 200433, P.R. China, e-mail: wh_tian@fudan.edu.cn

https://doi.org/10.1515/9783112216620-017

Keywords: Infectious disease, mass screening design, logistic difference, Bateman function, escalating outbreak waves, sugarloaf shape, rapid suppression, Bateman effectiveness, conversion from asymptomatic infection, cumulative incidence function, counting process

MSC 2020: 92C45, 92D45, 92D30

1 Introduction

A dynamic zero-COVID policy had been adopted for China in August 2021 to counter an expected increase of infections from the omicron variant already seen in South Africa [25, 40, 37]. Growing concern arose from persistent flare-ups and moderate to significant outbreaks across several provinces of China since mid-February 2022, driven by an omicron subvariant [37], possibly imported from the earlier Hong Kong omicron outbreak, which reached East China's Shanghai, a megacity of 25 million population, "vital to China's economy as its primary financial and industrial hub, home to major stock and futures exchanges, the world's busiest port, and a significant center for trade and innovation" [27]. Such threat—unprecedented in national and global significance—led to equally unprecedented drastic measures in size and determination when the rapidly accelerating surge in March 2022 could not be managed with localized lock-downs targeted to districts hit most [13]. In combination with repeated mass screening in central locations and immediate removal from risk of transmission of the test positives into cabin quarantine in shelter hospitals for two weeks [58], Shanghai's megacity close-off and lock-down in fact brought the spread down from a base reproduction number of $R_0 > 3.6$ to below one within two weeks [58].

In rapidly evolving outbreaks, day-to-day dynamics are of primary concern as early interventions aim to reduce speed and turn rise into peak, with subsequent persistent reduction in numbers of daily new infections. This is then in the first place an incidence rate problem of transitions $\boxed{\text{not infected}} \to \boxed{\text{infected}}$, while the desired shape of turning a steep rise to a peak and then into descent is reminiscent of a double-exponential Bateman function in one-compartment–blood plasma–pharmacokinetic (PK) models. These models formalize bioavailability of a drug in plasma as a function of administered dose, rate of absorption into the central compartment (blood plasma), and rate of elimination, which determine peak, time to peak, and total amount of drug absorbed into plasma as area under the curve (AUC) [18].

Despite some similarity, epidemic models, on the contrary, are based on mass-action equations of chemical kinetics [36] and used in forecasting extent and duration of infectious spread in order to plan resources for prolonged duration, but first to find the most effective parameters to influence the course of the epidemic to the better. This engages mechanistic models, e. g., [31, 3, 5, 7, 35, 30] of changing occupancy in several compartments—susceptibles, infected, infectious, hospitalized, remission, reinfected, immunes, immigration and emigration—within a population, and the more detailed, the more parameters will have to be estimated for relevant projections into the future, and accordingly the more time will elapse until specific data has been collected and processed for model fitting.

Therefore, the present authors focused on a simpler idea that would generate the same shape from physics concepts. The thought model is a container filled with fluid or gas—representing a reservoir of unnoticed infectious individuals—under increasing pressure until the encasement breaks open, the stored potential energy from the high pressure is rapidly converted into kinetic energy, causing the fluid or gas to exit at a high velocity. As the fluid or gas escapes, the pressure inside the container or reservoir decreases significantly. Because the driving pressure is falling, there is less energy to convert into kinetic energy, and the velocity of the outflow slows down. This is an illustration of Toricelli's law and a statement of energy conservation: the total energy (a combination of pressure energy, kinetic energy, and potential energy) remains constant. As the pressure energy decreases, the kinetic energy (and hence velocity) must also decrease.

The present authors therefore had applied a Bateman function to three major SARS-nCoV 2020 outbreaks in China (Hubei, Zhejiang Province, and Shanghai) [47], and demonstrated that Bateman functions not only fitted the ascending part with peak, but also—with the associated power series approximation up to power ten—the steeply descending part in the shape of daily numbers of new infected cases unexpectedly well, after a logarithmic transformation. In particular, the approach lends itself to measure the effectiveness of antiepidemic interventions in terms of AUCs. We therefore explored the same approach in the 2022 context.

In addition, we used a descriptive log-linear approach to monitor the daily arising numbers of new cases in Shanghai, and for contrast, in mainland China without Shanghai.

Section 2 explains the data model and the two kinds of analyses, the log-linear graphical analyses of day-to-day data in 2.3 and Bateman functions fitting for Shanghai incidences in 2.4. Section 3 presents the outcomes from both kinds, and Section 4 discusses our results and their distinction from other approaches. The Appendix collects necessary supplementary information about the raw data used in its Section A, mass screening design and analysis in Section B, main mechanistic epidemic models in Section C, and transition intensities from the data model of Section 2 in Section C.4.2.

2 Material and methods

2.1 Data source

New cases of confirmed asymptomatic infections (nAIC) and of primary clinically symptomatic COVID-19 without prior registration as AIC (nCD-19*) are detected daily by population screens (cf. Figure B.1) according to theoretical transition intensities λ and ϕ in the state diagram in Figure 12.1, respectively.

Their daily numbers were reported to the China National Health Commission [50], from all provinces, autonomous regions, and megacities directly under the Central Government Administration and are listed in condensed form from March 1 to May 16, 2022, in Table A.1. For our data analysis, the systematic seven days a week mass screening design in Figure B.1 was adopted.

From March 1 to May 16, 2022, 35,181 "primary," i. e., without known prior asymptomatic infection, clinically symptomatic COVID-19 cases, briefly denoted as CD-19* cases, 22,079 "secondary," i. e., with known prior asymptomatic infection, clinically symptomatic COVID-19 cases, reported as conversions and briefly denoted here as CONV cases, and 585,732 AIC were reported for Shanghai, compared to 47,624 CD-19* cases, 3,772 CONV cases, and 71,115 AIC for Mainland China without Shanghai, Table 12.1.

Table 12.1: Total numbers of daily new cases from March 1 to May 16, 2022, in Shanghai megacity and in Mainland China without Shanghai (Source [50]).

Diagnostic Classification	Abbreviation used in text	Shanghai megacity	Mainland China without Shanghai
Asymptomatic Infected	AIC	585,732	71,115
COVID-19 without prior AIC	CD-19*	35,181	47,624
COVID-19 with prior AIC	CONV	22,079	3,772

2.2 Data model

An observable epidemic population process with four states is adopted and transitions from one state to the next occur according to time-dependent conditional probability rates—so-called hazards—conditional on numbers currently being at risk, denoted by ϕ, λ, ψ for departure states in Figure 12.1. Note that conversions can occur only among active AIC and for at most 14 days by Shanghai epidemic protocol [62].

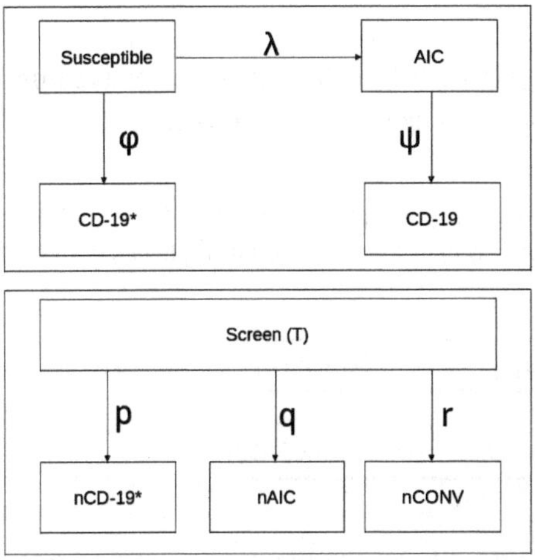

Figure 12.1: State diagram (top) prior to screening examination T (cf. Figure B.1): Any person is either not infected, (true "susceptible"), or is infected without clinical symptoms (true "asymptomatic infected case," AIC), or is clinically symptomatic COVID-19 without previous diagnosis of virus infection (true nCD-19*), or is clinically symptomatic COVID-19 after previously documented status of AIC (denoted by CONV, as a case of conversion). Greek letters for theoretical rates of change, i. e., intensities of possible progressions from one state to next, Latin letters for pretest probabilities of new symptomatic cases, theoretically detectable in screen *T*. The notation is collected in Table 12.2.

Table 12.2: Synopsis of notation in Figure 12.1.

State	Abbreviation	from		to	rate of change
Susceptible	S	S	→	AIC	λ
Susceptible	S	S	→	CD-19*	ϕ
Asymptomatic Infected	AIC	AIC	→	CD-19	ψ
COVID-19 without prior AIC	CD-19*	CD-19*	→	clinical treatment	
COVID-19 with prior AIC	CONV	CONV	→	clinical treatment	

2.3 Time series of daily new numbers

2.3.1 Log-linear day-by-day comparisons

Quantitative analysis is based on daily reported numbers of new cases that arise in Shanghai from daily screens T_i, $i = 1, 2, 3, \ldots$, and are considered—for unbiasedness—in pairs of diagnostic classes nAIC, nCD-19*, and nCONV; cf. Figure B.1 and Table 12.3 for the combination nAIC and nCD-19*.

Daily new AIC and new CD-19*

For any k screens from those T_i, $i = 1, 2, 3, \ldots$ (Figure B.1) in Shanghai, respective numbers of new cases, N_i(nAIC) and N_i(nCD-19*), are arranged as in Table 12.3.

Table 12.3: Two-by-k table of daily counts N of new PCR-detected cases of asymptomatic infected (nAIC) and of clinically detected cases of COVID-19 without previously detected virus infection (nCD-19*) from $k + 1$ screens T_{i+j}, $j = 0, 1, \ldots, k$.

Daily numbers of new cases	Screen T_i	Screen T_{i+1}	⋯	Screen T_{i+k}
Asymptomatic Infected (AIC)	N_i(nAIC)	N_{i+1}(nAIC)	⋯	N_{i+k}(nAIC)
COVID-19 without prior AIC	N_i(nCD-19*)	N_{i+1}(nCD-19*)	⋯	N_{i+k}(nCD-19*)

The ratios N_{i+j}(nAIC):N_{i+j}(nCD-19*) are actually empirical *odds*, since N_{i+j}(nAIC) + N_{i+j}(nCD-19*) equals the total number of *new* cases among the clinically and PCR-negative in the current population of SARS-CoV-2 susceptibles just prior to screen T_{i+j}, $j = 0, 1, \ldots, k$ (cf. (12.1)),

$$
{}_{i+j}\widehat{\text{odds}}^{\,\text{AIC}}_{\text{CD}-19*} = \frac{N_{i+j}(\text{nAIC})}{N_{i+j}(\text{nCD}-19*)} \tag{12.1}
$$

Empirical odds in (12.1) can then be interpreted as factors by which N_{i+j}(nAIC), the number of new PCR-detected cases of AIC in screen T_{i+j}, exceeds N_{i+j}(nCD-19*), the number of new cases of clinically detected COVID-19 without previously known SARS-CoV-2 infection in the same screen T_{i+j}, tantamount to the daily number of new AIC per one new case nCD-19*, for all $j = 0, 1, \ldots, k$.

As distances are more easily visually compared than relations, it is helpful to take logarithms on both sides in (12.2) to obtain empirical *logits*,

$$
{}_{i+j}\widehat{\text{logit}}^{\,\text{AIC}}_{\text{CD}-19*} = \log N_{i+j}(\text{nAIC}) - \log N_{i+j}(\text{nCD}-19*), \tag{12.2}
$$

$j = 0, 1, \ldots, k$. These now appear as daily vertical distances between the two time series

$$
\left.\begin{array}{llll} \log N_i(\text{nAIC}), & \log N_{i+1}(\text{nAIC}), & \ldots, & \log N_{i+k}(\text{nAIC}) \\ \log N_i(\text{nCD}-19*), & \log N_{i+1}(\text{nCD}-19*), & \ldots, & \log N_{i+k}(\text{nCD}-19*) \end{array}\right\} \tag{12.3}
$$

of log-transforms of daily numbers of new cases as depicted for Shanghai from March 1 to June 19, 2022, in Figure 12.2A and for mainland without Shanghai from March 1 to June 29, 2022, in Figure 12.2B.

Technically, logarithms spread the interval $]0, \infty[$ of values that odds can take to $]-\infty, \infty[$ for logits; a logit of zero means odds of one, negative logits correspond to odds less, positive logits to odds larger than one. Logits lend themselves to calculation of *average* vertical distances with *confidence intervals* for chance adjustments, during time periods of interest, as statistical distributions of logits can be approximated by a Gaussian distribution. Then comparisons of averages are possible with Student's t test.

Daily new AIC and new CONV

The methods used for nAIC and nCD-19* carry over when nCD-19* is replaced by nCONV, the daily number of new conversions from asymptomatically infected to clinically manifest COVID-19.

For any k screens from those T_i, $i = 1, 2, 3, \ldots$ (Figure B.1) in Shanghai, respective numbers of new cases, N_i(nAIC) and N_i(nCONV), are arranged as in Table 12.4.

Table 12.4: Two-by-k table of daily counts N of new PCR-detected cases of asymptomatic infected (nAIC) and of clinically detected cases of COVID-19 after previously detected virus infection (nCONV) from $k+1$ screens T_{i+j}, $j = 0, 1, \ldots, k$.

Daily numbers of new cases	Screen T_i	Screen T_{i+1}	…	Screen T_{i+k}
Asymptomatic Infected (AIC)	N_i(nAIC)	N_{i+1}(nAIC)	…	N_{i+k}(nAIC)
COVID-19 after prior AIC	N_i(nCONV)	N_{i+1}(nCONV)	…	N_{i+k}(nCONV)

The ratios N_{i+j}(nAIC):N_{i+j}(nCONV) are actually *empirical odds*, since N_{i+j}(nAIC) + N_{i+j}(nCD-19*) equals the total number of *new* cases among the clinically and PCR-negative in the current population of SARS-CoV-2 susceptibles just prior to screen T_{i+j}, $j = 0, 1, \ldots, k$; cf. (12.4),

$$ {}_{i+j}\widehat{\text{odds}}^{\text{AIC}}_{\text{CONV}} = \frac{N_{i+j}(\text{nAIC})}{N_{i+j}(\text{nCONV})} \tag{12.4} $$

Empirical odds in Equation (12.4) can then be interpreted as factors by which N_{i+j}(nAIC), the number of new PCR-detected cases of AIC, exceeds N_{i+j}(CONV), the number of new cases of clinically detected COVID-19 after previously known SARS-CoV-2 infection, tantamount to the daily number of new AIC per one new case nCONV. Take again logarithms on both sides, and obtain *empirical logits* in (12.5),

$$ {}_{i+j}\widehat{\text{logit}}^{\text{AIC}}_{\text{CONV}} = \log N_{i+j}(\text{nAIC}) - \log N_{i+j}(\text{nCONV}), \tag{12.5} $$

$j = 0, 1, \ldots, k$. These now appear as daily vertical distances between the two time series

$$ \left.\begin{array}{llll} \log N_i(\text{nAIC}), & \log N_{i+1}(\text{nAIC}), & \ldots, & \log N_{i+k}(\text{nAIC}) \\ \log N_i(\text{nCONV}), & \log N_{i+1}(\text{nCONV}), & \ldots, & \log N_{i+k}(\text{nCONV}) \end{array}\right\} \tag{12.6} $$

of log-transforms of daily numbers of new cases as depicted for Shanghai from March 1 to May 16, 2022, in Figure 12.3A, and for the Mainland without Shanghai from March 1 to May 16, 2022, in Figure 12.4A.

Daily new CD-19* and new CONV

This is a change of notation, replacing nAIC by nCD-19* in the preceding contrast; we present it for clarity without repeating explanations. Table 12.4 becomes Table 12.5 and (12.4) becomes (12.7) for *empirical odds*.

$$ {}_{i+j}\widehat{\text{odds}}_{\text{CONV}}^{\text{CD-19}*} = \frac{N_{i+j}(\text{nCD-19}*)}{N_{i+j}(\text{nCONV})}, \tag{12.7} $$

and after taking logarithms, one obtains respective *empirical logits*

$$ {}_{i+j}\widehat{\text{logit}}_{\text{CONV}}^{\text{CD-19}*} = \log N_{i+j}(\text{nCD-19}*) - \log N_{i+j}(\text{nCONV}), \tag{12.8} $$

$j = 0, 1, \ldots, k$. These appear as daily vertical distances between the two time series

$$ \left.\begin{array}{lll} \log N_i(\text{nCD-19}*), & \log N_{i+1}(\text{nCD-19}*), & \ldots, \log N_{i+k}(\text{nCD-19}) \\ \log N_i(\text{nCONV}), & \log N_{i+1}(\text{nCONV}), & \ldots, \log N_{i+k}(\text{nCONV}) \end{array}\right\} \tag{12.9} $$

of log-transforms of daily numbers of new cases as depicted in Figure 12.5A and B for Shanghai and for the Mainland without Shanghai, respectively.

Table 12.5: Two-by-k table of daily counts N of new cases of COVID-19 without prior PCR-detected asymptomatic infection (nCD-19*) and of clinically detected cases of COVID-19 after previously detected virus infection (nCONV) from $k+1$ screens T_{i+j}, $j = 0, 1, \ldots, k$.

Daily numbers of new cases	Screen T_i	Screen T_{i+1}	...	Screen T_{i+k}
COVID-19 without prior AIC	N_i(nCD-19*)	N_{i+1}(nCD-19*)	...	N_{i+k}(nCD-19*)
COVID-19 after prior AIC	N_i(nCONV)	N_{i+1}(nCONV)	...	N_{i+k}(nCONV)

2.3.2 *Post hoc* graphical regression

One application of the graphical method to reduce variability of daily vertical distances between the $\log N$(nAIC) and $\log N$(nCONV) curves is a time shift of the logN(nCONV) curve backwards to match the shape of the $\log N$(nAIC) curve more closely.

To compare occurrences of new cases of conversions to those of new cases of asymptomatic infected, nCONV of $t + \tau$ to nAIC of t, we consider time shifts τ that take the two time series in (12.6) to closest match across all days of reporting new cases, $i = \tau+1, \ldots, \ell$, say

$$ \left.\begin{array}{lll} \log N_{i-\tau}(\text{nAIC}), & \log N_{i-\tau+1}(\text{nAIC}), & \ldots, \log N_{i-\tau+k}(\text{nAIC}) \\ \log N_i(\text{nCONV}), & \log N_{i+1}(\text{nCONV}), & \ldots, \log N_{i+k}(\text{nCONV}) \end{array}\right\} \tag{12.10} $$

per least squares of vertical distances

$$\sum_{i=\tau+1}^{\ell} [\log N_{i-\tau}(\text{nAIC}) - \log N_i(\text{nCONV})]^2 = \min_{\tau}! \tag{12.11}$$

across the range of eligible values of τ, from zero to difference in peak daily number of new cases, $0 \le \tau \le (N_{\max}(\text{nAIC}) - N_{\max}(\text{nCONV}))$.

As an orientation without statistical confirmation seems sufficient, we obtained matching times series in (12.10) with manual shifts to visual congruence instead of algorithmic least squares solutions of (12.11) in Figures 12.3(B,D) for Shanghai and 12.4(B,D) for the Mainland without Shanghai.

2.3.3 Comparisons between Shanghai and Mainland China without Shanghai

Technically, this involves a comparison of two empirical odds as in (12.1), one from Shanghai data, denoted by (a), the other from Mainland without Shanghai data, denoted by (b), in terms of empirical *odds ratios* [9],

$${}^{\text{a}}_{\text{b}}\widehat{\text{OR}}^{\text{AIC}}_{\text{CD}-19*} = \frac{{}_{\text{a}}\widehat{\text{odds}}^{\text{AIC}}_{\text{CD}-19*}}{{}_{\text{b}}\widehat{\text{odds}}^{\text{AIC}}_{\text{CD}-19*}} \tag{12.12}$$

Inserting obtained daily numbers of new cases by (12.1), yields the representation by a cross-product familiar from 2-by-2 contingency tables,

$${}^{\text{a}}_{\text{b}}\widehat{\text{OR}}^{\text{AIC}}_{\text{CD}-19*} = \frac{{}_{\text{a}}N(\text{nAIC})_{\text{b}}N(\text{nCD}-19*)}{{}_{\text{b}}N(\text{nAIC})_{\text{a}}N(\text{nCD}-19*)} \tag{12.13}$$

Technically, more convenient is the *logistic difference* as the logarithm of the empirical odds ratio, denoted as logOR,

$$\left.\begin{aligned} {}_{\text{ab}}\widehat{\text{logOR}}^{\text{AIC}}_{\text{CD}-19*} &= \log({}^{\text{a}}_{\text{b}}\widehat{\text{OR}}^{\text{AIC}}_{\text{CD}-19*}) \\ &= {}_{\text{a}}\widehat{\text{logit}}^{\text{AIC}}_{\text{CD}-19*} - {}_{\text{b}}\widehat{\text{logit}}^{\text{AIC}}_{\text{CD}-19*} \end{aligned}\right\} \tag{12.14}$$

2.4 Bateman function fits

For the SARS-nCoV outbreaks in China 2020, the Bateman function of pharmacokinetics was used to fit the numbers of daily new cases surging to peak and quickly fading out after rigid counterepidemic interventions [47].

2.4.1 Bateman function

Assume constant rates k_a and k_e of *absorption* and *elimination* to represent the two phases of outburst and its containment, respectively,

$$B(t; D, k_a, k_e) = D\frac{k_a}{k_a - k_e}[e^{-k_e t} - e^{-k_a t}] \tag{12.15}$$

where $D > 0$ represents the size D of *hidden* infected cases in a population, on a log-scale [47].

2.4.2 Power series approximation

To guess parameter values in (12.15) as starting values for a nonlinear least-squares fit, an approximation proved useful in applications [47]. From the full power series representation of (12.15),

$$B(t) = D\frac{k_a}{k_a - k_e}\sum_{n=0}^{\infty}\frac{(-1)^n}{n!}[k_a^n - k_e^n]t^n, \tag{12.16}$$

only terms up to power ten,

$$B_{10}(t) = D\frac{k_a}{k_a - k_e}\sum_{n=0}^{10}\frac{(-1)^n}{n!}[k_a^n - k_e^n]t^n, \tag{12.17}$$

are used for approximations as long as $B_{10}(t) > 0$. Its shape of an asymmetrical cap (∩)—reminiscent of a "sugarloaf"—had indeed suggested that it may already cover most of very quickly contained outbreaks while ignoring late occurring flare-ups, in the 2020 data [47].

2.4.3 The modified fitting procedure for the omicron outbreak in 2022

While the "sugarloaf" shapes [53] generated by the wild-type virus in 2020 permitted single-curve fits [47], the 2022 outbreak of the SARS-CoV-2 omicron variant, however, required a more nuanced approach assuming several escalating "waves" of distinct contagiousness: first, a precursory moderate though persistently increasing occurrence of cases; second, some strong rapid-fire out-bursts generating highly surging numbers of cases in quick succession, and third, final culminating ones, which set the peak from which a drastic descent down to residual flare-ups occurred. The pattern is reminiscent of earthquakes that evolve in form of a precursory scatter of minor earthquakes, a (series of) strong major outburst(s), and a drastic drop of severity fading out in aftershocks [59].

Accordingly, the procedure used for SARS-nCoV in [47] could not be applied. Instead, a first Bateman function (12.15) would be fitted to the first-phase ascent by educated guessing, a second would be found by fitting first a closely attaching "sugarloaf" power-series approximation (12.17) by educated guessing and then inserting this one's parameters into (12.15) for a Bateman function identical in the ascending part. Notwithstanding a need to accommodate further outbursts that drive the total number to even higher levels, one final "wave" will mark the top at which the whole outburst culminates and shrinks in numbers of newly infected cases according the rapid descent seen in the 2020 data [47] under the rules of (12.17).

When functions (12.15) and (12.17) have been obtained, respective areas under the curve (AUC), AUC_B and $\mathrm{AUC}_{B_{10}}$, respectively, can be obtained by integration in R software. Then call

$$B_{\mathrm{eff}} = 1 - \frac{\mathrm{AUC}_{B_{10}}}{\mathrm{AUC}_B} \qquad (12.18)$$

the *Bateman effectiveness* of epidemic control of an outbreak as it represents that proportion of the area under the Bateman function that was prevented by control measures.

2.5 Technical data analysis

Analysis is based on comparisons of daily reported numbers of nCD-19*, nAIC, and nCONV on every single day, collected in time series on the log-scale for investigation of emerging epidemic dynamics. Calculations and graphical displays were done in an open-source spreadsheet, LibreOffice Calc (Linux Mint 17.1 Cinnamon 32-bit operating system). Bateman functions were plotted with R software.

3 Results

Structure of Section 3

3.1 Day-to-day comparisons of new cases. The purpose is to show the feasibility of exploratory monitoring of evolving epidemic spread in Shanghai and in Mainland China without Shanghai, concurrently with daily briefings by the China National Health Commission, without fitting mechanistic epidemic models.

3.1.1 Introduction. An overview of the analysis segments.

3.1.2: New asymptomatic infected cases (nAIC) vs. new clinical COVID-19 previously asymptomatic (nCD-19*), cf. page 254, with Figure 12.2 on page 255 and Table 12.6 on page 256. The purpose is to obtain an estimate of the relationship between otherwise hidden infected and clinically manifest cases in an evolving outbreak.

3.1.3: New asymptomatic infected cases (nAIC) vs. new clinical COVID-19 by conversion from previously asymptomatic (nCONV) cf. page 256, with Figure 12.3 on page 257, for Shanghai, Figure 12.4 on page 258 for the Mainland without Shanghai, and Table 12.7 on page 257 for comparison of the two regions. The purpose is to obtain information about the time delay in clinical manifestation of COVID-19 after an asymptomatic infection (AIC), in either epidemic context.
3.1.4: New clinical COVID-19 without prior asymptomatic infection (nCD-19*) and new clinical manifestation of COVID-19 after prior asymptomatic infection (nCONV), cf. page 258, with Figure 12.5 on page 259. The purpose is the confirmation of time delays seen in 3.1.3.
3.2: Curve fitting with escalating Bateman functions, page 260.
Appendix C.4.2: Transition intensities in four-state diagram, page 284.

3.1 Day-to-day comparisons of new cases

3.1.1 Introduction

From April 8 to 30, 2022, Shanghai cases accounted for most of the daily new cases of COVID-19 without prior asymptomatic infections (nCD-19*) in Mainland China because of epidemic outbreaks in several provinces in North-East China during March.

Since March 25, however, Shanghai's daily numbers of new asymptomatic infections (nAIC) began to outnumber the daily nAIC in the Mainland without Shanghai consistently from a factor of four on March 30 to a factor of 27 on April 30; see Table A.1.

While absolute numbers of daily new cases in both categories, nCD-19* and nAIC, are rather variable, their daily ratios N(nAIC):N(nCD-19*) show little variation around moderate values of one in March and of five in April.

Cumulative hazard functions of primary infections as either AIC or CD-19* (cf. Figure C.2) show the rate of detection in their slopes, e. g., between March 28 and April 6, the daily detection rate of AIC increases sharply, is fairly constant until April 22, and then declines until mid-May to a detection rate already close to zero.

The cumulative hazard function of secondary infection as the conversion from previously detected AIC must be seen in relation to the build-up of its risk set of AIC: daily detection rate of conversions can increase quickly when the number of daily nAIC during last 14 days is sufficiently large; cf. Figure C.2, page 285.

3.1.2 Daily new numbers of asymptomatic infected cases (AIC) and of clinically symptomatic COVID-19 without prior AIC (CD-19*)

Daily occurrences of new asymptomatic infected cases (AIC) and of concurrently reported cases of new COVID-19 cases without prior asymptomatic infections (CD-19*) are graphically represented in terms of logarithms of observed daily total numbers of each

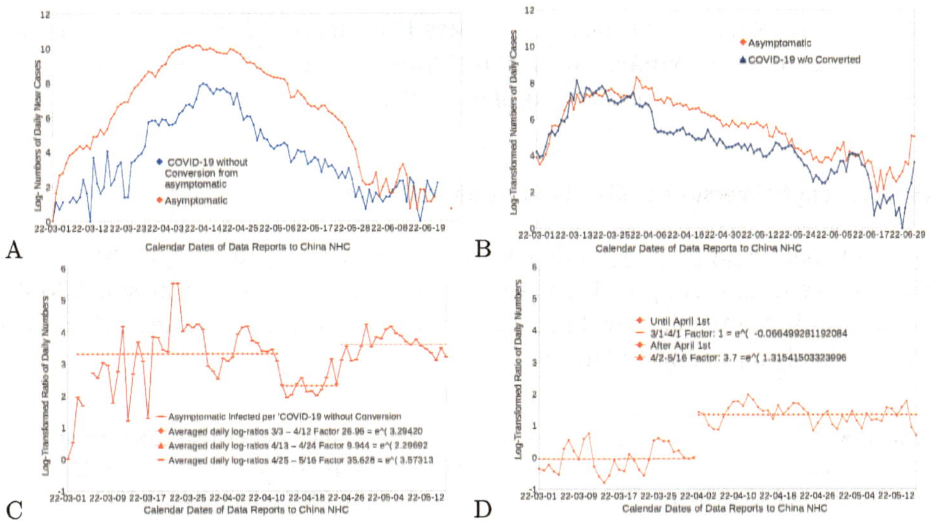

Figure 12.2: Log-transformed numbers of daily new cases of asymptomatic PCR-detected infections (nAIC) [red] and of new cases of clinically detected COVID-19 without prior detection of SARS-CoV-2 infection (nCD-19*) [blue] in the Shanghai megacity (Displays A, C) and in Mainland China without Shanghai (Displays B, D) during March 3 and June 29th, 2022. Note the different scales for calendar dates in top and bottom displays, due to lack of significant data after mid-May. Raw data in Table A.1, p. 272.

diagnostic examination outcome, $\log N(\text{nAIC})$ and $\log N(\text{nCD}{-}19{*})$, from March 1 to July 1, 2022, for Shanghai data in Figure 12.2(A,C) and for the Mainland without Shanghai in Figure 12.2(B,D).

(3.1.2) Shanghai

The time series of daily $\log N(\text{nAIC})$ in Figure 12.2(A) reveals a straight positive trend at a rate of 0.2132/d in daily $\log N(\text{nAIC})$ between March 3 and April 6. It then indicates a plateau, and from April 22 onwards, there is a rather straight decline at a rate of −0.1328/d.

From March 3 to April 6 and from April 22 to May 16 daily logN(nAIC) evolve fairly linearly. The volatility in Figure 12.2, top left, reflects high day-to-day variability in $\log N(\text{nCD-19*})$, and in daily vertical distances between $\log N(\text{nAIC})$ and $\log N(\text{nCD-19*})$, in Figure 12.2 (bottom, left).

(3.1.2) Mainland without Shanghai

Daily numbers of nAIC and nCD-19* rise until mid-March, then reach a plateau, and jointly decline (cf. Figure 12.2(B)) mainly with constant factors corresponding to fairly constant vertical distances from April 2 to May 16 in Figure 12.2(D). Specifically, before April 1, daily numbers, N(nAIC) and N(nCD-19*) are at par, and after April 1, both decline persistently at an almost constant rate of 95 nAIC/d.

(3.1.2) Shanghai versus the Mainland without Shanghai

Daily numbers of nAIC are approximately 36-fold (95 %CI:18;70) those of nCD-19*, on average, in Shanghai after April 24t, and approximately 4-fold (95 %CI:2;7) those of nCD-19*, on average, in the Mainland without Shanghai from April 2 to May 16; the difference is statistically significant by the Student's t test ($P < 0.0001$).

Table 12.6: Number of daily new cases of asymptomatic infections N(nAIC) per one daily new case of COVID-19 without prior asymptomatic infection with SASR-CoV-2 (nCD-19*), in Shanghai (Figure 12.2, Display C) and in Mainland China without Shanghai (Figure 12.2, Display D), March 1 to May 16, 2022.

From – to	average vertical distance	standard deviation	approx. 95 % confid. Interv.	average factor	approx. 95 % confid. Interv.
Fig. 12.2, C		Shanghai megacity			
Mar 03–Apr 12	3.294	0.950	[1.43, 5.16]	27.0	[4.2, 173]
Apr 13–Apr 24	2.297	0.333	[1.64, 2.95]	9.9	[5.1, 19.1]
Apr 25–May16	3.573	0.345	[2.89, 4.25]	35.6	[18.1, 70]
Figure 12.2 D		Mainland without Shanghai			
Mar 1-Apr 01	−0.067	0.422	[−0.89, 0.76]	0.94	[0.41, 2.14]
Apr 2–May 16	1.314	0.304	[0.72, 1.91]	3.73	[2.05, 6.76]

3.1.3 New asymptomatic infected cases (AIC) and new COVID-19 by conversion from prior asymptomatic (CONV)

(3.1.3) Shanghai

In Shanghai before March 24, conversions from AIC had only been reported sporadically; cf. Table A.1. Time series of daily log N(nCONV) are therefore moved backward in calendar time by 9 days, such that N(nCONV) reported for March 13 is related to N(nAIC) reported for March 4, etc.; cf. Figure 12.3 and Table 12.7.

(3.1.3) The Mainland without Shanghai

Conversions had been reported daily since March 1. The time series of daily log N(nCONV) is moved backward in calendar time by 10 days for congruent shape with that of the daily log N(nAIC). Accordingly, N(nCONV) reported for March 13 is related to N(nAIC) reported for March 4, etc.; cf. Figure 12.4B. Despite irregular large variation within each

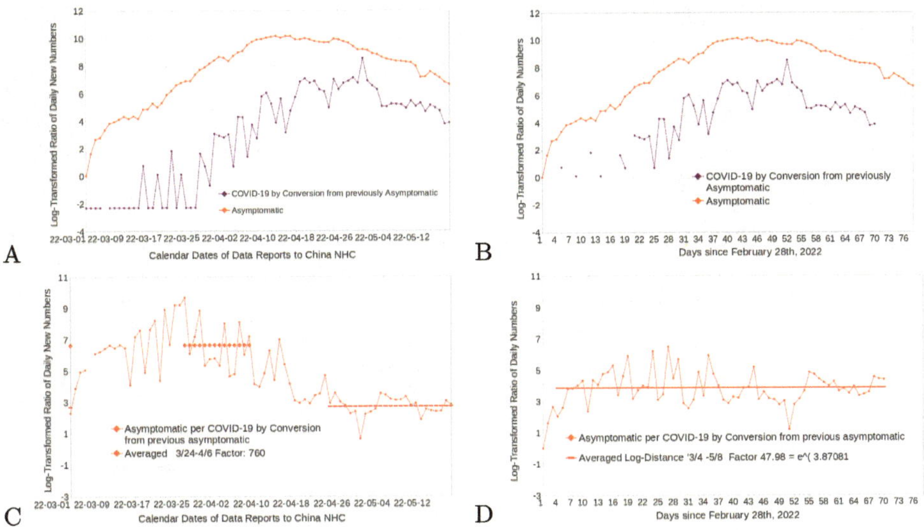

Figure 12.3: Shanghai megacity. A: Log-transformed numbers of daily new cases of asymptomatic PCR-detected infections (nAIC) [red] and of new cases of clinically detected COVID-19 after prior detection of SARS-CoV-2 infection (nCONV) [dark magenta] during March 3 to May 16, 2022. B: Same data, now with a time shift of nCONV 9 days backwards. C: Daily logistic difference of curves in A, with averages along time intervals of fairly equal distance. D: Same analysis, now for time shifted data for nCONV. Raw data in Table A.1, p. 272.

Table 12.7: *N*(nAIC) per one nCONV 9d later in Shanghai (cf. Figure 12.3D), and 10d later in the Mainland without Shanghai; cf. Figure 12.4D.

From – to	average vertical distance	standard deviation	approx. 95 % confid. Interv.	average factor	approx. 95 % confid. Interv.
Figure 12.3D		Shanghai megacity			
Mar 4–May 08	3.871	0.995	[1.92, 5.82]	47.98	[6.8, 337]
Figure 12.4D		Mainland without Shanghai			
Mar 13–May 07	3.222	0.620	[2.01, 4.44]	25.08	[7.4, 84.5]

time series, vertical distances can be represented fairly well by a single average vertical distance of 3.222 (STD = 0.620; 95 %CI:2.01;4.44) throughout, between March 13 and May 7; cf. Figure 12.4D, tantamount to an average factor of 25.08 (95 %CI:7.4;84.5) for *N*(nAIC) per one nCONV reported 10 days later.

(3.1.3) Shanghai versus the Mainland without Shanghai

Comparing the two regions, Shanghai megacity and the Mainland without Shanghai in Table 12.7, with respect to factors by which new cases of AIC occur more frequently then

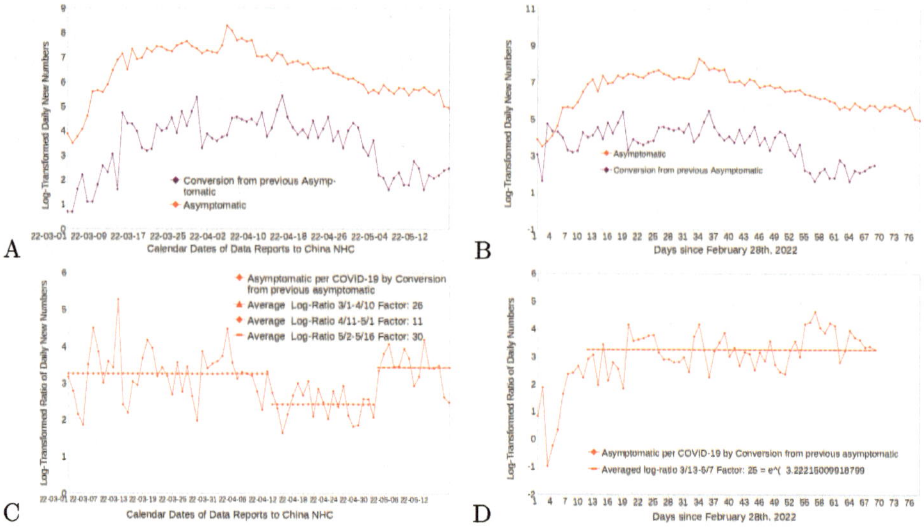

Figure 12.4: The Mainland without Shanghai. A: Log-transformed numbers of daily new cases of asymptomatic PCR-detected infections (nAIC) [red] and of new cases of clinically detected COVID-19 after prior detection of SARS-CoV-2 infection (nCONV) [dark magenta] during March 3 to May 16, 2022. B: Same data, now with a time shift of nCONV 9 days backwards. C: Daily logistic difference of curves in A, with averages along time intervals of fairly equal distance. D: Same analysis, now for time shifted data for nCONV. Raw data in Table A.1, p. 272.

those of new conversions on a day-by-day basis with a time shift of 9–10 days for congruent shapes of time profiles of $\log N(\text{nAIC})$ and $\log N(\text{nCONV})$, one finds the difference of their average factors of 48.0 and 25.08 are statistically significant by the Student's t-test ($P < 0.00005$).

Understanding $N(\text{nCONV})$ as emerging from previously reported $N(\text{nAIC})$ (cf. Fig. 12.3), the inverse relation, an average daily proportion of N(nCONV) per new AIC 9 days earlier, is of interest: by an average vertical distance of −3.871 (95 %CI:-5.81;-1.95), the average proportion is 2.1 % (95 %CI:0.3 %;14.2 %) from March 13 to May 8.

3.1.4 New clinical COVID-19 without prior asymptomatic infection (CD-19*) and new clinical manifestation of COVID-19 after prior asymptomatic infection (CONV)

(3.1.4) Shanghai

In Shanghai before March 24th, conversions from AIC had only been reported sporadically, cf. Table A.1, p. 272. Time series of daily $\log N(\text{nCONV})$ are therefore moved backward in calendar time by nine days, such that $N(\text{nCONV})$ reported for March 13th is compared to $N(\text{CD-19*})$ as reported for March 4th, etc. This shifts the time series of daily

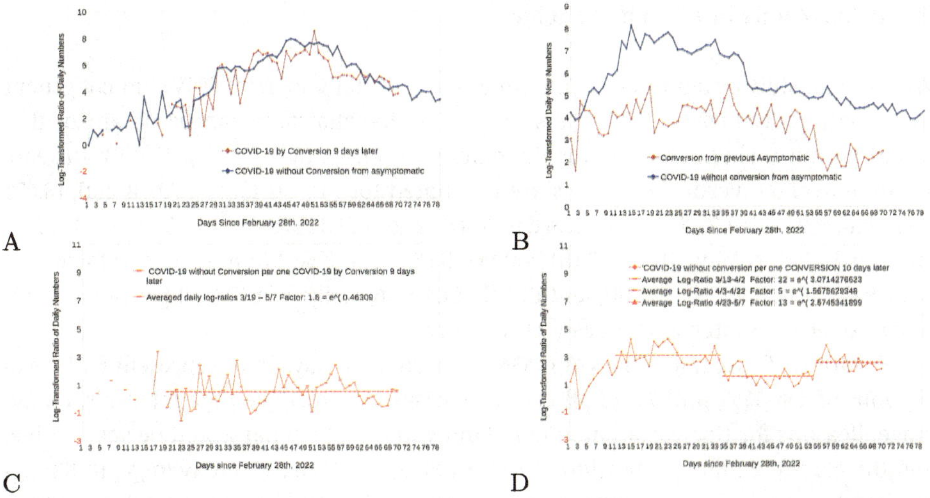

Figure 12.5: Top row: Log transformed numbers of daily new cases of COVID-19 without [blue] and with [magenta] previously detected asymptomatic infection, CD-19* and CONV, during first 80 days since 01 March 2022, respectively. Left: Shanghai, with *N*(nCONV) shifted 9 days back; right: mainland China without Shanghai, *N*(nCONV) shifted 10 days back. Bottom row: Daily differences of curves in top row (until mid-May 2022, only), with averages along time intervals of fairly equal distance. C: Shanghai data, D: Mainland China without Shanghai data. Raw data in Table A.1, p. 272.

log*N*(nCONV) to fairly good congruence with the time series of daily log*N*(CD-19*), as seen in Fig. 12.5(A).

Despite high volatility, a single average vertical distance of 0.584 (STD=1.158; 95%CI:1.68;2.85) fits throughout from March 5th to May 8th, cf. Fig. 12.5(C); the average daily factor is near two *N*(nCD-19*) per one nCONV reported nine days later; cf. Tab. 12.8. Taking the two diagnoses together, 36% of their daily sum of new cases, *N*(nCD-19*) and *N*(nCONV) reported nine days later, are conversions – and that applies fairly throughout, from March 5th to May 8th, 2022.

Table 12.8: *N*(nCD-19*) per one nCONV 9d later in Shanghai, cf. Fig. 12.5C, and 10d later in Mainland without Shanghai, cf. Fig. 12.5D.

From - to	average vertical distance	standard deviation	approx. 95% confid.Interv.	average factor	approx. 95% confid.Interv.
Fig. 12.5C		Shanghai Megacity			
Mar 05 - May 08	0.584	1.158	[1.68 , 2.85]	1.794	[0.186 , 17.3]
Fig. 12.5D		Mainland without Shanghai			
Mar 13 - Apr 02	0.071	0.674	[1.75 , 4.39]	21.6	[5 , 81]
Apr 03 - Apr 22	1.568	0.547	[0.50 , 2.64]	4.8	[1.5 , 14]
Apr 23 - May 07	2.575	0.511	[1.57 , 3.58]	13.1	[4.8 , 36]

(3.1.4) The Mainland without Shanghai

With a time shift of ten days for the time series of daily log*N*(nCONV), no congruent shape with daily log*N*(nCD-19*) as seen in the Shanghai data can be obtained, that would permit a single average for the daily differences here, cf. Fig. 12.5(B). Instead, separate average vertical distances are calculated for March 13th to April 2nd, (3.071 (STD=0.674; 95%CI:1.75;4.39)), April 3rd to April 22nd (1.568 (STD=0.547; 95%CI:0.50;2.64)), and April 23rd to May 7th (2.575 (STD=0.511; 95%CI:1.57;3.58)) tantamount to factors of 22 (95%CI:5;81), 5 (95%CI:1.5;14), and 13 (95%CI:4.8;36) daily *N*(nCD-19*) per one nCONV reported ten days later, respectively, cf. Tab. 12.8.

Regarding factors, numbers of nCONV reported ten days later represent daily proportions of 4%, 17%, and 7% of all COVID-19 cases in respective time intervals, on average. Reasons for this variation in mainland without Shanghai would be speculative, and the corresponding proportion of 35.8% (95%CI:5.46%;84%), on average, in Shanghai stands out with chance-correction by a statistically significant Student's t test ($P < 0.00001$).

(3.1.4) Shanghai versus the Mainland without Shanghai

For congruent shapes of time profiles of log*N*(nCD-19*) and log*N*(nCONV), the latter was shifted nine days backwards for Shanghai and ten days for mainland without Shanghai. While daily numbers of nCD-19* and of nCONV nine days later were fairly at par in Shanghai, the relation was different in Mainland China without Shanghai: Average factors of 22, 5, and 13 were found for daily *N*(nCD-19*) per nCONV reported ten days later in mainland without Shanghai; compared to an average factor of 2 (95%CI:0.18;17.4) throughout in Shanghai, the difference is statistically significant in vertical distances by Student's test ($P < 0.00001$).

3.2 Curve fitting with Bateman functions

3.2.1 Fitting daily new cases of asymptomatically infected

Logarithms of daily numbers of new asymptomatically infected cases (AIC) are depicted with the blue curve in Figure 12.6. It did rise to a maximum within 26 days from the onset (set as March 1, 2022) with an escalating pattern of several "waves" of exacerbating outbreaks, the next superseding the already receding last.

The first (from left) Bateman function rose rapidly to its maximum, from day 1 to day 20, [orange curve], the second Bateman function fitted to the first exacerbation started 7 days later and reached its maximum 35 days later (April 12) [green], while the last one [red] aimed to fit the second exacerbation started on day 19 and rose sharply to take its

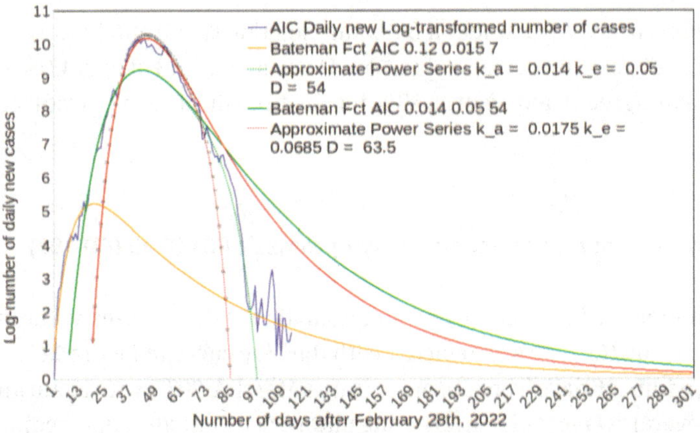

Figure 12.6: SHANGHAI: Logarithmically transformed numbers N(nAIC) of daily new cases of asymptomatic infection with SARS-CoV-2 (AIC) that had been recorded for 122 days from March 1, 2022, [50] and three separate Bateman function fits to address changes in velocity of ascent [orange and green] and the cap at the top [red].

Table 12.9: AIC in the outbreak of SARS-CoV-2 omicron in Shanghai from March 1, 2022: Parameter estimates from fitting the power-10 series approximation of Bateman function to log-transformed numbers of daily new cases of asymptomatically infected (AIC), used in Figure 12.6.

"Wave" Onset	D	k_a	k_e	B_{max}	t_{max}	t^*_{max}	AUC_{Bat}	AUC_{Bapp}	B_{eff}
1st (03/01)	7	0.12	0.0150	5.201	19.8	20	466.600	n. a.	n. a.
2nd (+7 d)	54	0.014	0.0500	9.216	35.4	43	1079.663	614.4973	0.4308
3rd (+19d)	63.5	0.0175	0.0685	10.16	26.8	48	926.973	502.1238	0.8461

Legend: D, virtual size prior to outbreak, k_a, absorption rate, k_e, elimination rate, all with respect to log-transformed numbers of cases, B_{max}: Bateman function (12.15) maximum, t_{max}: Bateman function (12.15) time maximum from "wave onset" time, $t^*_{max} = t_{max}+$ "wave onset" time, AUC_{Bat}: area under the Bateman function (12.15) $[0, \infty[$, AUC_{Bapp}: area under the first nonnegative part of Bateman power-series approximation (12.17), B_{eff} = Bateman effectiveness (12.18).

peak 26 days later in order to match the turning point of daily AIC incidences on day 48 (April 13). Details that characterize these Bateman functions are collected in Table 12.9.

Of note, the descent from peak of daily reported new AIC incidences did not need a new fit. In fact, the power-series approximations to the Bateman functions fitted to second [green] and third [red] exacerbations each fitted respective parts of a little "stuttering" descent surprisingly close.

Another point of interest is the observation that the long phasing-out of the first Bateman-function fit [orange] in Figure 12.6 has no role in fitting any part of the descent as the other two do, but still cuts right through the suddenly appearing "scribble" around day 100, both matching in their decline. Its occurrence may be ascribed to a subpopulation less attended for its lower profile and then "forgotten," calling "I'm still here"!

Considering the difference of AUCs under a Bateman function (12.15) and its power-series approximation (12.17) for effectiveness of antiepidemic management, it is 43 % in curbing the second "wave" [green] and even as 85 % for the final third "wave" [red]; cf. Table 12.9.

3.2.2 Fitting daily new cases of primary clinically symptomatic COVID-19 (CD-19*)

Logarithms of daily numbers of new clinically symptomatic COVID-19 cases without prior asymptomatic infection (CD-19*) are depicted with the blue curve in Figure 12.7. It did rise to a maximum within 46 days from the onset (set as March 1, 2022) with again an escalating pattern of several "waves" of exacerbating outbreaks, mutually superseding with little delay.

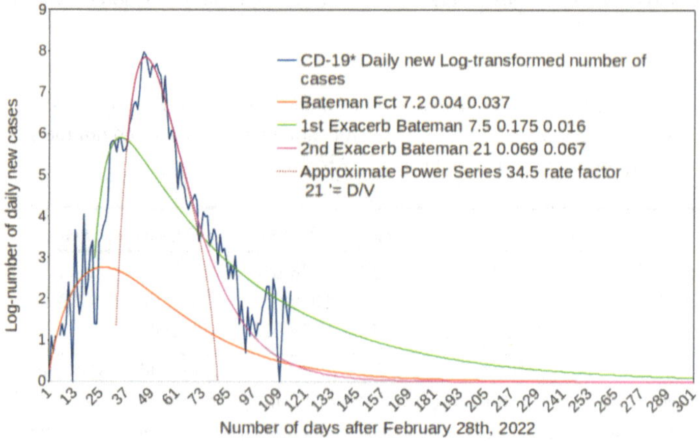

Figure 12.7: SHANGHAI: Logarithmically transformed numbers N(nCD-19*) of daily new cases of clinically symptomatic COVID-19 without prior asymptomatic infection with SARS-CoV-2 that had been recorded for 122 days from March 1, 2022, [50] and three separate Bateman function fits to address changes in velocity of ascent [orange and green] and the cap at the top [red].

The first (from left) Bateman function rises rapidly to its maximum, from day 1 to day 26 [orange curve], the second Bateman function fitted to the first exacerbation started 19 days later and reached its maximum another 15 days later (April 3) [green], while the last one [red] aimed to fit the second exacerbation started on day 31 and rose sharply to take its peak within 15 days in order to match the turning point of daily CD-19* incidences on day 46 (April 11). Details that characterize these Bateman functions are collected in Table 12.10.

Here again, the descent from the peak of daily reported new CD-19* incidences did not need a new fit, as in fact again, the power-series approximations to the Bateman

Table 12.10: CD-19* in Outbreak of SARS-CoV-2 omicron in Shanghai from March 1, 2022: Parameter estimates from fitting the power-10 series approximation of the Bateman function to log-transformed numbers of daily new cases of asymptomatically infected (AIC), used in Figure 12.7.

"Wave" Onset	D	k_a	k_e	B_{max}	t_{max}	t^*_{max}	AUC_{Bat}	AUC_{Bapp}	B_{eff}
1st (03/01)	7.2	0.04	0.037	2.752626	26.0	26	194.595	n. a.	n. a.
2nd (+19 d)	7.5	0.175	0.016	5.895438	15.0	34	468.715	n. d.	n. d.
3rd (+ 31 d)	63.5	0.0175	0.0685	7.839361	14.7	46	313.433	254.0318	0.1895

Legend: D, virtual size prior to outbreak, k_a, absorption rate, k_e, elimination rate, all with respect to log-transformed numbers of cases, B_{max}: Bateman function (12.15) maximum, t_{max}: Bateman function (12.15) time maximum from "wave onset" time, $t^*_{max} = t_{max} +$ "wave onset" time, AUC_{Bat}: area under the Bateman function (12.15) $[0, \infty[$, AUC_{Bapp}: area under the first nonnegative part of Bateman power-series approximation (12.17), B_{eff} = Bateman effectiveness (12.18).

functions fitted to third [red] exacerbation cut through half of the descent, when the tail of the second Bateman [green] took over.

Of note here, even the power-series approximations were not needed for the descent in any part, as exponential fading according to the second [green] and the third [red] cut through all "scribble," even when ragged—the red tail took over when the green missed out for a while. This unexpected phenomenology cannot be ascribed to the Bateman function, but to the design and realization of the mass screening program and antiepidemic measures.

Considering again the difference of AUCs under a Bateman function (12.15) and its power-series approximation (12.17) for effectiveness of antiepidemic management, it is 19 % in curbing the third "wave" [red]; cf. Table 12.10. The figure clearly then shows that successful management of the spread of spontaneously clinically manifest COVID-19, that is, without a prior phase of asymptomatic infection, had mainly evolved by the Bateman law of fading numbers of new cases in several involved source populations.

4 Discussion

Since the pandemic outbreak of SARS-nCoV beginning in January 2020, a burst of scientific articles has been published. Public focus was soon turned on the base reproduction number [19–23], a simple criterion to predict future development of epidemic spread in a population, given an appropriate mechanistic epidemic models, and hence not directly estimable without known close values of the adopted model's parameters [57, 17].

Classic are the SIR and SEIR models (see Appendix C), which have been extended to address challenges and features of interventions in the COVID-19 pandemic; see [32, 16, 26, 56] among others.

Many aspects of the Shanghai 2022 omicron outbreak have been covered extensively in the scientific literature, specifically the driving variant [37], transmission dynamics

[13, 11], interventions [58, 39], effects on public behavior [38, 61], have been considered, including considerations of consequences of China's new "dynamic zero-COVID" strategy [40] for the country's health burden from a nationwide omicron wave as a potential national threat [11].

Our data base

By the implemented epidemic and mass-screening diagnostic protocol in Figure B.1, Shanghai outbreak data promises significant insight even when individual follow-up data could not be used.

We had read the daily briefings by the China National Health Commission concurrently with the emerging outbreak, and hence compared the evolving time series of logarithms of numbers of new asymptomatic and of logarithms of numbers of new clinically symptomatic cases from the beginning, as their daily differences would be statistically independent observations and could be averaged over relevant time intervals, and also be compared with the times series arising in Mainland China without the Shanghai data. Differentiation in the Mainland would not be meaningful as the management of cases would not be comparable to that in Shanghai, because of smaller sizes of local outbreaks. Different from other authors, we had no information about numbers at risk in the screened populations, and no case follow-up information.

Synchronous analysis of evolving time series

The flow chart in Figure B.1 is a hypothetical design of systematic "combing" through the megacity's districts 7 days a week; cf. [42] for real timing. However, it still shows that bias will arise and can be avoided with the analysis of the odds ratios instead of reported numbers (cf. Appendix B) as long as different screening days are not mixed—as we do in shifting conversions. The methods are rooted in well-established log-linear statistical models for binary data; see B.2 [9]. All calculations are simple to perform, do not require professional software, are hence transparent, and results have immediate intuitive interpretations, as needed in synchronous monitoring.

Under these limitations, we investigated two aspects: (i) estimating the proportion of otherwise hidden asymptomatic infected in the population, which would now be revealed by systematic compulsory mass-screening, and (ii) an estimate of time from detected asymptomatic infection to clinical manifestation of COVID-19, as time to "conversion," according to the underlying four-state model with the diagram in Figure 12.1.

For asymptomatic infections detected from March 1 to May 16, 2022, averaged daily proportion of new asymptomatic cases (nAIC) among all new confirmed infections—AIC, COVID-19 without and with previous AIC—was 93.6 % (95 %CI: 67.7 %;99.0 %) in Shanghai and 64.3 % (95 %CI: 32.3 %;87.2 %) in the Mainland without Shanghai, where data arose from a mixture of flare-ups and smaller outbreaks that represents a broader spectrum of generally lower activity; this may explain closer agreement with 40.50 %

(95 %CI, 34.94 %-46.07 %) in a recent metaanalysis [44], notwithstanding even smaller 20 to 30 % in some earlier systematic reviews [12, 52, 60].

For time to conversion, the time series of daily log-transformed numbers of new cases from March 1 to May 16, 2022, were used. We shifted the time series of conversions for close match in shape (rise, peak, plateau, if any, and decline) to that for asymptomatic infected and to that for primary COVID-19 without previously detected asymptomatic infection. Our average time of 9 to 10 days to clinical expression of COVID-19 after asymptomatic infection is compatible with estimated means of 6 to 7 days (95 %CI: 5.6d; 8.8d), and upper 95th percentiles of 10 to 13 days, reported from fitting several survival-time distributions to Wuhan 2020 data [4].

This is consistent with two observations. First, it will need a certain size of the AIC subpopulation to generate visible numbers of conversions. In Shanghai, cases of COVID-19 and AIC had been reported every day in slowly increasing numbers since March 1, but the first case of a conversion was reported as late as March 13; cf. Table A.1. Second, average proportions of number of conversions 9 days later relative to current number of AIC and conversions combined, are very close in Shanghai and in the Mainland without Shanghai, about 2 % and 4 %, respectively, (cf. Figures 12.3(B,D) and 12.4 (B,D)), for Shanghai and for the Mainland without Shanghai, respectively, and the summary for each in Table 12.7,[1] which is less than 25 % reported in [52].

Post-hoc analysis of epidemic dynamics

Our analytical approach is exploratory descriptive to find expression of epidemic dynamics in the time series of reported numbers of daily new cases without assumptions about a specific infectious-disease model that was active in the population.

The four-state dynamics implied by Figure 12.1 cannot be analyzed without additional internal structure, as is characteristic for mechanistic infectious disease models. Such explanatory models are used to simulate the future course of an epidemic and need prior information about the level of immunity in the population, compliance of the people to stop transmission of an ambient virus by wearing masks, personal hygiene, frequent decontamination of "touch surfaces," and wide-sense social distancing to reduce person-to-person transmission opportunities wherever people gather.

Therefore, social dynamics, biological and physicochemical properties of a new virus must be taken up in mechanistic infectious-disease models in order to find best parameter combinations for reduction and eventual extinction of epidemic spread in a community. Prior estimates of parameters are needed to run the simulations. Accordingly, different countries may show different epidemic patterns [33].

The purpose for a purely phenomenological description of epidemic dynamics is different.

---

1 Use 1/(1+average-factor).

First note that descriptive data-fitting to capture the empirical shape of epidemic time series is well established in the scientific literature.

Advantages are (i) flexible fit to rise, peak, and decline in epidemic data without detailed mechanistic assumptions [15, 10], (ii) useful for monitoring and short-term forecasting [8, 45], (iii) require only few external parameters, hence computationally efficient [10], and (iv) especially valuable when mechanisms are unclear or data are limited [14, 48, 63].

Among the limitations of phenomenological models are that (i) they cannot provide inference about transmission dynamics [14], (ii) parameter interpretation and biological meaning are limited [41], (iii) determining which phase of the time series to fit may be difficult [43], (iv) cannot capture feedback, heterogeneity, or interventions [46], and (v) are prone to overfitting the shape and have limited predictive power when epidemic parameters change due to interventions or emergence of variants [49].

With this well in mind and phenomenologically inspired by volcano eruptions [29, 51], we adopted the Bateman function of pharmacokinetics (PK) [6, 24] as a flexible mathematical form to represent the transient pulse observed in many epidemic datasets, regardless of the underlying transmission dynamics. This is analogous to using exponential, Gaussian, or lognormal curves to fit time series data where capturing the pattern is more important than explaining the cause.

For estimation, the feathering/peeling method [18] is established in PK and the Gauss–Newton method [55] is the default option in program *nls*() in R [54]; for more general epidemic growth functions, see [43, 63]. However, to capture the characteristic sugarloaf shape [53] of China's major outbreaks in year 2020, we had instead used the power series (up to power ten) approximation to the Bateman function and obtained parameter estimates from its nonlinear least squares fit [47]; these parameter estimates were inserted into the Bateman function formula to generate the Bateman growth curve that matched the fitted approximation but for the long exponential elimination phase. The relative difference in area and the curve (AUC) between the two then lends itself to a measure of effectiveness of epidemic control management.

The phenomenon was different in the Shanghai 2022 outbreak. Instead of an volcanic-eruption type of outburst, the omicron spread occurred mainly in primarily "hidden" asymptomatic cases that flared up in some districts of the megacity, closer to trembling ground precursory to earthquakes [59] the peak size of which as well as the implication for conversions to clinically symptomatic cases are difficult to predict. This may explain a cautious approach with localized measures in the beginning (cf. actual diagnostic protocol in [42] Figure 2), hence the escalating waves that prevented the fitting of a single Bateman absorption parameter.

Instead, escalating waves in the Shanghai 2022 curves could easily be fitted with separate Bateman functions superimposed as soon as the previous one began to recede in amplitude. This time the parameters of the fitted Bateman functions were inserted into formulae of their respective approximations and these approximations showed an excellent fit to the almost abruptly descending part, of course, again piecewise according to the escalating Bateman absorption steps.

Even further, late flare-ups in ongoing descent surprisingly showed agreement with long-tailed elimination phases early-phase Bateman functions—as if parts of a forest fire had been suppressed but not definitely extinguished, while the fire line moved on.

5 Conclusions

Shanghai diagnostic protocol appears as a "golden standard" when estimation of hidden cases from a rapidly spreading pathogen is key to effective antiepidemic response.

For unbiasedness, epidemic time series of diagnostic classes must be analyzed in pairs; this is sufficient for synchronous monitoring, short-term forecasting, and summary assessments like average time from asymptomatic infection to clinical manifestation of COVID-19, and comparisons between regions.

The sugarloaf shape of daily incidences in China's unprecedented outbreaks, in Hubei 2020 and in Shanghai 2022, appear as a characteristic of the country's antiepidemic management; Bateman function dynamics are compatible with this shape in rising, peak, and descending numbers, even when easing out from the control regimen. Effectiveness of control was up to 85 % on asymptomatic cases and 20 % on COVID-19 without prior asymptomatic status in Shanghai 2022.

Bibliography

[1] O. O. Aalen. Nonparametric inference for a family of counting processes. *The Annals of Statistics*, 6:701–726, 1978.

[2] P. K. Andersen, Ø. Borgan, R. Gill, and N. Keiding. *Statistical Models Based on Counting Processes.* Springer, New York, 1 edition, 1993.

[3] R. M. Anderson and R. M. May. *Infectious Diseases of Humans: Dynamics and Control.* Oxford University Press, 1991.

[4] J. A. Backer, D. Klinkenberg, and J. Wallinga. The incubation period of 2019 novel coronavirus (2019-ncov) infections among travellers from Wuhan, China, 20–28 January 2020. *Euro Surveillance*, 25(5):2000062, 2020. https://doi.org/10.2807/1560-7917.ES.2020.25.5.2000062.

[5] N. T. Bailey. *The Mathematical Theory of Infectious Diseases and Its Applications.* Charles Griffin & Co. Ltd, London, 2 edition, 1975.

[6] H. Bateman. Solution of a system of differential equations occurring in the theory of radioactive transformations. *Proceedings of the Cambridge Philosophical Society*, 15:423–427, 1910.

[7] N. G. Becker. *Analysis of Infectious Disease Data.* Chapman and Hall/CRC, 1 edition, 1989.

[8] M. Betti and J. M. Heffernan. A simple model for fitting mild, severe, and known cases during an epidemic with an application to the current SARS-CoV-2 pandemic. *Infectious Disease Modelling*, 6:313–323, 2021. https://doi.org/10.1016/j.idm.2021.01.002.

[9] J. M. Bland and D. G. Altman. The odds ratio. *British Medical Journal*, 320(7247):1468, 2000. https://doi.org/10.1136/bmj.320.7247.1468.

[10] R. Bürger, G. Chowell, and L. Y. Lara-Dı́az. Comparative analysis of phenomenological growth models for modeling epidemics. *Mathematical Biosciences and Engineering*, 16(5):4250–4273, 2019. https://doi.org/https://doi.org10.3934/mbe.2019212.

[11] J. Cai, X. Deng, J. Yang, K. Sun, H. Liu, Z. Chen, C. Peng, X. Chen, Q. Wu, J. Zou, R. Sun, W. Zheng, Z. Zhao, W. Lu, Y. Liang, X. Zhou, M. Ajelli, and H. Yu. Modeling transmission of SARS-CoV-2 omicron in China. *Nature Medicine*, 28:1468–1475, 2022. https://doi.org/10.1038/s41591-022-01855-7.

[12] X. Chen, Z. Huang, J. Wang, S. Zhao, M. C.-S. Wong, K. C. Chong, D. He, and J. Li. Ratio of asymptomatic COVID-19 cases among ascertained SARS-CoV-2 infections in different regions and population groups in 2020: A systematic review and meta-analysis including 130 123 infections from 241 studies. *BMJ Open*, 11(12):e049752, 2021. https://doi.org/10.1136/bmjopen-2021-049752.

[13] Z. Chen, X. Deng, L. Fang, K. Sun, Y. Wu, T. Che, J. Zou, J. Cai, H. Liu, Y. Wang, T. Wang, Y. Tian, N. Zheng, X. Yan, R. Sun, X. Xu, X. Zhou, S. Ge, Y. Liang, L. Yi, J. Yang, J. Zhang, M. Ajelli, and H. Yu. Epidemiological characteristics and transmission dynamics of the outbreak caused by the SARS-CoV-2 omicron variant in Shanghai, China: A descriptive study. *The Lancet Regional Health. Western Pacific*, 29:100592, 2022. https://doi.org/10.1016/j.lanwpc.2022.100592.

[14] G. Chowell. Fitting dynamic models to epidemic outbreaks with quantified uncertainty: A primer for parameter uncertainty, identifiability, and forecasts. *Infectious Disease Modelling*, 2(3):379–398, 2017. https://doi.org/10.1016/j.idm.2017.08.001.

[15] G. Chowell, D. Hincapie-Palacio, J. Ospina, B. Pell, A. Tariq, S. Dahal, S. Moghadas, A. Smirnova, L. Simonsen, and C. Viboud. Using phenomenological models to characterize transmissibility and forecast patterns and final burden of Zika epidemics. *PLOS Currents*, 8:ecurrents.outbreaks.f14b2217c902f453d9320a43a35b9583, 2016. https://doi.org/10.1371/currents.outbreaks.f14b2217c902f453d9320a43a35b9583.

[16] S. Contreras, J. Dehning, M. Loidolt, J. Zierenberg, F. P. Spitzner, J. H. Urrea-Quintero, S. B. Mohr, M. Wilczek, M. Wibral, and V. Priesemann. The challenges of containing SARS-CoV-2 via test-trace-and-isolate. *Nature Communications*, 12:378, 2021. https://doi.org/10.1038/s41467-020-20699-8.

[17] P. L. Delamater, E. J. Street, T. F. Leslie, Y. Tony Yang, and K. H. Jacobsen. Complexity of the basic reproduction number (R_0). *Emerging Infectious Diseases*, 25(1):1–4, 2019.

[18] H. Derendorf and S. Schmidt. *Rowland and Tozer's Clinical Pharmacokinetics and Pharmacodynamics: Concepts and Applications*. Wolters Kluwer, Dordrecht, Netherlands, 5 edition, 2019.

[19] G. Freeguard. The story of the R number. How an obscure epidemiological figure took over our lives. Part 1: History. *Significance*, 21(1):6–10, 2024. https://doi.org/10.1093/jrssig/qmae004.

[20] G. Freeguard. The story of the R number. How an obscure epidemiological figure took over our lives. Part 2: Modelling. *Significance*, 21(2):6–10, 2024. https://doi.org/10.1093/jrssig/qmae027.

[21] G. Freeguard. The story of the R number. How an obscure epidemiological figure took over our lives. Part 3: Media. *Significance*, 21(3):26–29, 2024. https://doi.org/10.1093/jrssig/qmae043.

[22] G. Freeguard. The story of the R number. How an obscure epidemiological figure took over our lives. Part 4: The politics of R. *Significance*, 21(4):19–22, 2024. https://doi.org/10.1093/jrssig/qmae058.

[23] G. Freeguard. The story of the R number. How an obscure epidemiological figure took over our lives. Part 5: So what did we learn? *Significance*, 21(5):12–15, 2024. https://doi.org/10.1093/jrssig/qmae072.

[24] E. R. Garrett. Simplified methods for the evaluation of the parameters of the time course of plasma concentration in the one-compartment body model with first-order invasion and first-order drug elimination including methods for ascertaining when such rate constants are equal. *Journal of Pharmacokinetics and Biopharmaceutics*, 21:689–734, 1993.

[25] N. Garrett, A. Tapley, J. Andriesen, I. Seocharan, L. H. Fisher, L. Bunts, N. Espy, C. L. Wallis, A. K. Randhawa, M. D. Miner, N. Ketter, M. Yacovone, A. Goga, Y. Huang, J. Hural, P. Kotze, L. G. Bekker, G. E. Gray, L. Corey, and Ubuntu Study Team. High asymptomatic carriage with the omicron variant in South Africa. *Clinical Infectious Diseases*, 75(1):e289–e292, 2022. https://doi.org/10.1093/cid/ciac237.

[26] A. N. Gerasimov. The spread of SARS-CoV-2 in Russia and the evolution of the properties of the pathogen: Unfulfilled hopes, unpleasant surprises and mathematical models. In J. Mau, S. Mukhin, G. Wang and S. Xu, editors, *BIOKYBERNETIKA. Mathematics for Theory and Control in the Human and in Society*, pages 479–489. Walter de Gruyter GmbH, Berlin/Boston, Mass. USA, 12 2024. https://doi.org/10.1515/9783111341996-028.

[27] AI Google. What is the significance of Shanghai for economy, finance, and industry of China? https://www.google.com/search?client=firefox-b-lm&channel=entpr&q=what+is+the+significance+of+Shanghai+for+economy%2C+finance%2C+and+industry+of+China, 2025. Accessed 01 October 2025.

[28] A. Holborow, H. Asad, L. Porter, P. Tidswell, C. Johnston, I. Blyth, A. Bone, and B. Healy. The clinical sensitivity of a single SARS-CoV-2 upper respiratory tract RT-PCR test for diagnosing COVID-19 using convalescent antibody as a comparator. *Clinical Medicine*, 20(6):e209–e211, 2020. https://doi.org/10.7861/clinmed.2020-0555.

[29] C. Jaupart. Physical models of volcanic eruptions. *Chemical Geology*, 128(1–4):217–227, 1996. https://doi.org/10.1016/0009-2541(95)00175-1.

[30] M. J. Keeling and P. Rohani. *Modeling Infectious Diseases in Humans and Animals*. Princeton University Press, Princeton, 2008. https://doi.org/10.1515/9781400841035.

[31] W. O. Kermack and A. G. McKendrick. Contributions to the mathematical theory of epidemics–i. *Bulletin of Mathematical Biology*, 53(1–2):33–55, 1927. 1991.

[32] C. C. Kerr, D. Mistry, R. M. Stuart, K. Rosenfeld, G. R. Hart, R. C. Núñez, J. A. Cohen, P. Selvaraj, R. G. Abeysuriya, M. Jastrzębski, L. George, B. Hagedorn, J. Panovska-Griffiths, M. Fagalde, J. Duchin, M. Famulare, and D. J. Klein. Controlling COVID-19 via test-trace-quarantine. *Nature Communications*, 12:2993, 2021. https://doi.org/10.1038/s41467-021-23276-9.

[33] N. L. Komarova, L. M. Schang, and D. Wodarz. Patterns of the COVID-19 pandemic spread around the world: Exponential versus power laws. *Journal of the Royal Society Interface*, 17(170), 2020. https://doi.org/10.1098/rsif.2020.0518.

[34] E. Kortela, V. Kirjavainen, M. J. Ahava, S. T. Jokiranta, A. But, A. Lindahl, A. E. Jääskeläinen, A. J. Jääskeläinen, A. J. P. Jokela, H. Kallio-Kokko, R. Loginov, L. Mannonen, E. Ruotsalainen, T. Sironen, O. Vapalahti, M. Lappalainen, H.-R. Kreivi, H. Jarva, and S. K. E. Kekäläinen. Real-life clinical sensitivity of SARS-CoV-2 RT-PCR test in symptomatic patients. *PLoS ONE*, 16(5):e0251661, 2021. https://doi.org/10.1371/journal.pone.0251661.

[35] M. Kretzschmar and J. Wallinga. Mathematical models in infectious disease epidemiology. In A. Krämer, M. Kretzschmar and K. Krickeberg, editors, *Modern Infectious Disease Epidemiology*, Statistics for Biology and Health, pages 209–221. Springer, New York, NY, 2009. https://doi.org/10.1007/978-0-387-93835-6_12.

[36] T.-M. Le and J.-P. Onnela. Connecting mass-action models and network models for infectious diseases. *PLoS Computational Biology*, 21(8):e1013373, 2025. https://doi.org/10.1371/journal.pcbi.1013373.

[37] Y. Ling, G. Lu, F. Liu, Y. Tan, X. Xu, D. Wei, J. Xu, S. Wang, S. Yu, F. Jiang, X. Zhang, S. Chen, S. Wang, X. Fan, and S. Chen. The omicron BA.2.2.1 subvariant drove the wave of SARS-CoV-2 outbreak in Shanghai during spring 2022. *Cell Discovery*, 8:97, 2022. https://doi.org/10.1038/s41421-022-00468-1.

[38] C. Liu, Y. Tian, Y. Shi, Z. Huang, and Y. Shao. An analysis of public topics and sentiments based on social media during the COVID-19 omicron variant outbreak in Shanghai 2022. *Computational Urban Science*, 4:19, 2024. https://doi.org/10.1007/s43762-024-00128-y.

[39] H. Liu, J. Cai, J. Zhou, X. Xu, M. Ajelli, and H. Yu. Assessing the impact of interventions on the major Omicron BA.2 outbreak in spring 2022 in Shanghai. *Infectious Disease Modelling*, 9(2):519–526, 2024. https://doi.org/10.1016/j.idm.2024.02.013.

[40] J. Liu, M. Liu, and W. Liang. The dynamic COVID-zero strategy in China. *CCDC Weekly*, 4(4):74–75, 2022. https://doi.org/10.46234/ccdcw2022.015.

[41] Y. R. Liyanage, G. Chowell, G. Pogudin, and N. Tuncer. Structural and practical identifiability of phenomenological growth models for epidemic forecasting. *Viruses*, 17(4):496, 2025.

[42] L. Lou, L. Zhang, J. Guan, X. Ning, M. Nie, Y. Wei, and F. Chen. Retrospective modeling of the omicron epidemic in Shanghai, China: Exploring the timing and performance of control measures. *Tropical Medicine and Infectious Disease*, 8:39, 2023. https://doi.org/10.3390/tropicalmed8010039.
[43] J. Ma. Estimating epidemic exponential growth rate and basic reproduction number. *Infectious Disease Modelling*, 5:129–141, 2020.
[44] Q. Ma, J. Liu, Q. Liu, L. Kang, R. Liu, W. Jing, Y. Wu, and M. Liu. Global percentage of asymptomatic SARS-CoV-2 infections among the tested population and individuals with confirmed COVID-19 diagnosis: A systematic review and meta-analysis. *JAMA Netw Open*, 4(12):e2137257, 2021. https://doi.org/10.1001/jamanetworkopen.2021.37257.
[45] B. Majeed, A. Li, J. Peng, and Y. Lin. A multi-period curve fitting model for short-term prediction of COVID-19 spread in the U.S. Metropolitans, *Frontiers in Public Health*, 9, 2021. https://doi.org/10.3389/fpubh.2021.809877.
[46] G. Marion, L. Hadley, V. Isham, D. Mollison, J. Panovska-Griffiths, L. Pellis, G. S. Tomba, F. Scarabel, B. Swallow, P. Trapman, and D. Villela. Modelling: Understanding pandemics and how to control them. *Epidemics*, 39:10588, 2022. https://doi.org/10.1016/j.epidem.2022.100588.
[47] J. Mau and W. Tian. Control of SARS-nCoV outbreaks in China 2020. A semiparametric analysis by first-order pharmacokinetics. In J. Mau, S. Mukhin, G. Wang and S. Xu, editors, *BIOKYBERNETIKA. Mathematics for Theory and Control in the Human and in Society*, pages 513–530. Walter de Gruyter GmbH, Berlin/Boston, Mass. USA, 12 2024. https://doi.org/10.1515/9783111341996-030.
[48] M. H. Nguyen, T. Braeye, N. Hens, and C. Faes. Multivariate phenomenological models for real-time short-term forecasts of hospital capacity for COVID-19 in Belgium from March to June 2020. *Epidemiology and Infection*, 150:e12, 2021. https://doi.org/10.1017/S0950268821002491.
[49] K. Nixon, S. Jindal, F. Parker, N. G. Reich, K. Ghobadi, E. C. Lee, S. Truelove, and L. Gardner. An evaluation of prospective COVID-19 modelling studies in the USA: From data to science translation. *The Lancet Digital Health*, 4(10):e738–e747, 2022. https://doi.org/10.1016/S2589-7500(22)00148-0.
[50] National Health Commission of the PRC. Daily Briefing, 2022.
[51] J. Oldfield and A. Stevenson. After the fire: An ecological, phenomenological exploration of resilience-building following the Fuego volcanic eruption in Guatemala. *American Journal of Community Psychology*, 74(1–2):86–99, 2024. https://doi.org/10.1002/ajcp.12748.
[52] D. P. Oran and E. J. Topol. The proportion of SARS-CoV-2 infections that are asymptomatic: A systematic review. *Annals of Internal Medicine*, 174(5):655–662, 2021. https://doi.org/10.7326/M20-6976.
[53] S. C. Owen. *Geological origins of sugarloaf mountains in eastern Brazil and their environmental significance as refugia for the Mata Atlantica rainforest*. PhD thesis, Department of Geology, 6 2014. https://figshare.le.ac.uk/articles/thesis/Geological_origins_of_sugarloaf_mountains_in_eastern_Brazil_and_their_environmental_significance_as_refugia_for_the_Mata_Atlantica_rainforest/10139930.
[54] R Core Team. *R: A Language and Environment for Statistical Computing*. R Foundation for Statistical Computing, Vienna, Austria, 2021. https://www.R-project.org/.
[55] L. Ringwelski and G. Lihs. On parametric estimation for the Bateman-function by means of the Gauss-Newton method. *Methods of Information in Medicine*, 15(4):246–251, 1976. https://doi.org/10.1055/s-0038-1635746.
[56] R. Tang, A. G. Makeev, and N. L. Semendyaeva. Agent-based modeling of epidemic spread via kinetic Monte Carlo method. In J. Mau, S. Mukhin, G. Wang and S. Xu, editors, *BIOKYBERNETIKA. Mathematics for Theory and Control in the Human and in Society*, pages 491–512. Walter de Gruyter GmbH, Berlin/Boston, Mass. USA, 12 2024. https://doi.org/10.1515/9783111341996-029.
[57] P. van den Driessche. Reproduction numbers of infectious disease models. *Infectious Disease Modelling*, 2(3):288–303, 2017. https://doi.org/10.1016/j.idm.2017.06.002.
[58] S.-T. Wang, L. Li, J. Zhang, Y. Li, X.-F. Luo, and G.-Q. Sun. Quantitative evaluation of the role of Fangcang shelter hospitals in the control of omicron transmission: A case study of the outbreak in Shanghai, China in 2022. *One Health*, 16:100475, 2023. https://doi.org/10.1016/j.onehlt.2022.100475.

[59] A. M. Wein, S. K. McBride, J. S. Becker, A. Christophersen, E. E. H. Doyle, M. C. Gerstenberger, and S. H. Potter. Long-term communication of aftershock forecasts: The Canterbury earthquake sequence in New Zealand. *International Journal of Disaster Risk Reduction*, 114:104878, 2024. https://doi.org/10.1016/j.ijdrr.2024.104878.

[60] M. Yanes-Lane, N. Winters, F. Fregonese, M. B. S. Perlman-Arrow, J. R. Campbell, and D. Menzies. Proportion of asymptomatic infection among COVID-19 positive persons and their transmission potential: A systematic review and meta-analysis. *PLoS ONE*, 15(11):e0241536, 2020. https://doi.org/10.1371/journal.pone.0241536.

[61] J. Zhang, S. Tan, C. Peng, X. Xu, M. Wang, W. Lu, Y. Wu, B. Sai, M. Cai, A. G. Kummer, Z. Chen, J. Zou, W. Li, W. Zheng, Y. Liang, Y. Zhao, A. Vespignani, M. Ajelli, X. Lu, and H. Yu. Heterogeneous changes in mobility in response to the SARS-CoV-2 Omicron BA.2 outbreak in Shanghai. *PNAS*, 120(42):e2306710120 2023. https://doi.org/10.1073/pnas.230671012.

[62] Z. Zhang. Diagnosis and treatment protocol for COVID-19 patients (trial version 9). *Health Care Science*, 1(1):14–28, 2022. https://doi.org/10.1002/hcs2.1.

[63] A. M. Ćmiel and B. Ćmiel. A simple method to describe the COVID-19 trajectory and dynamics in any country based on Johnson cumulative density function fitting. *Scientific Reports*, 11:17744, 2021. https://doi.org/10.1038/s41598-021-97285-5.

Appendix A. Daily briefings by the China National Health Commission

Table A.1: Daily updated reported numbers of new cases in Mainland China and in Shanghai megacity from March 1 to May 16, 2022, obtained from the National Health Commission of People's Republic of China [50].

Day from Feb 28	Date of received report	Mainland			Shanghai			Mainland w/o Shanghai		
		nCD-19	nAIC	nCONV	nCD-19	nAIC	nCONV	nCD-19	nAIC	nCONV
0	1	2	3	4	5	6	7	8	9	10
1	Mar 1	71	48	2	1	1	0	70	47	2
2	Mar 2	54	37	2	3	5	0	51	32	2
3	Mar 3	61	56	5	2	14	0	59	42	5
4	Mar 4	102	73	9	3	16	0	99	57	9
5	Mar 5	175	127	3	0	28	0	175	99	3
6	Mar 6	214	312	3	3	45	0	211	267	3
7	Mar 7	175	330	6	4	51	0	171	279	6
8	Mar 8	233	322	13	3	62	0	230	260	13
9	Mar 9	402	435	10	4	76	0	398	359	10
10	Mar 10	397	703	21	11	64	0	386	639	21
11	Mar 11	476	1048	5	5	78	0	471	970	5
12	Mar 12	1807	1315	114	1	64	0	1806	1251	114
13	Mar 13	1337	788	77	41	128	2	1296	660	75
14	Mar 14	3507	1647	73	9	130	0	3498	1517	73
15	Mar 15	1860	1194	53	5	197	0	1855	997	53
16	Mar 16	1226	1206	28	8	150	1	1218	1056	27
17	Mar 17	2388	1742	24	57	203	0	2331	1539	24
18	Mar 18	2157	1713	26	8	366	0	2149	1347	26
19	Mar 19	1656	2177	76	17	492	6	1639	1685	70
20	Mar 20	1947	2384	54	24	734	0	1923	1650	54
21	Mar 21	2281	2313	61	31	865	1	2250	1448	60
22	Mar 22	2591	2346	94	4	977	0	2587	1369	94
23	Mar 23	2010	2722	50	4	979	0	2006	1743	50
24	Mar 24	1301	3489	121	29	1582	0	1272	1907	121
25	Mar 25	1280	4320	72	38	2231	5	1242	2089	67
26	Mar 26	1217	4333	124	45	2631	2	1172	1702	122
27	Mar 27	1219	4996	217	50	3450	0	1169	1546	217
28	Mar 28	1228	5658	48	96	4381	21	1132	1277	27
29	Mar 29	1565	7090	66	326	5656	18	1239	1434	48
30	Mar 30	1803	6651	56	355	5298	16	1448	1353	40
31	Mar 31	1787	5442	56	358	4144	20	1429	1298	36
32	Apr 1	2086	7789	44	260	6051	2	1826	1738	42
33	Apr 2	1455	11691	118	438	7788	73	1017	3903	45
34	Apr 3	1366	11771	163	425	8581	71	941	3190	92
35	Apr 4	1173	15239	100	268	13086	4	905	2153	96
36	Apr 5	1383	19089	127	311	16766	40	1072	2323	87

Table A.1 (continued)

Day from Feb 28	Date of received report	Mainland			Shanghai			Mainland w/o Shanghai		
		nCD-19	nAIC	nCONV	nCD-19	nAIC	nCONV	nCD-19	nAIC	nCONV
37	Apr 6	1286	21711	95	322	19660	15	964	2051	80
38	Apr 7	1540	22561	412	824	20398	323	716	2163	89
39	Apr 8	1334	23737	491	1015	22609	420	319	1128	71
40	Apr 9	1318	25037	306	1006	23937	191	312	1100	115
41	Apr 10	1164	26345	90	914	25173	47	250	1172	43
42	Apr 11	1251	23295	335	994	22348	273	257	947	62
43	Apr 12	1500	26420	152	1189	25141	23	311	1279	129
44	Apr 13	2999	26318	344	2573	25146	114	426	1172	230
45	Apr 14	3472	20694	404	3200	19872	307	272	822	97
46	Apr 15	3867	20813	985	3590	19923	922	277	890	63
47	Apr 16	3504	22512	1224	3238	21585	1177	266	927	47
48	Apr 17	2723	20639	910	2417	19831	853	306	808	57
49	Apr 18	3297	18187	1015	3084	17332	974	213	855	41
50	Apr 19	2753	17066	616	2494	16407	533	259	659	83
51	Apr 20	2830	16552	500	2634	15861	459	196	691	41
52	Apr 21	2119	16383	202	1931	15698	143	188	685	59
53	Apr 22	2971	21355	1217	2736	20634	1120	235	721	97
54	Apr 23	1566	20230	577	1401	19657	541	165	573	36
55	Apr 24	2666	17528	899	2472	16983	846	194	545	53
56	Apr 25	1908	15816	995	1661	15319	968	247	497	27
57	Apr 26	1818	12404	1308	1606	11956	1253	212	448	55
58	Apr 27	1494	9791	934	1292	9330	858	202	461	76
59	Apr 28	5646	9942	5125	5487	9545	5062	159	397	63
60	Apr 29	1410	9293	1013	1249	8932	985	161	361	28
61	Apr 30	916	7340	703	788	7084	683	128	256	20
62	May 1	846	6895	566	727	6606	529	119	289	37
63	May 2	368	5647	164	274	5395	155	94	252	9
64	May 3	353	5075	159	260	4722	151	93	353	8
65	May 4	360	4678	190	261	4390	185	99	288	5
66	May 5	356	4272	189	245	4024	181	111	248	8
67	May 6	345	4275	185	253	3961	175	92	314	10
68	May 7	319	4065	141	215	3760	135	104	305	6
69	May 8	401	3859	236	322	3625	230	79	234	6
70	May 9	349	3077	172	234	2780	156	115	297	16
71	May 10	302	1545	210	228	1259	198	74	286	12
72	May 11	222	1630	111	144	1305	106	78	325	5
73	May 12	312	2140	176	227	1689	167	85	271	9
74	May 13	253	1726	148	194	1487	140	59	239	8
75	May 14	226	1492	120	166	1203	111	60	289	9
76	May 15	140	1019	53	69	869	42	71	150	11
77	May 16	162	887	58	77	746	46	85	141	12
78	May 17	227	1000	79	96	759	56	131	241	23
79	May 18	191	825	57	82	637	48	109	188	9
80	May 19	176	939	82	88	770	71	88	169	11

Table A.1 (continued)

Day from Feb 28	Date of received report	Mainland			Shanghai			Mainland w/o Shanghai		
		nCD-19	nAIC	nCONV	nCD-19	nAIC	nCONV	nCD-19	nAIC	nCONV
81	May 20	181	951	66	84	784	49	97	167	17
82	May 21	157	667	45	52	570	29	105	97	16
83	May 22	174	628	56	55	503	30	119	125	26
84	May 23	141	498	56	58	422	39	83	76	17
85	May 24	102	418	39	44	343	32	58	75	7
86	May 25	104	356	44	48	290	31	56	66	13
87	May 26	80	274	36	45	219	33	35	55	3
88	May 27	71	193	20	39	131	18	32	62	2
89	May 28	54	161	21	29	93	18	25	68	3
90	May 29	20	102	3	6	61	2	14	41	1
91	May 30	28	69	3	9	22	2	19	47	1
92	May 31	22	46	3	5	10	1	17	36	2
93	Jun 1	18	43	4	5	8	3	13	35	1
94	Jun 2	20	54	2	8	8	2	12	46	0
95	Jun 3	21	55	3	5	9	2	16	46	1
96	Jun 4	33	55	3	6	16	1	27	39	2
97	Jun 5	25	61	1	4	4	0	21	57	1
98	Jun 6	39	85	14	3	7	0	36	78	14
99	Jun 7	44	80	1	4	11	0	40	69	1
100	Jun 8	53	111	10	4	5	0	49	106	10
101	Jun 9	30	43	2	6	5	0	24	38	2
102	Jun 10	65	73	1	7	9	0	58	64	1
103	Jun 11	122	74	52	10	19	0	112	55	52
104	Jun 12	69	74	2	11	26	1	58	48	1
105	Jun 13	60	67	2	3	14	0	57	53	2
106	Jun 14	54	65	9	13	2	1	41	63	8
107	Jun 15	42	38	4	9	7	0	33	31	4
108	Jun 16	23	41	2	2	2	0	21	39	2
109	Jun 17	11	41	1	1	6	0	10	35	1
110	Jun 18	5	22	0	3	6	0	2	16	0
111	Jun 19	14	10	0	10	3	0	4	7	0
112	Jun 20	9	25	0	6	3	0	3	22	0
113	Jun 21	10	12	0	4	4	0	6	8	0
114	Jun 22	13	18	0	9	0	0	4	18	0
115	Jun 23	18	26	13	0	0	0	18	26	13
116	Jun 24	7	15	1	0	0	0	7	15	1
117	Jun 25	2	12	0	0	0	0	2	12	0
118	Jun 26	5	18	0	2	2	0	3	16	0
119	Jun 27	1	21	0	0	0	0	1	21	0
120	Jun 28	3	36	0	0	0	0	3	36	0
121	Jun 29t	8	31	1	0	0	0	8	31	1
122	Jun 30	12	151	0	0	0	0	12	151	0
123	Jul 1	38	145	1	0	0	0	38	145	1
124	Jul 2	75	310	2	0	0	0	75	310	2

Daily numbers for the Mainland without Shanghai and daily numbers of new COVID-19 cases without conversion from previous AIC, nCD-19*, were obtained by respective subtractions.

Appendix B. Mass screening design and analysis

B.1 Design of mass screening

At each specific screening day T_i, every person from the defined screen population of Shanghai megacity will be in exactly one of the four states. By the rapidly evolving dynamics of the epidemic, the current state may have changed from the previous screening day's status: for a previously susceptible person in state "S" to new state "nAIC" or "nCD-19*" when now found asymptomatically infected or found already clinically manifest with COVID-19, respectively. Or, an individual had previously been tested positive for SARS-CoV-2 infection, though without clinical symptoms of COVID-19 at that time, and is now seen with clinical manifestation of COVID-19, hence changing from previous state "AIC" to new state "CONV."

At every day of mass screening, different districts had been selected leading to variable subpopulations involved—denote these screens as sets T_i of individuals, number i for the different days, and denote their cardinality (size) as T_i, $T_i = |T_i|$. Denote individual post-testing outcomes as "labels"; see Table B.1.

Table B.1: Screening test outcomes imply "labels" assigned to every screened individual according to the State Diagram in Figure 12.1.

True Pretest State	Population prevalence	Possible Post-test Labels			
		"NON"	"CD-19*"	"AIC"	"CD-19"
Susceptible	π_0	cNON	nCD-19*	nAIC	nCONV
COVID-19 without prior AIC	π_1	fNON	cCD-19*	–	–
Asymptomatic Infected	π_2	fNON	–	cAIC	–
COVID-19 with prior AIC	π_3	fNON	–	–	cCONV

Legend: Prefix "n" denotes diagnostic classification as "new case," prefix "c" denotes no change from previous testing, "f" denotes false negative test outcome, "-" practically not possible (except record documentation mistakes).

Denote the true proportions of the four categories in screen T_i by Latin p_0^i, p_1^i, p_2^i, p_3^i, according to district. Each individual tested in Shanghai on a specific day will have a post-testing status as either test-negative, denoted as nNON, or as test-positive and classified as either a new case of an asymptomatic infection with SARS-CoV-2, nAIC, or a new case of clinical COVID-19, nCD-19, either without previously tested positive, nCD-19*, or as new cases of conversion, nCONV, from previously detected asymptomatic infected case; cf. data-raising flow diagram in Figure B.1.

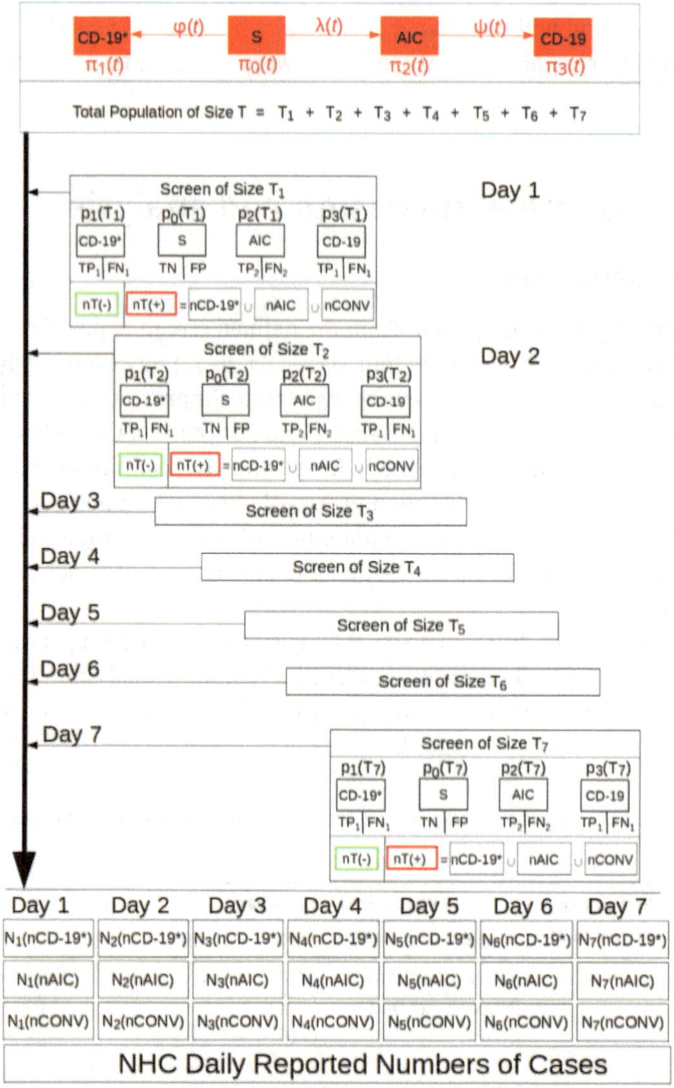

Figure B.1: Data-collection flow chart for mass screening in Shanghai during 7 days, with different samples in daily screens. CD-19*: previously asymptomatic infected status COVID-19; S: not infected (PCR negative); AIC: asymptomatic infected (PCR positive); CD-19: COVID-19 from previous AIC; $\phi(t), \lambda(t), \psi(t)$ intensities of status change in short time interval $]t, t + dt]$; $\pi_i(t), i = 0, 1, 2, 3$ prevalences of S, CD-19*, AIC, CD-19, respectively, in screened population at time t; $p_i(T_j)$ respective prevalences in screen T_j on day j; TP_i true-positive, FN_i false-negative, FP false-positive, TN true-negative as diagnostic test-outcome possibilities; n(T-) and n(T+): daily number of new test-negative and test-positive cases; N(nCD-19*), N(nAIC), N(nCONV) daily number of new diagnoses of CD-19*, AIC, and CONV, respectively.

Table B.2: Concepts of true and false "labeling" by diagnostic tests in mass screening T_i (cf. Figure B.1) with probabilities of occurrence according to each tested individual's unknown true state.

	Screenee's true state by State Diagram Figure 12.1			
	CD-19*	**Susceptible**	**AIC**	**CD-19**
Population prevalences in screen T_i	$p_1(T_i)$	$p_0(T_i)$	$p_2(T_i)$	$p_3(T_i)$
Diagnostic test outcome positive ("T+")	TP	FP	TP	TP
Probability of "T+" given true state	se_1^i	$1 - sp_0^i$	se_2^i	se_3^i
Diagnostic test outcome negative ("T-")	FN	TN	FN	FN
Probability of "T-" given true state	$1 - se_1^i$	sp_0^i	$1 - se_2^i$	$1 - se_3^i$

Legend: TP = true positive, FP = false-positive, FN = false-negative, TN = true-negative diagnostic finding for an individual's test result. Diagnostic sensitivities se and specificities sp defined as conditional probabilities, given a true state, defined in (B.1).

B.2 Analysis of daily numbers

The data-raising flow diagram in Figure B.1 exhibits a bias problem: population prevalences $\pi_0, \pi_1, \pi_2, \pi_3$ of states NON, CD-19*, AIC, CONV, respectively, will vary from day-to-day, by the dynamic spread of virus infections in the population as a whole, and in the residential districts selected for mass screening in weekly rotation, as well.

Any partial screen T_i can catch only from the T_i-predominant presampling prevalences $p_0(T_i)$, $p_1(T_i)$, $p_2(T_i)$, $p_3(T_i)$ of its assigned residential district. For visual simplicity, change to a notation that indicates the i-th screen T_i with superscript i: $p_0^i = p_0(T_i)$, $p_1^i = p_1(T_i)$, $p_2^i = p_2(T_i)$, $p_3^i = p_3(T_i)$, here and in general. Note that screen T_i sets a name, while its size is a variable, hence T_i denotes a number that equals the size of T_i.

As diagnostic tests have their well-known errors, also in SARS-CoV-2 detection, post-test classification of screened individuals as either NON, CD-19*, AIC, or CONV depends on detectable virus load, cooperation, and manual swab-practice, which render nominal sensitivities and specificities daily at variance from screen-to-screen [28, 34].

With (diagnostic) *sensitivity* as the conditional probability that any specific case in Figure 12.1, CD-19*, AIC, and CD-19 is tested positive with the applied test, and with (diagnostic) *specificity* for the conditional probability that any susceptible "S" is correctly tested as negative, with $\mathbb{P}$ for probability and $\mathbb{P}\{\ |A\}$ for conditional probability given realization of an event A. Use TP for "true positive" and TN for "true negative" test outcomes:

$$\left.\begin{aligned} se_1 &= \mathbb{P}\{\text{test–positive} \mid \text{nCD–19}*\} \\ se_2 &= \mathbb{P}\{\text{test–positive} \mid \text{nAIC}\} \\ se_3 &= \mathbb{P}\{\text{test–positive} \mid \text{nCD–19}\} \\ sp_0 &= \mathbb{P}\{\text{test–negative} \mid \text{S}\} \end{aligned}\right\} \tag{B.1}$$

Hence, daily reported numbers of new clinically symptomatic cases of COVID-19, CD-19*, according to the population background on day i arise with expected values by screen size T_i, current prevalence p_1^i of CD-19* in the i-th screen and diagnostic sensitivity se_1^i for CD-19* by swab sampling

$$\mathbb{E}\{N^i(\text{nCD}-19*)\} = p_1^i\,\mathrm{se}_1^i\,T_i \tag{B.2}$$

and analogously, for daily reported numbers of new asymptomatic infected cases (AIC)

$$\mathbb{E}\{N^i(\text{nAIC})\} = p_2^i\,\mathrm{se}_2^i\,T_i \tag{B.3}$$

and for daily reported numbers of new conversions (CONV) from a previously asymptomatic infected case (AIC)

$$\mathbb{E}\{N^i(\text{nCONV})\} = p_3^i\,\mathrm{se}_3^i\,T_i; \tag{B.4}$$

correspondingly for the daily number of new "noncases," which consists of true-negative from susceptibles S and false-positives from true CD-19*, AIC, and CD-19; cf. Tables B.1 and B.2,

$$\left.\begin{aligned}&\mathbb{E}\{N^i(\text{NON})\}\\ &\quad = T_i\{p_0^i\,\mathrm{sp}_0^i + p_1^i(1-\mathrm{sp}_1^i) + p_2^i(1-\mathrm{sp}_2^i) + p_3^i(1-\mathrm{sp}_3^i)\}\end{aligned}\right\} \tag{B.5}$$

Taking ratios N(nAIC) over N(nCD-19*), e. g., eliminates screen sizes T_i as a factor,

$$\frac{\mathbb{E}\{N^i(\text{nAIC})\}}{\mathbb{E}\{N^i(\text{nCD}-19*)\}} = \frac{p_2^i\mathrm{se}_2^i}{p_1^i\mathrm{se}_1^i} \tag{B.6}$$

$$\frac{\mathbb{E}\{N^i(\text{nCD}-19*)\}}{\mathbb{E}\{N^i(\text{nCONV})\}} = \frac{p_1^i\mathrm{se}_1^i}{p_3^i\mathrm{se}_3^i} \tag{B.7}$$

Under assumption of screen-to-screen, $i < j$, say, invariance for ratios of sensitivities $\mathrm{se}_2/\mathrm{se}_1$,

$$\frac{\mathrm{se}_2^j}{\mathrm{se}_1^j} = \frac{\mathrm{se}_2^i}{\mathrm{se}_1^i} \tag{B.8}$$

the ratio of prevalence odds

$$\frac{p_2^j}{p_1^j} : \frac{p_2^i}{p_1^i} \tag{B.9}$$

becomes suitable for screen-to-screen comparisons of prevalence odds nAIC to nCD-19* according to (B.6). Taking logarithms on the ratios,

$$\log \frac{\mathbb{E}\{N^j(\mathrm{nAIC})\}}{\mathbb{E}\{N^j(\mathrm{nCD{-}19}*)\}} - \log \frac{\mathbb{E}\{N^i(\mathrm{nAIC})\}}{\mathbb{E}\{N^i(\mathrm{nCD{-}19}*)\}} = \log \frac{p_2^j}{p_1^j} - \log \frac{p_2^i}{p_1^i} \tag{B.10}$$

and substituting the observable (sampled) daily numbers of new cases for their expectations, like $N^j(\mathrm{nAIC})$ for $\mathbb{E}\{N^j(\mathrm{nAIC})\}$, yields the difference of empirical (sample) *logits* in numerator and denominator,

$$\log \frac{N^j(\mathrm{nAIC})}{N^j(\mathrm{nCD{-}19}*)} - \log \frac{N^i(\mathrm{nAIC})}{N^i(\mathrm{nCD{-}19}*)} \approx \log \frac{p_2^j}{p_1^j} - \log \frac{p_2^i}{p_1^i}. \tag{B.11}$$

One further step yields the *logistic differences* for comparison of logits between screen T_i and T_j,

$$\left.\begin{aligned} \log N^j(\mathrm{nAIC}) - \log N^j(\mathrm{nCD{-}19}*) \\ - \log N^i(\mathrm{nAIC}) + \log N^i(\mathrm{nCD{-}19}*) \\ \approx \log\left(\frac{p_2^j}{p_1^j}\right) - \log\left(\frac{p_2^i}{p_1^i}\right) \end{aligned}\right\} \tag{B.12}$$

which reveals, that taking differences between logarithms of numbers of daily new cases of AIC and of numbers of daily new cases of CD-19* in screen T_i and doing this again in screen T_j, leads to valid comparisons of differences between respective logits of underlying population prevalences. Though estimation of individual prevalences, i. e., p_1^i, p_2^i, is not possible in the Shanghai 2022 sampling design, the relation in (B.12) still justifies logarithmic transformations of daily numbers of new cases and comparisons of vertical distances on the linear scales between curves in Figure 12.2. The intuitive reasoning remains valid for T_i and T_j from different regions as, e. g., in contrasts of Shanghai to Mainland China without Shanghai.

The same reasoning is valid for comparison of occurrences new CD-19* and new CD-19 (conversions from previous AIC) according to (B.7).

Remark B.1. The multinomial screen outcomes are NON, nAIC, nCD-19*, nCD-19 (nCONV) for screen negative and screen positive new cases of AIC, COVID-19 without prior AIC, and conversions from AIC, with unknown probabilities $0 < p_0 < 1$, $0 < p_1 < 1$, $0 < p_2 < 1$, $0 < p_3 < 1$, respectively, and $p_0 + p_1 + p_2 + p_3 = 1$.

Multinomial odds are commonly defined with respect to a reference category, here NON:

$$\mathrm{odds}_{\mathrm{NON}}^{\mathrm{AIC}} = \frac{p_2}{p_0} \wedge \mathrm{odds}_{\mathrm{NON}}^{\mathrm{CD{-}19}*} = \frac{p_1}{p_0} \Rightarrow \frac{p_2}{p_1} = \mathrm{OR}_{\mathrm{CD{-}19}*}^{\mathrm{AIC}} \tag{B.13}$$

revealing the prevalence odds in (B.9) as *multinomial odds ratios* with respect to reference category NON of screen negative examination outcome. However, as screen sizes $T = \text{NON} + \text{nAIC} + \text{nCD–19}* + \text{nCONV}$ are unknown to us, we withdraw from the multinomial data model for a stratification by changes of state, the transitions in the state diagram Figure 12.1, and argue conditionally on the selected pairing of screen outcomes, with *prevalent odds* justified by

$$\frac{p_2}{p_1} = \frac{p_2}{p_2 + p_1} : \frac{p_1}{p_2 + p_1} \tag{B.14}$$

Appendix C. Models

C.1 The Pharmacokinetic model equations

The one-compartment pharmacokinetic (PK) model with first-order kinetics

[18]

$$\boxed{\text{D}} \xrightarrow{k_a} \boxed{\text{C}} \xrightarrow{k_e} \boxed{\text{E}}$$

Definitions

$D(t)$ Depot: amount of drug available for absorption at time t,
$C(t)$ Circulation: amount of drug in blood circulation at time t,
$E(t)$ Elimination: amount of drug eliminated from circulation by time t.

Kinetic parameters

k_a constant absorption rate for passage from depot to circulation
k_e constant elimination rate for passage from circulation to other body compartments

Dynamics in force

$$\left.\begin{aligned} \dot{D}(t) &= -k_a D(t) \\ \dot{C}(t) &= k_a D(t) - k_e C(t). \end{aligned}\right\} \tag{C.1}$$

Solving these for an initial depot size $D = D(0)$,

$$C(t) = D\frac{k_\text{a}}{k_\text{a} - k_\text{e}}[e^{-k_\text{e}t} - e^{-k_\text{a}t}] \tag{C.2}$$

gives the Bateman function in (12.15).

C.2 The SIR epidemic equations

$$\boxed{\mathrm{S}} \xrightarrow{\alpha} \boxed{\mathrm{I}} \xrightarrow{\mu} \boxed{\mathrm{R}}$$

see Table C.1 for mass balances.[3]

Table C.1: Mass balances in the SIR model.

"Compartment"	"Flux In"	"Flux Out"	"Rate Parameters"
"S"	–	$-\alpha SI$	α
"I"	αSI	$-\mu I$	μ
"R"	μI	–	–

Definitions

$S(t)$: Number of susceptible individuals, not yet exposed to the pathogen, at time t,
$I(t)$: Number of infectious individuals (capable of transmitting disease) at time t,
$R(t)$: Number of individuals removed from infectious status by recovery at time t.

Kinetic parameters

α constant infection rate for a susceptible
μ constant recovery rate for an infected

Dynamics in force

$$\left.\begin{aligned} \dot{S}(t) &= -\alpha S(t)I(t), \\ \dot{I}(t) &= \alpha S(t)I(t) - \mu I(t), \\ \dot{R}(t) &= \mu I(t). \end{aligned}\right\} \tag{C.3}$$

The system (C.3) is a mass-action dynamic system from chemical kinetics and appropriate under a "well mixing" assumption, when infected move freely among the susceptibles and every susceptible has the same probability to get into contact with an infected for transmission of the pathogen; hence, never applicable in a human society framework.

For more realistic scenarios, however, more structure is introduced with a decomposition of the attack rate $\alpha = cp$ (cf. Figure C.1) such that new cases occur in $]t, t + \mathrm{d}t]$ according to

$$\mathrm{d}I(t) = cpI(t)S(t)\mathrm{d}t, \tag{C.4}$$

with c for contact rate of infectious individual with a susceptible and p for the conditional probability of transmission upon a contact.

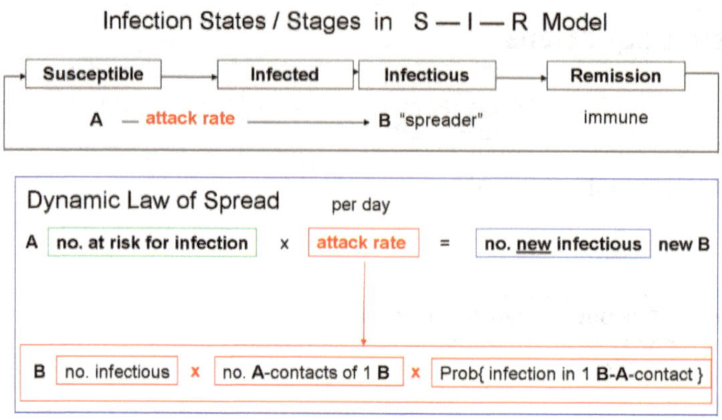

Figure C.1: The SIR model with three states, "S" for immunologically naïve "susceptibles" who may or may not enter state "I" of infected and—after an incubation period—become infectious with possibility to infect one or more susceptibles, before they finally are "removed" or in "remission" by immunity against reinfection. Dynamics of transitions between the states are driven by an "attack rate" that involves an (expected) number of contacts of susceptibles per one infectious individual and the probability of (an effective) transmission from one to the other. From [47].

Under special conditions of permanent mass screening and immediate isolation of infected individuals, (C.3) reduces to (C.1) such that it can be solved explicitly

$$I(t) = S_0 \frac{\mu}{\mu - \alpha}\left[e^{-\mu t} - e^{-\alpha t}\right] \tag{C.5}$$

for an initial number $S_0 = S(0)$ of susceptibles, yielding (C.2).

C.3 The SEIR epidemic equations

$$\boxed{\text{S}} \xrightarrow{\beta} \boxed{\text{E}} \xrightarrow{\sigma} \boxed{\text{I}} \xrightarrow{\gamma} \boxed{\text{R}}$$

see Table C.2 for mass balances. [30]

Table C.2: Mass balances in the SEIR model.

"Compartment"	"Flux In"	"Flux Out"	"Rate Parameters"
"S"	–	$-\beta SI$	β
"E"	βSI	$-\sigma E$	σ
"I"	σE	$-\gamma I$	γ
"R"	γI	–	–

Definitions

$S(t)$: Number of susceptible individuals, not yet exposed to the pathogen, at time t,
$E(t)$: Number of individuals exposed to the pathogen (infected but not yet infectious) at time t,
$I(t)$: Number of infectious individuals (capable of transmitting disease) at time t,
$R(t)$: Number of individuals removed from risk set by recovery or immunity at time t.

Dynamics in force

$$\left.\begin{aligned} \dot{S}(t) &= -\beta S(t)I(t), \\ \dot{E}(t) &= \beta S(t)I(t) - \sigma E(t) \\ \dot{I}(t) &= \sigma E(t) - \gamma I(t) \\ \dot{R}(t) &= \gamma I(t). \end{aligned}\right\} \tag{C.6}$$

This system cannot be solved analytically, in general.

C.4 The four-state model

The State Diagram in Figure 12.1 (top) for analysis of reported data of daily new cases listed in Table A.1 is

$$\boxed{I^*} \xleftarrow{\phi} \boxed{S} \xrightarrow{\lambda} \boxed{E} \xrightarrow{\psi} \boxed{I}$$

see Table C.3 for mass balances.

Table C.3: Mass balances in the four-state model.

"Compartment"	"Flux In"	"Flux Out"	"Rate Parameters"
"S"	–	$-(\phi + \lambda)S$	ϕ, λ
"I*"	ϕS	–	ϕ
"E"	λS	$-\psi I$	λ, ψ
"I"	ψE	–	–

Definitions

$S(t)$ Number of susceptible individuals, not yet exposed to the pathogen, at time t,
$I^*(t)$ number of infected individuals with clinical COVID-19 by time t, without prior diagnosis of infection,
$E(t)$ number of "Asymptomatically Infected Cases" (AIC) by time t,
$I(t)$ number of individuals with clinical COVID-19 by time t, after prior diagnosis of infection.

Kinetic parameters

- ϕ primary elimination rate for direct transitions from S to I,
- λ infection rate for transition from S to E,
- ψ secondary elimination (conversion) rate for transition from E to I by time t, given in status E just before.

Dynamics in force (Kolmogorov forward equations)

$$\left.\begin{aligned} \dot{S}(t) &= -(\phi + \lambda)S(t), \\ \dot{E}(t) &= \lambda S(t) - \psi E(t) I(t), \\ \dot{I}(t) &= \phi S(t) + \psi E(t), \end{aligned}\right\} \tag{C.7}$$

with initial conditions $S(0) = N, E(0) = 0, I(0) = 0$.

C.4.1 State-occupancy oriented parametric model

$$\left.\begin{aligned} S(t) &= S_0 e^{-(\phi+\lambda)t}, \\ E(t) &= \lambda S_0 \frac{1}{\psi - (\phi + \lambda)} [e^{-(\phi+\lambda)t} - e^{-\psi t}], \\ I(t) &= S_0 - S(t) - E(t), \end{aligned}\right\} \tag{C.8}$$

for $\psi \neq (\phi + \lambda)$.

C.4.2 Transition intensities of the four-state diagram

When evolving transition intensities $\phi(t), \lambda(t), \psi(t)$ are of interest, nonparametric analysis of their cumulative intensity (or hazard) functions is particularly simple using the daily numbers of new cases, listed in Table A.1, for the increments in $]t, t + dt]$:

$$\left.\begin{aligned} \Phi(t) &= \int_0^t \phi(s)\mathrm{d}s, \quad \mathrm{d}\hat{\Phi}(t) = \frac{N(\text{nCD19*})(t)}{Y_\phi(t-)}, \\ \Lambda(t) &= \int_0^t \lambda(s)\mathrm{d}s, \quad \mathrm{d}\hat{\Lambda}(t) = \frac{N(\text{nAIC})(t)}{Y_\lambda(t-)}, \\ \Psi(t) &= \int_0^t \psi(s)\mathrm{d}s, \quad \mathrm{d}\hat{\Psi}(t) = \frac{N(\text{nCONV})(t)}{Y_\psi(t-)}, \end{aligned}\right\} \tag{C.9}$$

where $Y_\nu(t-)$ denotes the pre-t current number at risk for transition $\nu \in \{\phi, \lambda, \psi\}$. For $Y_\phi(t-)$, $Y_\lambda(t-)$, these numbers are obtained by successive daily subtraction of reported numbers of new cases from total number at risk; for $Y_{\psi}(t-)$, this number depends on prior AIC; it rises with accumulating numbers and shrinks with decumulating numbers of AIC available.

For interpretation of results in Figure C.2 right, understand t-current risks as the t-current slopes to the curves in Figure C.2; see [2] for details of this and other statistical tools.

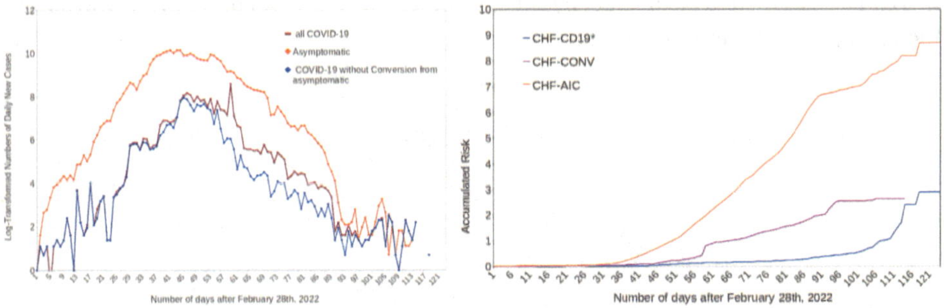

Figure C.2: Left: Log-transformed daily numbers of new cases. Right: Cumulative intensities of status change in the four-state diagram in Figure 12.1 from March 1 to May 26, 2022, in the Shanghai 2022 omicron outbreak. Middle [magenta] curve for intensity ψ of transition AIC $\longrightarrow$ CD-19 of conversions from previous asymptomatic, lower curve [blue] for intensity ϕ of transition Susceptible $\longrightarrow$ CD-19* of primary COVID-19 without previous asymptomatic infection, and the upper [red] curve of transition susceptible $\longrightarrow$ AIC of asymptomatic infections. Comparison of slopes at any common time point corresponds to the comparison of current risks of transition.

With the daily reported numbers of new cases, the four-state diagram in Figure 12.1 does not lend itself immediately to statistical methods from life-table analyses, because numbers of individuals at risk are not known. We therefore take the total number of detected cases during the Shanghai 2022 omicron outbreak as the size of the reservoir, that unknown set of susceptibles who could be revealed as a case.

Such state diagrams with a "source" state, one transient state, the AIC, and two absorbing states, clinical signs of COVID-19, either with or without previously detected asymptomatic infection, can be analyzed with methods based on the theory of counting processes [1]. The simplest among these is the robust estimation of the cumulative transition intensity between two adjacent states, e. g.,

$$\Psi(t) = \int_0^t \psi(s) \mathrm{d}s \tag{C.10}$$

with increments

$$\mathrm{d}\Psi(t) = \psi(t)\mathrm{d}t \tag{C.11}$$

for interpretation of ψ at time t as the t-current slope of the magenta color curve in Figure C.2; see [2] for details of this and other statistical tools.

The predominant information to be drawn from the three estimated cumulative transition intensities are the differences in rates: for conversion, the rate is fairly constant after mid-April (day 46); the same holds for the occurrence of new cases of asymptomatic infections from about the same date until day 90 when the rate drops. The rate of new cases of primary COVID-19, nCD-19*, rises sharply after 112 days, mid-May 2022; then this can be ascribed to an artefact by small numbers at risk, from mid-May onwards, less then 500 detected cases remained.

Part II-B.2: Civilization

Part II-B.3: Geography

Rationale: 6

Part II-B.2: **Civilization**

This class includes
- urbanization and industrialization
- occupational hazards
- ocean, wetland, farmland, forest economies
- exploitation of physical natural resources
- pollution

In this volume:
- A model of territorial development of cities as autowave self-organization of conjugated natural and anthropogenic subsystems
- Game-theoretic optimization in spatial control of invasive plants
- Enhancing maximum sustainable yield in multisite single-species and predator–prey fisheries

In former volume BIOKYBERNETIKA:
- Environmental modeling of pesticide exposure

Part II-B.3: **Geography**

This class includes habitual exposures to
- air
- water
- soil
- seas and oceans
- altitude
- radiation

In this volume: No contributions are available
In former volume BIOKYBERNETIKA: No contributions are available.

https://doi.org/10.1515/9783112216620-018

Alla Sidorova and Natalia Levashova

A model of territorial development of cities as autowave self-organization of conjugated natural and anthropogenic subsystems

Abstract: We propose a spatiotemporal model of urban ecosystem development that represents a city as a superposition of conjugated active media—the anthropogenic (activator) and natural (inhibitor) subsystems—explicitly accounting for spatial heterogeneity in both. Urban ecosystems are characterized by high population growth and density due to the concentration of residential, industrial, commercial, and transport infrastructure. These conditions lower the "buffer capacity" of natural subsystems, increase nonlinearity, and hence amplify the instability of system processes. The mathematical core of the model is a FitzHugh–Nagumo-type reaction–diffusion system modified to incorporate spatially heterogeneous activator–inhibitor interactions. Analytical results for such systems, together with numerical solutions in two spatial dimensions on rectangular domains with homogeneous Neumann boundary conditions, allow us to reproduce propagating autowave fronts and stationary contrast structures at interfaces between built-up areas and natural geobiocenoses. The model is validated against statistical and cartographic data for New Moscow and Shanghai (2017–2024). The developed model will enable the creation of optimal conditions for the formation of ecocities for human habitation and the preservation of essential biodiversity.

Keywords: Active media, autowave self-organization, urban ecosystem, reaction–diffusion, FitzHugh–Nagumo

MSC 2020: 35B36, 92D40, 35Q92

1 Introduction

1.1 Models of urban ecosystem development

By the mid-20th century, a serious global problem had emerged—the progressive growth of cities. According to [33], in 2010 the level of urbanization exceeded 50 %, and by 2050,

Acknowledgement: This study was conducted under the state assignment of Lomonosov Moscow State University

Alla Sidorova, Department of Biphysics Faculty of Physics, Lomonosov Moscow State University, Leninskie gory 1-2, 119991 Moscow, Russia, e-mail: sky314bone@mail.ru
Natalia Levashova, Department of Mathematics Faculty of Physics, Lomonosov Moscow State University, Leninskie gory 1-2, 119991 Moscow, Russia, e-mail: levashovant@physics.msu.ru

https://doi.org/10.1515/9783112216620-019

70 % of the Earth's population will live in urbanized areas [34]. This process is primarily associated with the formation of structural spatial heterogeneities in the distribution of population size and density, and consequently, with anthropogenically transformed urban geobiocenoses (construction, surface and underground utilities, roads, etc.).

For many years, the primary task in modeling urban ecosystems has been to consider cities as socioeconomic systems. The first recognized model of urban development is Forrester's simulation model [10], which analyzes the influence of population dynamics on investment and capital stock levels. In recent decades, socioeconomic models have been developed to assess the impact of urban growth on ecosystem services [19, 26, 7] such as waste water treatment, groundwater reserves, urban vegetation cover and recreational loads, land use, urban infrastructure, and landscape features.

Cellular automata models are widely used for spatiotemporal modeling of urban ecosystems. Using this method, predictive models have been developed for the growth of various cities worldwide, e. g., the territories of the Isfahan metropolis, Iran [1], Xiamen on the south-east coast of China [13], Toronto [35], Changshacity [40]. In a number of studies, urban growth modeling combines cellular automata theory with neural networks, e. g., to study the dynamics of Dongguan City, China, with forecasts up to 2024 [42, 24]. Overall, these methods describe urban development in terms of spatial and socioeconomic structure, but they do not account for the nonlinear effects arising from the interaction of multiple anthropogenic and natural factors within the system boundaries.

Most studies emphasize that the main drivers of urbanization affecting landscape changes, biogeochemical and hydrological cycles, trophic networks, and population dynamics are population size and density, as a result of complex socio-economic processes of urban development [14, 37, 17]. In most studies, population size and density are considered as the main driving forces of urbanization, influencing landscape changes, biogeochemical and hydrological cycles, food webs and population dynamics as a result of complex socioeconomic processes of urban development.

However, the lack of consideration of feedback effects between the city's social system and the natural ecosystem has a negative impact on modeling processes characteristic of cities [26]. Therefore, it is evident that the key concept in modeling urban ecosystems should be based on self-organization studies of urban ecosystems as coupled natural and anthropogenic subsystems.

1.2 Features of the interaction of anthropogenic and natural factors within the boundaries of urban ecosystems

An urban ecosystem (UES) is the ecosystem of a city, including urban geobiocenoses and anthropogenic components associated with residential and industrial development. The study of urban ecosystem development patterns makes it possible to create new methods for organizing urban space, ensuring the efficient and balanced development of all spheres of life within the UES, preserving the biodiversity necessary for the sustainable development of urban geobiocenoses, and predicting the evolution of these areas.

Let us give some examples of the interaction of anthropogenic and natural factors within the boundaries of urban ecosystems.

1.2.1 Anthropogenic factors affecting the lithosphere and hydrosphere in urban areas

UESs are characterized by changes in the physicochemical and biological parameters of geobiocenoses, increased fragmentation of anthropogenically transformed geobiocenoses, and significant alterations to trophic networks, which are due to multidirectional, overlapping anthropogenic impacts generated by spatially connected industrial and residential facilities.

Noise and vibration fields in some cases generate irreversible processes, as they penetrate the upper lithospheric layers. Also, these factors intensify karst-suffosis processes and reduce the biochemical activity of the biota.

Thermal loads

A characteristic feature of large urban agglomerations is the alteration of the thermal regime down to depths of 100–300 m: the temperature of rocks and groundwater often exceeds the background by 10–40 °C, which disrupts the physicochemical properties of the environment and reduces the rate of biodegradation (an increase in the temperature of the upper lithospheric layer by 10 °C accelerates the decomposition of organic matter by bacteria, fungi, and protozoa by a factor of 1.5–2). If the heat source is located near the groundwater level, an expansion of thermal anomaly zones by up to 30–50 % is observed.

Static loads

Subsidence of the upper lithospheric horizon (as a result of the construction of multistory buildings) can lead to tectonic movement of soil layers, irreversible soil compaction, and a decrease in groundwater levels (e. g., the rate of subsidence of the surface lithosphere layer for clay soils can reach 3 mm per year), which can lead to the formation of hydrodynamic, temperature, hydrochemical, and hydrobiological anomalies [31]. Changes in soil density lead to variations in the height of capillary rise, and consequently, to shifts in the species composition of vegetation. Disruption of the physicochemical properties of soils (movement of the saline groundwater solution through ion exchange, alternation of sorption–desorption processes of soil-dissolved substances caused by temperature changes, etc.) is highly likely to cause both qualitative and quantitative imbalances in trophic networks.

Municipal solid waste and wastewater

The radius of negative impacts on soils and groundwater in areas near solid waste landfills is 2–3 km, leading to changes in the thermal balance (to depths of up to 0.05 km), properties of natural geochemical barriers (pH, gley, sorption), and the composition and density of bottom sediments. This problem, on the one hand, is associated with unbalanced local growth of population and metabolic products, and on the other hand, with the need for "competent" waste disposal.

Wastewater increases chemical pollution of groundwater, leads to soil corrosion, soil erosion, and reduces the biochemical activity of biota.

Water basin

Groundwater accounts for 14 % of the world's freshwater reserves, and in some countries, up to 70–100 %. The scarcity of these resources is due to three factors: natural and man-made pollution, and the irrational use of aquifers. Groundwater level fluctuations in urban areas are significantly higher than in equilibrium and slightly disequilibrium natural ecosystems and are generally not subject to seasonal changes. As anthropogenic pressures on the water basin increase, hydrospheric feedback responses sharply increase an abrupt or gradual decline in groundwater quality and level, including hydrodynamic, temperature, hydrochemical, and hydrobiological anomalies. Soil moisture requirements within the UES boundaries are higher than for other ecosystem types, as soil moisture potential in these areas decreases sharply and soil density increases. Under these conditions, hydrobiological processes are disrupted and the activity of soil bacteria is reduced.

Chemical pollution

Most chemical pollution in cities is generated by industrial and thermal power plants, as well as by transport. These substances inevitably enter the upper lithosphere, groundwater, and surface water via atmospheric flows. A significant proportion of pollutants in megacities comes from motor vehicles—up to 90 % of total emissions (including sulfur, nitrogen, carbon, lead, and hydrocarbon compounds). Unlike stationary industrial sources of chemical pollutants, motor vehicles are dispersed throughout the entire UES. Furthermore, while industrial gases emitted into the atmosphere at altitudes above 300 meters are dispersed by wind currents, vehicle pollutants accumulate in the ground layer (tens of meters), creating a ground-level temperature inversion and a concentration of harmful chemical elements. As a result, vertical air exchange is impeded, as the threshold height of the mixing layer is approximately 1.5 meters. Constant exposure to sulfur dioxide, one of the most common chemicals, leads to the accumulation of toxic substances in plants, causing irreversible processes (damage to the structure of chloroplast membranes, and suppression of photosynthesis, which begins after just 20–30 minutes of exposure to this toxicant). Chemical pollution affects the chemical hydrosphere and the upper lithosphere, increases the corrosion of underground structures, karst-suffosite processes, soil erosion, and the negative impact of wastewater, which ultimately negatively impacts the biota of all environments in the city's natural ecosystems.

Electromagnetic fields of industrial frequencies

Population size and density in UESs, and thus building density, are associated with the formation of heterogeneous industrial-frequency electromagnetic fields (EMFs) generated by numerous household electrical appliances. The wavelengths of industrial-frequency electromagnetic fields (EMFs) extend to thousands of kilometers (approximately 6,000 km in vacuum and air, and 1,000–3,000 km in soil). EMF radiation from terrestrial sources is partially reflected from the Earth's surface. However, a significant portion of the EMF energy, depending on the frequency range, penetrates into the upper layers of the lithosphere (the active medium) to a significant depth—up to tens and hundreds of meters (unlike high-frequency fields, which are characterized by the skin effect). Importantly, at such wavelengths, EMFs can affect ecosystems virtually independently of their electric and magnetic components. As a result, densely populated industrial and residential areas create a highly inhomogeneous electromagnetic background with large gradients in the electric and magnetic components of the EMF. Scientific literature typically examines the impact of EMFs on ecosystems in the air, although most of the biosphere's biomass is concentrated in the upper layers of the lithosphere. All "targets" of EMF impact (magnetic, electrical, thermal, chemical, etc.), interacting nonlinearly, exhibit the property of nonadditivity [31]. We would like to specifically note that we are talking, first and foremost, about EMF sources directly connected to objects equipped with power supply networks, transformers, electric motors, switching systems, household and office appliances, etc. The size of typical zones of EMF intensity heterogeneity corresponds to the dimensions of UES biocenoses (from 30 to 100 m, depending on building height), and the average absorption of low-frequency EMF power emitted by surface sources by soil–ground environments reaches up to 20 %. Low-frequency EMFs can propagate through soils over considerable distances—tens to hundreds of meters from the source—so in EMF-affected zones, not only the quantitative but also the qualitative composition of biocenoses may change significantly.

Microorganisms

Microorganisms are a key factor in soil formation and can develop intensively only under certain environmental parameters: temperature in the range [+20—35 °C], humidity, the CO_2/O_2 ratio in soils, pH values (certain microorganism species are evolutionarily adapted to different pH values—from pH = 2.5 for highly oxidized soils to pH = 9.5 for saline soils), and redox potential (Eh) (beyond the permissible values $Eh_{min} \leq Eh \leq Eh_{max}$, a disruption in the ratio of dissolved and undissolved forms of certain minerals, migrating substances, and the number of anaerobes occurs).

Soil fauna plays an important role in nutrient cycling and water infiltration [27]. Changes in the physical, chemical, and biological properties of soil affect soil moisture, oxygen availability, and chemical composition. For example, soil decomposition rates correlate with nitrogen content, electrical conductivity, and pH [23]; changes in the CO_2/O_2 ratio in soils and groundwater alter pH, biological productivity, and species composition of trophic networks. Moreover, due to the consistency of trophic webs in all ecosystem environments, a violation of their quantitative and qualitative composition in soils inevitably leads to a change in trophic webs in all natural environments of urban ecosystems.

Fragmentation of ecosystems

Urban expansion over areas of natural geobiocenoses leads to ecosystem fragmentation, resulting in local isolation of populations and species, reduced biodiversity, degradation of key ecosystem functions due to biomass loss, altered resource availability [12], changes in trophic network structure [6], and modifications of migration routes, reproduction rates, and life spans. Data from 76 long-term studies (over 35 years) show that ecosystem fragmentation reduces biodiversity by 20–75 % [11, 8]. The sustainability of highly fragmented urban biocenoses is linked to the synergistic impacts of various factors on the ecosystem: the area of the biocenoses, their quantitative and qualitative composition, climatic conditions, anthropogenic factors, population size and density, etc., which limits the buffering capacity of the natural subsystem of the UES. These conditions disrupt the dynamic equilibrium of energy, matter, and information flows, reduce the "buffer capacity" of natural subsystems, and increase nonlinearity—and thus instability—of systemic processes.

1.3 Self-organization in urboecosystems

It is known that self-organization manifests itself through the formation of autowave dissipative structures [39]. According to our theory, a city should be regarded as a UES composed of conjugate anthropogenic and natural subsystems. A necessary condition for self-organization in a UES is the synergetic correspondence of natural (biotic and abiotic) factors to anthropogenic impacts, which is determined by the structural and functional complexity of the ecosystem, the intensity of anthropogenic processes, and the potential ability of natural subsystems to mitigate these processes ("buffer capacity"). The driving parameters of this process are natural and anthropogenic factors (excitable elements) that determine the wavelength and shape of the autowave [31, 29, 15, 30]. Under certain conditions, the values of the activator and inhibitor intensity functions at the periphery of the considered regions undergo abrupt changes. Such solutions are referred to as step-type contrast structures: in the vicinity of a transition point (or curve in the two-dimensional case), the solution can vary sharply [4].

2 Methods

Under certain conditions, the values of the intensity activator and inhibitor functions on the periphery of the studied regions can undergo sharp changes [4].

The aim of the modeling is to obtain the spatiotemporal distribution of the activator and inhibitor, taking into account the presence of sharp transition layers at UES boundaries. One of the mathematical models yielding solutions of this type is the FitzHugh–Nagumo system [9], which belongs to the reaction–diffusion class. However, this system does not account for medium heterogeneity and, therefore, cannot describe stationary distributions with large gradients. The general form of such systems and the conditions for the existence of moving and stationary front-type solutions are discussed in [3].

This type of ecosystem is characterized by a high rate of population growth and density due to the concentration of residential, industrial, commercial, and other facilities, as well as communication networks. These conditions disturb the dynamic equilibrium of energy, matter, and information flows, reduce the "buffer capacity" of natural subsystems, and increase the nonlinearity and, consequently, the instability of systemic processes. Being conjugate active media, UESs are nonlinear dynamic macrostructures consisting of natural and anthropogenic subsystems that are coupled in both space and time. The UES contains a distributed resource (population, industrial and residential facilities, transport, geobiocenoses), and in the presence of perturbations in neighboring cells that alter the resource, the process can become autocatalytic. As a result of increasing population size and density, along with the expansion of residential, commercial, and industrial development, traffic flows and communication networks grow, and the boundaries of the UES expand. Consequently, growing cities tend to merge into common systems, which reduces and fragments geobiocenoses [31, 29, 15, 30]. This inevitably leads to changes in the quantitative and, more importantly, the qualitative composition of trophic networks.

In dimensionless form, the spatiotemporal model of UES development is given by activator-inhibitor system of equations based on the modified FitzHugh–Nagumo system [31]:

$$\left.\begin{aligned}\frac{\partial u}{\partial t}-\varepsilon^2 D_u\left(\frac{\partial^2 u}{\partial x^2}+\frac{\partial^2 u}{\partial y^2}\right)&=-u(u-\alpha(x,y))(u-1)-uv,\\ \frac{\partial v}{\partial t}-\varepsilon D_v\left(\frac{\partial^2 v}{\partial x^2}+\frac{\partial^2 v}{\partial y^2}\right)&=-\gamma(x,y)v+\beta(x,y)u,\end{aligned}\right\} \tag{13.1}$$

where $u(x,y,t)$ is the function of intensity of anthropogenic processes (activator), and $v(x,y,t)$ is the function of intensity of natural processes (inhibitor). As the rates of anthropogenic processes exceed those of natural processes by at least an order of magnitude; therefore, anthropogenic processes are modeled as activators (u) and natural processes as inhibitors (v) of systemwide dynamics. In the system, $\alpha(x,y)$ is the activation parameter of the system ($0 < \alpha \leq 1$), $\gamma(x,y)$ is the kinetic decay parameter of the inhibitor ($\gamma > 0$), $\beta(x,y)$ is the kinetic interaction parameter between activator and inhibitor ($\beta > 0$, and D_u, D_v are the diffusion coefficients of the activator and inhibitor ($0 < D_u \leq 1$, $0 < D_v \leq 1$). The term uv extends the ability to analyze cross-interactions between activator and inhibitor. The functions u and v are nonnegative, and the initial distributions are assumed to be known. The governing parameters are natural and anthropogenic factors (excitable elements) that define the length and shape of the autowaves.

In the context of urban ecosystem development, the solutions of system (13.1) of particular interest are those in the form of propagating autowave fronts and stationary contrast structures, i. e., functions that undergo sharp changes in narrow zones (transition layers) between areas of urban development and natural biocenoses. The transition layer width is typically on the order of several hundred meters, whereas the characteristic size of real UESs is on the order of tens of kilometers. Therefore, the system (13.1) contains a small parameter ε, ($0 < \varepsilon \ll 1$) equal to the ratio of the transition layer width to L, the linear size of the considered domain. The multiplier ε^2 before the diffusion coefficient in the first equation (compared to the multiplier ε in the second equation) reflects the marked significant difference in the rates of change of the activator and the inhibitor.

The problem is solved in a quadratic domain $(x,y) \in [-1,1] \times [-1,1]$ with a given initial distribution and Neumann boundary conditions on all edges of the square:

$$\left.\frac{\partial u}{\partial x}\right|_{x=\mp 1}=\left.\frac{\partial v}{\partial x}\right|_{x=\mp 1}=\left.\frac{\partial u}{\partial y}\right|_{y=\mp 1}=\left.\frac{\partial v}{\partial y}\right|_{y=\mp 1}=0. \tag{13.2}$$

Let us find out under what conditions the model (13.1) can describe small cities with radial symmetry, the boundaries of which do not change for a long time. To do this, we study stationary solution of the one-dimensional system (13.1), (in which there is no dependence on the variable y) i. e., the solution to system

$$\left.\begin{aligned} \varepsilon^2 D_u u_{xx} - u(u-\alpha(x))(u-1) - uv = 0, \\ \varepsilon D_v v_{xx} - \gamma(x)v + \beta(x)u = 0, \end{aligned}\right\} \tag{13.3}$$

on the interval $x \in (-1,1)$ with Neumann boundary conditions.

The equilibrium positions of this system, depending on the value of the parameter x, are defined as solutions of the so-called reduced system of equations

$$\left.\begin{aligned} u(u-\alpha(x))(u-1) + uv = 0, \\ -\gamma(x)v + \beta(x)u = 0, \end{aligned}\right\} \tag{13.4}$$

The first equation has three roots:

$$\varphi_1 = 0, \quad \varphi_{2,3}(v,x) = \frac{1}{2}(\alpha(x) + 1 \mp \sqrt{(\alpha(x)-1)^2 - 4v}).$$

If the expression under the radical takes positive values, then all three roots of the equation are real. In this case, $\varphi_1(v,x) = 0$ and $\varphi_3(v,x)$ relate to saddle-type rest points of system (13.3) and a solution transition from $\varphi_1(v,x)$ to $\varphi_3(v,x)$ is possible [3]. For small ε, this transition is rapid and occurs in a small vicinity of the transition layer point x^* that is determined from equality $u(x^*) = \varphi_2(v(x^*), x^*)$.

System (13.4) has at least one real solution $(u,v) = (0,0)$. In order to describe a bistable medium, it must have another real solution $(u,v) = (\varphi_3(v_3(x),x), v_3(x))$. To obtain this solution, let us substitute $u = \varphi_3(v,x)$ into the second equation of system (13.4):

$$\gamma(x)v - \beta(x)\frac{1}{2}(\alpha(x) + 1 + \sqrt{(\alpha(x)-1)^2 - 4v}) = 0. \tag{13.5}$$

A real solution $v(x)$ to the resulting equation exists if the inequality holds

$$v - \frac{\beta(x)}{2\gamma(x)}(\alpha(x)+1) \geq 0 \tag{13.6}$$

and has a form

$$v_3(x) = \frac{1}{2}\left(\frac{\beta(x)}{\gamma(x)}(\alpha(x)+1) - \frac{\beta^2(x)}{\gamma^2(x)} + \frac{\beta(x)}{\gamma(x)}\sqrt{\left(\alpha(x)+1-\frac{\beta(x)}{\gamma(x)}\right)^2 - 4\alpha(x)}\right). \tag{13.7}$$

Substituting this expression into inequality (13.6), we obtain the following condition on the functions $\alpha(x)$, $\beta(x)$, and $\gamma(x)$ the fulfillment of which is necessary for the existence of two pairs of stable solutions in a reduced system:

$$\alpha(x) \leq 1 + \frac{\beta(x)}{\gamma(x)} - \sqrt{\frac{\beta^2(x)}{\gamma^2(x)} + 4\frac{\beta(x)}{\gamma(x)}}. \tag{13.8}$$

For the function $\varphi_3(v_3(x), x)$, we get

$$\varphi_3(v_3(x), x) = \gamma(x)\frac{v_3(x)}{\beta(x)}$$
$$= 0.5\left(\alpha(x) + 1 - \frac{\beta(x)}{\gamma(x)} + \sqrt{\left(\alpha(x) + 1 - \frac{\beta(x)}{\gamma(x)}\right)^2 - 4\alpha(x)}\right). \quad (13.9)$$

In order for the function to take real values, it is necessary to perform the inequality

$$\alpha(x) \leq 1 + \frac{\beta(x)}{\gamma(x)} - 2\sqrt{\frac{\beta(x)}{\gamma(x)}}. \quad (13.10)$$

The results of the analysis of the reduced system are presented in Figure 13.1 as the state diagram of the active medium as a function of the parameters α, β, γ. A sufficient condition for the formation of an autowave front of the activator is that the parameters α, β, γ correspond to the excitable (bistable) region below the red curve defined by the equality in expression (13.8).

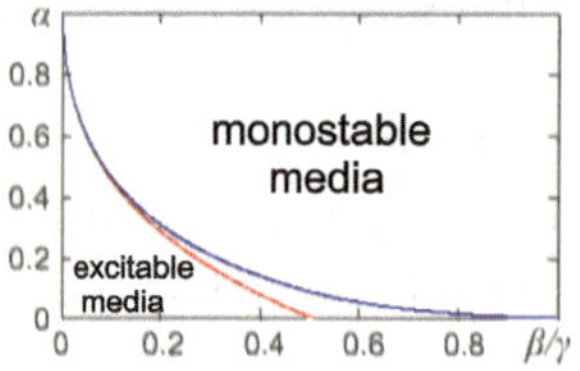

Figure 13.1: Schematic state diagram of the active medium in the (α, β, γ)-space: above the blue curve—nonexcitable; below the red curve—excitable/bistable.

If the parameter values fall within the area above the blue curve, then the medium is nonexcitable (monostable). Any external disturbance of this environment fades with time. An undisturbed environment may be, e. g., a large body of water. The blue curve is given by the equality in the expression (13.10).

In the case of spatial inhomogeneity of excitable and nonexcitable zones, which is typical for UESs, abrupt changes occur in the activator and inhibitor intensity functions at the boundaries of these zones. In this case, internal transition layers are formed.

It is well established that stationary contrast structures arise in media that are spatially inhomogeneous. In the present model, two primary factors determine spatial heterogeneity: (i) the spatial distribution of population density and (ii) the presence of barriers represented by urban geobiocenoses. Such inhomogeneity is reflected in the dependence of the parameters α, β, and γ on spatial coordinates, where this dependence becomes discontinuous in the presence of barriers. The conditions for the existence of stationary solutions of system (1) were derived in [4], while the conditions under which stationary solutions with internal transition layers form at barrier boundaries are given in [25].

For small towns, the activation parameter is related to population density, at the same time β and γ are constants. The coordinate of the stationary transition layer (urban suburb) is determined from the equation [31]

$$\frac{\gamma}{\beta}[v_3(\alpha(x^*))]^2 - (\alpha(x^*)+1)[v_3(\alpha(x^*)) - v^*(\alpha(x^*))]$$
$$+\frac{1}{6}[((\alpha(x^*)-1)^2 - 4v_3(\alpha(x^*)))^{3/2} - ((\alpha(x^*)-1)^2 - 4v^*(\alpha(x^*)))^{3/2}] = 0,$$

where v_3 is given in Equation (13.7) and

$$v^*(\alpha(x^*)) = \frac{9(\alpha-1)^2 - (\alpha+1)^2}{36}$$

is the value of the inhibitor function in the transition layer.

The most characteristic population density distribution for a small town is the Gaussian distribution. A single-peaked population density function has the following form [29]:

$$\rho(x) = \frac{\rho_0 \exp[-\lambda(x-x_0)^2]}{2}, \tag{13.11}$$

where

$$\lambda = \frac{2\ln(\rho_0/\rho_d)}{d^2}, \tag{13.12}$$

x_0 is the town center, $\rho_0 = \rho(x_0)$ is the population density at x_0, and $\rho_d = \rho(x_0 \pm d)$ is the population density at a distance d from the center.

If we assume $\alpha(x) = 1/\rho(x)$, solution (10) exists when $\rho(x) = \rho^*$ where ρ^* is the critical population density parameter in the transition layer. In the case of a Gaussian distribution, the density at the location of the largest gradient within the transition layer is given by $\rho^* = \rho_0/\sqrt{e}$.

The population density function for two neighboring cities in the two-dimensional case is given by the superposition of two Gaussian distributions:

$$\rho(x,y) = \sum_{i=1}^{2} \rho_i \exp(-\lambda_i((x-x_i)^2 + (y-y_i)^2)), \tag{13.13}$$

where (x_i, y_i), $i = 1,2$ are the coordinates of town centers, ρ_i are population densities in the points (x_i, y_i), respectively, $\lambda_i = d_i^{-2}\ln(\rho_i/\rho_{d_i})$, ρ_{d_i} is the population density at a distance d_i from the i-th center.

Numerical experiments for cities in the Moscow region have shown that, in the case of territorial merging of two cities of different diameters with different population density distributions, the formation of the autowave front and transition zones is determined by the population density function and by the ratio of the coefficients γ/β.

Figure 13.2 Display (a) shows the location of two cities in the Moscow region, Mytishchi and Korolev on the map. The approximate borders of the central parts of these cities are marked with black ovals. Urban development forms a united front and there is no pronounced boundary in the form of building changes between them.

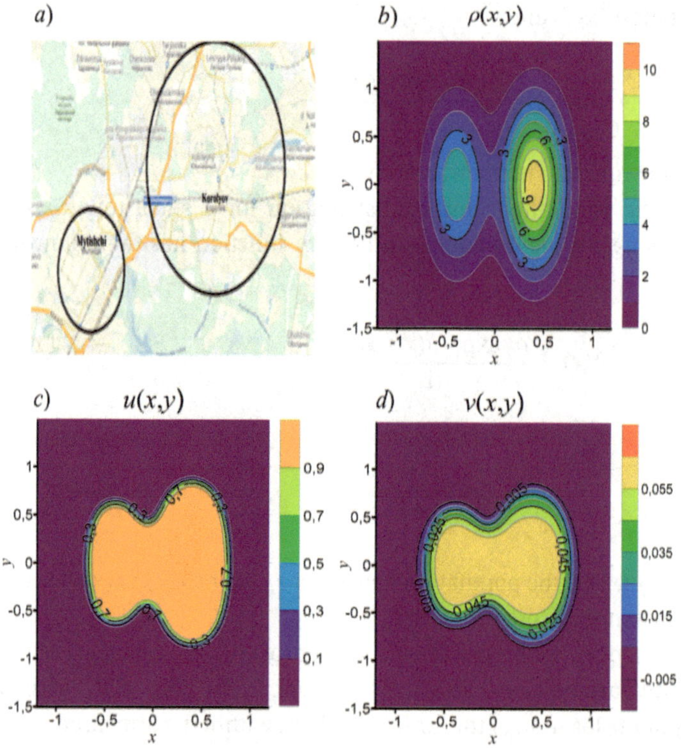

Figure 13.2: (a) Geographic location of Mytishchi and Korolyov, Moscow Region. (b) Population density distribution $\rho(x, y)$ for two neighboring cities, modeled as a superposition of two Gaussian functions. (c) Spatial distribution of the activator $u(x, y)$ obtained from numerical simulations. (d) Spatial distribution of the inhibitor $v(x, y)$ obtained from numerical simulations. The model predicts the formation of a common autowave front due to the short intercity distance, absence of landscape barriers, and presence of developed infrastructure.

Figure 13.2 Display (b) represents a population density distribution $\rho(x, y)$ for two neighboring cities, modeled as a superposition of two Gaussian functions of type (13.13).

Figure 13.2 Displays (c) and (d) present the results of numerical calculations for two cities of different diameters: spatial distribution of the activator $u(x, y)$ and the inhibitor $v(x, y)$. Due to the short distance between them, the absence of landscape barriers, and the presence of developed infrastructure, the merging of these cities into a single urban ecosystem is possible. The expansion of the city boundaries occurs at the expense of the territory of the natural ecosystem, reducing its potential as an inhibitor of anthropogenic processes. As a result, an ecosystem with a common autowave front may form. Similarly, through the annexation of adjacent territories, megacities are formed.

The rate and spatial extent of autowave propagation in a heterogeneous medium are influenced by the spatial distribution and size of local nonexcitable, weakly excitable, and highly excitable zones. In the context of urban ecosystems, these patterns are associated with the location and size of geobiocenoses and anthropogenically transformed areas, which are, in turn, related to population size and density—effectively, the presence of barriers. To model such a barrier, we employ discontinuous functions $\alpha(x)$, $\beta(x)$, and $\gamma(x)$ (see Sections 3.1 and 3.2). In our case, the ratio of the coefficients α, β, and γ within the barrier zone must correspond to the nonexcitable region in the state diagram of Figure 13.1.

3 Two applications

3.1 Autowave model of the territorial development of New Moscow (2017–2024)

Moscow is a typical example of the expansion of an urban ecosystem through the incorporation of adjacent territories. Between 1902 and 2002, the area of Moscow increased by a factor of 13, the population by a factor of 17, while the population density increased by a factor of 1.2. In 2012, certain territories of the Moscow region were annexed to the city of Moscow.

According to development plans, the territory of New Moscow is divided into three functional zones:

- Administrative, business, and residential zone—90 % building density.
- Historical, recreational, and medical complexes—70–80 % building density.
- Agroindustrial complex—50 % building density.

By 2035, the population of New Moscow is projected to increase by a factor of 6.5, while the real estate stock will increase by a factor of 4.2.

Based on cartographic data and urban development plans, the authors developed a mathematical model for predicting the territorial development of New Moscow up to the year 2024 [32]:

$$\left.\begin{aligned} &\frac{\partial u}{\partial t} - D_u\left(\frac{\partial^2 u}{\partial x^2} + \frac{\partial^2 u}{\partial y^2}\right) = -\frac{u(u-N_B K\alpha(x,y))(u-N_B)}{K_u^2 T^*} - \frac{1}{K_u T^*} uv, \\ &\frac{\partial v}{\partial t} - D_v\left(\frac{\partial^2 v}{\partial x^2} + \frac{\partial^2 v}{\partial y^2}\right) = \frac{1}{T^*}(-v + \beta\gamma u), \quad 0 < x,y < L,\ t_1 < t < t_2, \\ &0 = \left.\frac{\partial u}{\partial x}\right|_{x=0,\,x=L} = \left.\frac{\partial u}{\partial y}\right|_{y=0,\,y=L} = \left.\frac{\partial v}{\partial x}\right|_{x=0,\,x=L} = \left.\frac{\partial v}{\partial y}\right|_{y=0,\,y=L} \\ &u(x,y,t_1) = u_{\text{init}}(x,y), \\ &v(x,y,t_1) = v_{\text{init}}(x,y). \end{aligned}\right\} \tag{13.14}$$

Here u (km^2) denote the built-up area, and v (km^2) the area of destroyed geobiocenoses. According to the adopted urban planning standards, the estimated area of biocenoses in the territory of New Moscow should be at least 25 % of the building area [20]. In the model, v acts as the inhibitor of urban development.

The parameters D_u and D_v represent the diffusion coefficients of the activator and inhibitor, respectively: D_u (km^2/year) is the rate of building area growth, and D_v (km^2/year) is the rate of geobiocenosis loss, while L is the linear dimension of the study domain. Considering the pace of New Moscow's territorial development and the rate of geobiocenosis destruction during the construction of residential, administrative, and industrial facilities (including subsequent expansion of infrastructure due to population growth), the ratio $D_v/D_u = 1/10$ is adopted, corresponding to $D_u = 5$ km^2/year and $D_v = 0.5$ km^2/year.

To ensure dimensional consistency, two additional coefficients are introduced:

$$T^* = 1\,\text{year}, \quad K = 10^{-2}\,\text{km}^2, \tag{13.15}$$

where T^* is a time scale and K is the characteristic area, corresponding to an average single-family house with a garden.

The parameter γ denotes the fraction of destroyed geobiocenosis area per square kilometer. The maximum number of buildings per square kilometer is $N_B = 80$, depending on the building type (dense housing, cottage, industrial). The normalizing factor is $\beta = 100/N_B$, representing the ratio of the total area to the local building area.

The land-use function

$$\alpha(x,y) = e^{-0.05K_\rho(x,y)}$$

characterizes the designated land purpose (availability and type of development). The planned population density ρ (km^{-2}) is assigned according to land use [20]:
- $\rho = 4000$ people/km^2, $K_\rho = 40$ – for dense housing and industrial zones,
- $\rho = 2000$ people/km^2, $K_\rho = 20$ – for cottage housing,
- $\rho = 0$, $K_\rho = 0$ – for parks and natural geobiocenoses.

To determine the spatial distribution and numerical value of the kinetic parameter γ, an analysis of representative park areas, industrial zones, dense housing, and cottage development was conducted. The analysis was performed using a C++ application developed by the authors, which generates text files from image-based data (Figure 13.3 Display (a)). Based on cartographic data, the following average proportions of geobiocenoses were obtained:

- Typical Moscow park territories: $\approx$ 10 %,
- Industrial areas: $\approx$ 15 %,
- Cottage development: $\approx$ 50 %,

accordingly, γ in the New Moscow development model was set to:

- $\gamma = 0$ for natural geobiocenoses,
- $\gamma = 0.1$ for parks,
- $\gamma = 0.5$ for cottage development,
- $\gamma = 0.75$ for dense housing (minimum 25 % biocenoses per total area),
- $\gamma = 0.85$ for industrial zones.

To test the model, the Novomoskovskiy Administrative District (NAD) was considered; cf. Figure 13.3 Display (b). Initial conditions were derived from aerial imagery and cartographic data using the C++ application; cf. Figure 13.3 Display (a). The initial distribution of destroyed geobiocenosis area per square kilometer is shown in Figure 13.3 Display (c)—the reduction in the area of natural geobiocenoses is shown in colors: from dark green (maximum proportion of vegetation) to white—area without vegetation (minimum proportion). The initial distribution of building density fraction per square kilometer (from raster data) are shown in Figure 13.3 Display (e)—the increase in the share of built-up area is shown in colors: blue is undeveloped area; brown (minimum share of built-up area from the total area) to yellow (maximum share of built-up area from the total area). The simulated distributions for 2024 are shown in Figure 13.3 Displays (d) and (f).

Numerical solutions were obtained using the authors' code based on the method of lines, employing a Rosenbrock scheme with coefficient $\frac{1}{2}$ and spatial factorization.

A comparison of construction maps of New Moscow for 2024 with model results (Figure 13.3) shows:

- **Official data** [22]: built-up area – 52.3 %, forests, parks, and wastelands – 47.7 %.
- **Model results**: built-up area – 53 %, geobiocenoses – over 25 %.

Statistical and cartographic analyses of typical Moscow residential complexes show that areas for social infrastructure (commercial zones, clinics, schools, kindergartens, roads, etc.) have expanded at the expense of natural geobiocenoses. Similar patterns are reproduced by the model.

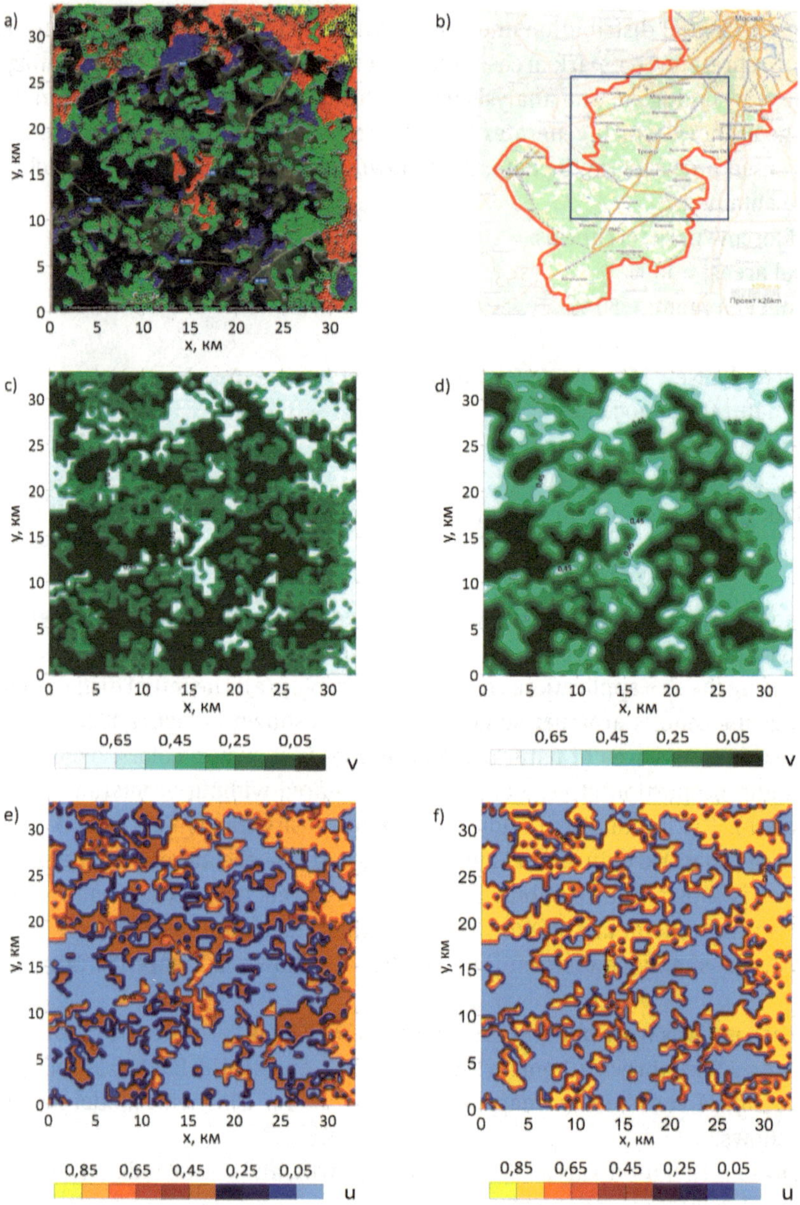

Figure 13.3: Spatial data and simulation results for the Novomoskovskiy Administrative District (NAD). (a) Rasterized land-use classification obtained from satellite imagery: red marks the dense housing (population density 4000 people/km^2), bright green marks cottage developments (population density 2000 people/km^2), blue marks industrial complexes, yellow marks geobiocenoses on the old Moscow territory, dark green color marks the territory of forests and parks in the territory of NAD. (b) Geographic boundaries of NAD (red contour), from [41] and simulation domain (blue rectangle). (c) Initial distribution of the destroyed geobiocenosis fraction v (from raster data). (d) Simulated distribution of v in 2024. (e) Initial distribution of building density fraction u. (f) Simulated distribution of u in 2024. Color bars indicate the fraction of the corresponding quantity per unit area.

3.2 Autowave model of territorial development of Shanghai (2017–2024)

Over the past three decades, China has experienced an intense wave of urbanization. Overall, the expansion of urban areas in China during 2000–2005 amounted to 40.42 %, which is significantly higher compared to the period 2005–2010 (14.86 %), a difference associated with the country's demographic policy [13]. It is projected that by 2030, 65 % of the country's population will reside in cities [2]. In Shanghai, both the population size and density steadily increases [2], inevitably leading to the expansion of its boundaries. At present, the directions of Shanghai's expansion are predominantly to the southeast, southwest, and south, which can be explained by the geographical location of the metropolis and the conditions of industrial and economic development [17] (Figure 13.4 Displays (a), (b), and (c)).

In parallel with population growth between 2000 and 2010, the demand for improved quality of life in Shanghai increased, causing more intensive use of natural geobiocenoses to meet these needs [5, 36]. Consequently, the structure and functions of urban geobiocenoses and adjacent areas undergo substantial changes, including alterations in landscape, biogeochemical and hydrological cycles, trophic networks, and population dynamics [28, 38]. Thus, the process of urbanization becomes autocatalytic.

The central core of Shanghai—its historically established center—is surrounded by a belt of inner suburbs with numerous satellite towns. The third ring of the agglomeration consists of rural peripheries, where population density is almost an order of magnitude lower than in the central district [2] (Figure 13.4 Display (b)). In 2015, in accordance with Shanghai's development plans, the area of land available for construction was restricted, leading to an increase in housing prices. In response, the Shanghai municipal government is currently implementing measures aimed at limiting both total and local population density by 2030.

The model of Shanghai's development is based on cartographic data, housing prices, and urban planning documents [16]. It is governed by the following system of reaction–diffusion equations:

$$\frac{\partial u}{\partial t} - D_u\left(\frac{\partial^2 u}{\partial x^2} + \frac{\partial^2 u}{\partial y^2}\right) = -\frac{1}{T^*}u(u - a(x,y))(u-1) - \frac{1}{T^*}uv,$$

$$\frac{\partial v}{\partial t} - D_v\left(\frac{\partial^2 v}{\partial x^2} + \frac{\partial^2 v}{\partial y^2}\right) = \frac{1}{T^*}(-v + \gamma(x,y)u), \quad 0 < x,y < L,\ t_1 < t < t_2$$

with the same type of initial and boundary conditions as in Equation (13.14).

Here, u is the *activator* of the system, representing the fraction of the built-up area per square kilometer (dimensionless), and v is the *inhibitor*, corresponding to the housing price per square meter expressed as a fraction of its maximum possible value (dimensionless). The parameter T^* is the characteristic time scale (1 year), while γ is a

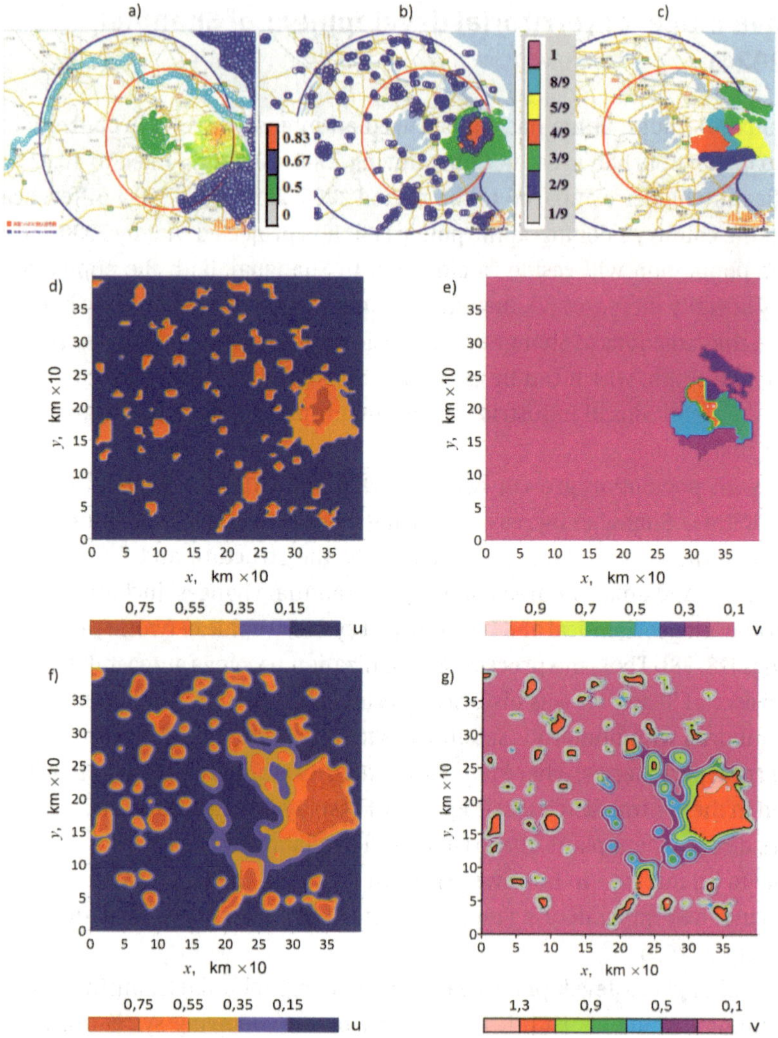

Figure 13.4: (a) Processing the 2017 data for Shanghai and surrounding areas (400 × 400 km), taking into account the building density and landscape features [21] the zone inside the red circle is the near suburbs, while the belt from the red circle to the blue circle is distant suburbs. The marked are areas of large bodies of water: Taihu Lake is green, the East China Sea is blue, and the Yangtze River is light blue. (b) Cartographic data of building density at the beginning of the calculation period (2017), the values on the legend are given in building shares per square kilometer: red color marks the highest density in the central part of Shanghai, blue color marks lower density in different areas of Shanghai. (c) Distribution of the relative housing price within the metropolitan area. The values on the legend are given in fractions of the housing price in the core of Shanghai in 2017: the purple color is the highest price in the central part of Shanghai, and then the price decreases as the economic benefit for developing areas of Shanghai decreases (light blue, yellow, red, green, and dark blue colors) until the price reaches zero in the area where no development is taking place (gray color). (d)–(f) Results of the simulation of Shanghai territorial development for the period 2017–2024. (d) Initial distribution of the built-up area u (activator). (e) Initial distribution of the relative housing price v (inhibitor). (f) Simulated distribution of the built-up area u in 2024. (g) Simulated distribution of the relative housing price v in 2024.

dimensionless parameter describing the dependence of housing prices on the development zone. The coefficients D_u and D_v represent the rates of change of the built-up area (km^2/year) and housing price (km^2/year), respectively.

Data analysis showed that in 2017, compared to 2016, housing prices increased by 22 % ([21]). Therefore, the living space that could be bought for the same price decreased by 22 % over the year. Hence, we set $D_v = 0.22D_u$. The function α is the activation parameter. Analysis of the corresponding point system revealed that for $\alpha = 0.3$, the system describes a bistable medium, while for $\alpha = 1$ it is nonexcitable (Figure 13.1). In the eastern direction, $\alpha = 0.3$ corresponds to a bistable zone (numerical value defined by regulations preserving at least 30 % of agricultural land from total territory); in the nonexcitable zone of the East China Sea and Lake Taihu, $\alpha = 1$; the Yangtze River acts as a barrier on the boundary between bistable and nonexcitable zones, thus $\alpha = 1$ (Figure 13.4 Display (a)).

The computational implementation of the Shanghai development model (2017–2024) based on building dynamics and housing prices is presented in Figure 13.4.

Comparison of cartographic data and the results of the Shanghai territorial development model [18] demonstrated that in 2024, in terms of both population density and the geographical direction of urban expansion (Figure 13.4 Displays (a)–(c)), the relative error of the model is 13.3 %. Between 2017 and 2024, the actual average housing price per square meter increased by a factor of 1.5, the same result is given by the model. The locations of local built-up areas in the model closely match those observed in reality.

4 Discussion

To predict the territorial development of urban ecosystems, a mathematical model of auto-wave self-organization from spatially and temporally coupled natural and anthropogenic subsystems was developed.

The classical FitzHugh–Nagumo reaction–diffusion model was modified to include spatially heterogeneous activator–inhibitor interactions.

As factors that determine the spatial heterogeneity have been considered: spatial barriers (city development and natural bioclimatic zones) and the distribution of population depending on the spatial coordinates.

The model was then validated with statistical and cartographic data for New Moscow and Shanghai (2017–2024).

A comparison between the development maps of New Moscow and the model results for 2024 showed that the relative error of numerical modeling is 4.8 %. According to official data [22], the built-up area accounts for 52.3 %, while the model predicts 53 %, implying that geobiocenoses occupy more than 25 % of the territory, in accordance with regulatory requirements to maintain the minimum necessary biodiversity.

A comparison between cartographic data and the Shanghai territorial development model demonstrated that, in 2024, the relative error in built-up density and geographical expansion direction is 13.3 %. Between 2017 and 2024, the actual average housing price per square metre increased by a factor of 1.5 as predicted by the model. Local development patterns in the simulation closely match those observed. The model's results are consistent with the Chinese government's plans and land-use norms.

5 Conclusion

Urban ecosystems were considered as spatially and temporally coupled natural and anthropogenic subsystems.

For the territorial development of New Moscow and Shanghai during the period 2017–2024, the results of the model simulations are in full agreement with the observed dynamics of real urban ecosystems.

The developed model will enable the creation of optimal conditions for the formation of ecocities to ensure the survival of human populations and the maintenance of the required biodiversity of species in urban eco-systems.

Bibliography

[1] N. Bihamta, A. Soffianian, S. Fakherran, and M. Gholamalifard. Using the SLEUTH urban growth model to simulate future urban expansion of the Isfahan Metropolitan Area, Iran. *Journal of the Indian Society of Remote Sensing*, 43:407–417, 2015.

[2] Urbanization trends in China. https://cn.nytimes.com/china/20150422/c22study. Accessed: 2025-08-12.

[3] V. F. Butuzov, N. T. Levashova, and A. A. Melnikova. Steplike contrast structure in a singularly perturbed system of equations with different powers of a small parameter. *Computational Mathematics and Mathematical Physics*, 52(11):1526–1544, 2012.

[4] V. F. Butuzov, N. T. Levashova, and A. A. Mel'nikova. A steplike contrast structure in a singularly perturbed system of elliptic equations. *Computational Mathematics and Mathematical Physics*, 53(9):1239–1254, 2013.

[5] W. Chen, J. Gao, and W. Chen. Urban land expansion and the transitional mechanisms in Nanjing, China. *Habitat International*, 53:274–283, 2016.

[6] K. F. Davies, C. R. Margules, and J. F. Lawrence. A synergistic effect puts rare, specialized species at greater risk of extinction. *Ecology*, 85(1):265–271, 2004.

[7] R. de Groot, L. Brander, S. van der Ploeg, et al. Global estimates of the value of ecosystems and their services in monetary units. *Ecosystem Services*, 1(1):50–61, 2012.

[8] G. Ferraz, G. J. Russell, P. C. Stouffer, R. O. Bierregaard, S. L. Pimm, and T. E. Lovejoy. Rates of species loss from Amazonian forest fragments. *Proceedings of the National Academy of Sciences*, 100(24):14069–14073, 2003.

[9] R. FitzHugh. Mathematical models of threshold phenomena in the nerve membrane. *The Bulletin of Mathematical Biophysics*, 17:257–278, 1955.

[10] J. W. Forrester. *Urban Dynamics*. MIT Press, Cambridge, MA, 1969.
[11] A. Gonzalez. Community relaxation in fragmented landscapes: The relation between species richness, area and age. *Ecology Letters*, 3(6):441–448, 2000.
[12] N. M. Haddad, L. A. Brudvig, J. Clobert, K. F. Davies, A. Gonzalez, R. D. Holt, et al. Habitat fragmentation and its lasting impact on Earth's ecosystems. *Science Advances*, 1(2):e1500052, 2015.
[13] L. Hua, L. Tang, Sh. Cui, and K. Yin. Simulating urban growth using the SLEUTH model in a coastal peri-urban district in China. *Sustainability*, 6(6):3899–3914, 2014.
[14] N. Larondelle and S. Lauf. Balancing demand and supply of multiple urban ecosystem services on different spatial scales. *Ecosystem Services*, 22:18–31, 2016.
[15] N. Levashova, A. Mel'nikova, A. Semina, and A. Sidorova. Autowave mechanisms of structure formation in urban ecosystems as the process of self-organization in active media. *Communications on Applied Mathematics and Computation*, 31(1):32–40, 2017.
[16] N. Levashova, A. Sidorova, A. Semina, and M. Ni. A spatio-temporal autowave model of Shanghai territory development. *Sustainability*, 11(13):3669, 2019.
[17] J. Li, W. Fang, T. Wang, S. Qureshi, J. M. Alatalo, and Y. Bai. Correlations between socioeconomic drivers and indicators of urban expansion: Evidence from the heavily urbanised Shanghai Metropolitan Area, China. *Sustainability*, 9(7):1199, 2017.
[18] Global Property Guide LLC. China's residential property market analysis. https://www.globalpropertyguide.com/asia/china/price-history, 2025.
[19] X. Luo, P. Jiang, J. Yang, J. Jin, and J. Yang. Simulating pm2.5 removal in an urban ecosystem based on the social-ecological model framework. *Ecosystem Services*, 47:101234, 2021.
[20] MoscowInvestment. Moscow investment portal. Available at: https://investmoscow.ru/city-projects/aip/ (accessed 22 November 2018.).
[21] National Bureau of Statistics of China. Official Statistics. http://www.stats.gov.cn/english/Statisticaldata/AnnualData, 2018. Accessed: 15 July 2018.
[22] Nedvio.com. What will new Moscow become? (in Russian), 2024. https://nedvio.com/kakoy-stanet-novaya-moskva/.
[23] D. A. Neher, T. R. Weicht, and M. E. Barbercheck. Linking invertebrate communities to decomposition rate and nitrogen availability in pine forest soils. *Applied Soil Ecology*, 54:14–23, 2012.
[24] O. Odebiri, O. Mutanga, and J. Odindi. Deep learning-based national-scale soil organic carbon mapping with sentinel-3 data. *Geoderma*, 411:115695, 2022.
[25] N. N. Nefedov, B. V. Tishchenko, and N. T. Levashova. An Algorithm for Construction of the Asymptotic Approximation of a Stable Stationary Solution to a Diffusion Equation System with a Discontinuous Source Function. *Algorithms*, 16(8):359, 2023.
[26] X. Ouyang and X. Luo. Models for assessing urban ecosystem services. *Sustainability*, 14(8):4725, 2022.
[27] H. R. P. Phillips et al. Global changes and environmental stressors impact soil biodiversity: A meta-analysis. *iScience*, 27(9):110540, 2024.
[28] S. Qureshi and D. Haase. Compact, eco-, hybrid or teleconnected? Novel aspects of urban ecological research seeking compatible solutions to socio-ecological complexities. *Ecological Indicators*, 42:1–4, 2014.
[29] A. E. Sidorova, N. T. Levashova, A. A. Mel'nikova, N. N. Deryugina, and A. E. Semina. Autowave self-organization in heterogeneous natural–anthropogenic ecosystems. *Moscow University Physics Bulletin*, 71(6):562–568, 2016.
[30] A. E. Sidorova, N. T. Levashova, A. A. Mel'nikova, and A. E. Semina. Model of the structure formation of urban ecosystems as a process of autowave self-organization in active media. *Mathematical Biology and Bioinformatics*, 12(1):186, 2017, in Russian.
[31] A. E. Sidorova, N. T. Levashova, A. A. Mel'nikova, and L. V. Yakovenko. A model of a human-dominated urban ecosystem as an active medium. *Biophysics*, 60(3):466–473, 2015.
[32] A. E. Sidorova, N. T. Levashova, A. E. Semina, and A. A. Mel'nikova. The application of a distributed model of active media for the analysis of urban ecosystems development. *Mathematical Biology and Bioinformatics*, 13: 454–465, 2018.

[33] Department of Economic United Nations and Population Division Social Affairs. World urbanization prospects: The 2011 revision. United Nations, New York, 2011.
[34] Department of Economic United Nations and Population Division Social Affairs. World urbanization prospects: The 2014 revision, highlights (st/esa/ser.a/352). United Nations, New York, 2014.
[35] E. Vaz and J. J. Arsanjani. Predicting urban growth of the Greater Toronto Area—coupling a Markov cellular automata with document meta-analysis. *Journal of Environmental Informatics*, 25(2):71–80, 2015.
[36] Y. D. Wei. Zone fever, project fever: Economic transition, development policy, and urban expansion in China. *Geographical Review*, 105(2):156–177, 2015.
[37] L. H. Williams, P. Montagna, D. Yoskowitz, and D. Scholz. Stakeholder perceptions of coastal habitat ecosystem services. *Estuaries and Coasts*, 38:67–80, 2013.
[38] G. Yang, L. C. Bowling, K. Cherkauer, and B. Pijanowski. The impact of urban development on hydrologic regime from catchment to basin scales. *Landscape and Urban Planning*, 103(2):237–247, 2011.
[39] Y. E. Yelkin. Autowave processes. *Mathematical Biology and Bioinformatics*, 1(1):27–62, 2006.
[40] Ch. Yin, D. Yu, H. Zhang, Sh. You, and G. Chen. Simulation of urban growth using a cellular automata-based model in a developing nation's region. *International Journal of Engineering Science and Advanced Technology*, 2(3):453–460, 2012.
[41] Zabir.ru. Map of new Moscow. https://26km.ru/new_news/2011/aug/new_map_moscow.jpg, 2023. Accessed on 12 August 2025.
[42] X. Zhang. Urban growth modeling using neural network simulation: A case study of Dongguan City, China. *Journal of Geographic Information System*, 8:317–329, 2016.

Emmanuelle Augeraud-Véron, Arnaud Ducrot, and Ousmane Seydi

Game-theoretic optimization in spatial control of invasive plants

Comparison of decentralized and centralized strategies

Abstract: We study the spatial control of an invasive plant population structured over a finite number of discrete habitat patches, each characterized by site-specific ecological conditions and connected through seed dispersal. We investigate two control strategies: a decentralized strategy and a centralized one. The former is modeled as a noncooperative differential game in which all sites aim to minimize their respective local costs independently. In the latter, a central planner can allocate resources across all sites to minimize the total costs. We show that the stationary equilibrium of a spatially extended invasive-species control game can be characterized as the fixed point of a globally monotone reaction operator. By using Browder's theorem, together with the spectral properties of the underlying dispersal kernel, we establish existence, uniqueness, and stability of the stationary control equilibrium under weak assumptions. Crucially, our fixed-point algorithm converges in practice even for systems comprising hundreds of spatial patches, thereby overcoming the "curse of dimensionality" inherent in shooting-based solutions of partial differential equation frameworks. This offers two economically significant advantages, the full scope of intersite externalities is captured in a single algebraic formulation, and closed-form expressions for marginal externality payments lend themselves to the design of spatially differentiated fiscal transfers that precisely internalize social costs and benefits. Impacts of spatial heterogeneity on invasion outcomes and of control strategies on management outcomes are illustrated by simulations.

Keywords: Ordinary differential equations, monotone dynamical systems, game theory, population dynamics, invasive species

MSC 2020: 34C12, 34D05, 91A10, 92D40

1 Introduction

Ecological threat

Invasive plant species are a major threat to biodiversity and ecosystem functioning worldwide. Characterized by local reproduction and long-range dispersal, they exhibit

Emmanuelle Augeraud-Véron, Bordeaux Sciences Économiques, Université de Bordeaux, 16 Av. Léon Duguit, 33600 Pessac, France, e-mail: emmanuelle.augeraud@u-bordeaux.fr
Arnaud Ducrot, Ousmane Seydi, Université Le Havre Normandie, Normandie Université, LMAH UR 3821, 76600 Le Havre, France, e-mails: arnaud.ducrot@univ-lehavre.fr, ousmane.seydi@univ-lehavre.fr

https://doi.org/10.1515/9783112216620-020

spatial dynamics that enable persistence and expansion despite localized suppression efforts [10, 12]. These dynamics pose a substantial challenge for land managers and policymakers, who must decide when and where to intervene under spatial externalities and resource constraints.

The ecological mechanisms driving these invasions are well documented: seeds accumulate in local soil banks, disperse through vectors such as wind or animals, and establish new populations that compete with native flora [13, 9]. In systems where managers operate at different spatial locations (e. g., landowners, municipalities), decisions are made in a decentralized manner, typically without internalizing the impact of seed dispersal on neighboring sites. This generates a classic case of uncoordinated control under spatial externalities, where the effort of one manager benefits others, potentially leading to underinvestment in control [23, 12].

Economical burden

Global estimates place the annual economic burden of biological invasions well above hundreds of billions of dollars, when accounting for lost agricultural output, infrastructure damage, human health impacts, and ecosystem restoration costs [6, 25]. In particular, invasive seed-bearing plants—whose dissemination through seed banks and long-distance dispersal mechanisms amplifies both direct crop yield losses and the expenses associated with habitat rehabilitation—contribute a sizeable fraction of these losses. For example, global synthesis data report cumulative costs of at least USD 8.9 billion for plant invasions between 1970 and 2017, despite under-reporting in many regions, while country-level studies (e. g., New Zealand) have found that tree and shrub invasions alone can amount to nearly 1 % of national GDP in some years [6, 2]. Such figures underscore the critical need to integrate species-specific externalities—especially those arising from seed-bearing taxa—into spatially explicit bioeconomic models to inform more efficient allocation of control resources.

Policy frameworks

In many jurisdictions, invasive species management is governed by spatially uniform policies—broad subsidies for mechanical removal, chemical treatments, or pan-regional quarantines—that neglect the ecological and economic heterogeneity of landscapes. Such blanket interventions frequently misallocate resources to low-risk “sink” areas while failing to target upstream “source” regions whose seed exports drive downstream spread, resulting in persistent inefficiencies and elevated control costs [1, 10]. Moreover, the absence of transnational coordination exacerbates the problem for invasive seed-bearing plants, which disperse across administrative boundaries and undermine purely domestic control schemes [6].

Modeling intent
To formalize this situation, we adopt the framework of a noncooperative differential game over a discrete spatial network, inspired by the approach developed by Dockner et al. [7] and surveyed by Van Long and Ngo [21]. Each manager controls the removal rate of the invasive species at their site, trading off between direct control costs and damages from local abundance. We focus on stationary (time-invariant) equilibria, which correspond to long-run patterns of species abundance and management effort. The ecological dynamics, including seed dispersal and germination, are modeled following the biological assumptions found in recent work on optimal spatial control of biological invasions [15, 26].

Structure of content
The ecological model is described in Section 2, control strategy modeling in Section 3. Specifically, Section 2 introduces the ecological and economic structure for the model in (2.1), obtains extinction and invasion dynamics in (2.2), and closes with a numerical illustration of the invasion (2.3). Section 3 characterizes the stationary Nash equilibrium under decentralized control in (3.1) and derives the centralized optimum strategy in (3.2); the two regimes are compared and incentive mechanisms are proposed in (3.3). Section 4 presents a broad discussion, and Section 5 draws the conclusion.

2 Material and methods

Our contribution is threefold. First, we extend the analysis of dynamic environmental games with spatial externalities to a setting with nonlinear ecological feedbacks and discrete space. Second, we quantify the cost of noncoordination and propose implementable incentive schemes to mitigate it. Third, we provide a microfoundation for the functional forms used in the cost structure, linking the parameters to interpretable economic quantities such as labor and restoration costs.

2.1 Spatially dynamic model of an invasive plant population

We consider a spatially structured model describing the dynamics of an invasive plant population distributed over N discrete sites indexed by $i \in \Omega := \{1, \ldots, N\}$. At each site, i, $n_i(t)$ denotes the local density of adult plants, and $s_i(t)$ the density of seeds in the seed bank at time $t \geq 0$. To capture the effect of local management interventions, we introduce a constant control variable $c_i \in [0, 1]$ representing the intensity of removal efforts at site i, with $c_i = 1$ corresponding to the maximum feasible intervention. The control acts directly on the adult population at rate $c_i n_i(t)$. The adult population $n_i(t)$ declines due to natural mortality at rate μ_i, and increases through germination from the seed bank at rate $\gamma_i s_i(t) H_i(n_i(t))$, where $H_i(\cdot)$ captures local saturation effects. Seeds at site i are produced via dispersal from other sites, with incoming seed flux governed by the kernel $K(i,j)$, which determines the fraction of seeds produced at site j that reach site i. The seed input is proportional to the seed production rate r_j and local adult density $n_j(t)$. The seed bank decays at rate δ_i. The resulting controlled system is given by

$$\left.\begin{aligned} \dot{n}_i(t) &= -\mu_i\, n_i(t) + \gamma_i\, s_i(t)\, H_i(n_i(t)) - c_i\, n_i(t), \\ \dot{s}_i(t) &= \sum_{j\in\Omega} K(i,j)\, r_j\, n_j(t) - \delta_i\, s_i(t), \end{aligned}\right\} \tag{14.1}$$

for all $i \in \Omega$ subjected to the initial condition $n_i(0) = n_{0i} \geq 0$ and $s_i(0) = s_{0i} \geq 0$ and

$$H_i(h) = 1 - \kappa_i h, \quad \forall h \geq 0.$$

To simplify notation and analysis, we reformulate the system in compact vector form. Define the nonlinear function $F : \mathbb{R}^N \times \mathbb{R}^N \to \mathbb{R}^N$ by

$$F_i(n, s) := s_i\, H_i(n_i), \quad i \in \Omega,\ (n, s) \in \mathbb{R}^N \times \mathbb{R}^N$$

and $L : \mathbb{R}^N \to \mathbb{R}^N$ the linear map given by

$$L_i[n] := \sum_{j\in\Omega} K(i,j)\, r_j\, n_j, \quad i \in \Omega,\ \forall n \in \mathbb{R}^N.$$

Next, let M, Γ, C, and D denote the diagonal matrices:

$$M = \operatorname{diag}(\mu_i), \quad C = \operatorname{diag}(c_i), \quad \Gamma = \operatorname{diag}(\gamma_i) \quad \text{and} \quad D = \operatorname{diag}(\delta_i)$$

so that (14.1) rewrites as

$$\left.\begin{aligned} \dot{n}(t) &= -Mn(t) + \Gamma F(s(t), n(t)) - Cn(t), \\ \dot{s}(t) &= L[n(t)] - Ds(t), \end{aligned}\right\} \tag{14.2}$$

with initial condition $n(0) = n_0 \in \mathbb{R}^N_+$ and $s(0) = s_0 \in \mathbb{R}^N_+$. Next, we define the biologically relevant set

$$\Theta := \{(n_0, s_0) \in \mathbb{R}^N_+ \times \mathbb{R}^N_+ : 0 \leq n_{0i} \leq 1/\kappa_i,\ \forall i \in \Omega\}. \tag{14.3}$$

The assumptions on the parameters of the model (14.1) are as follows.

Assumption 2. *Assume that:*
1. *r_i, δ_i, γ_i, κ_i, and μ_i are positive for all $i \in \Omega$.*

2. $K(i,j) = K(j,i) > 0$ *for all* $i,j \in \Omega$, *and*

$$\sum_{j\in\Omega} K(i,j) \leq 1, \quad \forall i \in \Omega.$$

Lemma 2.1. *Let Assumption 2 be satisfied. Then for each initial condition* $(n_0, s_0) \in \Theta$, *the system* (14.2) *admits a unique global nonnegative solution* $(n(t), s(t)) \in \Theta$ *for all* $t \geq 0$. *Furthermore, the system is monotone in* Θ, *i. e., for each* $(n_0, s_0), (\bar{n}_0, \bar{s}_0) \in \Theta$ *if*

$$n_0 - \bar{n}_0 \in \mathbb{R}_+^N \quad \textit{and} \quad s_0 - \bar{s}_0 \in \mathbb{R}_+^N,$$

then the corresponding solutions satisfy

$$n(t) - \bar{n}(t) \in \mathbb{R}_+^N \quad \textit{and} \quad s(t) - \bar{s}(t) \in \mathbb{R}_+^N \quad \textit{for all } t \geq 0.$$

Proof. First, we note that the vector field of the ODE system (14.2) is smooth in $\mathbb{R}^{2N}$. Next, observe that the right-hand side of $\dot{s}_i(t)$ is nonnegative for $s_i = 0$ and $n \in \mathbb{R}_+^N$. Furthermore the right-hand side $\dot{n}_i(t)$ is nonnegative of $n_i = 0$ and $s \in \mathbb{R}_+^N$ and it is nonpositive for $n_i = 1/\kappa_i$ and $s \in \mathbb{R}_+^N$. Consequently, we infer from the Nagumo–Brezis theorem [24, Proposition 4.5] that for each initial condition $(n_0, s_0) \in \Theta$ there exists a unique maximally defined nonnegative solution in Θ. Consequently the n-component is always bounded so that the s-component is also bounded by using the second equation of (14.2). This proves that the solutions are nonnegative and globally defined in times for initial condition in Θ. To prove monotonicity, we observe that the nonlinear function $F_i(n, s) = s_i H_i(n_i)$ satisfies in Θ,

$$\frac{\partial F_i}{\partial n_j} \geq 0 \quad \text{and} \quad \frac{\partial F_i}{\partial s_j} \geq 0 \quad \text{for } i \neq j.$$

Furthermore, the matrices appearing in (14.2) all have nonnegative off-diagonal entries. Therefore, the Kamke condition [8, Chapter 8] is satisfied in Θ, and the system is monotone on Θ. □

In the following, to determine the invasion threshold, we note that near the invasion-free state, the dynamic of (14.1) is approximated by a linear system driven by the following matrix with positive off-diagonal entries:

$$\mathbf{V} = \begin{pmatrix} -M - C & \Gamma \\ L & -D \end{pmatrix} \tag{14.4}$$

Consequently, the spectral bound $\mathrm{s}(\mathbf{V}) = \{\mathrm{Re}(\lambda) : \lambda \text{ is an eigenvalue of } \mathbf{V}\}$ is an eigenvalue of $\mathbf{V}$ and determines whether the invading population initially grows ($\mathrm{s}(\mathbf{V}) > 0$) or decays ($\mathrm{s}(\mathbf{V}) < 0$). To link this to a threshold quantity, we define the next generation operator $\mathcal{G}(c) : \mathbb{R}^N \to \mathbb{R}^N$ as

$$\mathcal{G}(c)[s] := L(M+C)^{-1}\Gamma D^{-1}s, \tag{14.5}$$

which acts componentwise as

$$\mathcal{G}(c)[s]_i = \sum_{j\in\Omega} K(i,j)\frac{\gamma_j r_j}{\delta_j(\mu_j+c_j)}s_j, \quad i\in\Omega \tag{14.6}$$

and we define the invasion threshold $\mathcal{T}_0(c)$ as the spectral radius of $\mathcal{G}(c)$, i. e.,

$$\mathcal{T}_0(c) := \mathrm{r}(\mathcal{G}(c)). \tag{14.7}$$

The operator $\mathcal{G}(c)$ admits a natural ecological interpretation as a *next generation operator* describing the full life cycle of an invading plant population. In this framework, $n_i(t)$ denotes the density of established plants (e. g., juveniles or adults) at location i, while $s_i(t)$ represents the local seed density. The action of $\mathcal{G}(c)$ on a given seed distribution s accounts for the expected number of new seeds produced through the following process: seeds germinate and give rise to plants at rate γ_i, which survive on average for a time $(\mu_i+c_i)^{-1}$ under natural mortality and control. Each plant reproduces at rate r_j and produces seeds that disperse across space according to the kernel $K(i,j)$. The resulting seeds then accumulate at location i, modulated by their survival time δ_i^{-1}. Therefore, $\mathcal{G}(c)[s]_i$ captures the *total contribution of the seed population at site* i after one generation. As such, the spectral radius $\mathcal{T}_0(c) = \mathrm{r}(\mathcal{G}(c))$ serves as an *invasion threshold*: if $\mathcal{T}_0(c) < 1$, each generation of seeds produces fewer viable offspring than itself, leading to eventual extinction of the invading species. Conversely, if $\mathcal{T}_0(c) > 1$, the population can grow and establish, signifying successful invasion despite the control effort. Economically, the index

$$\mathcal{T}_0(c) = r(\mathcal{G}(c))$$

measures the net reproduction rate of the invasive population per unit of control effort: when $\mathcal{T}_0(c) > 1$, each additional dollar spent on control yields more than a dollar in avoided future damages (justifying increased effort), whereas $\mathcal{T}_0(c) \le 1$ indicates that the marginal cost of effort exceeds the marginal benefit of damages avoided, thus defining the optimal resource allocation threshold.

2.2 On extinction and invasion dynamics

Here, we demonstrate that the threshold $\mathcal{T}_0(c)$ plays a central role in determining whether extinction or invasion occurs in the long run. Specifically, we show that the system (14.1) (or equivalently (14.2)) exhibits such threshold-dependent dynamics. As a first step, we establish the following lemma.

Lemma 2.2. *Let Assumption 2 be satisfied. Then* $\mathsf{s}(\mathbf{V}) < 0$ *(resp.,* $\mathsf{s}(\mathbf{V}) = 0$*) if and only if* $\mathcal{T}_0(c) < 1$ *(resp.,* $\mathcal{T}_0(c) = 1$*).*

Proof. First, note that

$$\mathbf{V} = \begin{pmatrix} -M - C & \Gamma \\ L & -D \end{pmatrix} = \underbrace{\begin{pmatrix} 0 & \Gamma \\ L & 0 \end{pmatrix}}_{\mathbf{N}:=} - \underbrace{\begin{pmatrix} M + C & 0 \\ 0 & D \end{pmatrix}}_{\mathbf{M}:=}, \tag{14.8}$$

from where $\mathsf{r}(\mathbf{N}\mathbf{M}^{-1}) - 1$ and $\mathsf{s}(\mathbf{V})$ have the same sign. The result follows by using the fact that $\mathsf{r}((\mathbf{N}\mathbf{M}^{-1})^2) = \mathsf{r}(\mathbf{N}\mathbf{M}^{-1})^2 = \mathcal{T}_0(c)$. □

Lemma 2.3. *Let Assumption 2 be satisfied. If* $\mathcal{T}_0(c) \le 1$*, then the invading species goes extinct for every initial condition* $(n_0, s_0) \in \Theta$*.*

Proof. Recalling that $H_i(h) \le 1$ for all $h \ge 0$, it follows from the system (14.1) (equivalently, (14.2)) that the solution satisfies the differential inequality

$$\begin{pmatrix} \dot{n}(t) \\ \dot{s}(t) \end{pmatrix} \le \mathbf{V} \begin{pmatrix} n(t) \\ s(t) \end{pmatrix}, \quad \forall t \ge 0.$$

Let $\lambda > 0$ be sufficiently large such that $\lambda - \delta_i > 0$ and $\lambda - (\mu_i + c_i) > 0$ for all $i \in \Omega$. Then

$$(\lambda I + \mathbf{V})^2 \ge \begin{pmatrix} \Gamma L & \Gamma(\lambda I - D) \\ (\lambda I - D)L & L\Gamma \end{pmatrix},$$

which shows that $(\lambda I + \mathbf{V})^2$ is a positive irreducible matrix. Therefore, by the Perron–Frobenius theorem, there exists a componentwise positive left eigenvector $W \in \mathbb{R}_+^{2N}$ such that $W^\top \mathbf{V} = \mathsf{s}(\mathbf{V}) W^\top$. Since $\mathsf{s}(\mathbf{V}) \le 0$, it follows that

$$\frac{\mathrm{d}}{\mathrm{d}t}\left(W^\top \begin{pmatrix} n(t) \\ s(t) \end{pmatrix} \right) = W^\top \begin{pmatrix} \dot{n}(t) \\ \dot{s}(t) \end{pmatrix} \le \mathsf{s}(\mathbf{V})\, W^\top \begin{pmatrix} n(t) \\ s(t) \end{pmatrix} \le 0, \quad \forall t \ge 0. \tag{14.9}$$

In addition, note that the n-component is uniformly bounded by $1/\kappa_i$ in Θ. Using this in the second equation of (14.2), we obtain for each $i \in \Omega$,

$$\dot{s}_i(t) \le \sum_{j \in \Omega} K(i,j)\, r_j \frac{1}{\kappa_j} - \delta_i\, s_i(t),$$

which yields

$$\limsup_{t \to \infty} s_i(t) \le \frac{1}{\delta_i} \sum_{j \in \Omega} K(i,j)\, r_j \frac{1}{\kappa_j},$$

uniformly with respect to initial conditions in bounded sets of $\mathbb{R}_+^N$. Hence, the system is bounded dissipative, and by standard theory [16, 20], it admits a compact global attractor $\mathcal{A} \subset \Theta$. By (14.9),

$$\mathcal{V}(t) := W^\top \begin{pmatrix} n(t) \\ s(t) \end{pmatrix}$$

is a Lyapunov function and the only possible omega limit point is the origin, i. e., $\mathcal{A} = \{(0,0)\}$. Thus, the solution converges to the extinction state, completing the proof. □

The preceding lemma guarantees that the system (14.2) does not admit any nontrivial equilibrium when $\mathcal{T}_0(c) \leq 1$.

In what follows, we focus on the case where $\mathcal{T}_0(c) > 1$. More precisely, we are interested in the existence of a stationary state $(n^*, s^*) \in \Theta$ such that $(n^*, s^*) \neq (0,0)$ and

$$-\mu_i n_i^* + \gamma_i s_i^* H_i(n_i^*) - c_i n_i^* = 0, \tag{14.10a}$$

$$\sum_{j=1}^{N} K(i,j) r_j n_j^* - \delta_i s_i^* = 0\,. \tag{14.10b}$$

Next, observe that if $n_i^* > 0$ for some i, then equation (14.10b) implies $s_j^* > 0$ for all $j \in \Omega$, and then (14.10a) implies $n_j^* > 0$ for all $j \in \Omega$. Conversely, if $s_i^* > 0$ for some i, then (14.10b) guarantees the existence of some j with $n_j^* > 0$, which again implies $n_i^* > 0$ and $s_i^* > 0$ for all $i \in \Omega$. Hence, any nontrivial equilibrium necessarily satisfies

$$n_i^* > 0 \quad \text{and} \quad s_i^* > 0 \quad \text{for all } i \in \Omega.$$

Next, solving (14.10a) we obtain

$$n_i^* := \frac{\gamma_i s_i^*}{\mu_i + c_i + \kappa_i \gamma_i s_i^*}, \qquad i \in \Omega \tag{14.11}$$

so that (14.2) admits a nontrivial equilibrium if and only if there exists $s^* = (s_1^*, \dots, s_N^*) \in \mathbb{R}_+^N$, $s_i^* > 0$ for all $i \in \Omega$, such that

$$s_i^* = \frac{1}{\delta_i} \sum_{j=1}^{N} K(i,j)\, r_j \frac{\gamma_j s_j^*}{\mu_j + c_j + \kappa_j \gamma_j s_j^*}, \qquad i \in \Omega. \tag{14.12}$$

Proposition 2.1. *Let Assumption 2 hold. Let $c \in [0,1]^N$ be given. If $\mathcal{T}_0(c) > 1$, then system (14.2) admits a unique nontrivial equilibrium $(n^*(c), s^*(c)) \in \Theta$, with $n_i^*(c) > 0$ and $s_i^*(c) > 0$ for all $i \in \Omega$. Moreover, if $c, \bar{c} \in [0,1]^N$ such that $c \geq \bar{c}$, $\mathcal{T}_0(c) > 1$, and $\mathcal{T}_0(\bar{c}) > 1$, then we have*

$$n^*(\bar{c}) \geq n^*(c) \quad \textit{and} \quad s^*(\bar{c}) \geq s^*(c). \tag{14.13}$$

Proof. Define the operator $\mathcal{N}(c) : \mathbb{R}_+^N \to \mathbb{R}_+^N$ by

$$\mathcal{N}(c)[s]_i := \frac{1}{\delta_i} \sum_{j=1}^{N} K(i,j)\, r_j \,\frac{\gamma_j s_j}{\mu_j + c_j + \kappa_j \gamma_j s_j}, \qquad i \in \Omega,\ s \in \mathbb{R}_+^N. \tag{14.14}$$

Existence: Clearly, for all $s \in \mathbb{R}_+^N \setminus \{0\}$ and $i \in \Omega$, we have $\mathcal{N}(c)[s]_i > 0$. Moreover, the function $s \mapsto \mathcal{N}(c)[s]$ is monotone increasing and maps the compact convex set $[0, \kappa_0]^N$ into itself, where

$$\kappa_0 := \max_{i\in\Omega} \frac{1}{\delta_i} \sum_{j=1}^{N} K(i,j)\, r_j \,\frac{1}{\kappa_j}. \tag{14.15}$$

Next, observe that the Jacobian matrix $D\mathcal{N}(c)(0)$ is equal to the next generation matrix $\mathcal{G}(c)$. Since $\mathcal{G}(c)$ is a positive irreducible matrix and $\mathcal{T}_0(c) := \mathrm{r}(\mathcal{G}(c)) > 1$, the Perron–Frobenius theorem implies the existence of a strictly positive eigenvector $u \in \mathbb{R}_+^N$ such that $\mathcal{G}(c)[u] = \mathcal{T}_0(c)\, u$. Hence, for sufficiently small $\varepsilon_0 > 0$, we have

$$\mathcal{N}(c)[\varepsilon u] \geq \varepsilon u, \forall \epsilon \in (0, \epsilon_0)$$

componentwise. By monotonicity, the sequence $(\mathcal{N}(c)^k[\varepsilon u])_{k\geq 0}$ is increasing and bounded in $[0, \kappa_0]^N$. Thus, it converges to a nontrivial fixed point $s^*(\varepsilon) \in [0, \kappa_0]^N$ satisfying $\mathcal{N}(c)[s^*(\varepsilon)] = s^*(\varepsilon)$. Similarly, letting $v_i = \kappa_0$, $i \in \Omega$, we have $\mathcal{N}(c)[v] \leq v$, so the sequence $(\mathcal{N}(c)^k[v])_{k\geq 0}$ is decreasing and converges to a fixed point $s^*(\kappa_0) \in [0, \kappa_0]^N$ with $s^*(\varepsilon) \leq s^*(\kappa_0)$.

Uniqueness: We now prove uniqueness. Let $\bar{s}$ be any fixed point of Ω. Since $0 < \bar{s}_i \leq \kappa_0$ and $u_i > 0$ for all $i \in \Omega$, there exists $\varepsilon > 0$ such that $\varepsilon u \leq \bar{s} \leq v$. Then by monotonicity, $\mathcal{N}(c)^k[\varepsilon u] \leq \mathcal{N}(c)^k[\bar{s}] = \bar{s} \leq \mathcal{N}(c)^k[v]$ for all k. Taking the limit when $k \to +\infty$, we obtain $s^*(\epsilon) \leq \bar{s} \leq s^*(\kappa_0)$. Assume by contradiction that there exists some $i_0 \in \Omega$ such that $s_{i_0}^*(\epsilon) < s_{i_0}^*(\kappa_0)$ and set

$$\ell := \min_{i\in\Omega} \frac{s_i^*(\epsilon)}{s_i^*(\kappa_0)} \in (0, 1).$$

Then we have

$$\begin{aligned} s_i^*(\epsilon) - \ell s_i^*(\kappa_0) &= \frac{1}{\delta_i} \sum_{j=1}^{N} K(i,j)\, r_j \left[\frac{\gamma_j s_j^*(\kappa_0)}{\mu_j + c_j + \kappa \gamma_j s_j^*(\kappa_0)} - \frac{\gamma_j \ell s_j^*(\varepsilon)}{\mu_j + c_j + \kappa_j \gamma_j s_j^*(\varepsilon)} \right] \\ &\geq \frac{1}{\delta_i} \sum_{j=1}^{N} K(i,j)\, r_j \left[\frac{\gamma_j s_j^*(\kappa_0)}{\mu_j + c_j + \kappa_j \gamma_j s_j^*(\kappa_0)} - \frac{\gamma_j \ell s_j^*(\varepsilon)}{\mu_j + c_j + \kappa_j \gamma_j \ell s_j^*(\varepsilon)} \right] \end{aligned}$$

and since $s_{i_0}^*(\kappa_0) > s_{i_0}^*(\epsilon)$, it follows that $s_i^*(\epsilon) - \ell s_i^*(\kappa_0) > 0$ for all $i \in \Omega$. This is a contradiction with the definition of ℓ. This completes the proof of uniqueness.

Monotonicity: Let $c, \bar{c} \in [0,1]^N$ be given such that $c \geq \bar{c}$, $\mathcal{T}_0(c) > 1$, and $\mathcal{T}_0(\bar{c}) > 1$. Let $s^*(c)$ and $s^*(\bar{c})$ be the unique solution to

$$\mathcal{N}(c)[s^*(c)] = s^*(c) \quad \text{and} \quad \mathcal{N}(\bar{c})[s^*(\bar{c})] = s^*(\bar{c}).$$

Since $c \geq \bar{c}$ it follows that

$$s^*(c) = \mathcal{N}(c)[s^*(c)] \leq \mathcal{N}(\bar{c})[s^*(c)]$$

so that $(\mathcal{N}^k(\bar{c})[s^*(c)])_{k\geq 0}$ is an increasing sequence which converges to the unique fixed point of $\mathcal{N}(\bar{c})$. Consequently, $s^*(c) \leq s^*(\bar{c})$ so that $n^*(c) \leq n^*(\bar{c})$. □

As a consequence of Proposition 2.1, we have the corollary.

Corollary 2.1. *Let Assumption 2 hold. Assume in addition that $\mathcal{T}_0(c) > 1$ for all $c \in [0,1]^N$. Then there exists a continuous nonincreasing map $\Phi : [0,1]^N \to \operatorname{Int}(\mathbb{R}_+^N)$ such that (s^*, n^*, c) with $(s^*, n^*) \in \Theta \setminus \{(0,0)\}$ satisfies* (14.1) *if and only if*

$$s_i^*(c) = \Phi_i(c) \quad \text{and} \quad n_i^*(c) = \frac{\gamma_i \Phi_i(c)}{\mu_i + c_i + \gamma_i \kappa_i \Phi_i(c)}. \tag{14.16}$$

Proposition 2.2. *Let Assumption 2 hold. Let $c \in [0,1]^N$ be given such that $\mathcal{T}_0(c) > 1$. Then the positive equilibrium $(n^*(c), s^*(c))$ is globally asymptotically stable in $\Theta \setminus \{(0,0)\}$.*

Proof. Let $(n_0, s_0) \in \Theta \setminus \{(0,0)\}$ be arbitrary. The proof relies on monotonicity arguments. We first construct a family of initial conditions whose corresponding solutions converge to (n^*, s^*) as $t \to +\infty$.

From the proof of Lemma 2.3, there exists a vector $(\bar{n}, \bar{s})^\top \in \mathbb{R}_+^{2N}$ with $\bar{n}_i > 0$ and $\bar{s}_i > 0$ for all $i \in \Omega$ such that

$$\mathbf{V}(\bar{n}, \bar{s})^\top = \mathrm{s}(\mathbf{V})(\bar{n}, \bar{s})^\top.$$

Since $\mathcal{T}_0(c) > 1$, we have $\mathrm{s}(\mathbf{V}) > 0$. For each $\varepsilon > 0$, define $(\bar{n}_\varepsilon, \bar{s}_\varepsilon) = (\varepsilon\bar{n}, \varepsilon\bar{s})$. Then

$$\left.\begin{aligned} -M\bar{n}_\varepsilon + \Gamma F(\bar{s}_\varepsilon, \bar{n}_\varepsilon) - C\bar{n}_\varepsilon &= \varepsilon(-M\bar{n} + \Gamma\bar{s} - C\bar{n} - \varepsilon\bar{F}(\bar{n}, \bar{s})) \\ &= \varepsilon(\mathrm{s}(\mathbf{V})\bar{n} - \varepsilon\bar{F}(\bar{n}, \bar{s})), \\ L[\bar{n}_\varepsilon] - D\bar{s}_\varepsilon = \varepsilon(L[\bar{n}] - D\bar{s}) &= \varepsilon\mathrm{s}(\mathbf{V})\bar{s}, \end{aligned}\right\}$$

where $\bar{F}_i(\bar{n}, \bar{s}) = \kappa_i \bar{s}_i \bar{n}_i$ for all $i \in \Omega$. Therefore, there exists $\varepsilon_0 > 0$ such that for all $\varepsilon \in (0, \varepsilon_0)$, both right-hand sides are componentwise positive. By monotonicity and dissipativity of the system (14.1), the solution with initial condition $(\bar{n}_\varepsilon, \bar{s}_\varepsilon)$ is increasing in time and converges to the unique nontrivial steady state (n^*, s^*).

Next, for each $\eta \geq \kappa_0$ (with κ_0 defined in (14.15)), define $(\bar{n}_\eta, \bar{s}_\eta)$ as the vector with value $(1/\min \kappa_j, \eta)$ for each $i \in \Omega$. Then

$$\left.\begin{aligned}-M\bar{n}_\eta + \Gamma F(\bar{s}_\eta, \bar{n}_\eta) - C\bar{n}_\eta = -M\bar{n}_\eta - C\bar{n}_\eta \in \mathbb{R}^N_-,\\ L[\bar{n}_\eta] - D\bar{s}_\eta \in \mathbb{R}^N_-.\end{aligned}\right\}$$

Hence, the solution with initial condition $(\bar{n}_\eta, \bar{s}_\eta)$ is decreasing and converges to the same positive equilibrium (n^*, s^*).

To complete the proof, we use the irreducibility and positivity of the operator L. By standard arguments, for any $(n_0, s_0) \in \Theta \setminus \{(0,0)\}$, there exists $t_0 > 0$ such that $n_i(t_0) > 0$ and $s_i(t_0) > 0$ for all $i \in \Omega$. Choosing $\varepsilon > 0$ small enough and $\eta \geq \kappa_0$ large enough, we can ensure that

$$\bar{n}_\varepsilon \leq n(t_0) \leq \bar{n}_\eta, \quad \bar{s}_\varepsilon \leq s(t_0) \leq \bar{s}_\eta.$$

By comparison principles, it follows that the solution $(n(t), s(t))$ converges to the unique positive equilibrium (n^*, s^*) as $t \to +\infty$. □

2.3 Numerical illustration of the invasion

We now provide a numerical illustration of the invasion dynamics under the proposed control strategy. This serves to complement the analytical results by highlighting how the spatial distribution and intensity of the invader evolve over time, depending on the value of the threshold $\mathcal{T}_0(c)$. The numerical simulations are performed on a spatial grid composed of $N = 20$ sites arranged in a 4×5 layout, with each cell representing a square. Each site $i = 1, \ldots, N$ is assigned a suitability index $q_i \in [0,1]$, representing the favorability of local environmental conditions for the establishment and growth of the invasive species. This index is constructed to capture both spatial structure and randomness that influence local ecological dynamics. More precisely, $q_i \in [0,1]$ is obtained as a combination of a uniform random component, a spatially correlated sinusoidal variation, and Gaussian noise

$$q_i = \frac{z_i - \min_j z_j}{\max_j z_j - \min_j z_j},$$

with

$$z_i = 0.6U_i + 0.3\sin\left(\frac{x_i}{20{,}000}\right)\cos\left(\frac{y_i}{30{,}000}\right) + 0.1\mathcal{N}_i,$$

where $U_i \sim \mathcal{U}(0,1)$ are i. i. d. random variables uniformly distributed on the interval $[0,1]$, and $\mathcal{N}_i \sim \mathcal{N}(0, 0.2)$ are i. i. d. Gaussian random variables with mean zero and variance 0.2. The coordinates (x_i, y_i) denote the centroids of the square grid cells associated with site i. The natural mortality is set as $\mu_i = 1 - q_i^3$, decreasing with suitability to reflect higher survival in favorable environments. The germination rate is $\gamma_i = 0.01 + (1 - 0.01)q_i^2$, and the seed production rate is $r_i = 50 + 50q_i^3$, both increasing with q_i to model enhanced reproduction in suitable sites. The saturation coefficient is $\kappa_i = 1/(50 + 50 \cdot q_i)$, decreasing with q_i to account for more gradual saturation in rich environments. The seed decay rate is fixed at $\delta_i = 0.1$. Control costs are assumed to increase with site suitability, reflecting the idea that managing the invader is more difficult and resource-intensive in favorable environments (e. g., due to higher biomass, or faster growth). To capture this, the site-specific weights for the cost of intervention and the cost of invasion are defined, respectively, as $C_i = 1 + 3 \cdot q_i$ and $L_i = 1 + 2 \cdot q_i$. The discount factor is set to $\rho = 0.01$.

Simulations are run over a time horizon of $T = 1{,}100$ time units. To model spatial interactions, we use an exponentially decaying dispersal kernel defined by

$$K(i,j) = \frac{\exp(-d_{ij}/\lambda)}{\max_{i\in\Omega}(\sum_{k=1}^{N} \exp(-d_{ik}/\lambda))}, \tag{14.17}$$

where d_{ij} is the Euclidean distance between patches i and j, and $\lambda = 1$ controls dispersal range. The parameters are summarized in Table 14.1.

Table 14.1: Model parameters [Par], values, and their descriptions used in simulations.

Par.	Value	Description
N	20	Number of sites in the spatial grid (here, 4×5)
q_i	$0 \le q_i \le 1$	Suitability index of site i,
μ_i	$1-q_i^3$	Natural mortality rate of the invader at site i (decreases with q_i^3)
γ_i	$0.01+(1-0.01)q_i^2$	Germination rate at site i (increases with q_i^2)
r_i	$50+50q_i^3$	Seed production rate at site i (increases with q_i^3)
κ_i	$1/(50+50q_i)$	Saturation coefficient at site i (decreases with q_i)
δ_i	0.1	Decay rate of seeds at site i (fixed)
ρ	0.01	Discount factor for cost functions (fixed)
$K(i,j)$	(14.17)	Dispersal kernel from site j to site i (Exponentially decaying)
C_i	$1 + 3q_i$	Local cost coefficient for control effort in centralized strategy
L_i	$1 + 2q_i$	Local cost coefficient for control deviation in decentralized strategy
c_i	1	Intensity of removal efforts at site i
T	1,100	Final simulation time for numerical simulations
$n_i(t)$	outcome	Plant density at site i and time t
$s_i(t)$	outcome	Seed density at site i and time t

Figure 14.1 illustrates the convergence of the system (14.1) toward a unique equilibrium distribution under maximum control effort $c = (1, \ldots, 1)$, starting from different initial conditions. All trajectories lead to the same stationary distribution, confirming the uniqueness and global attractivity of the equilibrium state. The equilibrium distribution is numerically computed using (14.11) and (14.12), while the transient dynamics are obtained through numerical simulation of the model.

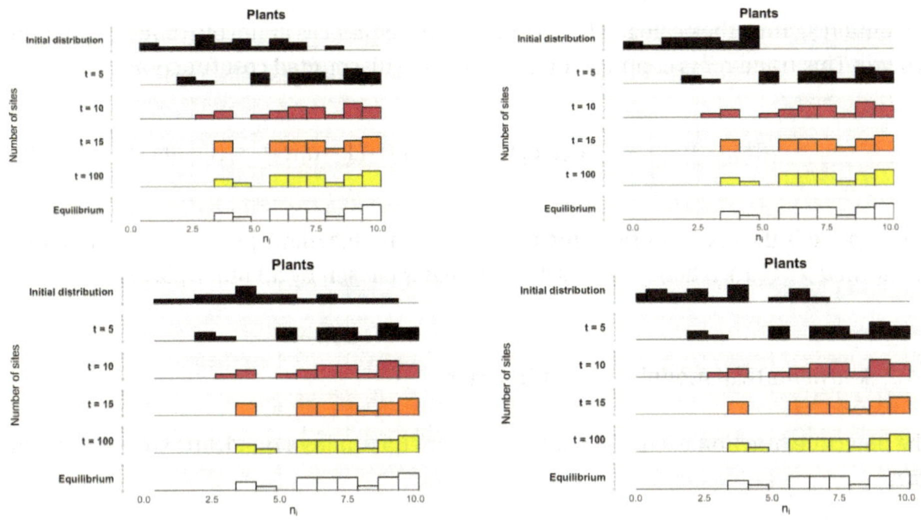

Figure 14.1: Each panel illustrates the convergence toward the same equilibrium state under maximum control effort ($c_i = 1$), starting from different initial conditions.

3 Control strategies

In this section, we consider first the decentralized control through local control regimes, then optimization of a centralized control strategy, and finally compare the two.

3.1 Decentralized control strategy

Here, we focus on the case where $\mathcal{T}_0(c) > 1$ for all $c \in [0,1]^N$, corresponding to situations where the invasive species persists despite the most intense feasible local interventions. This ecological regime reflects highly aggressive or resilient species that cannot be eradicated through local control alone.

3.1.1 Problem setting and motivation

We consider N local actors engaged in a noncooperative differential game, each responsible for controlling the dynamics of an invasive plant population at a specific site. These actors, indexed by the set $\Omega := \{1, \dots, N\}$, act independently to manage local invasion processes over time. At each location $i \in \Omega$, the local planner selects a control function $d_i : \mathbb{R}_+ \to [0,1]$, which represents the intensity of intervention (e.g., removal, treatment, or containment) applied to the invasive plant population. The goal of player i is to minimize the long-term cost associated with the invasion, by balancing the cost of intervention against the ecological or economic consequences of uncontrolled population growth. This trade-off is captured by the following discounted cost functional:

$$J_i(d_i; d_{-i}) = \int_0^{\infty} e^{-\rho t} [C(i)\, d_i^2(t)\, n_i^2(t) + L(i)\,(1 - d_i(t))^2\, n_i^2(t)]\, \mathrm{d}t, \tag{14.18}$$

where $\rho > 0$ is the discount rate, and $C(i), L(i) > 0$ are location-specific cost parameters. The term $d_{-i} := (d_j)_{j \neq i}$ denotes the set of strategies chosen by all other players.

3.1.2 Mathematical model and cost function

The objective function used throughout this paper assumes a quadratic structure of the form:

$$C(i)d_i^2 n_i^2 + L(i)(1 - d_i)^2 n_i^2,$$

where $d_i \in [0, 1]$ represents the fraction of biomass removed through local control at site i, and n_i is the standing biomass of the invasive species.

Interpretation of the cost terms

This functional form captures two intuitive features. First, both the control cost and the ecological damage are increasing and convex in biomass: controlling denser infestations is more costly, and leaving higher biomass unchecked leads to more severe ecological or economic impacts. Second, the marginal cost of effort rises with its intensity: doubling the fraction of biomass removed more than doubles the effort required, consistent with observed increasing returns to scale in invasive species management [11, 22].

Quadratic cost functions have become standard in dynamic bioeconomic models due to their analytical tractability and ability to capture increasing marginal costs or damages without imposing arbitrary bounds [4, 3]. In our context, the term $C(i)d_i^2 n_i^2$ represents effort-based costs, such as labor, machinery, and herbicide use, which scale nonlinearly with the size and aggressiveness of the infestation. The term $L(i)(1 - d_i)^2 n_i^2$ captures the residual damages caused by uncontrolled biomass, including loss of native biodiversity, degradation of ecosystem services, and reduced land usability. These are assumed to rise quadratically with density, following the "impact-per-abundance" relationships documented in ecological field studies [29].

While we do not calibrate $C(i)$ and $L(i)$ to empirical data in this theoretical analysis, their interpretation is aligned with practical management considerations. For instance, $C(i)$ can reflect the marginal cost of physical removal (labor and fuel), while $L(i)$ corresponds to an implicit valuation of ecological damages or opportunity costs of lost land use. In applications, these parameters could be estimated from field trials, economic valuation studies, or expert surveys [17].

3.1.3 Dynamics and strategic coupling

The dynamics at each location are governed by a system of differential equations given by

$$\dot{n}_i(t) = -\mu_i\, n_i(t) + \gamma_i\, s_i(t)\, H_i(n_i(t)) - d_i(t)\, n_i(t), \tag{14.19}$$

$$\dot{s}_i(t) = \sum_{j=1}^{N} K(i,j)\, r_j\, n_j(t) - \delta_i\, s_i(t), \tag{14.20}$$

with initial conditions $n_i(0) = n_i^0$, $s_i(0) = s_i^0$. The decentralized game is thus defined by the set of individual optimization problems (14.18), coupled through the dynamics (14.19)– (14.20). The strategic interaction among players arises from these shared dynamics, even though each player optimizes independently.

3.1.4 Optimality conditions and Nash equilibrium

Pontryagin's maximum principle

We introduce the costate processes $\lambda_i^n(t)$ and $\lambda_i^s(t)$ for each player i. The current-value Hamiltonian is

$$\begin{aligned}\mathcal{H}_i^c(n, s, d, \lambda^n, \lambda^s) = &-\left[C(i)\, d_i^2\, n_i^2 + L(i)\,(1-d_i)^2\, n_i^2\right] \\ &+ \lambda_i^n\left[-\mu_i\, n_i + \gamma_i\, s_i\, H_i(n_i) - d_i\, n_i\right] \\ &+ \lambda_i^s\left[\sum_j K(i,j)\, r_j\, n_j - \delta_i\, s_i\right].\end{aligned}$$

An open-loop Nash equilibrium must satisfy, for each i, the following first-order conditions:

(a) Control stationarity:

$$\frac{\partial \mathcal{H}_i^c}{\partial d_i} = -2\,C(i)\, d_i\, n_i^2 + 2\,L(i)\,(1-d_i)\, n_i^2 - \lambda_i^n\, n_i = 0.$$

(b) State equations: The dynamics (14.19)– (14.20) hold.

(c) Costate equations:

$$\dot{\lambda}_i^n = \rho\,\lambda_i^n - \frac{\partial \mathcal{H}_i^c}{\partial n_i}, \quad \dot{\lambda}_i^s = \rho\,\lambda_i^s - \frac{\partial \mathcal{H}_i^c}{\partial s_i}.$$

(d) Transversality:

$$\lim_{t\to\infty} e^{-\rho t}\,\lambda_i^n(t) = 0, \quad \lim_{t\to\infty} e^{-\rho t}\,\lambda_i^s(t) = 0.$$

Let $(n_i^*, s_i^*, d_i^*, \lambda^{n*}, \lambda^{s*})$ for all $i \in \Omega$ denote the global equilibrium state and control profile. The dynamics must satisfy, for each $i \in \Omega$,

$$0 = -\mu_i n_i^* + \gamma_i s_i^* H_i(n_i^*) - d_i^* n_i^*, \tag{14.21a}$$

$$0 = \sum_{j=1}^{N} K(i,j) r_j n_j^* - \delta_i s_i^*. \tag{14.21b}$$

Moreover, the local optimality conditions for planner i then require that the following hold:

$$\left.\begin{aligned} 0 &= -2C(i)\, d_i^* n_i^{*2} + 2L(i)(1-d_i^*)\, n_i^{*2} - \lambda_i^{n*} n_i^*, \\ 0 &= \rho\lambda_i^{n*} + 2\big[C(i)d_i^{*2} + L(i)(1-d_i^*)^2\big] n_i^* \\ &\quad + \lambda_i^{n*}(\mu_i - \gamma_i s_i^* H_i'(n_i^*) + d_i^*) - \lambda_i^{s*} K(i,i)\, r_i, \\ 0 &= (\rho + \delta_i)\lambda_i^{s*} - \gamma_i H_i(n_i^*)\lambda_i^{n*}. \end{aligned}\right\} \tag{14.22}$$

By Corollary 2.1, (s^*, n^*, d^*) with $(s^*, n^*) \in \Theta \setminus \{(0,0)\}$ satisfies (14.21) if and only if

$$s_i^* := s_i^*(d^*) = \Phi_i(d^*) \quad \text{and} \quad n_i^* := n_i^*(d^*) = \frac{\gamma_i \Phi_i(d^*)}{\mu_i + d_i^* + \gamma_i \kappa_i \Phi_i(d^*)} \tag{14.23}$$

for all $i \in \Omega$.

Furthermore, we also have for each $i \in \Omega$,

$$\left.\begin{aligned} \Phi_i(d^*) &= \frac{1}{\delta_i} \sum_{j=1}^{N} K(i,j) \frac{r_j \gamma_j \Phi_j(d^*)}{\mu_j + d_j^* + \kappa_j \gamma_j \Phi_j(d^*)} \\ &= \frac{1}{\delta_i} \sum_{j=1}^{N} K(i,j)\, r_j n_j^* \end{aligned}\right\} \tag{14.24}$$

and

$$\mu_i - \gamma_i \Phi_i(d^*) H_i'(n_i^*) + d_i^* = \frac{\gamma_i}{n_i^* \delta_i} \sum_{j=1}^{N} K(i,j)\, r_j n_j^*. \tag{14.25}$$

Consequently, the existence of an equilibrium profile $(n_i^*, s_i^*, d_i^*, \lambda^{n*}, \lambda^{s*})$ is equivalent of the existence of $d^* \in [0,1]^N$, and λ^{n*} such that for all $i \in \Omega$,

$$n_i^* = \frac{\gamma_i \Phi_i(d^*)}{\mu_i + d_i^* + \gamma_i \kappa_i \Phi_i(d^*)}, \tag{14.26a}$$

$$0 = -2C(i)\, d_i^* n_i^{*2} + 2L(i)(1-d_i^*)\, n_i^{*2} - \lambda_i^{n*} n_i^*, \tag{14.26b}$$

$$\begin{aligned} 0 &= \rho\lambda_i^{n*} + 2\big[C(i)d_i^{*2} + L(i)(1-d_i^*)^2\big] n_i^* \\ &\quad + \lambda_i^{n*}(\mu_i - \gamma_i \Phi_i(d^*) H_i'(n_i^*) + d_i^*) \end{aligned} \tag{14.26c}$$

$$- \frac{\gamma_i H_i(n_i^*)}{\rho + \delta_i} \lambda_i^{n*} K(i,i) r_i,$$

$$\lambda_i^{s*} = \frac{\gamma_i H_i(n_i^*)}{\rho + \delta_i} \lambda_i^{n*}. \tag{14.26d}$$

Next, using (14.26b), we obtain

$$d_i^* = \frac{L(i)}{C(i) + L(i)} - \frac{\lambda_i^{n*}}{2n_i^*} \frac{1}{C(i) + L(i)}, \quad i \in \Omega. \tag{14.27}$$

Economic interpretation of control rule

Equation (14.27) admits a clear economic interpretation: the baseline term

$$\frac{L(i)}{C(i) + L(i)}$$

corresponds to the relative valuation of residual damages versus control costs, and thus represents the effort one would exert in the absence of strategic considerations. The corrective term

$$- \frac{\lambda_i^{n*}}{2 n_i^* \, (C(i) + L(i))}$$

reflects the opportunity cost of control as measured by the shadow price λ_i^{n*} on local biomass: when the marginal value of reducing biomass to downstream sites is high, λ_i^{n*} increases and the optimal effort d_i^* is further adjusted downward to account for diminishing returns on additional removal. In this way, Equation (14.27) balances the private marginal cost of effort with the social marginal benefit of reduced spread, yielding an equilibrium rule that is both economically intuitive and directly implementable in spatially explicit management frameworks.

Plugging (14.25) into (14.26c), it becomes

$$\lambda_i^{n*} = \widehat{\Psi}_i(\rho, d^*) \lambda_i^{n*} - \mathsf{R}_i(\rho, d^*) \tag{14.28}$$

with n_i^* defined in (14.23), and for all $i \in \Omega$, $\widehat{\Psi}_i$ and R_i are given by

$$\widehat{\Psi}_i(\rho, d^*) := \frac{\frac{\gamma_i}{\delta_i} \frac{\delta_i}{\rho + \delta_i} H_i(n_i^*) K(i,i) \, r_i n_i^*}{\rho n_i^* + \frac{\gamma_i}{\delta_i} \sum_{j=1}^N K(i,j) \; r_j n_j^*} \tag{14.29}$$

and

$$\mathsf{R}_i(\rho, d^*) := \frac{2[C(i) d_i^{*2} + L(i)(1 - d_i^*)^2] n_i^{*2}}{\rho n_i^* + \frac{\gamma_i}{\delta_i} \sum_{j=1}^N K(i,j) \; r_j n_j^*}. \tag{14.30}$$

Recalling that $0 < H_i(n_i^*) < 1$, it follows that $\widehat{\Psi}_i(\rho, d^*) \in (0,1)$ for all $i \in \Omega$ so that (14.28) can be solved to obtain

$$\Psi_i(\rho, d^*) := -\frac{\lambda_i^{n*}}{2n_i^*} = \left(1 - \widehat{\Psi}_i(\rho, d^*)\right)^{-1} \frac{\mathrm{R}_i(\rho, d^*)}{2n_i^*}, \quad i \in \Omega \tag{14.31}$$

with n_i^* given by (14.26a).

3.1.5 Characterizing equilibria: Fixed-point system

Next, by combining (14.27) and (14.31) we end-up with the following fixed-point problem:

$$\left.\begin{aligned} n_i^* &= \frac{\gamma_i \Phi_i(d^*)}{\mu_i + d_i^* + \gamma_i \kappa_i \Phi_i(d^*)}, \quad i \in \Omega \\ d_i^* &= \frac{L(i)}{C(i) + L(i)} + \frac{\Psi_i(\rho, d^*)}{C(i) + L(i)}, \quad i \in \Omega. \end{aligned}\right\} \tag{14.32}$$

Note that for each $i \in \Omega$ and $\rho > 0$, the map $d \in [0,1]^N \mapsto \Psi(\rho, d)$ is continuous and bounded with

$$0 \le \Psi_i(\rho, d) \le \frac{\max(C(i), L(i))}{\rho + \frac{\rho}{\rho+\delta_i} \frac{\gamma_i}{\delta_i} K(i,i)\, r_i}, \quad i \in \Omega.$$

Next, we set

$$\Psi_i^+(\rho) := \sup_{d \in [0,1]^N} \Psi_i(\rho, d), \quad i \in \Omega. \tag{14.33}$$

Consequently, we obtain the following existence result.

Proposition 3.1. *Let Assumption 2 be satisfied. Assume in addition that $\mathcal{T}_0(c) > 1$ for all $c \in [0,1]^N$, and that*

$$0 \le \Psi_i^+(\rho) \le C(i), \quad \forall i \in \Omega. \tag{14.34}$$

Then there exists $d^* \in [0,1]^N$ solution to the system

$$d_i^* = \frac{L(i)}{C(i) + L(i)} + \frac{\Psi_i(\rho, d^*)}{C(i) + L(i)}, \quad i \in \Omega.$$

Proposition 3.1 follows from standard fixed-point arguments. Specifically, the continuity of $\Psi_i(\rho, d)$ with respect to $d \in [0,1]^N$ and the condition (14.34) ensure that the map

$$d \in [0,1]^N \mapsto \frac{L(i)}{C(i)+L(i)} + \frac{\Psi_i(\rho,d)}{C(i)+L(i)}$$

is continuous and maps the compact convex set $[0,1]^N$ into itself, allowing the application of Brouwer's fixed-point theorem to guarantee the existence of a solution.

Moreover, condition (14.34) is satisfied if the discount rate $\rho > 0$ fulfills

$$\rho + \frac{\rho}{\rho+\delta_i}\frac{\gamma_i}{\delta_i}K(i,i)\,r_i \geq \frac{\max(C(i),\,L(i))}{C(i)}, \quad \forall i \in \Omega, \tag{14.35}$$

which provides a simple and conservative criterion for ensuring boundedness of $\Psi_i(\rho,d)$ within the admissible control range.

3.2 Centralized control strategy

We now consider a centralized optimization framework in which a single global planner coordinates control efforts across all locations with the objective of minimizing the total long-term cost of managing the invasive species. This framework represents a coordinated management regime, where all decisions are made collectively to account for spatial interactions and externalities resulting from the dispersal of the invasive population.

3.2.1 Centralized control problem statement

The planner selects control functions $c_i : \mathbb{R}_+ \to [0,1]$ for each site $i \in \Omega := \{1,\ldots,N\}$, where $c_i(t)$ denotes the intensity of intervention (e. g., mechanical removal, chemical treatment, or habitat management) applied at location i at time t. The aim is to determine a control strategy that minimizes the aggregate discounted cost incurred from both intervention efforts and the persistence of the invasive species whenever $\mathcal{T}_0(c) > 1$ for all $c \in [0,1]^N$.

This trade-off is captured by the following global cost functional:

$$J_P := \sum_{i=1}^{N}\int_0^{\infty} e^{-\rho t}\big[C(i)\,c_i^2(t)\,n_i^2(t) + L(i)\,(1-c_i(t))^2\,n_i^2(t)\big]\mathrm{d}t, \tag{14.36}$$

where $\rho > 0$ is the discount rate, and $C(i), L(i) > 0$ are site-specific cost coefficients associated respectively with control actions and invasion-induced damages.

Global cost functional

The dynamics of the system evolve according to the same spatially coupled differential equations as in the decentralized setting:

$$\dot{n}_i(t) = -\mu_i\, n_i(t) + \gamma_i\, s_i(t)\, H_i(n_i(t)) - c_i(t)\, n_i(t), \tag{14.37}$$

$$\dot{s}_i(t) = \sum_{j=1}^{N} K(i,j)\, r_j\, n_j(t) - \delta_i\, s_i(t), \tag{14.38}$$

with initial conditions $n_i(0) = n_i^0$, $s_i(0) = s_i^0$.

3.2.2 Coupled population dynamics

In this centralized setting, the planner simultaneously optimizes the full control profile $(c_1, \dots, c_N)$, fully internalizing the ecological interdependencies among sites. This allows the global strategy to exploit synergies and avoid the inefficiencies that typically arise in noncooperative settings, especially when the invasive species exhibits strong spatial coupling across locations.

3.2.3 Centralized Hamiltonian and costate variables

As in the decentralized setting, we introduce costate variables $\lambda_i^n(t)$ and $\lambda_i^s(t)$ to characterize the necessary optimality conditions. The current-value Hamiltonian associated with the centralized control problem is given by

$$\mathcal{H}_P^c := \sum_{i=1}^{N} \mathcal{H}_i^c(n, s, c, \lambda^n, \lambda^s), \tag{14.39}$$

where for each $i \in \Omega$,

$$\begin{aligned}
\mathcal{H}_i^c(n, s, c, \lambda^n, \lambda^s) &= -\left[C(i)\, c_i^2\, n_i^2 + L(i)\,(1 - c_i)^2\, n_i^2\right] &(14.40)\\
&\quad + \lambda_i^n\left[-\mu_i n_i + \gamma_i s_i H_i(n_i) - c_i n_i\right] &(14.41)\\
&\quad + \lambda_i^s\left[\sum_{j=1}^{N} K(i,j)\, r_j\, n_j - \delta_i s_i\right]. &(14.42)
\end{aligned}$$

The necessary conditions for optimality are as follows:

(a) Control stationarity:

$$\begin{aligned}
\frac{\partial \mathcal{H}_P^c}{\partial c_i} &= -2C(i)\, c_i\, n_i^2 + 2L(i)\,(1 - c_i)\, n_i^2 - \lambda_i^n\, n_i \\
&= 0.
\end{aligned}$$

(b) State dynamics: for all $i \in \Omega$,

$$\dot{n}_i = -\mu_i n_i + \gamma_i s_i H_i(n_i) - c_i n_i,$$

$$\dot{s}_i = \sum_{j=1}^{N} K(i,j)\, r_j\, n_j - \delta_i s_i,$$

(c) Costate equations: for all $i \in \Omega$,

$$\dot{\lambda}_i^n = \rho\lambda_i^n - \frac{\partial \mathcal{H}_P^c}{\partial n_i},$$
$$\dot{\lambda}_i^s = \rho\lambda_i^s - \frac{\partial \mathcal{H}_P^c}{\partial s_i}.$$

(d) Transversality conditions:

$$\lim_{t\to\infty} e^{-\rho t}\lambda_i^n(t) = 0, \quad \lim_{t\to\infty} e^{-\rho t}\lambda_i^s(t) = 0.$$

3.2.4 Equilibrium states and controls

Derivation of equilibrium controls

Following the same approach as in the decentralized case, we find that the equilibrium values (n_i^*, s_i^*) satisfy

$$s_i^* = \Phi_i(c^*), \quad n_i^* = \frac{\gamma_i \Phi_i(c^*)}{\mu_i + c_i^* + \gamma_i \kappa_i \Phi_i(c^*)}, \quad i \in \Omega, \tag{14.43}$$

where $\Phi_i(\cdot)$ is defined as in Corollary 2.1. Solving the stationarity condition yields the equilibrium control ·

$$c_i^* = \frac{L(i)}{C(i) + L(i)} - \frac{\lambda_i^{n*}}{2n_i^*(C(i) + L(i))}, \quad i \in \Omega. \tag{14.44}$$

3.2.5 Algebraic structure of costates

The equilibrium costate equations then reduce to the following algebraic system:

$$\begin{aligned} 0 = {} & \rho\lambda_i^{n*} + 2[C(i)(c_i^*)^2 + L(i)(1 - c_i^*)^2]n_i^* \\ & + \lambda_i^{n*}(\mu_i - \gamma_i s_i^* H_i'(n_i^*) + c_i^*) \\ & - \sum_{k=1}^{N} \lambda_k^{s*} K(k,i)\, r_i, \quad i \in \Omega, \end{aligned} \tag{14.45a}$$

$$0 = (\rho + \delta_i)\,\lambda_i^{s*} - \gamma_i H_i(n_i^*)\,\lambda_i^{n*}, \quad i \in \Omega. \tag{14.45b}$$

Substituting (14.45b) into the first equation and using similar arguments as in the decentralized case, we obtain for all $i \in \Omega$,

$$\lambda_i^{n*} = -\frac{2[C(i)(c_i^*)^2 + L(i)(1-c_i^*)^2](n_i^*)^2}{\rho n_i^* + \frac{\gamma_i}{\delta_i}\sum_{j=1}^N K(i,j)\, r_j\, n_j^*} + \frac{n_i^* r_i}{\rho n_i^* + \frac{\gamma_i}{\delta_i}\sum_{j=1}^N K(i,j)\, r_j\, n_j^*} \times \sum_{k=1}^N K(k,i)\,\frac{\gamma_k}{\delta_k}\,\frac{\delta_k}{\rho+\delta_k}\, H_k(n_k^*)\,\lambda_k^{n*}. \tag{14.46}$$

To express (14.46) more compactly, we define for each $i \in \Omega$,

$$\mathsf{R}_i^P[c^*,\rho] := \frac{2[C(i)(c_i^*)^2 + L(i)(1-c_i^*)^2](n_i^*)^2}{\rho n_i^* + \frac{\gamma_i}{\delta_i}\sum_{j=1}^N K(i,j)\, r_j\, n_j^*},$$

and the linear operator $\widehat{\Psi}^P[c^*,\rho] : \mathbb{R}^N \to \mathbb{R}^N$ given by

$$\widehat{\Psi}_i^P[c^*,\rho](v) := \frac{n_i^* r_i}{\rho n_i^* + \frac{\gamma_i}{\delta_i}\sum_{j=1}^N K(i,j)\, r_j\, n_j^*} \times \sum_{k=1}^N K(k,i)\,\frac{\gamma_k}{\delta_k}\,\frac{\delta_k}{\rho+\delta_k}\, H_k(n_k^*)\, v_k, \quad v \in \mathbb{R}^N,\ i \in \Omega.$$

Fixed-point operator representation

With the above notation, (14.46) becomes

$$\lambda_i^{n*} = \widehat{\Psi}_i^P[c^*,\rho](\lambda^{n*}) - \mathsf{R}_i^P[c^*,\rho], \quad i \in \Omega. \tag{14.47}$$

3.2.6 Existence and boundedness conditions

Lemma 3.1. *Let Assumption 2 be satisfied. Let $c^* \in [0,1]^N$ be given such that $\mathcal{T}_0(c^*) > 1$. Let n_i^* be given by* (14.43). *Then the linear operator $I - \widehat{\Psi}^P : \mathbb{R}^N \to \mathbb{R}^N$ is invertible with $(I - \widehat{\Psi}^P[c^*,\rho])^{-1}(\mathbb{R}_+^N \setminus \{0\}) \subset \operatorname{Int}(\mathbb{R}_+^N)$. Moreover, we have*

$$\sup_{c\in[0,1]^N} \left\| (I - \widehat{\Psi}^P[c,\rho])^{-1} \mathsf{R}^P[c,\rho] \right\| < \infty. \tag{14.48}$$

Proof. To prove the first part of our lemma, we prove that the operator $\widehat{\Psi}^P(c^*,\rho)$ has spectral radius, $\mathrm{r}(\Psi^P(c^*,\rho))$, strictly less than one. To this end, consider the vector

$$v_k := \frac{\delta_k}{\gamma_k} r_k n_k^*, \quad k \in \Omega \tag{14.49}$$

and note that for each $i \in \Omega$,

$$\widehat{\Psi}_i^P(c^*,\rho)[v] = \frac{n_i^* r_i}{\rho n_i^* + \frac{\gamma_i}{\delta_i}\sum_{j=1}^N K(i,j)\, r_j n_j^*} \times \sum_{k=1}^N K(k,i)\,\frac{\delta_k}{\rho+\delta_k} H_k(n_k^*)\, r_k n_k^*.$$

Recalling that $H_k(n_k^*)$, $\frac{\delta_i}{\rho+\delta_i} \in (0,1)$, it follows that

$$\widehat{\Psi}_i^P(c^*,\rho)[v] < \frac{v_i}{\rho n_i^* + \frac{\gamma_i}{\delta_i}\sum_{j=1}^N K(i,j)\, r_j n_j^*}\frac{\gamma_i}{\delta_i} \times \sum_{k=1}^N K(k,i)\, r_k n_k^* < v_i, i \in \Omega.$$

Thus, using the Perron–Frobenius we obtain $\mathrm{r}(\Psi^P(c^*,\rho)) < 1$. To complete the proof, we note that since $c \mapsto \mathrm{r}(\Psi^P(c^*,\rho))$ is continuous and $[0,1]^N$ is compact, there exists $\bar{c} \in [0,1]^N$ such that

$$\mathrm{r}(\Psi^P(c,\rho)) \le \mathrm{r}(\Psi^P(\bar{c},\rho)) < 1, \quad \forall c \in [0,1]^N.$$

To obtain (14.48), we first note that $c \in [0,1]^N \to (I - \widehat{\Psi}^P[c,\rho])^{-1} \in \mathcal{L}(\mathbb{R}^N)$ is continuous so that

$$\sup_{c\in[0,1]^N} \|(I - \widehat{\Psi}^P[c,\rho])^{-1}\|_{\mathcal{L}(\mathbb{R}^N)} < \infty.$$

Next, by Corollary 2.1, we know that $\inf_{c\in[0,1]^N} n_i^*(c) > 0$ and $\sup_{c\in[0,1]^N} n_i^*(c) \le 1/\kappa_i$, for all $i \in \Omega$ so that

$$\sup_{c\in[0,1]^N} \mathrm{R}_i[\rho,c] < \infty, \quad \forall i \in \Omega.$$

The proof is completed. □

Recursive solution scheme

Thanks to Lemma (3.1) and (14.44), we can formulate the existence of control as the following fixed-point problem

$$\left.\begin{aligned} n_i^* &= \frac{\gamma_i \Phi_i(c^*)}{\mu_i + c_i^* + \gamma_i\kappa_i\Phi_i(c^*)}, \\ \Psi^P[\rho,c^*] &:= -\lambda^{n*} \\ &= (I - \widehat{\Psi}^P(\rho,c^*))^{-1}\mathrm{R}[\rho,c^*], \\ c_i^* &= \frac{L(i)}{C(i)+L(i)} + \frac{\Psi_i^P[\rho,c^*]}{2n_i^*(C(i)+L(i))}. \end{aligned}\right\} \tag{14.50}$$

3.2.7 Existence of centralized control solution

Consequently, by using Browder fixed-point theorem, we obtain the following result.

Proposition 3.2. *Let Assumption 2 be satisfied. Assume in addition that* $\mathcal{T}_0(c) > 1$ *for all* $c \in [0,1]^N$*, and that*

$$0 \leq \frac{\Psi^P[\rho, c]}{2n_i^*(c)} \leq C(i), \quad \forall i \in \Omega,\ c \in [0,1]^N. \tag{14.51}$$

Then there exists $c^* \in [0,1]^N$ *solution to*

$$c_i^* = \frac{L(i)}{C(i) + L(i)} + \frac{\Psi_i^P[\rho, c^*]}{2n_i^*(c^*)(C(i) + L(i))}, \quad \forall i \in \Omega.$$

3.3 Comparison of decentralized and centralized strategies, and incentive mechanisms

This section compares the decentralized equilibrium strategies derived in Section 3.1 with the socially optimal solution of the centralized problem in Section 3.2. We quantify the performance gap between these two configurations and propose an incentive mechanism that aligns decentralized decisions with the social optimum by internalizing spatial externalities. All the parameters used in this section are described in Section 2.3.

Both the decentralized equilibrium and the centralized optimal control are computed using an iterative fixed-point algorithm. Starting from an initial guess in the admissible control space $[0,1]^N$, the algorithm alternates between solving for the stationary state of the system and updating the control variables using explicit formulas derived from first-order optimality conditions (Proposition 3.2 and Proposition 3.1). Convergence is declared when the sup-norm of the difference between successive iterates satisfies $\|u^{(k+1)} - u^{(k)}\|_\infty < \varepsilon$, where $u^{(k)} \in [0,1]^N$ is the control at iteration k and $\varepsilon > 0$ is a prescribed tolerance. To ensure robustness and verify uniqueness, the algorithm is executed from multiple random initializations, and pairwise discrepancies between the resulting controls are examined (see Figures 14.2–14.3).

3.4 Benefits of centralized control and coordination costs

The centralized planner systematically outperforms the decentralized agents for two primary reasons. First, the planner internalizes spatial spillovers: the seed production at one location affects all connected sites through dispersal. In contrast, local agents minimize their private costs, disregarding the ecological consequences of their actions on others. Formally, this discrepancy manifests in the adjoint equations: the centralized

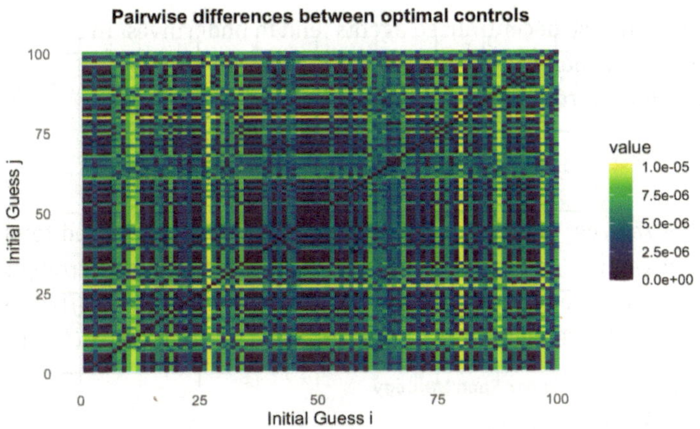

Figure 14.2: Pairwise maximum absolute differences $\max_{k\in\Omega}|d_k^{(i)*} - d_k^{(j)*}|$ between optimal decentralized controls obtained from 1000 independent runs of the fixed-point algorithm. Each run is initialized with a random vector in $[0,1]^N$. The consistently small values confirm the numerical stability and apparent uniqueness of the decentralized equilibrium.

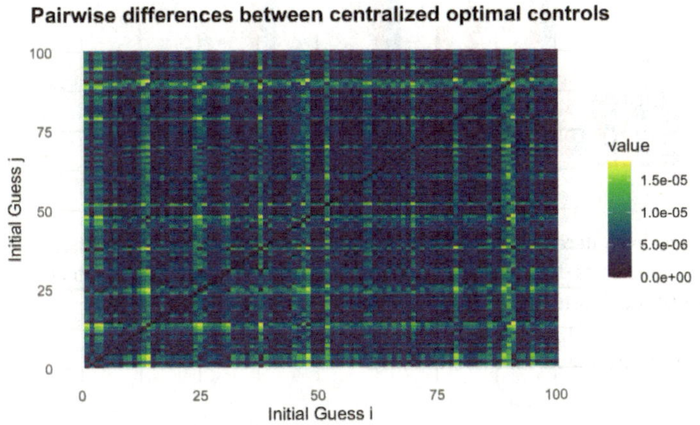

Figure 14.3: Pairwise maximum absolute differences $\max_{k\in\Omega}|c_k^{(i)*} - c_k^{(j)*}|$ between optimal centralized controls computed over 1000 randomized initializations. The low variability supports the robustness and uniqueness of the centralized optimal control.

formulation includes the aggregated effect

$$\sum_{k=1}^{N} \lambda_k^{S*} K(k,i) r_i,$$

whereas the decentralized problem only accounts for the self-effect $-\lambda_i^{S*} K(i,i) r_i$.

Second, the centralized solution ensures an efficient spatial allocation of control effort by equalizing the marginal social cost of reducing biomass across locations. High-dispersal “source” sites receive intensified efforts to mitigate their contribution to regional spread, while “sink” sites, which are susceptible but weak emitters, are subject to less intense control. In contrast, decentralized agents tend to underinvest in sources, since the benefits of reduced dispersal are external to their objective.

We define the cost of noncoordination as the difference in total cost:

$$\Delta J := J_P^{\text{dec}} - J_P^{\text{cent}} \geq 0,$$

where J_P^{cent} is the centralized cost and $J_P^{\text{dec}} = \sum_{i=1}^{N} J_i(d_i^*, d_{-i}^*)$ is the decentralized total. This gap is strictly positive when $K(i,j) > 0$ for some $i \neq j$, and increases with the strength of spatial coupling $\max_{i,j} K(i,j)$, heterogeneity in local cost ratios $|C(i)/L(i) - C(j)/L(j)|$, and asymmetries in dispersal (see Figures 14.4–14.5). Numerical simulations (not re-

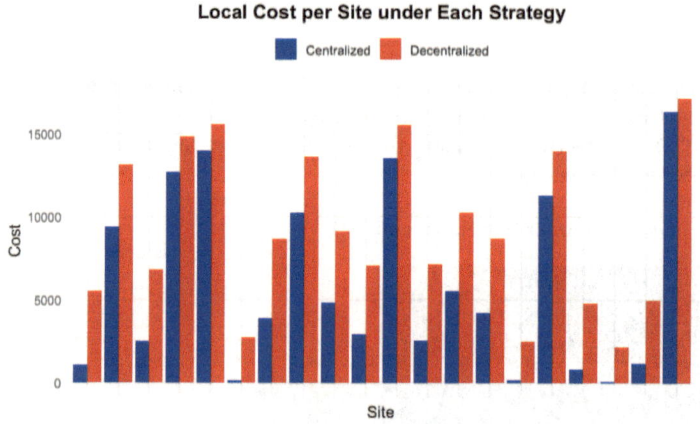

Figure 14.4: Local costs per site under decentralized (red) and centralized (blue) strategies. While both strategies yield heterogeneous outcomes due to spatial variation in suitability, the centralized planner systematically achieves lower or equal costs per site by internalizing externalities.

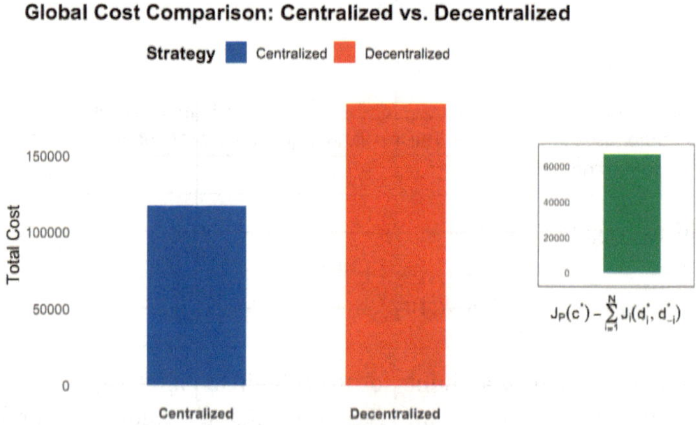

Figure 14.5: Total cost incurred under decentralized and centralized control strategies. The centralized strategy results in a strictly lower aggregate cost, illustrating the inefficiency of uncoordinated local decisions.

ported here) show that in highly connected networks, the coordination gap ΔJ can exceed 25 % of total cost.

3.5 Fiscal transfers for spatial externality internalization

To mitigate the inefficiencies of decentralization, we propose a dynamic transfer scheme that rewards local agents for reducing seed dispersal to others. The transfer $\tau_i(t)$ paid to agent i at time t modifies its objective to

$$\tilde{J}_i = \int_0^{\infty} e^{-\rho t}\left[C(i)d_i^2 n_i^2 + L(i)(1-d_i)^2 n_i^2 - \tau_i(t)\right]dt.$$

The optimal transfer $\tau_i^*(t)$ is derived by explicitly identifying and quantifying the spatial externalities generated by agent i's decisions.

Step 1: Identifying the marginal externality

In the centralized problem, the adjoint equation associated with the seed stock s_k includes the term:

$$-\sum_{k=1}^{N} \lambda_k^{s*} K(k,i) r_i,$$

which captures the marginal impact of plant density n_i at site i on seed inflows at all sites k, weighted by the marginal cost of seeds λ_k^{s*}. In contrast, the decentralized agent i internalizes only the term $-\lambda_i^{s*} K(i,i) r_i$, and ignores the effect on $k \neq i$. The resulting externality is therefore

$$\mathcal{E}_i(t) := \sum_{k \neq i} \lambda_k^{s*}(t) K(k,i) r_i n_i(t).$$

Step 2: Marginal impact of control effort

Control reduces plant biomass n_i according to the dynamic $\dot{n}_i = \cdots - d_i n_i$. The marginal effect of effort d_i on the externality is

$$\frac{\partial \mathcal{E}_i}{\partial d_i} = \sum_{k \neq i} \lambda_k^{s*}(t) K(k,i) r_i \frac{\partial}{\partial d_i}(n_i d_i).$$

Assuming a short-term approximation where n_i is quasiconstant with respect to d_i, we have

$$\frac{\partial}{\partial d_i}(n_i d_i) \approx n_i(t),$$

and hence

$$\frac{\partial \mathcal{E}_i}{\partial d_i} \approx \sum_{k \neq i} \lambda_k^{s*}(t) K(k,i) r_i n_i(t).$$

Step 3: Transfer design

The fiscal transfer should reward agents proportionally to the marginal social benefit generated by their effort. This gives the optimal transfer:

$$\tau_i^*(t) = \phi \cdot \left(\frac{\partial \mathcal{E}_i}{\partial d_i} \cdot d_i(t) \right) = \phi \sum_{k \neq i} \lambda_k^{s*}(t) K(k,i) r_i n_i(t) d_i(t),$$

where $\phi \in [0,1]$ allows for partial or full internalization.

Step 4: Ecological interpretation

The term $K(k,i) r_i n_i d_i$ represents the reduction in seed export from site i to k, achieved by controlling biomass at i. Multiplying by the marginal social cost λ_k^{s*} quantifies the ecological benefit accrued at k. Summing over all $k \neq i$ yields the total external benefit attributable to agent i's effort.

Step 5: Final expression and implementation

Using consistent notation with previous sections (index j for recipient sites), we obtain

$$\tau_i^*(t) = \phi \sum_{j \neq i} \lambda_j^{s*}(t) K(j,i) r_i n_i(t) d_i(t).$$

In steady state, the implementation relies on optimal values $(\lambda_j^{s*}, n_i^*, d_i^*)$ from the centralized solution:

$$\tau_i^* = \eta \sum_{j \neq i} \lambda_j^{s*} K(j,i) r_i n_i^* d_i^*, \quad \text{with } \eta \in [0,1].$$

When real-time monitoring is costly, simplified schemes may rely on observed reductions in plant or seed densities (n_i, s_i).

4 Discussion

We have shown that the stationary equilibrium of a spatially extended invasive-species control game can be characterized not through conventional first-order conditions but as the fixed point of a globally monotone reaction operator. By using Browder's theorem together with the spectral properties of the underlying dispersal kernel, we established existence, uniqueness and stability of the stationary control equilibrium under weak assumptions. Crucially, our fixed-point algorithm converges in practice even for systems comprising hundreds of spatial patches, thereby overcoming the "curse of dimensionality" inherent in shooting-based solutions of partial differential equation frameworks.

Our modeling approach is motivated by empirical studies showing that surveillance and treatment strategies must be spatially coordinated to be cost-effective [10, 19]. Moreover, the cost structure we consider—quadratic in both effort and abundance—is commonly used in the environmental economics literature for its tractability and for capturing the increasing marginal costs of control and damages [14, 27]. The quadratic cost terms can be interpreted as stylized approximations of real-world components such as labor input, equipment degradation, and nonlinear valuation of ecosystem loss.

Cobourn et al. (2018), demonstrated that bilateral negotiations—even when theoretically optimal—frequently fail in practice due to transaction costs and power asymmetries, highlighting the need for a centralized institutional mechanism capable of ensuring the longevity of cooperative agreements [5]. By contrast, the mechanism proposed in this chapter implements a spatially explicit transfer scheme that compensates land managers according to the marginal external benefits of their control efforts. This design internalizes spatial externalities, aligns private incentives with social welfare, and facilitates coordinated action across jurisdictions, thereby offering an innovative and effective alternative to existing policy frameworks.

In the absence of coordination, we show that decentralized equilibria systematically under-control invasive species in key dispersal hubs (or "sources"), resulting in higher overall costs. This inefficiency motivates the design of incentive-compatible mechanisms that align private and social interests. Building on the theory of Pigouvian taxation [28] and recent applications to ecosystem service provision [18], we propose a spatially explicit transfer scheme that compensates upstream managers for their role in reducing seed dispersal to neighboring sites. We derive the optimal tax formula by internalizing the marginal externalities, and show that it relies only on local observable quantities in the stationary regime.

From an economic perspective, this methodological shift affords two significant advantages. First, it captures in a single algebraic formulation the full scope of intersite externalities: each agent's optimal effort instantaneously internalizes the ecological feedback arising from all other agents. Second, it yields closed-form expressions for marginal externality payments, enabling the design of spatially differentiated fiscal transfers that precisely internalize social costs and benefits. These features render our approach not only theoretically rigorous but also operationally tractable for policymakers confronting heterogeneous territories and high-resolution monitoring constraints.

While theoretically grounded, this mechanism assumes the existence of a central authority with full knowledge of ecological parameters and sufficient monitoring capacity. It also requires legitimacy in redistributing intersite payments. Nonetheless, it exemplifies how the principles of Payment for Ecosystem Services can be adapted to dynamic, spatially structured ecological games, offering a promising tool for decentralized invasive species control. When real-time monitoring is costly, simplified schemes may rely on observed reductions in plant or seed densities.

Looking forward, our framework suggests several promising avenues for extension. Incorporating stochastic environmental shocks or endogenous network formation into the operator fixed-point structure may further enhance realism, while empirical calibration using emerging spatiotemporal datasets could validate and refine the transfer mechanisms proposed. Ultimately, by uniting global mathematical properties with economic incentive design, this work paves the way for robust, scalable interventions in spatial public-goods management.

5 Conclusion

The operator fixed-point framework provides a rigorous mathematical basis for characterizing equilibrium solutions in spatially structured, game-theoretic control problems, lending itself to iterative computation in systems with complex interactions.

Specifically, in ecological invasive plant management, the proposed mechanism implements a spatially explicit transfer scheme that compensates land managers according to the marginal external benefits of their control efforts. This design internalizes spatial externalities, aligns private incentives with social welfare, and facilitates coordinated action across jurisdictions, thereby offering an innovative and effective alternative to existing policy frameworks.

Bibliography

[1] H. J. Albers, C. Fischer, and J. N. Sanchirico. Invasive species management in a spatially heterogeneous world: Effects of uniform policies. *Resource and Energy Economics*, 32(4):483–499, 2010.

[2] W. Born, F. Rauschmayer, and I. Bräuer. Economic evaluation of biological invasions—a survey. *Ecological Economics*, 55(3):321–336, 2005.

[3] W. A. Brock and A. Xepapadeas. Pattern formation, spatial externalities and regulation in coupled economic–ecological systems. *Journal of Environmental Economics and Management*, 59(2):149–164, 2010.

[4] C. W. Clark. *Mathematical Bioeconomics: The Optimal Management of Renewable Resources*. Wiley-Interscience, 2nd edition, 1990.

[5] K. M. Cobourn, G. S. Amacher, and R. G. Haight. Cooperative management of invasive species: A dynamic Nash bargaining approach. *Environmental & Resource Economics*, 72(4):1041–1068, 2018.
[6] C. Diagne, B. Leroy, A.-C. Vaissière, R. E. Gozlan, D. Roiz, I. Jarić, J.-M. Salles, C. J. A. Bradshaw, and F. Courchamp. High and rising economic costs of biological invasions worldwide. *Nature*, 592:571–576, 2021.
[7] E. J. Dockner, S. Jørgensen, N. V. Long, and G. Sorger. *Differential Games in Economics and Management Science*. Cambridge University Press, 2000.
[8] A. Ducrot, Q. Griette, Z. Liu, and P. Magal. *Differential Equations and Population Dynamics I: Introductory Approaches*. Lecture Notes on Mathematical Modelling in the Life Sciences. Springer, Cham, 1 edition, 2022. https://doi.org/10.1007/978-3-030-98136-5.
[9] R. S. Epanchin-Niell. Cooperative management of invasive species: Theory and evidence. *American Journal of Agricultural Economics*, 96(5):1291–1315, 2014.
[10] R. S. Epanchin-Niell, R. G. Haight, L. Berec, J. M. Kean, and A. M. Liebhold. Optimal surveillance and eradication of invasive species in heterogeneous landscapes. *Ecology Letters*, 15(8):803–812, 2012.
[11] R. S. Epanchin-Niell and A. Hastings. Controlling established invaders: Integrating economics and spread dynamics to determine optimal management. *Ecology Letters*, 10(7):689–700, 2007.
[12] R. S. Epanchin-Niell and A. Hastings. Controlling established invaders: Integrating economics and spread dynamics to determine optimal management. *Ecology Letters*, 13(4):528–541, 2010.
[13] R. S. Epanchin-Niell and J. E. Wilen. Optimal spatial control of biological invasions. *Proceedings of the National Academy of Sciences*, 109(19):7681–7686, 2011.
[14] M. Finus. *Game Theory and International Environmental Cooperation: New Thinking for Old Problems*. Edward Elgar Publishing, 2001.
[15] R. G. Haight and S. Polasky. Optimal spatial control of biological invasions. *Ecological Economics*, 78:123–130, 2012.
[16] J. K. Hale. *Asymptotic Behavior of Dissipative Systems*, volume 25 of Mathematical Surveys and Monographs. American Mathematical Society, 1988.
[17] N. Hanley, J. F. Shogren, and B. White. *Introduction to Environmental Economics*. Oxford University Press, 2001.
[18] M. C. Jolejole-Foreman, F. Lupi, and M. D. Kaplowitz. Using incentive payments to encourage environmentally beneficial land use practices. *Ecological Economics*, 169:106505, 2020.
[19] B. Leung, D. M. Lodge, D. Finnoff, J. F. Shogren, M. A. Lewis, and G. Lamberti. An ounce of prevention or a pound of cure: Bioeconomic risk analysis of invasive species. *Proceedings of the Royal Society of London. Series B, Biological Sciences*, 269(1508):2407–2413, 2002.
[20] P. Magal and X.-Q. Zhao. Global attractors and steady states for uniformly persistent dynamical systems. *SIAM Journal on Mathematical Analysis*, 37(1):251–275, 2005.
[21] V. L. Ngo. *A Survey of Dynamic Games in Economics*. World Scientific Publishing Co Pte Ltd, Singapore, 2010. https://doi.org/10.1142/7577.
[22] L. J. Olson and S. Roy. The economics of controlling a stochastic biological invasion. *American Journal of Agricultural Economics*, 88(3):704–717, 2006.
[23] L. J. Olson and S. Roy. The economics of controlling a stochastic biological invasion. *American Journal of Agricultural Economics*, 89(3):678–692, 2007.
[24] N. H. Pavel and D. Motreanu. *Tangency, Flow Invariance for Differential Equations, and Optimization Problems*, volume 219. CRC Press, 1999.
[25] D. Pimentel, R. Zuniga, and D. Morrison. Update on the environmental and economic costs associated with alien-invasive species in the United States. *Ecological Economics*, 52(3):273–288, 2005.
[26] Z. Qin et al. Modeling long-term management of invasive species under uncertainty. *Nature Communications*, 11(1):1–11, 2020.
[27] A. Xepapadeas. Economic dynamics and spatial environmental externalities: Lessons for environmental policy. *Environment and Development Economics*, 10(6):679–698, 2005.
[28] A. Xepapadeas. Environmental policy and firm behavior: Abatement investment and location decisions under uncertainty and irreversibility. *Journal of Environmental Economics and Management*, 54(2):240–250, 2007.
[29] H. Yokomizo, H. P. Possingham, M. B. Thomas, and Y. M. Buckley. Managing the impact of invasive species: The value of knowing the density–impact curve. *Ecological Applications*, 19(2):376–386, 2009.

Pierre Auger, Ali Moussaoui, Tri Nguyen-Huu, and Doanh Nguyen-Ngoc

Enhancing maximum sustainable yield in multisite single-species and predator–prey fisheries

Mathematical analysis of spatial structure, migration and species interactions

Abstract: This paper investigates the impact of fast migration between two spatially distinct fishery patches on the Maximum Sustainable Yield (MSY). Several mathematical models are considered, including the single-species logistic model, as well as predator–prey dynamics governed by Lotka–Volterra and Holling type II functional responses. In the single-species case, we demonstrate that fast migration cannot increase the global MSY beyond the sum of isolated yields. However, in predator–prey systems, especially under conditions of ecological heterogeneity and asymmetric migration, fast connectivity can significantly enhance the overall fishery performance. Analytical derivations and numerical simulations show that the MSY of the connected system may exceed that of disconnected patches, highlighting the ecological and economic benefits of spatial management. This study emphasizes the importance of integrating spatial structure into fishery management strategies to optimize long-term yields.

Keywords: Maximum sustainable yield (MSY), multisite fishery, Schaefer fishery model, prey–predator fish community, spatial heterogeneity, increase in fish biomass and MSY by connectivity

MSC 2020: 34A34, 92B05, 92C42, 92F05

1 Introduction

There are many mathematical models describing the growth of a population in an environment where there is a resource that allows this population to survive and reproduce [26]. The most popular model is undoubtedly the logistic model. The logistic equation is

Pierre Auger, UMMISCO, Sorbonne Université, Institut de Recherche pour le Développement, IRD, Paris, France, e-mail: pierre.auger@ird.fr
Ali Moussaoui, Laboratory of Nonlinear Analysis and Applied Mathematics, University of Tlemcen, Faculty of Sciences, Department of Mathematics, 13000, Algeria, e-mail: moussaoui.ali@gmail.com
Tri Nguyen-Huu, UMMISCO, Sorbonne Université, Institut de Recherche pour le Développement, IRD, Paris, France; and IXXI, ENS Lyon, 46 allée d'Italie, 69007 Lyon, France, e-mail: tri.nguyen-huu@ird.fr
Doanh Nguyen-Ngoc, College of Engineering and Computer Science, and Center for Environmental Intelligence, VinUniversity, Da Ton, Gia Lam, Ha Noi, Vietnam

https://doi.org/10.1515/9783112216620-021

written as follows:

$$\frac{\mathrm{d}B}{\mathrm{d}\tau} = rB\left(1 - \frac{B}{K}\right) \tag{15.1}$$

where $B(t)$ represents the population biomass at time t, $r > 0$ is the growth rate of the population and $K > 0$ is the carrying capacity. This equation admits two equilibrium points, 0, which is unstable, and K which is stable. For any initial condition $B(0) > 0$, the solution tends towards the carrying capacity. Therefore, the carrying capacity is a parameter related to the quantity of locally available resource and determines the maximum population that can survive in this environment given the quantity of locally available resources. When it comes to a fish population, fish can be caught. We then add a negative term to the logistic equation that corresponds to the capture, leading to the following Schaefer model: [29]:

$$\frac{\mathrm{d}B}{\mathrm{d}\tau} = rB\left(1 - \frac{B}{K}\right) - qBE \tag{15.2}$$

where q is the catchability and E the fishing effort. The catch is assumed to be proportional to the fish biomass and the fishing effort. The corresponding catch is called a Schaefer catch [29]. This equation has two equilibrium points, 0 and $B^* = K\left(1 - \frac{qE}{r}\right)$. The latter equilibrium point is positive when we fish less quickly than the population reproduces, i. e., $qE < r$. In this case, the equilibrium 0 is unstable and the positive equilibrium B^* is stable for any positive initial condition. It is a case of a sustainable fishery. Otherwise, B^* is negative and has no meaning, 0 is stable and the population goes extinct in the long term. An important concept in fisheries is the Maximum Sustainable Yield (MSY) [9]. It corresponds to the maximum of the catch in the case of the positive equilibrium point B^*. The MSY has been the target to be achieved for many fisheries by researching the fishing effort needed to be deployed to reach it.

Another important aspect concerns multisite fisheries, which consist of several "sites" into and between which fish may move by migration. An example of a fishing site is a Fish Aggregating Device (FAD), [17]. FADs are typically floating objects that are deployed throughout the area because they attract large pelagic fish such as tuna, which remain there for a while before returning to the open sea. They allow fishermen to locate fish and catch them. Generally speaking, several fishing sites may exist with different resource levels that are visited by fish moving from one site to another by migration. In a network of sites connected by fish migrations, several questions arise regarding the total equilibrium population. Can the total equilibrium population in the network of sites exceed the sum of the carrying capacities of all sites? For multisite fisheries, can the optimal catch of the network with connectivity exceed the sum of local catches at MSY? The purpose of this chapter is to present a mini review of the works carried out on this subject, focusing primarily on two-site models.

2 Single-species models

In this section, we consider several regimes under which fish population may be managed in a fishery; of interest is the maximal yield.

This section analyzes a fish population split between two spatial patches with logistic growth and fast migration between them. The dynamics are modeled in two contexts: without and with harvesting. The goal is to understand whether fast migration between patches can enhance total population biomass and/or maximum sustainable yield (MSY).

2.1 Without harvesting

We consider two patches indexed by $i = 1, 2$, with fish biomass B_i on patch i. The growth in each patch follows a logistic model with intrinsic rate r_i and carrying capacity K_i. Fish migrate between patches at constant rates m_1 and m_2. The dynamics on a fast time scale τ are

$$\left.\begin{aligned}\frac{dB_1}{d\tau} &= m_2B_2 - m_1B_1 + \varepsilon r_1B_1\left(1-\frac{B_1}{K_1}\right)\\ \frac{dB_2}{d\tau} &= m_1B_1 - m_2B_2 + \varepsilon r_2B_2\left(1-\frac{B_2}{K_2}\right)\end{aligned}\right\} \tag{15.3}$$

Here, $\varepsilon \ll 1$ denotes the ratio of slow (growth) to fast (migration) time scales. Using aggregation of variables method [22, 30, 5, 3], a reduced model can be derived by substituting the fast subsystem's stable equilibrium into the full system. By setting $\varepsilon = 0$, we obtain the fast migration equilibrium:

$$\left.\begin{aligned}B_1^* &= uB\\ B_2^* &= (1-u)B\end{aligned}\right\} \tag{15.4}$$

where $u = m_2/(m_1+m_2)$ and $B = B_1+B_2$ is the total biomass. Substituting this equilibrium into the original system and rescaling time via $t = \varepsilon\tau$, we obtain the following aggregated model at the slow time scale. This approximation is valid when the reduced system is structurally stable and for sufficiently small $\varepsilon \ll 1$,

$$\frac{dB}{dt} = rB\left(1-\frac{B}{K}\right) \tag{15.5}$$

with

$$r = ur_1 + (1-u)r_2 \tag{15.6}$$

and

$$K = \frac{K_1K_2(ur_1 + (1-u)r_2)}{K_2r_1u^2 + K_1r_2(1-u)^2} \tag{15.7}$$

It is well known that the equilibrium fish biomass in a system of connected patches can exceed the sum of the equilibrium fish biomass of the isolated patches. In other words, $K > K_1 + K_2$. For example, setting $K_1 = K_2 = 50$, $r_1 = 10$, $r_2 = 1$, and $u = \frac{1}{3}$, we find $K = 128.57 > 100 = K_1 + K_2$. This result on the increase in total biomass through connectivity has been known since the 1970s. It was first shown in the case of fast and symmetrical migrations, $(m_1 = m_2)$, i. e., $(u = m_2/(m_1 + m_2) = 0.5)$; see [18, 20, 19] and [21].

In the homogeneous case, where $r_1 = r_2$ and $K_1 = K_2$, the expression for the global carrying capacity simplifies to

$$K_{\text{Hom}} = \frac{K_1}{u^2 + (1-u)^2} \tag{15.8}$$

For symmetric migration, this reduces to

$$K_{\text{Hom,Sym}} = 2K_1 \tag{15.9}$$

In the heterogeneous case with symmetric diffusion ($r_1 \neq r_2$ and, e. g., $K_1 > K_2$), the total carrying capacity is given by

$$K_{\text{Het,Sym}} = \frac{2K_1K_2(r_1 + r_2)}{K_2r_1 + K_1r_2} \tag{15.10}$$

It can be easily verified that under symmetric migration, the total population in the homogeneous case is always greater than in the heterogeneous case, i. e., $K_{\text{Het,Sym}} < K_{\text{Hom,Sym}}$, assuming $K_2 < K_1$.

Poggiale et al. [28] extended this study to the more general case where migration is asymmetric ($m_1 \neq m_2$). In this case and when $K_2 = K_1$, they showed that it is necessary to connect heterogeneous sites and gave a condition on growth and migration rates that allows for an increase in biomass at equilibrium through connectivity:

$$1 < \frac{m_1}{m_2} < \frac{r_1}{r_2} \tag{15.11}$$

To obtain a global carrying capacity greater than the sum of the local ones in isolation, both heterogeneity and asymmetric diffusion are thus required. There has been much interest more recently in networks of logistic sites connected by slow or fast as well as symmetric or asymmetric migrations and with any number of sites greater than 2, see [1, 2, 8, 7, 11, 13, 12, 16, 14, 15].

2.2 With harvesting

We now analyze the case where the fish population is harvested. Harvesting is modeled using the Schaefer model, which assumes a linear relationship between biomass and catch. The catch per unit time is given by qEB, where E is the fishing effort and q is the catchability coefficient. For simplicity, we take $q = 1$. Including harvesting in the aggregated model results in

$$\frac{\mathrm{d}B}{\mathrm{d}t} = rB\left(1 - \frac{B}{K}\right) - EB \tag{15.12}$$

This is a logistic equation with a linear harvest term. The equilibrium biomass B^* is found by setting $\mathrm{d}B/\mathrm{d}t = 0$:

$$B^* = K\left(1 - \frac{E}{r}\right) \tag{15.13}$$

This equilibrium is biologically meaningful (i. e., positive) only if $E < r$. The yield, or total harvest at equilibrium, is given by

$$Y^*(E) = EB^* = EK\left(1 - \frac{E}{r}\right) \tag{15.14}$$

Maximizing $Y(E)$ with respect to E gives the effort at MSY. This occurs at $E^* = \frac{r}{2}$ leading to the fact that $B^* = \frac{K}{2}$ and the MSY reads

$$Y^*_{\mathrm{MSY}} = E^*B^* = \frac{rK}{4} \tag{15.15}$$

Figure 15.1 shows a numerical example of the equilibrium catch as a function of the fishing effort E.

Now, consider the maximum sustainable yields from each patch when isolated (i. e., not connected by migration). In this case, each patch i ($i = 1, 2$) has its own MSY:

$$Y^*_{\mathrm{MSY},i} = \frac{r_iK_i}{4} \tag{15.16}$$

Summing the isolated yields

$$Y^*_{\mathrm{MSY},1} + Y^*_{\mathrm{MSY},2} = \frac{r_1K_1}{4} + \frac{r_2K_2}{4} \tag{15.17}$$

To compare this to the yield in the connected system, recall that the global MSY for the coupled model is $Y^*_{\mathrm{MSY}} = \frac{rK}{4}$. We aim to prove the following inequality:

$$\frac{rK}{4} \leq \frac{r_1K_1}{4} + \frac{r_2K_2}{4} \tag{15.18}$$

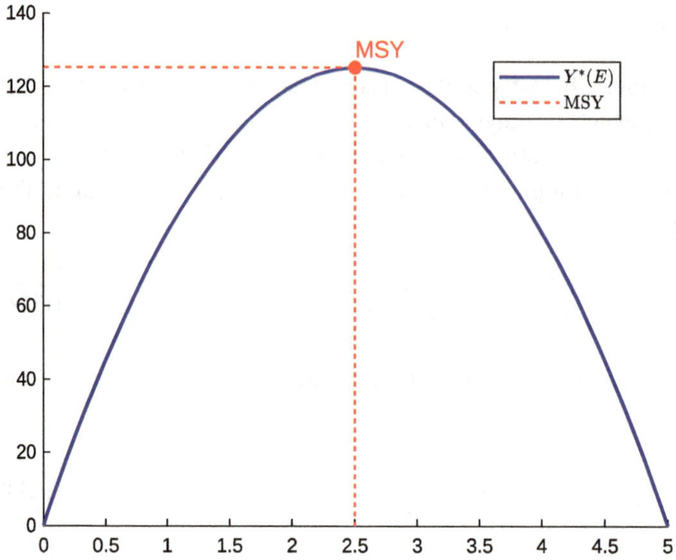

Figure 15.1: Equilibrium catch as a function of the fishing effort E: $r = 5$, $K = 100$, showing a maximum equal to $Y^*_{\text{MSY}} = 125$ when $E = 2.5$. At MSY, $Y^*_{\text{MSY}} = \frac{rK}{4} = 125$.

Substituting the expressions of r and K into inequality (15.18), we obtain

$$\frac{(r_1 u + r_2(1-u))^2 K_1 K_2}{r_1 K_2 u^2 + r_2 K_1 (1-u)^2} \leq r_1 K_1 + r_2 K_2 \tag{15.19}$$

This simplifies, for all $0 < u < 1$, to

$$(K_1(1-u) - K_2 u)^2 > 0, \tag{15.20}$$

which is always satisfied. This result was shown in [4] and extended to any number of sites in [6]. Therefore, although the total equilibrium biomass in the aggregated system may exceed the sum of local carrying capacities (i. e., $K > K_1 + K_2$), the MSY of the coupled system is always less than or equal to the sum of MSYs from isolated patches.

Figure 15.2 illustrates this result. Each logistic growth function is a downward-opening parabola, reaching its maximum when biomass is at half the carrying capacity. The MSY corresponds to this maximum. The slope at the origin equals the intrinsic growth rate. Since $r_1 < r < r_2$, the growth curve of the aggregated system lies between the two local growth curves, at least near the origin.

In the illustrated example, even though $K > K_1 + K_2$, we find that $\frac{rK}{4} = 3.73$ is still less than $\frac{r_1 K_1}{4} + \frac{r_2 K_2}{4} = 3.75$. For $u = 1$, the green curve coincides with patch 1 (blue); for $u = 0$, it coincides with patch 2 (red). For intermediate u, the aggregated curve lies between the other two.

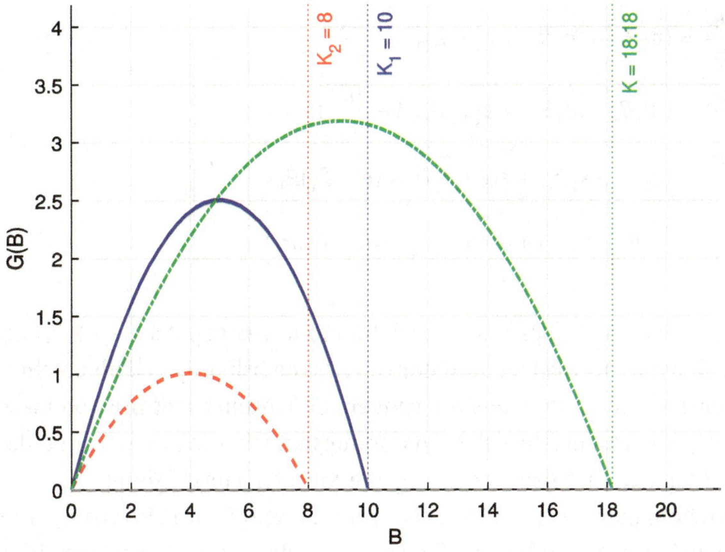

Figure 15.2: Logistic growth functions for patch 1 (blue), patch 2 (red), and the aggregated system (green). Parameters: $r_1 = 1, r_2 = 0.5, K_1 = 10, K_2 = 8, u = 0.4$, so that $r = r_1 u + r_2(1-u) = 0.7$ and $K = \frac{K_1 K_2 (r_1 u + r_2(1-u))}{K_2 r_1 u^2 + K_1 r_2 (1-u)^2} = 18.18$.

In conclusion, while spatial coupling can enhance total biomass capacity, it cannot increase the total MSY beyond the sum of isolated yields, due to the averaging effect on growth rates. This result has been extended to the case of any number of sites connected by fast asymmetric migrations, [6].

3 Results for predator–prey Lotka–Volterra model

We now consider a much more complex system of fish, i. e., a prey–predator fish community living in a two patches environment with predator harvesting and use the predator–prey Lotka–Volterra model.

3.1 Holling type I (linear) functional response

3.1.1 Model specification

This model was first studied in [4] and more in-depth in [27]. The complete model reads as follows:

$$\left.\begin{aligned}\frac{dB_1}{d\tau} &= (m_2B_2 - m_1B_1) + \varepsilon\left(r_1B_1\left(1 - \frac{B1}{K1}\right) - a_1B_1P_1\right)\\ \frac{dB_2}{d\tau} &= (m_1B_1 - m_2B_2) + \varepsilon\left(r_2B_2\left(1 - \frac{B2}{K2}\right) - a_2B_2P_2\right)\\ \frac{dP_1}{d\tau} &= (k_2P_2 - k_1P_1) + \varepsilon(ea_1B_1P_1 - (d + E_1)P_1)\\ \frac{dP_2}{d\tau} &= (k_1P_1 - k_2P_2) + \varepsilon(ea_2B_2P_2 - (d + E_2)P_2)\end{aligned}\right\} \tag{15.21}$$

where B_i and P_i are respectively fish prey and predator biomasses at patch i, $i = 1, 2$, and τ the fast time. For fish dynamic: the fish predator catches the fish prey in both patches. a_i is the predation parameter at site i. e is the conversion parameter of prey biomass into predator biomass. d represents the fish predator mortality rate. We assume that the predator is harvested at patch i at a fishing effort E_i. We consider a total fishing effort $E = E_1 + E_2$ distributed on both patches. The fishing efforts E, E_1, and E_2 are also considered as parameters: the level of fishing effort can be easily decided by buying/selling ships at the fishery level, or by setting up policies at the governmental level. We denote w the proportion of the total fishing effort, which is allocated to patch 1, such that $E_1 = wE$ and $E_2 = (1 - w)E$.

3.1.2 Model reduction

We will now take advantage of the time scales to obtain a reduced model governing total fish biomass, $B = B_1 + B_2$, $P = P_1 + P_2$, on a slow time scale. It is easy to show that there exists a stable fast equilibrium characterized by constant prey fish (u) and predatory fish (v) proportions on site 1:

$$u = \frac{m_2}{m_1 + m_2} \tag{15.22}$$

$$v = \frac{k_2}{k_1 + k_2} \tag{15.23}$$

The next step is to substitute the fast variables in terms of the fast equilibrium into the complete model, i. e., $B_1^* = uB$, and $P_1^* = vP$ to obtain the following reduced model or aggregated model at slow time t:

$$\left.\begin{aligned}\frac{dB}{dt} &= rB\left(1 - \frac{B}{K}\right) - aBP\\ \frac{dP}{dt} &= eaBP - (d + bE)P\end{aligned}\right\} \tag{15.24}$$

where $\tau = \varepsilon t$. The prey fish parameters of the aggregated model including the overall carrying capacity and the total growth rate are similar to the case of single fish model. These parameters read:

$$r = ur_1 + (1-u)r_2 \tag{15.25}$$

and

$$K = \frac{K_1K_2(r_1u + r_2(1-u))}{(K_2r_1u^2 + K_1r_2(1-u)^2)} \tag{15.26}$$

New global parameters are: the global predation parameter, which reads

$$a = a_1uv + a_2(1-u)(1-v), \tag{15.27}$$

and the fishing parameter is given by

$$b = vw + (1-v)(1-w). \tag{15.28}$$

3.1.3 Equilibria and yield analysis

The model has three equilibria

$$(0,0), (K,0) \text{ and } (B^*(E), P^*(E)) = \left(\frac{bE+d}{ea}, r\frac{eaK-(d+bE)}{ea^2K}\right). \tag{15.29}$$

Their asymptotic stability is summarized as follows:

- If $bE > eaK - d$, then the model has two positive equilibria: $(0,0)$, which is unstable, and $(K,0)$ is stable. In this case, the predator population becomes extinct. Therefore, there is no fishing output and the yield is $Y^*(E) = 0$.
- If $bE < eaK - d$, then the system has three equilibria: $(0,0)$ and $(K,0)$ are both unstable, and $(B^*(E), P^*(E))$ is stable. The yield is given by $Y^*(E) = bEP^*(E) = \frac{rbE}{a}(1 - \frac{d+bE}{eaK})$.

The catch at MSY for the Lotka–Volterra model with predator fishing was shown in [23]; see also [24]. The catch at the maximum sustainable yield, Y^*_{MSY}, of the system of the two patches connected by fast migration is given by

$$Y^*_{\text{MSY}} = \begin{cases} \frac{r}{4ea^2K}(eaK-d)^2, & \text{if } eaK > d \\ 0, & \text{otherwise,} \end{cases} \tag{15.30}$$

The MSYs can also be defined in the same manner for each disconnected patch and read for $i \in \{1, 2\}$:

$$Y^*_{\mathrm{MSY},i} = \begin{cases} \frac{r}{4ea_i^2 K_i}(ea_i K_i - d)^2, & \text{if } ea_i K_i > d \\ 0, & \text{otherwise.} \end{cases} \tag{15.31}$$

The condition $ea_i K_i > d$ indicates that the predator population is persistent at patch i when disconnected from the other patch. Under this condition, we notice patch i as *viable*. We will also indifferently refer to such a patch as *viable for fishing* or *viable for the predator*.

We then define the excess yield between the global model and the sum of the MSYs of the disconnected patches as

$$DY^*_{\mathrm{MSY}} = Y^*_{\mathrm{MSY}} - Y^*_{\mathrm{MSY},1} - Y^*_{\mathrm{MSY},2}, \tag{15.32}$$

Our question is still whether fast migration between patches can enhance MSY, i. e., whether $DY^*_{\mathrm{MSY}} > 0$. Below, we show that it is possible to increase MSY by connectivity via fast migration. We consider Y^*_{MSY} as a function of two variables u and v.

3.1.4 Illustration

We now illustrate our result based on the number of viable patches (when disconnected) and the number of local equilibria. The following figures illustrate the work already presented in [27].

Figure 15.3 illustrates case 1 where both patches are viable for fishing, i. e., $ea_1K_1 - d > 0$ and $ea_2K_2 - d > 0$ (or equivalently, $Y^*_{\mathrm{MSY},i} > 0$ for $i \in \{1, 2\}$). Figure 15.3a) corresponds to a subcase with one local (and global) maximum while Figure 15.3b) is related to the subcase with two local maxima. The green part of the surface corresponds to a positive excess yield: the yield of the system consisting of the two connected sites is greater than the sum of the yield of the two disconnected sites. This means that connecting the two patches with fast migration allows the catch to increase to a value that is impossible to reach without connectivity.

Figure 15.4 represents case 2 in which only one patch is viable. We assume patch 2 is viable while patch 1 is not. It means that we have the following conditions $ea_1K_1 < d$ and $ea_2K_2 > d$. Figure 15.4 shows configurations similar to the ones in case 1: one global maximum (Figure 15.4a) or two local maxima on a same connected component of the support domain (Figure 15.4b). These cases correspond respectively to the ones in Figure 15.3a and Figure 15.3b by keeping the same parameters, except for the mortality rate d, which has been increased until patch 1 becomes not viable. One can see that the values of these quantities have been decreased, respectively.

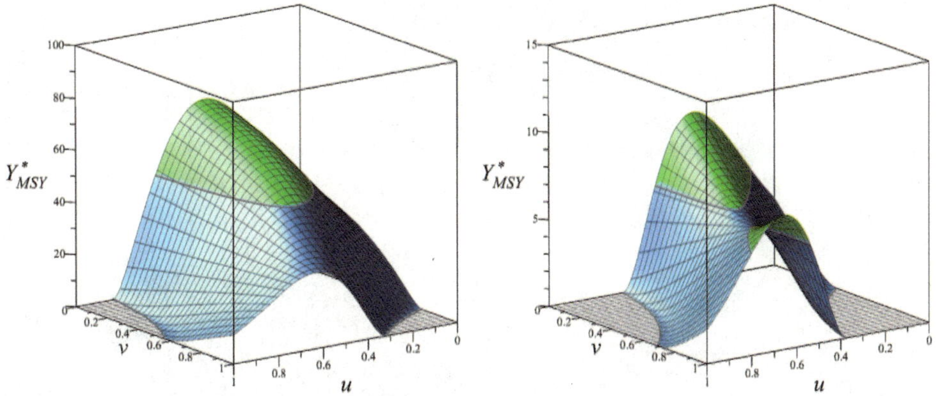

Figure 15.3: Case 1 (both patches are viable for fishing): catch at MSY with respect to the proportion u and v of prey and predators at patch 1 with parameters (a) $K_1 = 140, K_2 = 140, r_1 = 2, r_2 = 2, a_1 = 0.8, a_2 = 2, e = 0.8, d = 50$ and (b) $K_1 = 100, K_2 = 100, r_1 = 1, r_2 = 1.2, a_1 = 2, a_2 = 2.3, e = 0.8, d = 110$. The blue part of the surface corresponds to positive yield values, but no excess yield. The green part corresponds to a positive excess yield due to the connectivity between the two patches.

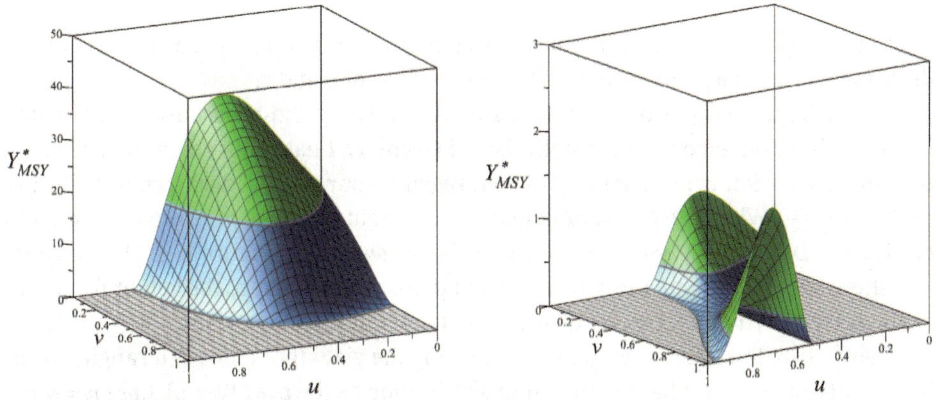

Figure 15.4: Case 2 (One viable patch): catch at MSY with respect to the proportion u and v of prey and predators in patch 1 with parameters (a) $K_1 = 100, K_2 = 100, r_1 = 2, r_2 = 2, a_1 = 0.2, a_2 = 2, e = 0.8, d = 60$ and (b) $K_1 = 100, K_2 = 100, r_1 = 0.5, r_2 = 1.2, a_1 = 2.1, a_2 = 2.5, e = 0.8, d = 180$. The blue part of the surface corresponds to positive yield values, but no excess yield. The green part corresponds to a positive excess yield due to the connectivity between the two patches.

Figure 15.5 shows an emergent case (case 3) where none of the sites are viable. This emergent case occurs when predators cannot survive at the equilibrium in the case of disconnected patches (i. e., $ea_1K_1 - d < 0$ and $ea_2K_2 - d < 0$) and, therefore, there is no yield. However, there exists a coexistence prey–predator equilibrium when the two patches are connected. In other words, there exist some prey and predator fish proportions $0 < u < 1$ and $0 < v < 1$ such that $eaK - d > 0$ for which a positive global yield

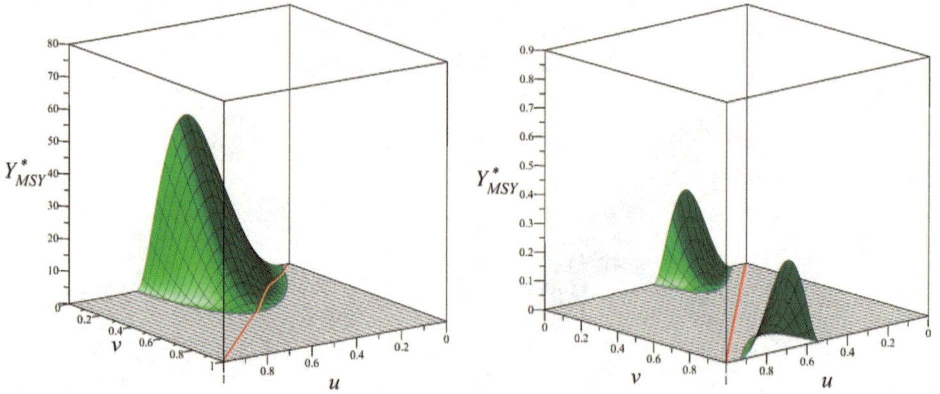

Figure 15.5: Case 3 (emergence case): catch at MSY with respect to the proportion u and v of prey and predators at patch 1 with parameters (a) $K_1 = 500$, $K_2 = 200$, $r_1 = 4$, $r_2 = 2$, $a_1 = 0.2$, $a_2 = 0.7$, $e = 0.8$, $d = 120$ and (b) $K_1 = 70$, $K_2 = 70$, $r_1 = 1$, $r_2 = 1$, $a_1 = 2$, $a_2 = 2$, $e = 0.8$, $d = 120$. The blue part of the surface corresponds to positive yield values, but no excess yield. The green part corresponds to a positive excess yield due to the connectivity between the two patches. The orange line corresponds to the predator population being at ideal free distribution ($u = v$).

exists. (see Figure 15.5). This is because the global carrying capacity may be larger than those of the sum of K_1 and K_2, for well-chosen values of u and v.

When the catch is positive, the excess yield is always positive because neither patch is viable. However, emergence is unlikely to be achieved naturally when the prey population distribution is proportional to its foraged resource, i. e., ideal free distribution (IFD) $u = \frac{K_1}{K_1+K_2}$. When the predator is spatially distributed like its prey, we consider it to be also at IFD, i. e., $u = v$, which is shown as the orange line of Figure 15.5. This suggests that there should be some human controlled patterns for the distribution of fish or an artificial way to choose a correct distribution to obtain a positive yield.

Figure 15.6 illustrates emergence by showing the yield for values of u ranging from 0 to 1, with the predator having the same distribution as the prey ($v = u$). Case $u = v = 1$ (resp., 0) corresponds to the entire population being located on patch 1 (resp., patch 2), each case representing the dynamics on each patch when they are not connected with fast migration. There is no catch for $u = v = 0$ or $u = v = 1$, as neither patch is viable. However, a positive yield appears for some values of u. This suggests that asymmetry in the distribution of the populations between the two patches (population significantly larger at one of the sites) is required to obtain a case of emergence.

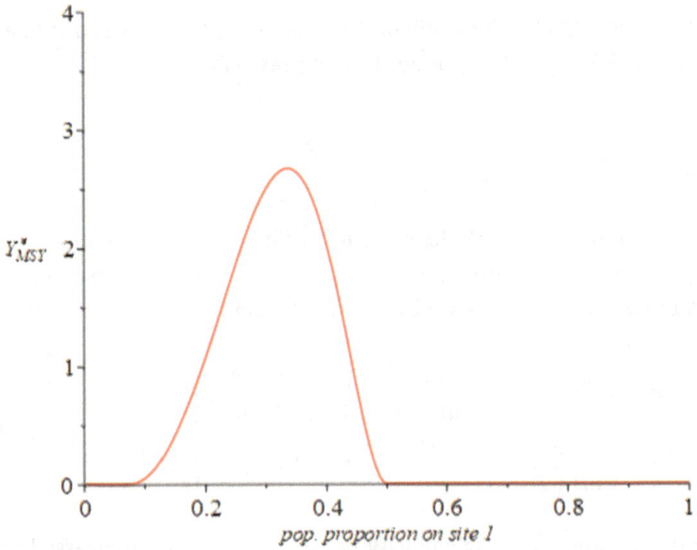

Figure 15.6: Emergence of a yield due to fast migration. For the set of parameters $K_1 = 500$, $K_2 = 200$, $r_1 = 4$, $r_2 = 2$, $a_1 = 0.2$, $a_2 = 0.7$, $e = 0.8$, $d = 120$, there is no capture in sites 1 and 2 when they are disconnected: the dynamics on each isolated patch can be obtained by choosing $u = v = 0$ or $u = v = 1$. However, when $0 < u, v < 1$, there is a domain for which there is a positive yield and a MSY with connected patches.

3.2 Holling type II (rectangular hyperbolic) functional response

3.2.1 Model specification

Now, we consider an extended version of the model in Section 2 by considering Holling type II functional response, and constant predator fishing effort E. The complete model reads

$$\left.\begin{aligned}
\frac{dB_1}{d\tau} &= m_2B_2 - m_1B_1 + \varepsilon r_1B_1\left(1 - \frac{B_1}{K_1}\right) - \varepsilon\frac{aB_1P_1}{B_1 + D}\\
\frac{dB_2}{d\tau} &= m_1B_1 - m_2B_2 + \varepsilon r_2B_2\left(1 - \frac{B_2}{K_2}\right) - \varepsilon\frac{aB_2P_2}{B_2 + D}\\
\frac{dP_1}{d\tau} &= k_2P_2 - k_1P_1 + \varepsilon\left(\frac{eaB_1P_1}{B_1 + D} - dP_1 - EP_1\right)\\
\frac{dP_2}{d\tau} &= k_1P_1 - k_2P_2 + \varepsilon\left(\frac{eaB_2P_2}{B_2 + D} - dP_2 - EP_2\right)
\end{aligned}\right\} \tag{15.33}$$

Here, r_i and K_i still denote the intrinsic growth rate and carrying capacity of the prey in patch i; a and D are the attack rate and handling time, assumed equal in both patches; e is the conversion efficiency; d is predator mortality; and m_i, k_i are prey and predator migration rates, respectively. The study of the Rosenzweig–MacArthur model was carried out in [4] and [25] for symmetric migration of prey fish.

3.2.2 Model reduction

Assuming fast migration dynamics ($\varepsilon \ll 1$), the system is reduced via aggregation. Let $u = \frac{m_2}{m_1+m_2}$ and $v = \frac{k_2}{k_1+k_2}$ represent the equilibrium proportion of prey and predator in patch 1. The aggregated model for total prey $B = B_1+B_2$ and predator $P = P_1+P_2$ becomes

$$\left.\begin{aligned} \frac{dB}{dt} &= rB\left(1-\frac{B}{K}\right) - aBP\left(\frac{uv}{uB+D}+\frac{(1-u)(1-v)}{(1-u)B+D}\right) \\ \frac{dP}{dt} &= eaBP\left(\frac{uv}{uB+D}+\frac{(1-u)(1-v)}{(1-u)B+D}\right) - (d+E)P \end{aligned}\right\} \tag{15.34}$$

where the global growth rate and carrying capacity parameters are still expressed as follows:

$$r = ur_1 + (1-u)r_2 \tag{15.35}$$

and

$$K = \frac{K_1K_2(r_1u + r_2(1-u))}{(K_2r_1u^2 + K_1r_2(1-u)^2)}. \tag{15.36}$$

3.2.3 Equilibrium and yield analysis

We consider the interior equilibrium (B^*, P^*), where the predator population satisfies

$$P^* = \frac{re}{d+E}B^*\left(1-\frac{B^*}{K}\right), \tag{15.37}$$

which is positive provided that $B^* < K$. The equilibrium biomass B^* corresponds to the positive root of the quadratic equation:

$$F(B) = \alpha_1B^2 + \alpha_2B - \alpha_3 = 0, \tag{15.38}$$

with the coefficients defined as

$$\begin{aligned} \alpha_1 &= (ea-(d+E))u(1-u), \\ \alpha_2 &= eaD[uv + (1-u)(1-v)] - D(d+E), \\ \alpha_3 &= (d+E)D^2. \end{aligned}$$

Since by assumption $a_1 > 0$, Descartes' Rule of Signs guarantees the existence of a unique positive root of equation (15.38), given by

$$B^* = \frac{-eaD[uv + (1-u)(1-v)] + D(d+E) + \sqrt{\Delta}}{2(ea - (d+E))u(1-u)}, \tag{15.39}$$

where the discriminant Δ of the polynomial F is

$$\begin{aligned}\Delta = &\left[eaD(uv + (1-u)(1-v)) - D(d+E)\right]^2\\ &+ 4(d+E)D^2(ea - (d+E))u(1-u).\end{aligned}$$

The corresponding yield at equilibrium, defined by $Y^*(E) = EP^*$, is expressed as

$$Y^*(E) = EP^* = re\frac{E}{d+E}B^*\left(1 - \frac{B^*}{K}\right). \tag{15.40}$$

3.2.4 Illustration

Using parameter values such as $K_1 = 70$, $K_2 = 10$, $r_1 = 0.1$, $r_2 = 5$, $a = 2$, $D = 15$, $e = 1$, $d = 0.4$, $u = 0.7$, $v = 0.9$, Figure 15.7 shows that the MSY of the aggregated (connected) system exceeds the sum of the MSYs of isolated patches by approximately 44.38 %.

This indicates that under conditions of strong heterogeneity and appropriate asymmetry in species distribution, spatial connectivity through fast migration enhances overall fishery productivity. The gain arises from the bottom-up control of the ecosystem, where mobile prey optimize their distribution for growth, thereby increasing the food availability for predators.

4 Discussion

We have shown that the total equilibrium population in a Schaefer network of two fishing sites can be greater than the sum of local carrying capacities. In contrast, the optimal catch still in the Schaefer fishery with connectivity is always less than or equal to the sum of optimal catches at MSY of each isolated site. Unexpectedly, when considering a marine prey–predator community with predator fishing, the opposite result is obtained. The total catch at MSY of the predator can be higher than the sum of local catches at MSY that would be obtained in the absence of connectivity. This result is obtained when two heterogeneous fishing sites are connected with asymmetric migrations as well as in the case of type I or II functional responses.

It is clear that marine communities are quite complex with several trophic levels. It appears necessary to generalize the study to the case of marine communities such as Phytoplankton Zooplankton Fish (PZF) and to take into account several fish populations

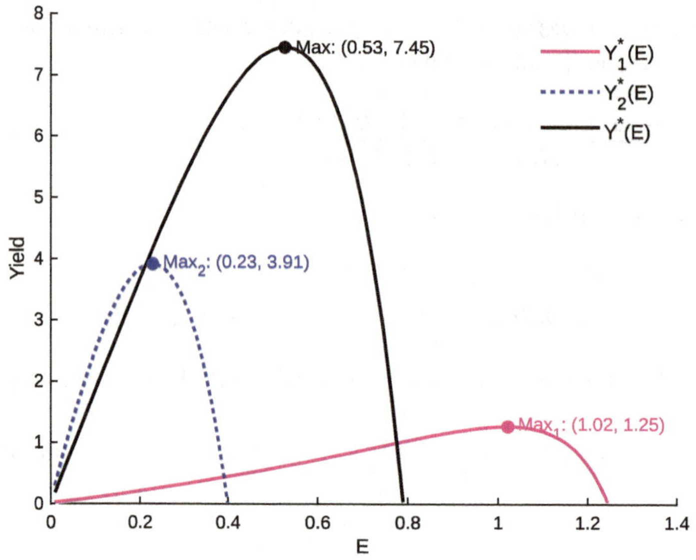

Figure 15.7: Catch at equilibium for the connected system $Y^*(E)$ (black) and for isolated sites 1 ($Y_1^*(E)$, blue) and 2 ($Y_2^*(E)$, pink) with respect to the fishing effort E, with parameter set: $K_1 = 70$, $K_2 = 10$ (carrying capacities on site 1 and 2, resp.), $D = 15$, $r_1 = 0.1$, $r_2 = 5$ (growth rates on sites 1 and 2, resp.), $a = 2$ (predator efficiency), $e = 1$ (biomass conversion factor for predation), $d = 0.4$ (fish predator mortality rate), $u = 0.7$ (proportion of prey in site 1) and $v = 0.9$ (proportion of predators in site 2). The fishing effort E and the yields are dimensionless quantities arising from the internal scaling of the theoretical model. Its scale is determined by the parameter values and is not intended to represent a specific empirical unit.

such as prey fish, predator fish, and top predator fish. It is also important to consider multiple fishing efforts on fish populations corresponding to different trophic levels and to study the MSY when they are exploited at the same time.

Finally, we mention the importance of taking into account economic aspects and in particular the operating costs of the fishery. We refer to a work corresponding to the Gordon–Schaefer model where the fishing effort is variable depending on benefice and costs. In this case, it has been shown that subject to similar costs on the fishing sites, it was possible to increase the catch at MSY by connectivity[10].

5 Conclusion

It is demonstrated that while fast migration and connectivity between patches do not increase maximum sustainable yield (MSY) in single-species fisheries beyond the sum of isolated sites, they can substantially enhance predator MSY in spatially structured predator–prey systems—particularly under ecological heterogeneity and asymmetric migration—highlighting the critical need to incorporate spatial structure, species interactions, and economic considerations into fishery management for optimizing both sustainability and profitability.

Bibliography

[1] R. Arditi, C. Lobry, and T. Sari. Is dispersal always beneficial to carrying capacity? New insights from the multi-patch logistic equation. *Theoretical Population Biology*, 106:45–59, 2015.

[2] R. Arditi, C. Lobry, and T. Sari. Asymmetric dispersal in the multi-patch logistic equation. *Theoretical Population Biology*, 120:11–15, 2018.

[3] P. Auger, R. Bravo de La Parra, J.-C. Poggiale, E. Sánchez, and L. Sanz. Aggregation methods in dynamical systems and applications in population and community dynamics. *Physics of Life Reviews*, 5(2):79–105, 2008.

[4] P. Auger, B. Kooi, and A. Moussaoui. Increase of maximum sustainable yield for fishery in two patches with fast migration. *Ecological Modelling*, 467:109898, 2022.

[5] P. Magal, and S. Ruan. *Structured Population Models in Biology and Epidemiology*, volume 1936. Springer, 2008.

[6] P. Auger, T. Nguyen-Huu, and D. Nguyen-Ngoc. On the impossibility of increasing the msy in a multisite Schaefer fishing model. *Mathematical Biosciences and Engineering*, 22(2):415–430, 2025.

[7] M. Benaïm, C. Lobry, T. Sari, and É. Strickler. Untangling the role of temporal and spatial variations in persistence of populations. *Theoretical Population Biology*, 154:1–26, 2023.

[8] M. Benaim, C. Lobry, T. Sari, and E. Strickler. When can a population spreading across sink habitats persist? *Journal of Mathematical Biology*, 88(2):19, 2024.

[9] C. W. Clark. *Mathematical Bioeconomics: The Optimal Management of Renewable Resources*. John Wiley & Sons, Inc. 2 edition, 1990.

[10] R. Bravo de la Parra, J.-C. Poggiale, and P. Auger. The effect of connecting sites in the environment of a harvested population. *Mathematical Modelling of Natural Phenomena*, 18:4, 2023.

[11] D. L. DeAngelis, C. C. Travis, and W. M. Post. Persistence and stability of seed-dispersed species in a patchy environment. *Theoretical Population Biology*, 16(2):107–125, 1979.

[12] D. L. DeAngelis, W.-M. Ni, and B. Zhang. Effects of diffusion on total biomass in heterogeneous continuous and discrete-patch systems. *Theoretical Ecology*, 9(4):443–453, 2016.

[13] D. L. DeAngelis and B. Zhang. Effects of dispersal in a non-uniform environment on population dynamics and competition: A patch model approach. *Discrete and Continuous Dynamical Systems. Series B*, 19(10), 2014.

[14] B. Elbetch. Effects of rapid population growth on total biomass in multi-patch environment. *Differential Equations & Applications*, 15(4):323–359, 2023. https://doi.org/10.7153/dea-2023-15-17.

[15] B. Elbetch. Generalized logistic equation on networks. *Comptes Rendus. Mathématique*, 361(G5):911–934, 2023.

[16] B. Elbetch and A. Moussaoui. Nonlinear diffusion in multi-patch logistic model. *Journal of Mathematical Biology*, 87(1):1, 2023.

[17] A. Fonteneau, J. Ariz, D. Gaertner, V. Nordstrom, and P. Pallares. Observed changes in the species composition of tuna schools in the gulf of Guinea between 1981 and 1999, in relation with the fish aggregating device fishery. *Aquatic Living Resources*, 13(4):253–257, 2000.

[18] H. I. Freedman, B. Rai, and P. Waltman. Mathematical models of population interactions with dispersal II: Differential survival in a change of habitat. *Journal of Mathematical Analysis and Applications*, 115(1):140–154, 1986.

[19] H. I. Freedman and Y. Takeuchi. Global stability and predator dynamics in a model of prey dispersal in a patchy environment. *Nonlinear Analysis: Theory, Methods & Applications*, 13(8):993–1002, 1989.

[20] H. I. Freedman and P. Waltman. Mathematical models of population interactions with dispersal. I: Stability of two habitats with and without a predator. *SIAM Journal on Applied Mathematics*, 32(3):631–648, 1977.

[21] R. D. Holt. Population dynamics in two-patch environments: Some anomalous consequences of an optimal habitat distribution. *Theoretical Population Biology*, 28(2):181–208, 1985.

[22] Y. Iwasa, S. A. Levin, and V. Andreasen. Aggregation in model ecosystems ii. Approximate aggregation. *Mathematical Medicine and Biology: A Journal of the IMA*, 6(1):1–23, 1989.

[23] T. Legović and S. Geček. Impact of maximum sustainable yield on independent populations. *Ecological Modelling*, 221(17):2108–2111, 2010.

[24] T. Legović, J. Klanjšček, and S. Geček. Maximum sustainable yield and species extinction in ecosystems. *Ecological Modelling*, 221(12):1569–1574, 2010.

[25] A. Moussaoui, P. Auger, and B. Elbetch. Enhancing maximum sustainable yield in a multi-patch Rosenzweig–Macarthur model with symmetrical prey and asymmetrical predator migration. *Mathematical Modelling of Natural Phenomena*, 20:13, 2025.

[26] J. D. Murray. *Mathematical Biology: I. an Introduction*, volume 17. Springer Science & Business Media, 2007.

[27] D. Nguyen-Ngoc, T. Nguyen-Huu, and P. Auger. Connectivity between two fishing sites can lead to an emergence phenomenon related to maximum sustainable yield. *Journal of Theoretical Biology*, 595:111913, 2024.

[28] J-C. Poggiale, P. Auger, D. Nerini, C. Manté, and F. Gilbert. Global production increased by spatial heterogeneity in a population dynamics model. *Acta Biotheoretica*, 53(4):359–370, 2005.

[29] M. B. Schaefer. Some considerations of population dynamics and economics in relation to the management of the commercial marine fisheries. *Journal of the Fisheries Board of Canada*, 14(5):669–681, 1957.

[30] A. N. Tikhonov. Systems of differential equations containing small parameters in the derivatives. *Matematičeskij Sbornik*, 73(3):575–586, 1952.

Part II-C: Person's sociosphere exposures

Rationale: 7

Part II-C: **Person's sociosphere exposures**

Conceptually, health impacts of opportunities and exposures in person's habitual social micro-context embedding and social macro-context of societal "framing," including structural disease prevention and protective interventions, health care delivery, consumer safety, occupational health, and epidemiology of adverse events.

Three classes are considered:
- economic factors
- social factors
- societal factors

Part II-C.1: **Economic factors**

This class includes:
- base-resources access (food, housing, mobility)
- education, economic self-sufficiency, welfare system participation
- health service access, health insurance, social security

In this volume:
- The effects of social support to frailty of elderly people in a longevity township of China: A community-based study

In former volume BIOKYBERNETIKA:
- Evolution of the health system in Shanghai, China, 2016–2020

https://doi.org/10.1515/9783112216620-022

Wenhua Tian

The effects of social support to frailty of elderly people in a longevity township of China: A community-based study

Abstract: Our study aims to analyze the association between social support and frailty of elderly people in a longevity township of China. We selected 1,721 participants from a community-based study in a longevity township of China as a sample. Frailty was defined by the frailty index (FI). Social support was measured by scales and a questionnaire from three aspects: objective support, subjective support, and utilization of support. Logistic regression models were used to examine the relationship between frailty and social support while adjusting for several sociodemographic variables. Our results revealed that 18.3 % (315/1721) of the participants were defined to be frail according to FI. Objective support (odds ratio (OR) = 1.123, 95 % confidence interval (CI): 1.030–1.223), subjective support (OR = 0.885, 95 % CI: 0.827–0.947), and utilization of support (OR = 1.089, 95 % CI: 1.002–1.184) were significantly associated with frailty. The subgroup analysis on participants with different genders showed different results. We found that the three aspects of social support might be linked with frailty. Women seemed to be more sensitive to social support and were more likely to be frail. Frailty is a dynamic condition. The prevalence of frailty may change if proper prevention and intervention policies of social support were applied in the future.

Keywords: Frailty index; social support, elderly people, longevity township, rating scales, Xiao Shuiyuan scales

MSC 2020: 62P20, 62-07, 92D30

1 Introduction

Frailty is a geriatric syndrome featured with decreased physiological activities as a result of aging [29]. Recently, more and more attention has been paid to frailty. Studies have shown that frailty may increase risks of adverse health outcomes like falls and overnight hospitalizations [17], Alzheimer's disease [28], depression, [19] and others. Frailty has an

Acknowledgement: The author is most grateful to Professor Xizhe Peng for providing the survey data for this study. Contributions by Junrui Long and Yujia Guo in data collation and analysis are gratefully acknowledged.

Wenhua Tian, School of Social Development and Public Policy, Fudan University, Shanghai, China, e-mail: wh_tian@fudan.edu.cn

https://doi.org/10.1515/9783112216620-023

important impact on the quality of life of elderly people, and available medical health care services. Researchers tend to reach a consensus that frailty is not part of the natural aging process. It is distinct from disability or multi-morbidity, although strictly related to them, and it is a complex, multifaceted, dynamic process [8, 9]. Therefore, numerous studies on frailty have begun to explore the risk factors and preventions medically and socially [1].

As for the risk factors of frailty, researchers have done much work and drawn attention to the potential impact of social support. Studies have suggested that social support influences physical and mental health and will provide care for frail older people [5, 27]. A study conducted in Brazil has shown social support would be associated with less-steep increases in frailty, and social support might protect against increases in frailty for people with progressive moderate frailty [23]. Another study has shown that older adults who had poor social support were more likely to develop frailty. And there might be interaction effects on frailty between social support and other social psychological factors [14]. A cohort study was conducted to evaluate the role of perceived social support, a specific component of social support, in the prediction of frailty development in myocardial infarction survivors after controlling for other psychosocial risk factors. The study indicated that lower level of perceived social support was associated with greater risk of frailty development in myocardial infarction survivors [18]. Studies have demonstrated that low social support plays a role in the process of becoming frail. One possible explanation for their results may be the effects of social on health behavior and adherence to medical treatment because social support also has an effect on specific diseases or mortality endpoints [4].

Studies on the effects of social support to frailty, especially in elderly Chinese people, have drawn increasing attention with the societal and economic intensification of aging. To investigate the effects of social support on frailty is particularly important among the elderly Chinese people. Since China is undergoing the stage of social changes as well as the pressure of population aging. The generation of elderly Chinese people have been influenced and witnessed great socioeconomic and epidemiological transitions during their lives. The experiences of these reforms and changes are embedded in their psychosocial pressure, resulting in variations in their lifespans and aging processes [2].

It is important to explore the impact of psychosocial factors on the development of frailty in older life. However, the association between social support, as an important aspect of psychosocial factors, and frailty has not been investigated thoroughly. It has been found that frailty can be reduced by participating in social activities and increasing social support [30]. And being active in participating in social leisure activities helped to prevent from being depression and lead to longer life expectancy in the elderly [15]. China is entering an aging society. Elderly Chinese people are at high risk of adverse health outcomes, which could contribute to frailty. It is significant to explore whether there are modifiable social factors related to frailty in elderly Chinese people.

2 Material and methods

Study population

We defined our study as a community-based study, which was conducted in 2014. The sample comes from a community-based study in a longevity township. Jiang'an is a middle-sized township in China. There were 11,198 elderly individuals aged 70–84 years according to the detailed register list from the Public Health Bureau of Jiang'an township. 1,960 elderly were selected among 31 villages considering the 5-year age and sex strata, randomly and equally. The detailed list of the targeted population was reviewed by the village doctors. The targeted population were well informed about the field investigation from the village doctors. One-thousand seven-hundred and eighty eight individuals formed the original study population. There were no substantial differences regarding age and sex ratios between the responders and nonresponders [17]. In our study, about 15 participants with severe deafness and blindness were excluded owing to missing data for all variables. Fifty-two participants were excluded owing to missing data in relevant variables, leaving a final study sample of 1,721.

Data collection

Data were collected by using the standardized questionnaire. Participants were helped to complete the questionnaire under the guidance of our trained investigators. The questionnaire included individuals and sociodemographic variables, indexes, and variables relating to frailty, social support, and so on.

Dependent variable

The definition of frailty is still controversial. In 2015, the World Health Organization summarized the background knowledge on frailty. And frailty was defined as a progressive age-related decline in physiological systems that results in decreased reserves of intrinsic capacity, which confers extreme vulnerability to stressors and increases the risk of a range of adverse health outcomes [21]. Methods have been developed to operationalize frailty for various purposes [10, 22, 24]. On the basis of Zhaoxue Liu's study, the measurement of frailty was based on the score of frailty index (FI) following an earlier research in this field [17]. We use FI to describe participants' state of frailty. Forty-five health deficits including symptoms, activities of daily living (basic and instrumental), comorbidities, cognitive and psychological function were used to construct a FI. The FI was calculated by summing all deficits and dividing by the total number of deficits ($n = 45$), with the FI ranging from 0 to 1. Participants were divided into two categories: "nonfrailty" ($FI < 0.25$), "frailty" ($FI \geq 0.25$).

Independent variables

For social support, the concept is multifaceted. It can be defined as the availability or provision of a relationship, information, or assistance that empower a person to manage their day-to-day life effectively in the presence or absence of crisis [6]. Also, there are numerous studies about the influences of social support on improved functional recovery and psychosocial health [7]. Researchers have applied scales and questionnaires to quantify the definition of social support, such as the famous Social Support Questionnaire (SSQ), developed by Sarason [20, 25]. The Interview Schedule for Social Interaction (ISSI), developed by Henderson [11]. But in consideration of culture differences, we used the Shuiyuan Xiao's Social Support Rating Scales (Chinese version) for reference and made some modifications to adapt to the study [31]. Xiao's Social Support Rating Scales were developed in 1986 and included ten items. In 1990, the scales were revised according to their application. In our study, we tried to measure participants' social support from three aspects, which were objective support, subjective support, and utilization of support, with two items for each aspect, as listed in Table 16.1. The total score of social support was calculated through the sum of the scores of six items.

Confounding factors

Several individual and demographic factors were questioned, they were: age, sex, BMI, educational level, marital status, smoking status, and drinking status. Body mass index was calculated as weight divided by height in square (kg/m2). According to the standard criteria for Asian-Pacific population, the BMI was divided into five groups. They were normal range (BMI: 18.5–22.9), underweight (BMI $<$ 18.5), preobese (BMI: 23.0–24.9), obese class I (BMI: 25.0–29.9), obese class II (BMI $\geq$ 30.0) [3]. The illiterate were defined as someone who had never went to school and not received formal education. Marital status was categorized as currently married and other status (spouse's loss, divorced, and unmarried). Smoking status was divided into three groups: the current smoker was someone who reported smoking every day at present; the former smoker was someone who had ever smoked continuously for more than 6 months. The nonsmoker was someone who had never smoked. Drinking status was divided into three groups: the current drinker was someone who reported drinking every day at present; the former drinker was someone who had ever drunk continuously for more than 6 months; the nondrinker was someone who had never drunk.

Statistical analyses

We used the Chi-square test or Fisher exact test for assessing gender differences in marital status, education, smoking status, drinking status, BMI, and frailty. Student *t*-test was used to compare the gender differences in scores of social support. The relationship between social support and frailty was examined by using the multiple logistic regression analysis (e. g., frailty vs. nonfrailty). In the process of logistic regression analysis, the social support was not only analyzed as a whole, but also as components. In Model 1, crude

Table 16.1: The questionnaire about social support.

No.	Question (Options)
1	How many close friends do you have, from whom you can get support and help? (choose only one option) (1) None (2) 1 or 2 (3) 3 to 5 (4) more than 6
2	For the nearly 1 year, (choose only one option) (1) You stay away from others and live alone in a room. (2) Most of the time you live with strangers and accommodation changes frequently. (3) You live with classmates, colleagues or friends. (4) You live with family members.
3	In the past, when you were in an emergency situation, you used to get help from the following sources: (Multiple choices are allowed) (1) No source (2) The following sources: A. the spouse; B. other family members; C. relatives; D. friends; E. colleagues; F. Work unit; G. Official or semiofficial organizations such as party, youth league and trade union; H. nongovernmental organizations such as religious and social groups; I. Others (please list) ______________
4	The way you talk about your troubles (choose only one option) (1) You never talk with anyone. (2) You only talk with one or two people who are very close to each other. (3) You will speak up if your friend asks you. (4) You take the initiative to complain about your troubles, in order to get support and understanding.
5	As for organizing activities for groups (such as party and youth league organizations, religious organizations, trade unions, student union, etc.): (choose only one option) (1) You never participate (2) You occasionally participate (3) You often participate (4) You actively participate and take an active part in activities
6	How much time have you spent with your friends, relatives or other people (whether you went to his place, he came to your place, or you went out together) in the past week? (choose only one option) (1) 1 day or more (2) 2 or 6 times (3) 1 time (4) no

odds ratios (OR) with 95 % confidence intervals (CI) were calculated. In Model 2, the odds ratios were adjusted for sex. In Model 3, the odds ratios were adjusted for several confounders. The following covariates were considered: age, sex, educational level, marital status, BMI, smoking status, and drinking status, and all aspects of social support were entered simultaneously. Finally, an additional analysis was conducted in a subgroup of different genders and age groups, because a significant interaction effect was observed between social support and gender in the relation with frailty and studies have shown that diseases of elderly people vary according to age [13, 16]. In Model 4, crude odds ratios with 95 % CI were calculated in men and in women, separately. In Model 5, the odds ratios were adjusted for several confounders where the following covariates were considered: age, educational level, marital status, BMI, smoking status, and drinking status, and all aspects of social support were entered. In Models 6 and 7, participants were divided into two groups: group A (age < 80) and group B (age ≥ 80). In Model 6, crude odds ratios with 95 % CI were calculated in groups A and B, separately. In Model 7, the odds ratios were adjusted for several confounders where the following covariates were considered: sex, educational level, marital status, BMI, smoking status, and drinking status, and all aspects of social support were entered simultaneously.

The models were screened for multicollinearity according to the calculation of variance inflation factors, which revealed no problems. ORs and 95 % CIs were calculated in all models. IBM SPSS Statistics 23.0 was used to perform all analyses. A P value of <0.05 (two-tailed) was considered as statistically significant without adjustment for multiple testing.

3 Results

3.1 Sociodemographic characteristics and gender differences

The total sample of $n = 1{,}721$ participants had a mean age of 75.3 years with ±3.9 years SD, and was composed of 46.1 % male ($n = 793$) and 53.9 % female ($n = 928$), statistically without an age difference between sexes, $P = 0.132$; see Table 16.2.

Two-thirds of the chort sample were currently married (65.8 %), with men significantly ($P < 0.001$) more likely to be married than women, 73.5 % and 59.3 %, respectively. Similarly, literacy rates were significantly ($P < 0.001$) higher among men with 80.0 % than among women with 18.3 %.

Regarding lifestyle risk factors, men reported as current smokers in 22.7 % and as current drinkers in 18.7 %, each of which is significantly higher ($P < 0.001$) than 1.1 % and 3.3 %, respectively, found for women. The proportion of participants with a normal body mass index was relatively low for both sexes, 38.2 % of men and 34.8 % of women, with $P = 0.005$ for difference.

The prevalence of frailty, as assessed by the frailty index, was significantly higher in women compared with men, 25.7 % *versus* 9.7 %, and statistically significant with $P <$ 0.001.

The overall sample exhibited a mean (± SD) total score on the social support rating scale (Table 16.1) of 14.69 (± 3.81), as a sum of component scores for mean objective support of 6.02 (± 1.76), mean subjective support of 4.65 (± 2.08), and mean support utilization of 4.01 (± 1.66). When stratified by sex, men reported a significantly ($P < 0.001$) higher mean total social support score than women, 15.17 (± 3.90) *versus* 14.28 (± 3.69).

Table 16.2: Demographics of the study participants.

	All (*n* = 1,721)	Men (*n* = 793)	Women (*n* = 928)	*P* value[a]
Age, mean ± SD	75.28 ± 3.89	75.13 ± 3.86	75.41 ± 3.91	0.132
Marital status				<0.001
Other status, *n* (%)	588 (34.17)	210 (26.48)	378 (40.73)	
Currently married, *n* (%)	1133 (65.83)	583 (73.52)	550 (59.27)	
Education level				<0.001
Literate, *n* (%)	804 (46.72)	634 (79.95)	170 (18.32)	
Illiterate, *n* (%)	917 (53.28)	159 (20.05)	758 (81.68)	
Smoking Status				<0.001
Nonsmoker, *n* (%)	1287 (74.78)	389 (49.05)	898 (96.77)	
Former smoker, *n* (%)	244 (14.18)	224 (28.25)	20 (2.16)	
Current smoker, *n* (%)	190 (11.04)	180 (22.70)	10 (1.08)	
Drinking Status				<0.001
Nondrinker, *n* (%)	1221 (70.95)	418 (52.71)	803 (86.53)	
Former drinker, *n* (%)	321 (18.65)	227 (28.63)	94 (10.13)	
Current drinker, *n* (%)	179 (10.40)	148 (18.66)	31 (3.34)	
BMI				0.005
18.5–22.9, *n* (%)	626 (36.37)	303 (38.21)	323 (34.81)	
<18.5, *n* (%)	62 (3.60)	28 (3.53)	34 (3.66)	
23.0–24.9, *n* (%)	395 (22.95)	190 (23.96)	205 (22.09)	
25.0–29.9, *n* (%)	544 (31.61)	246 (31.02)	298 (32.11)	
≥30.0, *n* (%)	94 (5.46)	26 (3.28)	68 (7.33)	
FI				<0.001
Frailty, *n* (%)	315 (18.30)	77 (9.71)	238 (25.65)	
Nonfrailty, *n* (%)	1406 (81.70)	716 (90.29)	690 (74.35)	
Social support	14.69 ± 3.81	15.17 ± 3.90	14.28 ± 3.69	<0.001
Objective, mean ± SD	6.02 ± 1.76	6.07 ± 1.85	5.98 ± 1.69	0.342
Subjective, mean ± SD	4.65 ± 2.08	4.96 ± 2.08	4.40 ± 2.05	<0.001
Utilization, mean ± SD	4.01 ± 1.66	4.15 ± 1.68	3.89 ± 1.64	0.001

Notes: *n*: absolute frequency. SD: standard deviation. [a]For categorical variables, the Chi-square test or Fisher's exact test were used, for continuous variables, Student *t*-test was used. BMI: Body Mass Index. FI: Frailty Index.

For objective support, scores were comparable between men and women (6.07 ± 1.85 *versus* 5.98 ± 1.69; $P = 0.342$), indicating little sex difference in the amount of tangible or instrumental support received. However, subjective support—reflecting perceived emotional and interpersonal support—was somewhat higher in men (4.69 ± 2.08) than in women (4.40 ± 2.05) with $P < 0.001$, and utilization of support was on average also scored somewhat higher in men, 4.15 ± 1.68, than in women, 3.89 ± 1.64, $P = 0.001$.

3.2 Association between social support and frailty

Here, we analyze the association of measurements on the social support rating scales in comparison with the frailty index, using logistic regression in three scenarios: without adjustments (Model 1), with adjustment for age and gender (Model 2), and with further adjustment for the other covariates (Model 3).

We first do the analysis in the total sample across gender in 3.2.1, then within gender in 3.2.2, and finally within two old-age groups in 3.2.3.

3.2.1 Association in total sample

The final results from fitting logistic regression models under the three scenarios in the total sample are presented in Table 16.3.

Table 16.3: Association between social support score on the Social Support Rating Scale (SSRS) and frailty index in $n = 1,721$ persons by logistic regression analysis in three scenarios, Model 1 to Model 3, of adjustment for covariates; estimated OR with 95%CI.

SSRS	Model 1	Model 2	Model 3
Total (Sum) Score	0.982 (0.951, 1.014)	0.998 (0.965, 1.032)	1.007 (0.972, 1.043)
P	0.263	0.091	0.699
Objective Subscore	1.107 (1.026, 1.195)	1.114 (1.029, 1.206)	1.123 (1.030, 1.223)
P	0.009	0.007	0.008
Subjective Subscore	0.856 (0.803, 0.913)	0.879 (0.824, 0.939)	0.885 (0.827, 0.947)
P	<0.001	<0.001	<0.001
Utilization Subscore	1.050 (0.970, 1.136)	1.063 (0.981, 1.153)	1.089 (1.002, 1.184)
P	0.228	0.136	0.046

Notes: SSRS = Social Support Rating Scale (Table 16.1). OR = odds ratio. CI = confidence interval. *P* = nominal level of statistical significance, *P* value. Model 1: Unadjusted for covariates. Model 2: Adjusted for only age and gender. Model 3: Further adjusted for marital status, education, smoking, drinking, Body Mass Index (BMI), and other covariates as described in the text.

The relationship between social support total score and frailty is robustly insignificant across the three scenarios with odds ratios from 0.982 (95 %CI: 0.951, 1.014) without any adjustments to 1.007 (95 %CI: 0.972, 1.043) after adjustment for all covariates, meaning that social support total score and frailty index mutually do not provide information about the other. The same finding applies to the utilization of support subscore, with odds ratios from 1.050 (95 %CI: 0.970, 1.136) without any adjustments to 1.089 (95 %CI: 1.002, 1.184) after adjustment for all covariates, irrespective of a nominal $P < 0.05$ that must be ascribed to random variation.

For the objective subscore of the SSRS—reflecting the amount of tangible or instrumental support received— there emerges some marginally significant association with the frailty index with odds ratios from 1.107 (95 %CI: 1.026, 1.195) without any adjustments to 1.123 (95 %CI: 1.030, 1.223) after adjustment for all covariates, by $P = 0.01$.

This is more pronounced in the subjective subscore of the SSRS—reflecting perceived emotional and interpersonal support—which shows a robustly significant

negative association with the frailty index by odds ratios from 0.856 (95 %CI: 0.803, 0.913) without any adjustments to 0.885 (95 %CI: 0.827, 0.947) after adjustment for all covariates, with consistent $P < 0.001$.

3.2.2 Association within men and within women

Within each of the sexes, there is no evidence of an association between Social Support Rating Scale (SSRS) scores and frailty index scores, regarding the total (sum) score and the utilization subscore, beyond random variation. Noteworthy is a statistically significant association between frailty and the objective social support subscore —reflecting the amount of tangible or instrumental support received—among women and between frailty and the subjective social support sub-score—reflecting perceived emotional and interpersonal support— among men.

Table 16.4 shows the detailed results of logistic regression analyses for men in the two scenarios of adjustments for covariates, i. e., adjusted for age under Model 4 and further adjusted for all other covariates under Model 5.

Estimated odds ratios (OR) of 0.941 (95 %CI: 0.886, 1.000) and 0.949 (95 %CI: 0.889, 1.013) for the total score of the Social Support Rating Scale (SSRS) do not support an association with frailty by $P = 0.048$ and $P = 0.113$ under Models 4 and 5, respectively.

Table 16.4: Association between social support score on the Social Support Rating Scale (SSRS) and frailty index in subgroup analysis by gender (men, $n = 793$; women, $n = 928$); separate logistic regression analyses, each in two scenarios, Model 4 and Model 5, of adjustment for covariates; estimated OR with 95%CI.

SSRS	Model 4		Model 5	
	Men	Women	Men	Women
Total (Sum) Score	0.941	1.024	0.949	1.034
	(0.886, 1.000)	(0.984, 1.066)	(0.889, 1.013)	(0.991, 1.079)
	$P = 0.048$	$P = 0.244$	$P = 0.113$	$P = 0.124$
Objective Subscore	0.978	1.192	0.966	1.214
	(0.857, 1.117)	(1.080, 1.316)	(0.834, 1.118)	(1.090, 1.352)
	$P = 0.746$	$P = 0.001$	$P = 0.642$	$P < 0.001$
Subjective Subscore	0.822	0.906	0.812	0.916
	(0.727, 0.929)	(0.839, 0.979)	(0.714, 0.924)	(0.845, 0.992)
	$P = 0.002$	$P = 0.012$	$P = 0.002$	$P = 0.032$
Utilization Subscore	1.075	1.056	1.136	1.070
	(0.927, 1.247)	(0.959, 1.163)	(0.972, 1.327)	(0.968, 1.182)
	$P = 0.341$	$P = 0.267$	$P = 0.110$	$P = 0.186$

Notes: SSRS = Social Support Rating Scale (Table 16.1). OR = odds ratio; CI = confidence interval. P = nominal level of statistical significance, P value. Model 4: adjusted for age; Model 5: further adjusted for marital status, education, smoking, drinking, Body Mass Index (BMI), and other covariates as described in the text.

Estimated odds ratios (OR) of 0.978 (95 %CI: 0.857, 1.117) and 0.966 (95 %CI: 0.834, 1.118) for the objective support subscore of the Social Support Rating Scale (SSRS) do not support an association with frailty by $P = 0.746$ and $P = 0.642$ under Models 4 and 5, respectively.

However, for the subjective support subscore of the Social Support Rating Scale (SSRS) the estimated odds ratios (OR) of 0.822 (95 %CI: 0.727, 0.929) and 0.812 (95 %CI: 0.714, 0.924) do indicate a **negative** association with frailty among men by $P = 0.002$ and $P = 0.002$ in both scenarios, Models 4 and 5.

For the social support utilization subscore of the Social Support Rating Scale (SSRS), the estimated odds ratios (OR) of 1.075 (95 %CI: 0.927, 1.247) and 1.136 (95 %CI: 0.972, 1.327) do not support an association with frailty among men, with $P = 0.341$ and $P = 0.110$ under Models 4 and 5, respectively.

For women, the detailed results of logistic regression analyses in the two scenarios of adjustments for covariates, i. e., adjusted for age under Model 4 and further adjusted for all other covariates under Model 5, are shown in Table 16.4, as well.

Estimated odds ratios (OR) of 1.024 (95 %CI: 0.984, 1.066) and 1.034 (95 %CI: 0.991, 1.079) for the total score of the Social Support Rating Scale (SSRS) do not support an association with frailty by $P = 0.244$ and $P = 0.124$ under Models 4 and 5, respectively.

Estimated odds ratios (OR) of 1.192 (95 %CI: 1.080, 1.316) and 1.214 (95 %CI: 1.090, 1.352) for the objective support subscore of the Social Support Rating Scale (SSRS) do support a robust **positive** association with frailty by $P = 0.001$ and $P < 0.001$ under Models 4 and 5, respectively.

However, for the subjective support subscore of the Social Support Rating Scale (SSRS), the estimated odds ratios (OR) of 0.906 (95 %CI: 0.839, 0.979) and 0.916 (95 %CI: 0.845, 0.992) do not support a statistically significant association with frailty among men by $P = 0.012$ and $P = 0.032$ in both scenarios, Models 4 and 5, beyond what can still be ascribed to random variation.

For the social support utilization subscore of the Social Support Rating Scale (SSRS), the estimated odds ratios (OR) of 1.056 (95 %CI: 0.959, 1.163) and 1.070 (95 %CI: 0.968, 1.182) also do not support an association with frailty among men, with $P = 0.267$ and $P = 0.186$ under Models 4 and 5, respectively.

3.2.3 Association within "younger-old" and within "older-old"

Within each of the two age groups—aged < 80 years, "younger-old," and aged ≥ 80 years, "older-old," there is no evidence of an association between Social Support Rating Scale (SSRS) scores and frailty index scores, regarding the total (sum) score and the utilization subscore, beyond random variation.

Noteworthy are a statistically significant association between frailty and the subjective social support subscore —reflecting perceived emotional and interpersonal support—among "younger-old" (aged < 80 years), and an only nominal association between frailty and the objective social support subscore—reflecting the amount of tangible or instrumental support received— among "older-old" (aged ≥ 80 years).

For the "younger-old" (aged < 80 years), the detailed results of logistic regression analyses in the two scenarios of adjustments for covariates, i. e., adjusted for age under Model 6 and further adjusted for all other covariates under Model 7, are shown in Table 16.5.

Table 16.5: Association between social support score on the Social Support Rating Scale (SSRS) and Frailty Index in subgroup analysis by age across sexes ("younger-old" (aged < 80 years), $n = 1,425$; "older-old" (aged ≥ 80 years), $n = 296$); separate logistic regression analyses, each in two scenarios, Model 6 & Model 7, of adjustment for covariates; estimated OR with 95%CI.

SSRS	Model 6		Model 7	
	Age < 80	Age ≥ 80	Age < 80	Age ≥ 80
Total (Sum) Score	0.981	1.012	0.996	1.044
	(0.946, 1.019)	(0.946, 1.083)	(0.956, 1.037)	(0.969, 1.124)
	$P = 0.321$	$P = 0.720$	$P = 0.844$	$P = 0.261$
Objective Subscore	1.095	1.197	1.088	1.260
	(1.002, 1.197)	(1.021, 1.404)	(0.987, 1.200)	(1.051, 1.509)
	$P = 0.045$	$P = 0.027$	$P = 0.089$	$P = 0.012$
Subjective Subscore	0.850	0.877	0.877	0.885
	(0.789, 0.915)	(0.769, 1.000)	(0.811, 0.947)	(0.746, 1.024)
	$P < 0.001$	$P = 0.050$	$P = 0.001$	$P = 0.102$
Utilization Subscore	1.062	1.090	1.081	1.144
	(0.969, 1.164)	(0.924, 1.286)	(0.982, 1.189)	(0.958, 1.367)
	$P = 0.195$	$P = 0.306$	$P = 0.112$	$P = 0.137$

Notes: SSRS = Social Support Rating Scale (Table 16.1). OR = odds ratio; CI = confidence interval. P = nominal level of statistical significance, P value. Model 6: adjusted for age; Model 7: further adjusted for marital status, education, smoking, drinking, Body Mass Index (BMI), and other covariates as described in the text.

Estimated odds ratios (OR) of 0.981 (95 %CI: 0.946, 1.019) and 0.996 (95 %CI: 0.956, 1.037) for the total score of the Social Support Rating Scale (SSRS) do not support an association with frailty by values of $P = 0.321$ and $P = 0.844$ under Models 6 and 7, respectively.

As well, for the objective support subscore of the Social Support Rating Scale (SSRS), estimated odds ratios (OR) of 1.095 (95 %CI: 1.002, 1.197) and 1.088 (95 %CI: 0.987, 1.200) do not support an association with frailty by values of $P = 0.045$ and $P = 0.089$ under Models 6 and 7, respectively.

However, for the subjective support subscore of the Social Support Rating Scale (SSRS), the estimated odds ratios (OR) of 0.850 (95 %CI: 0.789, 0.915) and 0.877 (95 %CI: 0.811, 0.947) do support a statistically significant **negative** association with frailty among "younger-old" (aged < 80 years) in both scenarios, Models 6 and 7, by values of $P < 0.001$ and $P = 0.001$, respectively.

For the social support utilization subscore of the Social Support Rating Scale (SSRS), the estimated odds ratios (OR) of 1.062 (95 %CI: 0.969, 1.164) and 1.081 (95 %CI: 0.982, 1.189) also do not support an association with frailty among "younger-old" (aged < 80 years) in both scenarios, Models 6 and 7, by values of $P = 0.195$ and $P = 0.112$, respectively. Table 16.5 also shows the detailed results of logistic regression analyses in the two scenarios of adjustments for covariates, i. e., adjusted for age under Model 6 and further adjusted for all other covariates under Model 7, for the "older-old" (aged ≥ 80 years).

Estimated odds ratios (OR) of 1.012 (95 %CI: 0.946, 1.083) and 1.044 (95 %CI: 0.969, 1.124) for the total score of the Social Support Rating Scale (SSRS) do not support an association with frailty by values of $P = 0.720$ and $P = 0.261$ under Models 6 and 7, respectively.

However, for the objective support sub-score of the Social Support Rating Scale (SSRS), the estimated odds ratios (OR) of 1.197 (95 %CI: 1.021, 1.404) and 1.260 (95 %CI: 1.051, 1.509) do indicate a statistically marginally significant **positive** association with frailty among "older-old" (aged ≥ 80 years) in both scenarios, Models 6 and 7, by values of $P = 0.027$ and $P = 0.012$, respectively, which can still be ascribed to random variation.

For the subjective support subscore of the Social Support Rating Scale (SSRS), estimated odds ratios (OR) of 0.877 (95 %CI: 0.769, 1.000) and 0.885 (95 %CI: 0.746, 1.024) do not support an association with frailty by values of $P = 0.050$ and $P = 0.102$ under Models 6 and 7, respectively.

For the social support utilization subscore of the Social Support Rating Scale (SSRS), the estimated odds ratios (OR) of 1.090 (95 %CI: 0.924, 1.286) and 1.144 (95 %CI: 0.958, 1.367) also do not support an association with frailty among "older-old" (aged ≥ 80 years) in both scenarios, Models 6 and 7, by values of $P = 0.306$ and $P = 0.137$, respectively.

4 Discussion

Our study aimed at exploring the association between social support and frailty. When social support was analyzed as a whole, we noticed that social support was not significantly related to frailty from the final adjusted ORs. It implies that there might not be a very strong link between social support and frailty. Since our study is a cross-sectional study. There is still much work to do to detect the relationship between social support and frailty. On the other hand, we cannot ignore the measurement of social support. We used the Shuiyuan Xiao's Social Support Rating Scales (Chinese version) for reference and made some modifications to adapt to the study [31]. Xiao's Social Support Rating Scales were first developed in 1986 and then revised in 1990. As time went by, maybe it is time to revised some items in the scales. And before the whole study was conducted, the reliability analysis and validity analysis of the questionnaire's social support part were not applied. So, in our future study, we need to study further to explore the association between social support and frailty. We might revise the questionnaire of social support or conduct a cohort study.

But when social support was classified as three aspects, the results were different. When all participants were included in the logistic regression analysis, we observed that objective support (OR = 1.123, 95 % CI: 1.030–1.223), subjective support (OR = 0.885, 95 % CI: 0.827–0.947) and utilization of support (OR = 1.089, 95 % CI: 1.002–1.184) were all significantly related to frailty. Participants with higher scores in objective support and utilization of support were more likely to be frail, but conversely in subjective support. Up to now, few researches on the effects of social support on frailty. According to the definition of social support that we have mentioned above [6], we might make a hypothesis that if someone gains more social support (with high scores in our study) in daily life, he or she is less likely to be frail. In fact, a cohort study conducted in older Mexican American adults supported that hypothesis [23]. There was also a study conducted in Taiwan indicating that infrequent social activities might be a risk factor for frailty in elderly people [1]. However, in our study the results of subjective support were in accordance with former studies. As for objective support and utilization of support, the results seemed to be contradicted. There is no denying that we still need to refine our research. But it may also imply that our recognition about social support needs to be improved. Maybe the interactions and contrasts of these three aspects need to be considered while judging an individual's status of social support.

But in consideration of subgroup analysis, there were some differences. In the subgroup analysis of different genders, objective support (OR = 0.966, 95 % CI: 0.834–1.118) and utilization of support (OR = 1.136, 95 % CI: 0.972–1.327) were not significantly related to frailty in men. And men with higher scores in subjective support less likely to be frail (OR = 0.812, 95 % CI: 0.714–0.924). While only utilization of support (OR = 1.070, 95 % CI: 0.968–1.182) was not significantly related to frailty in women. Women with higher scores in objective support (OR = 1.214, 95 % CI: 1.090–1.352) were more likely to be frail, however, women with higher scores in subjective support (OR = 0.916, 95 % CI: 0.845–0.992) were less likely to be frail. In the subgroup analysis of ages, objective support (OR = 1.088, 95 % CI: 0.987–1.200) and utilization of support (OR = 1.081, 95 % CI: 0.982–1.189) were not significantly related to frailty in group A. And people in group A with higher scores in subjective support less likely to be frail (OR = 0.877, 95 % CI: 0.811–0.947). While subjective support (OR = 0.885, 95 % CI: 0.746–1.024) and utilization of support (OR = 1.144, 95 % CI: 0.958–1.367) were not significantly related to frailty in group B. People with higher scores in objective support were more likely to be frail (OR = 1.260, 95 % CI: 1.051–1.509).

Based on the data from the Rugao Longevity and Ageing Study conducted from November 13, 2014, to December 21, 2014, at the beginning, women (25.65 %, 238/928) were more likely to be classified as frail than men (9.71 %, 77/793). And according to our results of subgroup analysis, women seemed to be more sensitive to social support than men, and were more likely to be affected by frailty. In consideration of ages, the oldest-old people seem more sensitive to objective support than the youngest-old people. While the youngest-old people are more sensitive to subjective support than the oldest-old people. Frailty is a dynamic condition. The prevalence of frailty may change if proper prevention and intervention polices and targets about social support were applied in the future. Therefore, to understand the differences and demands of men and women in social support and to further investigate the causes and associations, long-term data must be collected.

The data on social support in Table 16.2 suggest modest sex differences in only the perceived and actual use of social support among older adults in this sample. The patterns mirror previous research in older Chinese populations, where social support—particularly emotional and instrumental components—is shaped by household composition, marital status, and traditional family roles. Throughout late life, men may report higher overall support due in part to higher rates of marriage and partnership, while women, despite potentially broader informal networks, may have lower perceived and utilized support, especially in settings where spouse-based support predominates. The comparable objective support scores reflect the relatively equal access to tangible resources, yet nuanced differences in perception and utilization illustrate the complex interplay of gender, family structure, and cultural context typical of contemporary older adults in China [12, 32, 26].

5 Conclusion

Frailty of elderly people might be associated with objective support, subjective support and utilization of support in a longevity township. Women seemed to be more sensitive to social support and were more likely to be frail. The prevalence of frailty may change if proper prevention and intervention policies of social support were applied in the future.

Bibliography

[1] C.-Y. Chen, S.-C. Wu, L.-J. Chen, and B.-H. Lue. The prevalence of subjective frailty and factors associated with frailty in Taiwan. *Archives of Gerontology and Geriatrics*, 50(Suppl 1):S43–S47, 2010.

[2] A. Clegg, J. Young, S. Iliffe, M. O. Rikkert, and K. Rockwood. Frailty in elderly people. *Lancet*, 381:752–762, 2013.

[3] World Health Organization Expert Consultation. Appropriate body-mass index for Asian populations and its implications for policy and intervention strategies. *Lancet*, 363:157–163, 2004.

[4] R. M. DiMatteo. Social support and patient adherence to medical treatment: A meta-analysis. *Health Psychology*, 23(2):207–218, 2004.

[5] W. W. Dressler, K. S. Oths, and C. C. Gravlee. Race and ethnicity in public health research: Models to explain health disparities. *Annual Review of Anthropology*, 34(1):231–252, 2005.

[6] T. Elloker and A. J. Rhoda. The relationship between social support and participation in stroke: A systematic review. *African Journal of Disability*, 7:357, 2018.

[7] M. Fallahpour, K. Tham, M. T. Joghataei, and H. Jonsson. Perceived participation and autonomy: Aspects of functioning and contextual factors predicting participation after stroke. *Journal of Rehabilitation Medicine*, 43(5):388–397, 2011.
[8] L. Fried. Frailty: What's been done and what needs doing [presentation]. International Conference on Frailty & Sarcopenia Research, Miami, USA, 2018.
[9] L. Galluzzo, R. O'Caoimh, A. Rodriguez-Laso, N. Beltzer, A. H. Ranhoff, J. Van der Heyden, M. Lamprini-Koula, M. Ciutan, L. Lopez-Samaniego, and A. Liew. Incidence of frailty: A systematic review of scientific literature from a public health perspective. *Annali dell'Istituto Superiore di Sanità*, 54(3):239–245, 2018.
[10] J. Goldstein, R. E. Hubbard, P. Moorhouse, and M. K. Andrew. Feasibility of using information derived from a care partner to develop a frailty index based on comprehensive geriatric assessment. *Journal of Frailty & Aging*, 2(1):15–21, 2013.
[11] S. Henderson. Social network, support and neurosis. The function of attachment in adult life. *British Journal of Psychiatry*, 131:185–191, 1977.
[12] Y. Jing, W. Wang, W. Peng, M. Wang, X. Chen, X. Liu, P. Wang, F. Yan, Y. Yang, X. Wang, S. Sun, Y. Ruan, and Y. Wang. Social support and cognitive function in Chinese older adults who experienced depressive symptoms: Is there an age difference? *Frontiers in Aging Neuroscience*, 15:1175252, 2023. https://doi.org/10.3389/fnagi.2023.1175252.
[13] S.-B. Lee, J.-H. Oh, J.-H. Park, S.-P. Choi, and J.-H. Wee. Differences in youngest-old, middle-old, and oldest-old patients who visit the emergency department. *Clinical and Experimental Emergency Medicine*, 5(4):249–255, 2018.
[14] J. Li, D. Zhao, B. Dong, D. Yu, Q. Ren, J. Chen, Q. Qin, P. Bi, and Y. Sun. Frailty index and its associations with self-neglect, social support and sociodemographic characteristics among older adults in rural China. *Geriatrics & Gerontology International*, 18(7):987–996, 2018.
[15] L.-L. Li, H.-J. Chang, H.-I. Yeh, C.-J. Y. Hou, C.-H. Tsai, and J.-P. Tsai. Factors associated with leisure participation among the elderly living in long-term care facilities. *International Journal of Gerontology*, 4(2):69–74, 2010.
[16] Z. Liu, L. Han, X. Wang, Q. Feng, and T. M. Gill. Disability prior to death among the oldest-old in China. *The Journals of Gerontology. Series A, Biological Sciences and Medical Sciences*, 74(3):433, 2018.
[17] Z. Liu, Q. Wang, T. Zhi, Y. Zhu, Y. Wang, Z. Wang, J. Shi, X. Xie, X. Chu, X. Wang, and X. Jiang. Frailty index and its relationship to falls and overnight hospitalizations in elderly Chinese people: A population-based study. *The Journal of Nutrition, Health & Aging*, 20(5):561–568, 2016.
[18] I. Lurie, V. Myers, U. Goldbourt, and Y. Gerber. Perceived social support following myocardial infarction and long-term development of frailty. *European Journal of Preventive Cardiology*, 22(10):1346–1353, 2015.
[19] B. T. Mezuk, L. Edwards, M. Lohman, M. Choi, and K. Lapane. Depression and frailty in later life: A synthetic review. *International Journal of Geriatric Psychiatry*, 27(9):879–892, 2012.
[20] R. Moonesar, I. Sammy, P. Nunes, and J. Paul. Social support in older people: Lessons from a developing country. *Quality of Life Research*, 25(1):233–236, 2016.
[21] World Health Organization. World report on aging and health, 2015. Available online at http://www.who.int/ageing/publications/world-report-2015/en/. [Accessed 29 November 2018].
[22] N. Oubaya, R. Mahmoudi, D. Jolly, A. A. Zulfiqar, E. Quignard, C. Cunin, P. Nazeyrollas, J.-L. Novella, and M. Drame. Screening for frailty in elderly subjects living at home: Validation of the Modified Short Emergency Geriatric Assessment (SEGAm) instrument. *The Journal of Nutrition, Health & Aging*, 18(8):757–764, 2014.
[23] M. K. Peek, B. T. Howrey, R. S. Ternent, L. A. Ray, and K. J. Ottenbacher. Social support, stressors, and frailty among older Mexican American adults. *The Journals of Gerontology. Series B, Psychological Sciences and Social Sciences*, 67(6):755–764, 2012.
[24] K. Rockwood and A. Mitnitski. Frailty in relation to the accumulation of deficits. *The Journals of Gerontology. Series A, Biological Sciences and Medical Sciences*, 62(7):722–727, 2007.

[25] B. R. Sarason. Social support, social behavior, and cognitive processes. In R. Schwarzer, editor, *Self-Related Cognitions in Anxiety and Motivation*. Psychology Press, New York, 1 edition, 1986.

[26] M. A. Simon, R. Chen, and X. Dong. Gender differences in perceived social support in U.S. Chinese older adults. *Journal of Gerontology & Geriatric Research*, 3(4):163, 2014. https://doi.org/10.4172/2167-7182.1000163.

[27] P. A. Thoits. Stress, coping, and social support processes: Where are we? What next? *Journal of Health and Social Behavior*, 53–79, 1995. Special No.

[28] L. Wallace, O. Theou, K. Rockwood, and M. K. Andrew. Relationship between frailty and Alzheimer's disease biomarkers: A scoping review. *Alzheimer's & Dementia (Amsterdam)*, 10:394–401, 2018.

[29] X. Wang, C. Zhou, Y. Li, H. Li, Q. Cao, and F. Li. Prognostic value of frailty for older patients with heart failure: A systematic review and meta-analysis of prospective studies. *BioMed Research International*, 8739058, 2018.

[30] J. Woo, W. Goggins, A. Sham, and S. C. Ho. Public health significance of the frailty index. *Disability and Rehabilitation*, 28(8):515–521, 2006.

[31] S. Xiao. The theoretical basis and research application of social support rating scales. *Journal of Clinical Psychological Medicine*, 4(2):98–101, 1994.

[32] J. You, H. Fung, and P. Vitaliano. The pattern of social support seeking and its socio-demographic variations among older adults in China. *European Journal of Ageing*, 17(3):341–348, 2020. https://doi.org/10.1007/s10433-019-00550-w.

Part III: Technologies

Part III-A: **Engineering sciences**

Part III-B: **Information sciences**

Rationale: 8

Part III: **Technologies**

The “tool box” of technologies and methodologies needed to measure dynamics of the physiological processes and assess impact on internal dynamics from outside factors.

Three classes are considered:
- engineering sciences
- information sciences
- data-analytic sciences

Part III-A: **Engineering sciences**

This class includes
- engineered systems theory
- biometrology (measurement theory)
- biomedical technology

Part III-A.1: **Engineered systems theory**

In former volume BIOKYBERNETIKA:
- A method for automatically generating models of complex systems,
- Slow/fast dynamic models with applications to engineering problems

Part III-B: **Information sciences**

This class includes:
- computing technology
- imaging for visualization and pattern recognition
- artificial intelligence applications

Part III-B.2: **Imaging for visualization and pattern recognition**

In this volume:
- Solving the magnetoencephalography inverse problem for continuously distributed source models

In former volume BIOKYBERNETIKA:
- Numerical modeling of medical ultrasound using the grid-characteristic method
- The direct and the inverse magnetic encephalography problem

Part III-B.3: **Artificial intelligence applications**

In this volume:
- Classification of intracytoplasmic sperm injection videos

https://doi.org/10.1515/9783112216620-024

Arthur I. Sabirov and Tatyana V. Zakharova

Solving the magnetoencephalography inverse problem for continuously distributed source models

Abstract: The inverse problem of magnetoencephalography (MEG) and electroencephalography (EEG) is inherently ill-posed, particularly when modeling continuously distributed sources of cortical activity. Traditional distributed approaches such as low-resolution electromagnetic tomography (LORETA) provide smooth reconstructions but demand high computational resources and often lose spatial specificity, while dipole-based methods achieve focal localization yet fail in extended activity scenarios. We propose a hybrid algorithm that combines independent component analysis (ICA) with point-source localization to bridge these limitations. Under the assumption of independence between functional components, ICA decomposes MEG recordings into dominant source signals, which are then localized with single-dipole fitting methods. This strategy enables robust estimation of centers of neural activity even under high noise levels or correlated source configurations. To extend the approach, we integrate ICA-based localization with LORETA applied to local voxel grids surrounding each estimated center. This reduces dimensionality, improves spatial resolution, and significantly lowers computational cost while mitigating artifacts associated with smooth regularization. Tests on synthetic data with multiple independent sources demonstrate improved mapping accuracy, reduced background smoothing, and faster runtimes. These results establish the ICA-LORETA framework as a flexible and efficient tool for solving the MEG/EEG inverse problem, providing a foundation for future clinical and research applications.

Keywords: Magnetoencephalography, electroencephalography, inverse problem, source localization, independent component analysis, LORETA, distributed sources

MSC 2020: 92C55

1 Introduction

Magnetoencephalography (MEG) and electroencephalography (EEG) complement magnetic resonance imaging (MRI) by providing superior temporal resolution of neural activity. These techniques are particularly valuable for detecting abnormalities that occur only in high-frequency components of brain signals, offering diagnostic insights that

Arthur I. Sabirov, Tatyana V. Zakharova, Department of Mathematical Statistics, Faculty of Computational Mathematics and Cybernetics, Lomonosov Moscow State University, Leninskie Gory 52-1, 119991 Moscow, Russia, e-mails: a.i.sabirov@yandex.ru, tvzaharova@mail.ru

https://doi.org/10.1515/9783112216620-025

MRI alone cannot provide. However, their utility is limited by the difficulty of accurate localization of the neural activity sources. The MEG/EEG inverse problem—the reconstruction of the neural activity sources from the signals recorded on the scalp—is mathematically ill-posed and does not admit a unique solution.

Traditionally, the MEG inverse problem has been addressed using distributed source models such as low-resolution electromagnetic tomography (LORETA) [6] and its extensions, which formulate the task as an optimization problem that favors smooth spatial distributions of activity. While these methods are capable of reconstructing continuous maps of neural activation, they require substantial computational resources and may suffer from reduced spatial specificity. Alternative dipole-based approaches, such as the Multiple Signal Classification (MUSIC) [3]) algorithm, estimate source locations by exploiting the signal and noise subspaces of the sensor covariance matrix. MUSIC provides high localization accuracy for a small number of dipolar sources but is less effective in scenarios involving extended or distributed neural activity.

In our previous work, we introduced an algorithm capable of efficiently addressing the MEG/EEG inverse problem under the assumption of independent dipolar sources. In the present study, we extend this approach to demonstrate its applicability to a broader class of source models. In particular, we consider more realistic scenarios involving continuous or spatially distributed neural activity, expanding the range of problems for which our method may provide reliable source localization [9, 8, 5, 4].

2 Methods

Direct and inverse MEG problems with dipolar sources

MEG records magnetic fields at the scalp that are generated primarily by neural activity. However, the measured signals also contain contributions from external noise sources (e. g., environmental electromagnetic activity such as telecommunications) and internal physiological processes (e. g., cardiac and respiratory artifacts).

Under the dipolar source model with a spherical head volume, the direct (forward) problem can be described by the Biot–Savart law [4]:

$$\kappa(r,t) = -k \sum_{j=1}^{M} \frac{r}{|r - r_j'|^3} \cdot (Q_j(t) \times r_j') + \epsilon(t). \tag{17.1}$$

Here, $\kappa(r,t)$ denotes the signal recorded at sensor location r and time t; M is the number of dipolar sources; r_j' and $Q_j(t)$ are the location and dipole moment of source j, respectively; and $\epsilon(t)$ represents noise signals.

The corresponding inverse problem in this model is to estimate the number of active dipoles M, their spatial locations r_j', and their time-varying dipole moments $Q_j(t)$ from the observed sensor data. Even under this simplified formulation, the problem is ill-posed: multiple distinct configurations of dipoles can give rise to nearly identical sensor measurements, and a unique solution is therefore not guaranteed. Moreover, the associated optimization problem is highly nonconvex, making the search for accurate and stable solutions computationally challenging.

ICA-based dipole localization algorithm

We address the inverse problem using a two-stage algorithm based on independent component analysis (ICA) and dipole localization methods:

1. Source decomposition.
   The recorded MEG signals are first decomposed into M statistically independent components using ICA [1]. The number of components M is determined empirically by repeatedly applying ICA with increasing dimensionality and selecting the last value of M that yields a substantial reduction in the independence measure (e. g., non-Gaussianity). The resulting independent components approximate the normalized temporal signals of the underlying neural sources, $Q_j(t)$.
   In addition, this stage allows for the identification and rejection of noise and artifact components, which typically emerge among the last components in the decomposition. These components represent mixtures of useless signals $\epsilon(t)$.
   Applying the inverse ICA transformation to each retained independent component reconstructs the corresponding contribution in sensor space, effectively yielding a separate single-dipole inverse problem for each of the M sources.
2. Dipole localization.
   For each component j, we estimate the dipole location r_j using a suitable dipole localization method [4, 7, 2]. The tangential orientation and amplitude of the dipole moment $Q_j(t)$ are then approximated from the reconstructed temporal dynamics.

This algorithm is computationally efficient, requires only moderate resources, and can be applied rapidly even with high-density MEG data. A limitation of the approach is that it reconstructs at most M dipoles, where M is typically smaller than the number of available sensors.

3 Results

3.1 MEG inverse problem with correlated dipolar sources

The developed algorithm is not restricted to multidipole configurations. Because any extended neural activity can be approximated by a finite set of equivalent dipoles, the method is also applicable to distributed sources. The ICA-based decomposition tends to isolate the most dominant signals. In cases where bioelectrical activity consists of a primary source accompanied by weaker secondary contributions, the current implementation accurately reconstructs the principal component, while the weaker signals may be suppressed by noise or lost when the number of independent components is limited. Furthermore, when activity of a single physiological origin manifests across multiple overlapping regions, the ICA decomposition separates it into several components. Although such fragmentation may complicate neurophysiological interpretation, it simultaneously simplifies the task of source localization.

Robustness of ICA-based algorithm

We evaluated the robustness of the algorithm and demonstrated its stability under perturbations of different components of the forward model (Equation (17.1)), including the additive noise term $\epsilon(t)$, the dipole locations r_j, and the dipole moments $Q_j(t)$. As an example, consider perturbations applied to the dipole moment [9, 8, 5, 4]:

$$\kappa(r,t) = -k\sum_{j=1}^{M}\frac{r}{|r-r_j'|^3}\cdot((Q_j(t)+\epsilon_j(t))\times r_j') + \epsilon(t), \tag{17.2}$$

$$\kappa(r,t) = -k\sum_{j=1}^{M}\frac{r}{|r-r_j'|^3}\cdot(Q_j(t)\times r_j') + \left(-k\sum_{j=1}^{M}\frac{r\cdot\epsilon_j(t)\times r_j'}{|r-r_j'|^3}\right) + \epsilon(t), \tag{17.3}$$

$$\kappa(r,t) = -k\sum_{j=1}^{M}\frac{r}{|r-r_j'|^3}\cdot(Q_j(t)\times r_j') + \epsilon'(t) + \epsilon(t), \tag{17.4}$$

$$\kappa(r,t) = -k\sum_{j=1}^{M}\frac{r}{|r-r_j'|^3}\cdot(Q_j(t)\times r_j') + \epsilon''(t). \tag{17.5}$$

Thus, the perturbed model retains the same functional form as the original forward equation, with the perturbation absorbed into an effective noise term $\epsilon''(t)$. Consequently, the ICA-based decomposition remains equally applicable and demonstrates stable performance under such noise conditions.

Localization of dipole clusters

The results obtained for cases with noise in dipole coordinates and moments provide a foundation for addressing a more complex problem—the localization of correlated dipole clusters. We define a connected dipole cluster as a group of dipoles, potentially of varying spatial configuration, that collectively exhibit a dominant primary dipole moment. Within such clusters, nonzero correlations are assumed between any pair of dipoles. In this formulation, perturbations in the coordinates represent the spatial extent and variability of the cluster, while perturbations in the dipole moments capture the contributions of secondary sources arising from activity of the same physiological origin.

Let J_m denote the set of indices assigned to cluster m, and let $Q_m(t)$ represent the primary dipole moment of cluster m, which defines the dominant source of neural activity within that cluster. The contribution of cluster m to the signal recorded at sensor position r can be expressed as

$$\kappa_m(r,t) = -k \sum_{j \in J_m} \frac{r}{|r - r_j'|^3} \cdot ((Q_m(t) + Q_j(t)) \times r_j'), \tag{17.6}$$

$$\kappa_m(r,t) = -kr \cdot \sum_{j \in J_m} (Q_m(t) + Q_j(t)) \times \frac{r_j'}{|r - r_j'|^3}, \tag{17.7}$$

$$\kappa_m(r,t) = -k\left(r, Q_m(t), \sum_{j \in J_m} \frac{r_j'}{|r - r_j'|^3}\right) - kr \cdot \sum_{j \in J_m} Q_j(t) \times \frac{r_j'}{|r - r_j'|^3}, \tag{17.8}$$

$$\kappa_m(r,t) = -k(r, Q_m(t), R_m') - kr \cdot \xi_m(t). \tag{17.9}$$

Here, R_m' may be interpreted as the effective (offset) center of cluster m. Finally, the total measured signal at the scalp is given by the superposition of cluster contributions:

$$\kappa(r,t) = \sum_{m=1}^{M} \kappa_m(r,t) \tag{17.10}$$

$$\kappa(r,t) = -k \sum_{m=1}^{M} (r, Q_m(t), R_m') + \epsilon(t) \tag{17.11}$$

Notably, this expression retains the same functional form as the original forward model (Equation (17.1)), with each cluster represented by an equivalent dipole characterized by its effective center R_m' and primary moment $Q_m(t)$.

Thus, the proposed algorithm can be applied to more sophisticated cases, including configurations with correlated dipole clusters. In such cases, the method localizes the functional center of the cluster's activity. Depending on the spatial distribution of dipoles, this equivalent center may lie outside the geometric extent of the cluster (e. g., in the presence of internal voids). While the approach does not provide a continuous map of distributed activity, it yields point estimates corresponding to cluster centers, which can nevertheless serve as a basis for extensions toward more detailed source reconstructions.

3.2 MEG inverse problem with continuous source of activity

Applying the point-localization algorithm to continuously distributed sources of activity yields approximate centers of independent neural components. These centers are generally easier to estimate using optimization methods, as they are less affected by interference among overlapping signals. The resulting localization provides an effective basis for identifying regions where neural activity is most prominent. This framework can be particularly advantageous for LORETA, as it reduces the dimensionality of the problem and thereby decreases the computational resources required to generate distributed maps of neural activity.

The accuracy of LORETA critically depends on the number of voxels used for the spatial discretization of the solution space. In order to obtain an accurate continuous reconstruction, thousands of voxels are typically required along each dimension. When the number of voxels is substantially reduced—for example, to only a few hundred—the method becomes computationally efficient, but the resulting solutions are strongly affected by artifacts and lose physiological interpretability. To explore this trade-off, we tested such coarse approximations in localized regions of interest, using them as efficient surrogates for full-scale distributed mapping.

Gaussian distributed sources

The first thing we wanted to check was how LORETA copes with a similar single-dipole task with lowered number of voxels, but here, instead of a single dipole, Gaussian densities of equal width in each direction were used, large enough that they could not be modeled by a single dipole. Example 1. A continuous source of activity, defined by a Gaussian distribution, Q, is directed toward the Oy axis (green color in RGB). The coordinates of the center of the Gaussian beam are 0.75 of the head radius on the Oz axis Figure 17.1. The head radius is set to 10 cm.

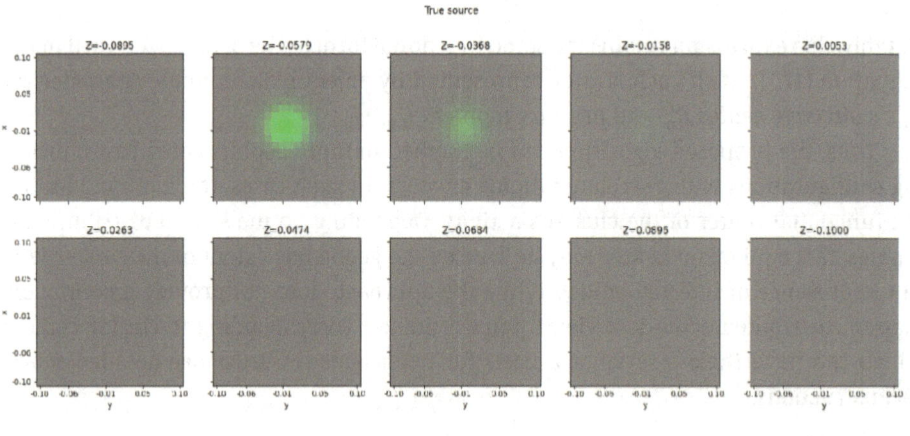

Figure 17.1: Gaussian beam activity source. Sliced along Oz axis, RGB channels show activity along X, Y, and Z directions. Activity directed towards the Oy axis.

In the reconstructed solution, in addition to the components along the Oy axis, activity is observed in all possible directions (which manifests itself in the full spectrum of color shades) Figure 17.2. This is explained by smooth regularization. Since the normal component of magnetic induction is reconstructed in the model, the final signal is a superposition of contributions from different spatial directions. When considering the projection onto the Oy axis Figure 17.3, this effect manifests itself in the smearing of the

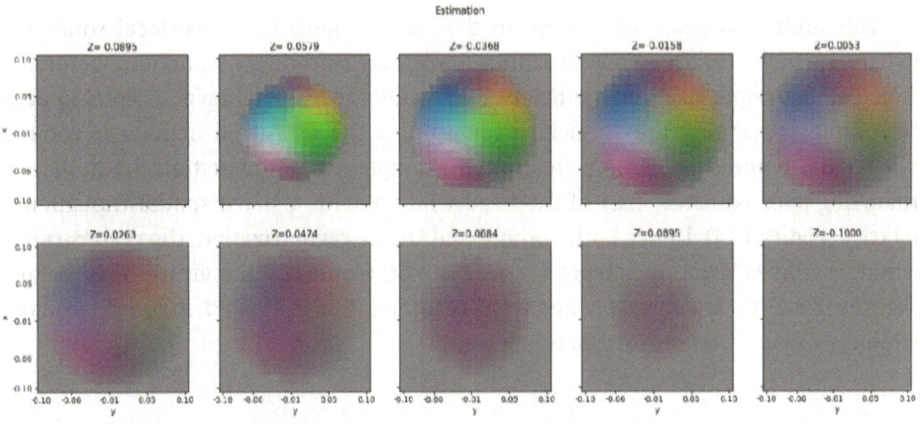

Figure 17.2: LORETA approximation of the Gaussian beam activity source. Regularization with low resolution leads to excessive smooth. Approximated activity is larger than true activity source.

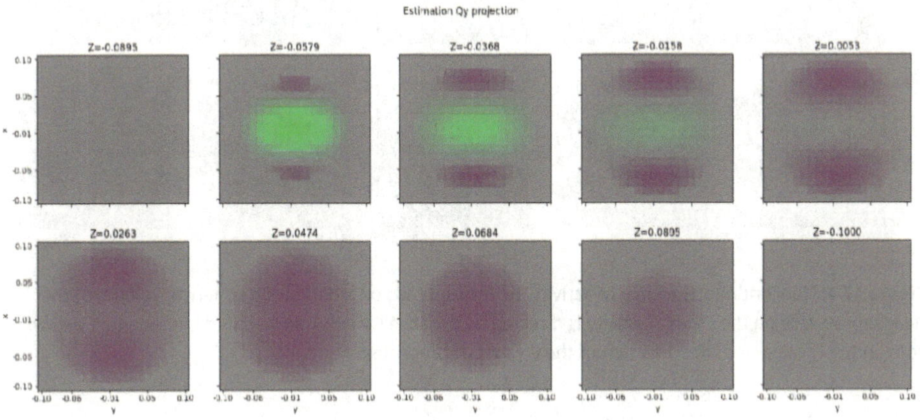

Figure 17.3: Oy projection of the LORETA approximation of the Gaussian beam. Regularization with low resolution smears activity source among the direction of the activity vector. Also, a zone of negative activity appears beyond the true source of activity.

activation zone along this axis and in the appearance of a negative background potential in volume, which is also a consequence of the regularized continuity of the solution.

Noncontinuous sources

In LORETA, the strength of the regularization parameter directly influences the spatial characteristics of the solution. Increasing the regularization coefficient enforces stronger smoothness constraints, yielding more continuous and spatially extended estimates of activity. However, excessive regularization also amplifies the appearance of spatially smeared artifacts and reduces the accuracy of focal source localization.

This limitation becomes evident in the case of noncontinuous, focal sources of activity. In such scenarios, regularization inevitably introduces smoothing errors: no choice of the regularization coefficient can simultaneously achieve acceptable accuracy while preventing the artificial expansion of the estimated activity beyond its true spatial boundaries. As an illustrative example, we simulated electrical activity occupying approximately half of the head volume, with a sharp spatial transition in polarity (Figure 17.4). Under both optimal and strong regularization, the reconstructed activity becomes severely smeared, extending across nearly the entire head volume (Figures 17.5, 17.6). In contrast, when weak regularization is applied, the reconstruction exhibits pronounced noise artifacts.

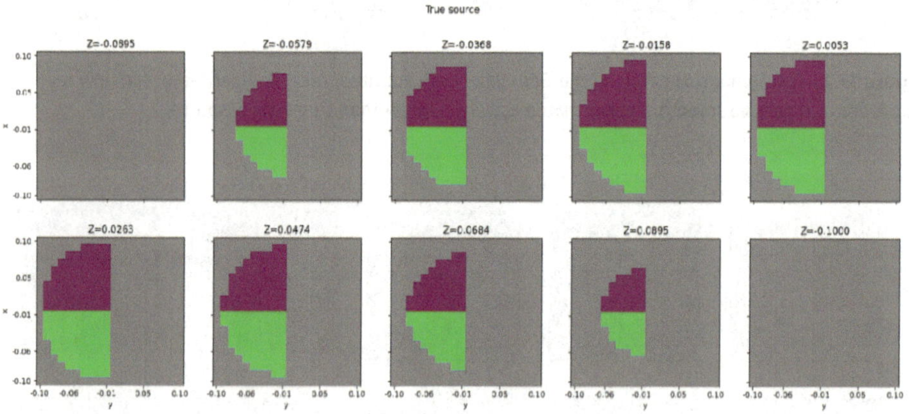

Figure 17.4: Noncontinuous source of activity. Activity spreads on a half of the spherical volume of the head, above zero on the Ox axis activity is directed against the Oy axis (represented in magenta), below zero on the Ox axis — in the direction of the Oy axis (represented in green).

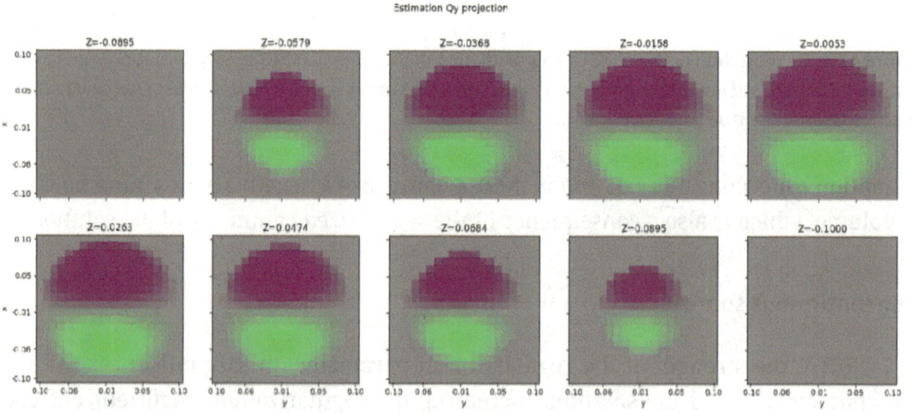

Figure 17.5: Oy projection of the LORETA approximation of the noncontinuous activity source with higher regularization parameter. Directions of activity remain same to original activity source, but they are smoothed on the border. Approximation spreads to the entire volume of the head.

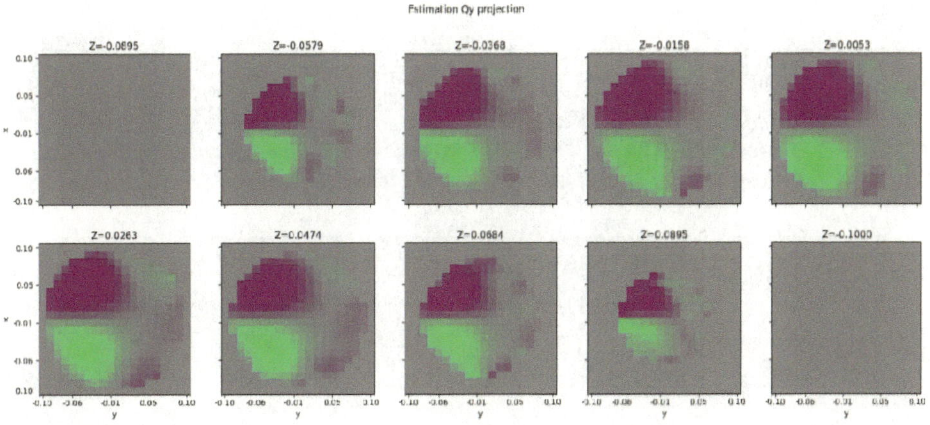

Figure 17.6: Oy projection of the LORETA approximation of the noncontinuous activity source with lower regularization parameter. Reconstruction exhibits pronounced noise artifacts and smoothing, but most of the activity zone remains same to original activity source.

Diffraction pattern sources

As a next test case, the activity was modeled as the diffraction pattern resulting from the superposition of two radially fluctuating signals (Figure 17.7). The method failed to recover the internal sources of activity for any value of the regularization coefficient (Figure 17.8). With insufficient regularization, the solution degenerated into noise, whereas excessive regularization yielded reconstructions restricted to superficial sources only.

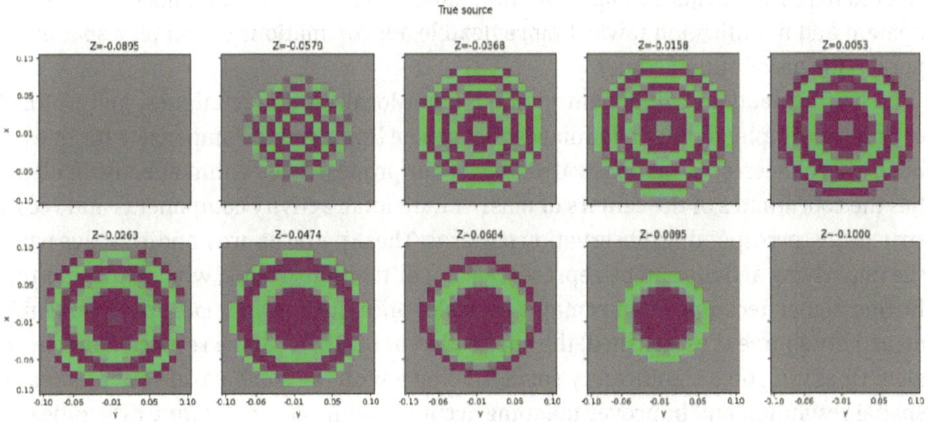

Figure 17.7: Fluctuating activity source. Superposition of two radially fluctuating signals..

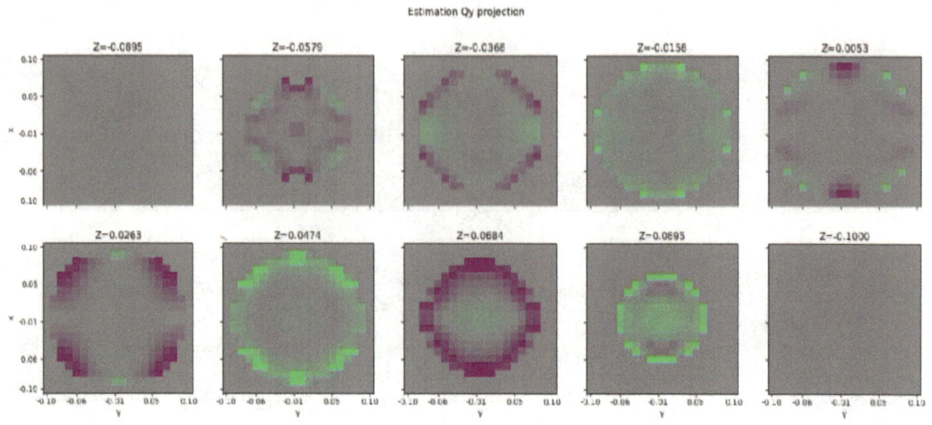

Figure 17.8: Oy projection of the LORETA approximation of the fluctuating activity source. Only superficial activity zone is relevant. Most of the activity degenerates to noise.

Limitations of LORETA regularization

It should be noted that the experiments described above were conducted on a relatively coarse voxel grid. This simplification was introduced deliberately in order to highlight a key limitation of the method. Because the algorithm attempts to localize activity across the entire volume simultaneously, no single universal value of the regularization parameter exists, it must be selected on a case-by-case basis. Additional difficulties arise when reconstructing discontinuous or rapidly fluctuating sources: smooth regularization is inadequate in such scenarios, while omission of the regularizing term renders the problem unstable and without a well-defined solution. Nevertheless, the approach considered here represents an important foundational tool, which warrants further development and modification toward more flexible approximations of complex spatiotemporal patterns of brain activity.

The independent component method, point localization techniques, and LORETA each have complementary advantages that can be leveraged to compensate for one another's weaknesses. As demonstrated above, our proposed algorithm effectively identifies the coordinates of the centers of mass of individual activity components and reconstructs the corresponding observation patterns. These patterns are nondiffractive when the underlying activity can be represented as a diffraction grating, which indicates that further signal decomposition remains possible. Once the approximate spatial extent of an activity source is established, the application of LORETA can be restricted to that region. Operating on a significantly smaller volume with the same voxel count increases spatial resolution and improves mapping accuracy, while also mitigating difficulties associated with cavities or toroidal activity distributions.

3.3 A new algorithm

On the basis of these considerations, we developed a generalized algorithm that combines ICA and LORETA. In its current form, the algorithm applies ICA and point localization to estimate the centers of mass of independent activity sources. Local voxel grids are then constructed around each center, within which the current distribution is reconstructed using LORETA. At present, this approach relies on a priori assumptions about the average spatial extent of activity sources; future work will be directed toward automatic estimation of these parameters.

The algorithm was evaluated on a synthetic dataset containing 50 independent activity sources (Figures 17.9, 17.10). The results demonstrated a marked improvement in reconstruction quality: background smoothing was confined to local grid regions, tran-

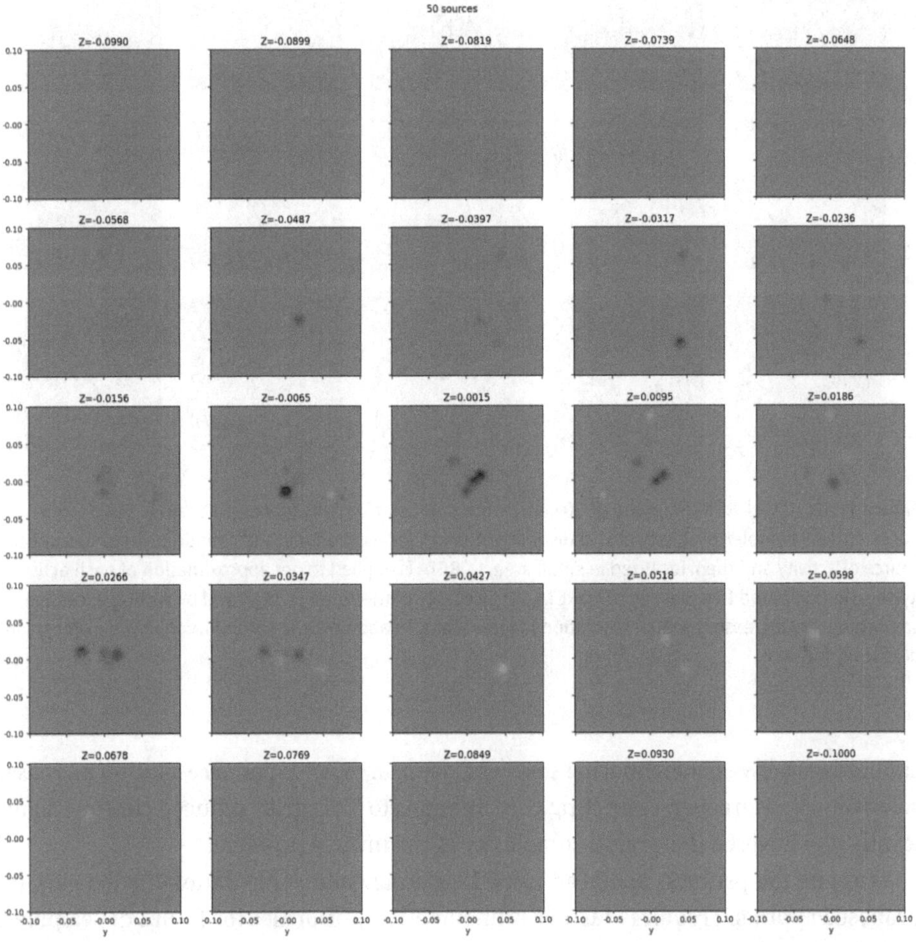

Figure 17.9: 50 independent continuous activity sources (Gaussian beams). Sources have different directions and coordinates.

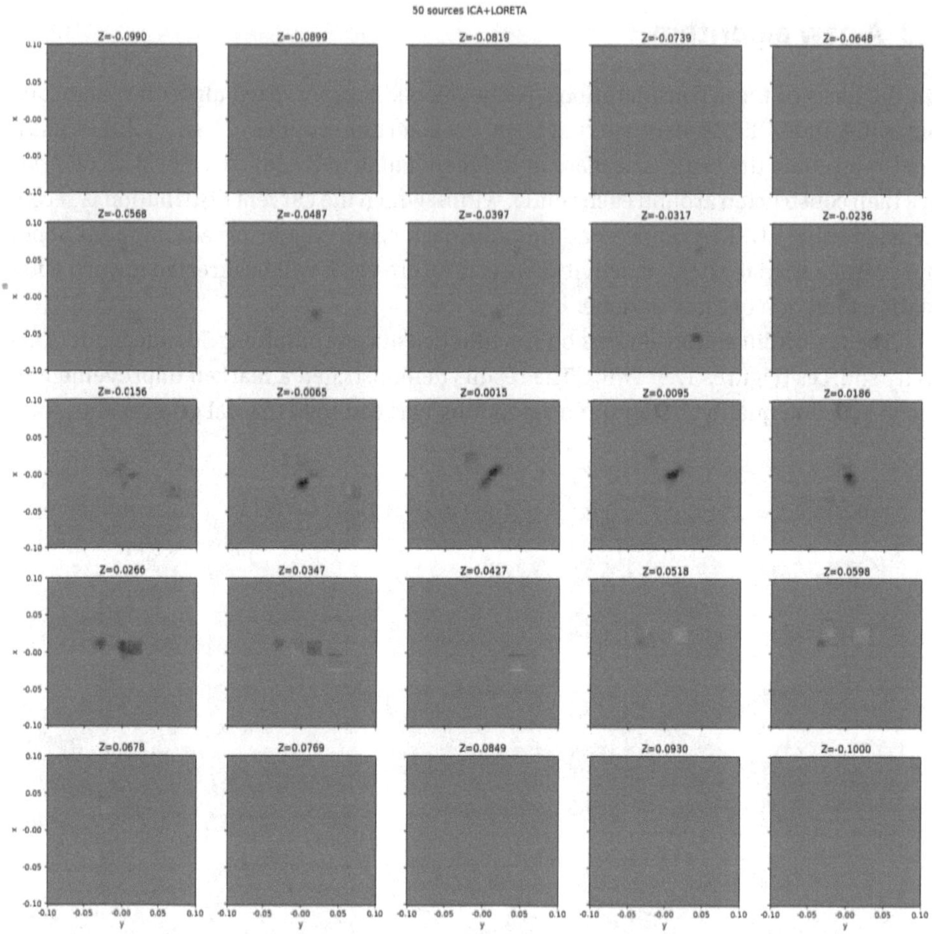

Figure 17.10: ICA–LORETA mapping of 50 continuous sources of activity (gaussian beams). ICA decomposes original complex problem into 50 distinct problems, localization methods obtain coordinates of each source of activity and then, localized in small area, LORETA is applied to get approximation of each activity source. Compared to usual use of LORETA, the accuracy of the result is improved by cutting excessive smoothing and increasing spatial resolution in local areas. In addition, the computational requirements are drastically lowered.

sition zones between neighboring sources disappeared, and spatial resolution increased substantially. Moreover, restricting computations to local grids reduced runtime significantly and lowered the overall complexity of the inverse problem.

Despite the progress achieved in both accuracy and computational efficiency, the proposed approach requires further refinement. Key priorities for future development include automated determination of the spatial scale of activity sources and the exploration of alternative regularization strategies. In particular, the smoothness matrix currently employed leads to excessive propagation of activity along current-flow directions and throughout the modeled volume.

4 Discussion

This study investigated methods for functional mapping of the cerebral cortex within the framework of the MEG/EEG inverse problem with distributed and continuous sources of activity. Point localization algorithms and the independent component method, previously developed for a multidipole model, were evaluated in more realistic settings. The methods demonstrated robustness to distortions and noise, supporting their applicability to problems with continuous source distributions. A critical analysis of the LORETA method on coarse voxel grids revealed characteristic features of distributed solutions shaped by smooth regularization, as well as inherent limitations of this approach. As a compensatory strategy, we proposed a hybrid method in which ICA-based localization constrains LORETA reconstructions to local voxel regions, thereby simplifying mapping and improving approximation accuracy.

In summary, the presented work bridges the gap between discrete and continuous formulations of the MEG/EEG inverse problem by combining the localization precision of ICA with the robustness of distributed regularization methods. The proposed solutions and identified development pathways provide a foundation for subsequent clinical and research applications in neuroscience.

Limitations

The ICA stage decomposes recorded MEG signals into dominant, statistically independent components that can be aligned with neural activity sources of different functional nature. These components can then be localized with dipole-fitting methods, yielding accurate estimates of functional centers even under correlated or distributed conditions. By restricting LORETA to local voxel grids around these estimated centers, we retained spatial detail while reducing the excessive smoothing characteristic of whole-volume reconstructions.

Despite these advantages, several limitations remain. ICA is sensitive to the number of independent components extracted, which may lead to two opposite problems: weak activity sources may be lost or absorbed into noise components due to the limited number of sensors, and a single physiological source may be fragmented into multiple components, requiring careful interpretation to reassemble them. In addition, the LORETA stage still depends on assumptions about the spatial extent of sources. If activity is widespread across the brain volume, the method inherits the same difficulties as standard LORETA and requires a fine voxel grid with substantial computational cost. Even within localized reconstructions, the definition of voxel grid size and location remains a manual step, limiting full automation.

Comparison with other approaches
Compared with classical distributed approaches such as LORETA and sLORETA, the proposed algorithm achieves higher effective resolution by focusing on localized reconstructions. In contrast to beamforming methods, which rely on adaptive spatial filters and are prone to signal cancellation in the presence of correlated sources, the ICA-based stage explicitly separates overlapping contributions. Subspace techniques such as MUSIC provide accurate localization for a small number of focal dipoles but degrade in performance when sources are extended or correlated, whereas the hybrid ICA–LORETA framework extends applicability to such scenarios while maintaining computational efficiency.

Directions for future work
Future priorities include the development of adaptive regularization schemes sensitive to the spatiotemporal structure of neurophysiological activity and the design of reliable algorithms for automatic estimation of source spatial scales. Another important direction is the transition to individualized head models, requiring recalculation of forward-model parameters with anatomical specificity. Finally, we highlight the potential of deep learning approaches to provide flexible approximating architectures capable of integrating these methodological advances into a unified framework for neuroimaging source localization.

5 Conclusion

These results establish the hybrid ICA–LORETA framework as an efficient tool for solving the MEG/EEG inverse problem with distributed sources. The method combines precise localization with scalable distributed mapping, outperforming classical LORETA in both accuracy and computational cost. Despite remaining challenges in developing adaptive schemes for individualized head models and diverse patterns of neurophysiological activity, the approach provides a promising basis for more reliable and practical applications of MEG/EEG in neuroscience.

Bibliography

[1] A. Hyvärinen, J. Karhunen, and E. Oja. *Independent Component Analysis*. Adaptive and Cognitive Dynamic Systems: Signal Processing, Learning, Communications and Control. Wiley, 2001.

[2] P. I. Karpov and T. V. Zakharova. Magnetoencephalography inverse problem in the spheroid geometry. *Journal of Inverse and Ill-Posed Problems*, 27(2):159–169, 2019.

[3] J. Mosher, R. Leahy, and P. Lewis. Eeg and meg: Forward solutions for inverse methods. *IEEE Transactions on Bio-Medical Engineering*, 46:245–259, 04 1999.

[4] I. P. Pakhnenko, T. V. Zakharova, and A. I. Sabirov. The direct and the inverse magnetic encephalography problem. In J. Mau, S. Mukhin, G. Wang and S. Xu, editors, *BIOKYBERNETIKA: Mathematics for Theory and Control in the Human and in Society*, pages 635–651. De Gruyter, Berlin, Boston, 2025.

[5] I. P. Pakhnenko, A. I. Sabirov, and T. V. Zakharova. The direct and the inverse magnetic encephalography problem. In J. Mau, S. Mukhin, G. Wang and S. Xu, editors, *BIOKYBERNETIKA: Mathematics for Theory and Control in the Human and in Society*, volume 9 of De Gruyter Series in Mathematics and Life Sciences, pages 635–650, De Gruyter, Berlin, New York, 2024.

[6] R. D. Pascual-Marqui, C. M. Michel, and D. Lehmann. Low resolution electromagnetic tomography: A new method for localizing electrical activity in the brain. *International Journal of Psychophysiology*, 18(1):49–65, 1994.

[7] T. V. Zakharova, P. I. Karpov, and V. M. Bugaevskii. Localization of the activity source in the inverse problem of magnetoencephalography. *Computational Mathematics and Modeling*, 28(2):148–157, 2017.

[8] T. V. Zakharova and A. I. Sabirov. The solution of the inverse meg problem in a multidipole model. *Bulletin of the Moscow University. Episode 15. Computational Mathematics and Cybernetics*, 15(3):27–32, 2023.

[9] T. V. Zakharova and A. I. Sabirov. Solving the inverse magnetoencephalography problem in the multidipole model. *Moscow University Computational Mathematics and Cybernetics*, 47(3):135–140, 2023.

Alexei Shishkin and Stepan Denisov

Classification of intracytoplasmic sperm injection videos

Abstract: Intracytoplasmic Sperm Injection (ICSI) procedures remain highly operator-dependent, with their success rate significantly influenced by subtle technical factors and lacking robust automated methods for objective assessment. This study explores the application of computer vision and machine learning techniques to classify and predict the outcomes of ICSI procedures using video data. The research addresses challenges such as noise and artifacts in raw ICSI videos by employing semantic segmentation models, specifically U-Net and Feature Pyramid Network (FPN) architectures, to isolate key components: the oocyte, micropipette, microinjection needle, and background. The FPN model achieved the highest segmentation accuracy, with an Intersection over Union (IoU) score of 0.976. To overcome limited data availability, a Denoising Diffusion Probabilistic Model (DDPM) was utilized to generate synthetic ICSI frames, further enhancing segmentation performance (IoU: 0.979). Six critical features—including oocyte deformation, needle insertion speed, and entry angle—were extracted from segmented frames and used to train a gradient boosting classifier for predicting ICSI success. The classifier achieved a classification F1-score of 0.833, with feature importance analysis highlighting oocyte deformation and needle speed as the most influential factors. This work demonstrates the potential of automated video analysis to improve ICSI procedural quality and outcomes, offering a scalable solution for clinical settings.

Keywords: Assisted reproductive technology, semantic segmentation, diffusion models, convolutional neural networks, computer vision

MSC 2020: 68T07, 68T45, 68U10

1 Introduction

Intracytoplasmic sperm injection (ICSI) is a method of assisted reproductive technology (ART) in which a single sperm cell is injected directly into the center of a female oocyte using micromanipulation tools (Figure 18.1). This technique was developed as an alternative to natural fertilization and is employed for couples with severe male factor infertility or those who have not succeeded with previous in vitro fertilization (IVF) attempts [16].

Alexei Shishkin, M.V.Lomonosov Moscow State University, Computational Mathematics & Cybernetics Lenin's Hills, 2nd academic building, 119991, Moscow, Russia, e-mail: shishkin@cs.msu.ru
Stepan Denisov, M.V.Lomonosov Moscow State University, Lenin's Hills, 2nd academic building, 119991, Moscow, Russia, e-mail: svdenrus@mail.ru

https://doi.org/10.1515/9783112216620-026

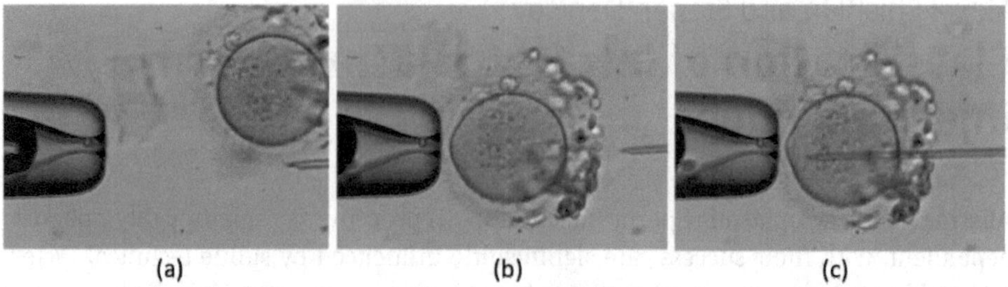

Figure 18.1: Schematic representation of the ICSI procedure stages captured in the analyzed videos. These phases define the key visual elements for semantic segmentation: (a) Initial oocyte capture; (b) fixation of the oocyte by a holding pipette; (c) injection of the sperm into the oocyte by a microinjector.

Despite the widespread use of ICSI in combination with IVF, certain technical aspects of the procedure remain insufficiently studied [16]. Moreover, the efficiency of ICSI may depend on the operator's skill level, and the clinical success rate varies between 50 % and 80 % [20, 3, 18, 10]. A number of studies have shown that the technique used during the ICSI procedure significantly affects its outcomes as well [20, 3, 18, 5]. For example, modifying the sperm injection technique one can obtain higher pregnancy success rates in women with prior unsuccessful ICSI attempts [5]. In [3], which analyzed over 500 video recordings of ICSI procedures, it was revealed that a specific method of inserting the microinjection pipette into the oocyte could substantially decrease the likelihood of successful fertilization.

In the past two decades, several studies [13, 4, 14, 1] have been conducted employing various machine learning algorithms and statistical models to predict the success of pregnancy following ICSI, based on patient data. These datasets usually were processed using attribute selection methods to identify the most relevant features for prediction. Female age was identified as the fundamental predictor across all these studies.

There are only a few works dedicated to ICSI image segmentation [1, 12, 17, 19, 21, 6, 8]. For instance, in [1] one of the objective of the work was aimed to automatically segment the endometrium to calculate its surface, thickness, and endometrial mean grey. Then authors predicted the outcome of ICSI using machine learning techniques with the new value.

In this work, we investigate the use of computer vision methods for predicting the outcome of the ICSI procedure based solely on the operation video, without relying on patient-specific data. The objective is to automate the quality control of ICSI procedures.

Unfortunately, raw video recordings of the ICSI process cannot be directly utilized for further analysis due to the presence of noise, artifacts, and foreign objects such as the culture medium surrounding the oocyte. This necessitates preliminary video preprocessing and noise removal.

One of the methods used to reduce noise and isolate important objects in the image is semantic segmentation. In [17], a gradient-weighted Hough transform was applied to detect oocyte contours, while [6] employed deep neural networks for the semantic segmentation of the micropipette, oocyte, and the microinjection needle tip. However, study [6] reported incomplete segmentation of the microinjection needle, attributed to its orientation and the injection technique, which substantially impacts the success of the procedure.

The goal of the present work is to develop a classification model capable of predicting ICSI outcomes using video data from the intracytoplasmic sperm injection procedure. To achieve this, a sequence of tasks must be performed, beginning with the preprocessing and analysis of the video data. The first stage involves removing the noise and artifacts that may degrade the quality of subsequent analysis from original video sequences.

Next, we develop a semantic segmentation model to extract four primary classes from the video: the oocyte, micropipette, microinjection needle, and background. From the segmented data, key features characterizing the ICSI procedure are extracted. These include the degree of oocyte deformation, the velocity of needle penetration and withdrawal, the entry angle, the number of puncture attempts, and oocyte size.

Given the limited volume of available surgical video data, a generative model will be developed to synthesize realistic representations of ICSI procedures. This will allow for augmentation of the training set for the segmentation model.

The final stage involves constructing a classification model based on ensemble machine learning methods to predict the outcome of the ICSI procedure using the identified features.

2 Materials and methods

2.1 Data

The dataset used in experiments consists of video sequences of ICSI procedures provided by the Reproductive Medicine Center "Liniya Zhizni" ("Line of Life"). The total number of raw videos was 350, each with a high resolution of 1024 × 768 pixels. However, the majority of frames contained artifacts and noise unrelated to the research objective.

For this study, 150 video sequences were randomly selected. From each video, 10–15 frames were extracted, evenly distributed throughout the entire ICSI procedure. Frames that lacked useful information or were of poor quality (e. g., blurred or missing relevant objects) were discarded. The remaining frames were converted from RGB color space to grayscale. As a result, approximately 1600 frames were obtained.

Each frame was manually annotated using an image editing tool. Every pixel was assigned a specific label depending on its class (see Table 18.1 and Figure 18.2).

Table 18.1: Color encoding used for manual semantic annotation of ICSI video frames. Each anatomical or tool-related object (oocyte, micropipette, microinjector, background) is assigned a unique RGB color code for mask generation, enabling supervised training of segmentation models.

Object	Color (R,G,B)
Background	Black (0,0,0)
Oocyte	Red (255,0,0)
Micropipette	Blue (0,255,0)
Microinjector	Green (0,0,255)

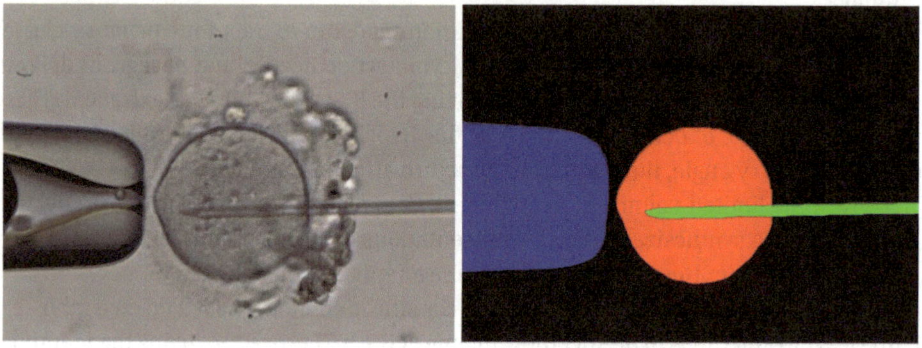

(a) (b)

Figure 18.2: Example of raw and annotated ICSI video frame used for training segmentation models. (a) Original grayscale frame from the ICSI video; (b) corresponding manually labeled mask where each class is assigned a unique RGB color (see Table 18.1).

2.2 Semantic segmentation

Currently, there are numerous convolutional neural network (CNN) architectures available for image segmentation tasks. In this work, we considered two of the most effective architectures for biomedical image segmentation: U-Net and Feature Pyramid Network (FPN) [15, 9].

Segmentation quality was evaluated using the Intersection over Union (IoU) metric, which is one of the most widely used performance measures in image segmentation tasks.

We have leveraged several network architectures for segmenting ICSI images. Table 18.2 presents the results of several segmentation experiments demonstrating the highest performance. Analysis of these results indicates that among the different UNet and FPN architectures, under identical experimental conditions, the FPN network with the encoder `resnext_101_32x8d` yielded the best segmentation quality. This encoder, ResNeXt with a depth of 101 layers and 32 parallel grouped convolutions, transforms the input image into a feature vector of fixed dimensionality.

Table 18.2: Performance comparison of deep learning models for semantic segmentation of ICSI frames. The table reports Intersection over Union (IoU) scores per object class and the overall mean IoU across U-Net and FPN architectures with different encoder backbones. The FPN model with ResNeXt-101 yields the best segmentation accuracy.

Model	Background	Oocyte	Microinjector	Micropipete	Average IoU
UNET resnext101_32x8d	0.990582	0.945986	0.787482	0.947283	0.917833
UNET xception	0.994633	0.965215	0.812562	0.965372	0.934445
Fpn resnext101_32x8d	**0.99843**	**0.99350**	**0.91807**	**0.99663**	**0.976453**
Fpn xception	0.996139	0.982156	0.870264	0.991512	0.960018

From Table 18.2, it can be seen that the classification of the microinjection needle was the least accurate. This is likely due to the fact that, in most images, the needle appears blurred, as the microscope's focus is primarily on the oocyte, resulting in varying degrees of blur for other objects in the image. Additionally, there are frequent cases where the needle moves rapidly, and it makes difficult to detect its contours and to label it accurately. This contributes to poor detection performance by the neural network in such cases.

In contrast, during the core ICSI procedure—namely, the insertion of the sperm into the oocyte—the model successfully predicts object positions and performs segmentation with high accuracy.

2.3 Augmentation model

To improve segmentation quality, we proposed augmenting the training dataset using synthetic image generation techniques. Under conditions of limited data availability, synthetic data can significantly enhance the diversity and representativeness of the training set. Generation is performed according to a distribution approximating the real data, preserving the semantic structure and visual features of the images, thereby improving the generalization ability of the segmentation model.

Among the various classes of generative models, Denoising Diffusion Probabilistic Models (DDPMs) [7] are gaining increasing popularity and outperform traditional approaches such as generative adversarial networks (GAN) and variational autoencoders (VAE) in image synthesis tasks. DDPMs are characterized by stable training, high-fidelity image generation, and the absence of typical artifacts found in alternative models. Due to these advantages, we selected DDPM as the generative module, adapted specifically for the task of synthetic data augmentation to boost segmentation performance.

The core component of DDPM is a convolutional network U-Net, which learns to predict the noise added at a given diffusion time step in a noised image.

The U-Net architecture consists of:

- *Downsampling path (encoder)*: Several blocks that progressively reduce spatial resolution while increasing channel depth. Each block also incorporates the temporal embedding via linear layers.
- *Bottleneck*: A central block that includes a self-attention mechanism.
- *Upsampling path (decoder)*: Symmetrically restores spatial resolution using skip-connections from the encoder.
- *Output layer*: A final convolutional layer that predicts noise with the same dimensions as the input.

Each block can be implemented using residual blocks, and attention mechanisms are employed to enhance global coordination—particularly important for high-resolution image synthesis. The model output is the predicted noise, which is used to reconstruct the previous state of the image.

To quantitatively evaluate the realism and diversity of synthetically generated images, we use the Fréchet Inception Distance (FID), a widely accepted metric for generative model assessment. Realism refers to how closely the generated images resemble real samples, while diversity measures how distinct the generated samples are. FID is known to be robust to noise and specific artifacts present in generated images. Lower FID scores indicate better performance.

2.4 Synthetic data generation

All image generation experiments were performed in a GPU-enabled environment, which enabled accelerated model training. We performed extensive hyperparameter tuning experiments. Parameters including learning rate, batch size, number of diffusion steps, and optimizer settings were systematically varied. Model performance was assessed using both the quantitative FID metric and visual inspection.

Table 18.3 presents the best generation results achieved by the DDPM model using a U-Net backbone, evaluated via the FID metric. It can be seen that among the tested configurations, the `ddpm_1000` architecture achieved the best FID scores.

Table 18.3: Optimization results for diffusion models used to generate synthetic ICSI video frames. Shows the effect of different DDPM configurations (number of diffusion steps, batch size, U-Net dimensions) on the Fréchet Inception Distance (FID), a metric for image quality and diversity. Lower FID indicates better realism.

DDPM	diff_steps	batch_size	Unet-dim	Unet-dim-mults	FID
ddpm_500	500	64	128	(1, 2, 4, 8)	31.4
ddpm_750	750	64	64	(1, 2, 4, 8)	23.5
ddpm_1000	1000	32	32	(1, 2, 4, 4)	**15.3**

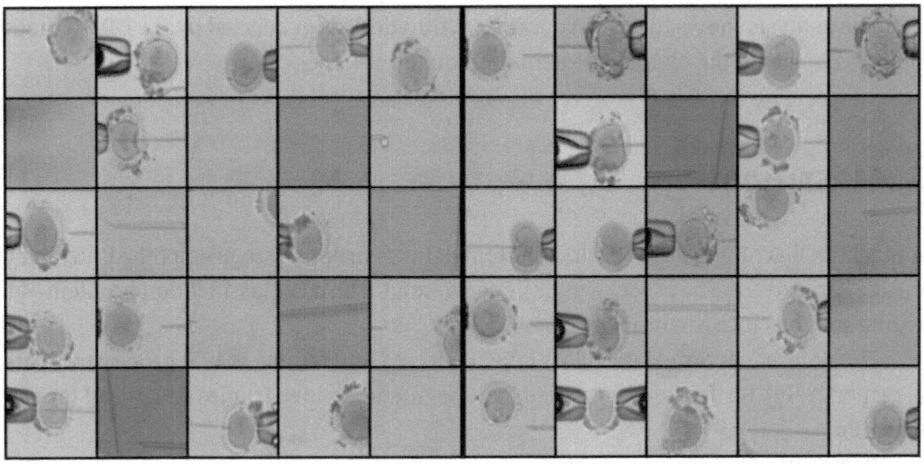

Figure 18.3: Examples of synthetic ICSI frames generated using the trained DDPM model. Images exhibit realistic anatomical and tool structures, closely resembling real surgical video frames. These were used for augmenting segmentation training data.

Examples of ICSI frames generated by the models are shown in Figure 18.3. Many synthetic frames are visually indistinguishable from real ones. However, some erroneous examples were also observed, such as frames containing duplicate objects like two pipettes or two needles. Such images were removed after careful visual inspection.

Using the developed DDPM model, a total of 3,000 new ICSI frames were generated. These frames were annotated using the previously trained semantic segmentation model described in the preceding section. During training a pseudo-label was added for the synthetic data to adjust the loss weighting such that the model relied less on synthetic data.

It is important to note that the test set consisted exclusively of real data. This was intentional, as the aim of testing was to evaluate the model's generalization ability on real (nonsynthetic) samples. Including synthetic samples in the test set could bias results due to the model potentially memorizing synthetic patterns, thus compromising the validity of the generalization assessment. No modifications were made to the existing segmentation model architecture or parameters. New training runs with the augmented datasets were conducted to assess the impact of synthetic data on overall segmentation performance. Training results, including per-class accuracy metrics, are reported in Table 18.4.

Table 18.4: Impact of synthetic data augmentation on segmentation model performance. Comparison of segmentation accuracy (IoU per class and mean IoU) with and without using synthetic ICSI frames generated by the DDPM model. Incorporating synthetic data improved performance across all object categories.

Data	Background	Oocyte	Microinjector	Micropipette	Mean IoU
No DDPM data	0.99843	0.99350	0.91807	0.99663	0.97465
With DDPM data	**0.99984**	**0.99731**	**0.92073**	**0.99721**	**0.97877**

The analysis shows that incorporating synthetic data generated by the DDPM model slightly improved the overall segmentation quality.

3 Classification

A machine learning-based classification model was proposed to predict the outcome of the ICSI procedure. The model was designed to solve a binary classification problem—dividing samples into predefined classes.

The dataset contained two class labels: success and failure of ICSI procedure (1 and 0, respectively). Six features extracted from segmented video frames were used for classification:

- Degree of oocyte deformation,
- Needle insertion speed,
- Needle insertion angle,
- Needle withdrawal speed,
- Number of penetration attempts,
- Oocyte size.

The degree of deformation was calculated as a relative difference between the oocyte area at rest A_{rest} (Figure 18.4(a)) and at maximum deformation A_{deformed} (Figure 18.4(b)):

$$\text{Deformation} = \frac{|A_{\text{deformed}} - A_{\text{rest}}|}{A_{\text{rest}}}$$

To extract insertion/withdrawal speeds and angles from the surgical video, first objects were detected on each video frame using the Haar Cascades algorithm [2]. Then

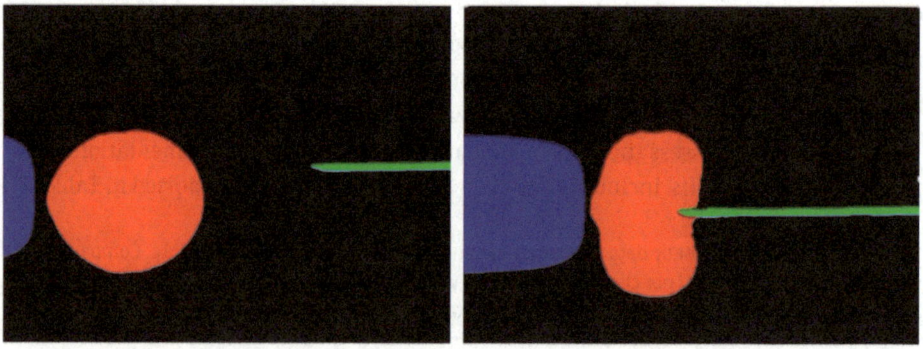

Figure 18.4: Oocyte deformation during the sperm injection stage in ICSI. (a) Oocyte in resting state; (b) maximum deformation at needle penetration. This dynamic feature is quantified and used as a key predictor in classification.

object position and motion were estimated using coordinates from consecutive frames. The Lucas–Kanade optical flow method [11] was used to compute velocities.

To achieve optimal results, a series of experiments was conducted to select the best ensemble method and the most effective combination of hyperparameters. Hyperparameters were selected using Bayesian optimization algorithms, such as the Thompson sampling algorithm, which build a probabilistic model of the hyperparameter space and then select the most promising set of hyperparameters to evaluate. This allows more efficient and faster optimization compared to brute-force grid search.

The best results shown in Table 18.5 were obtained using the gradient boosting method.

Table 18.5: Predictive results of gradient boosting model for ICSI outcome classification. Precision, recall, and F1-scores are reported for both outcome classes (success and failure), with weighted average metrics reflecting overall classifier quality.

Gradient Boosting	Precision	Recall	F1-score
Failure	0.594595	0.666667	0.628571
Success	0.905172	0.875000	0.889831
Weighted Avg.	0.838185	0.830065	0.833481

We estimated each feature's importance, i. e., their contribution to the predictive ability of the model. It can be seen from Figure 18.5 that the oocyte deformation degree and needle entry speed contribute most significantly, while the number of attempts and needle entry angle have lesser impact.

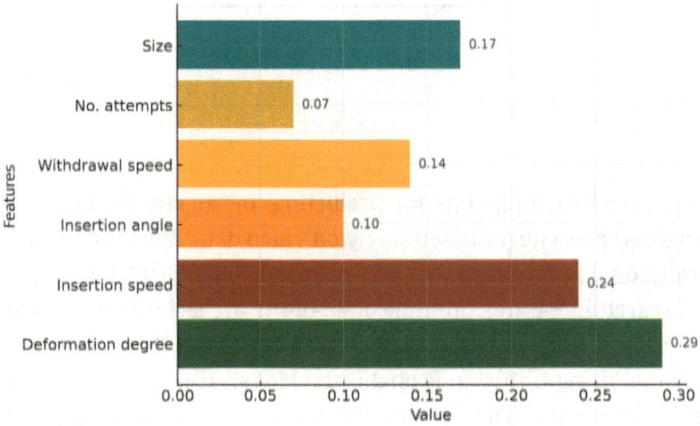

Figure 18.5: Feature importance analysis from gradient boosting classifier. Shows the relative contribution of each extracted video feature (e. g., oocyte deformation, needle speed) to the prediction of ICSI success. More important features significantly impact classification performance.

Based on these values, feature elimination experiments were performed. In each experiment, one or more features were removed from the dataset and model performance was assessed using quality metrics. These experiments help determine the significance of individual features and their impact on model accuracy.

Experimental results (Table 18.6) demonstrate that even features with relatively low importance still have a notable effect on overall model accuracy. For instance, eliminating the feature *number of attempts* reduces gradient boosting model accuracy to 0.808443; removing the *needle entry angle* leads to a more significant drop to 0.719723. Eliminating both reduces accuracy further to 0.709426. Other experiments showed similarly notable performance declines. Thus, each selected feature is important for prediction.

Table 18.6: Ablation study showing the impact of individual features on ICSI outcome classification. on ICSI outcome classification. The table lists classification F1-scores after selectively removing features such as oocyte deformation, insertion angle, or number of attempts. Even less important features contribute noticeably to model accuracy.

Deformation	Insertion Speed	Insertion Angle	Withdrawal Speed	No. Attempts	Oocyte Size	F1-score
+	+	+	+	+	+	0.833481
+	+	+	+	–	+	0.808443
+	+	–	+	+	+	0.719723
+	+	–	+	–	+	0.709426
-	+	+	+	+	+	0.653129
+	–	+	+	+	+	0.684210
-	–	+	+	+	+	0.586363
+	+	+	+	+	–	0.687695
+	+	+	–	+	+	0.717414

4 Discussion

This study presents a comprehensive pipeline for predicting the success of Intracytoplasmic Sperm Injection (ICSI) procedures based solely on video data, marking a significant departure from traditional approaches that rely primarily on clinical and patient metadata [13, 14]. The research integrates multiple state-of-the-art techniques, including semantic segmentation with deep convolutional neural networks (CNNs), synthetic data augmentation using Denoising Diffusion Probabilistic Models (DDPMs), and gradient boosting classifiers for prediction. The novelty lies in the complete automation of quality assessment of ICSI procedures through direct video interpretation.

One of the key strengths of this chapter is the effective use of semantic segmentation to extract high-fidelity, interpretable object masks from surgical video frames. The

Feature Pyramid Network (FPN) model with a ResNeXt-101 encoder achieved a mean Intersection-over-Union (IoU) of 0.976—a performance metric that surpasses previous studies such as [6], where segmentation of the injection needle remained a challenge. The segmentation quality was further improved with the use of a DDPM-based generative model, achieving an IoU of 0.978. This demonstrates not only the feasibility but also the superiority of diffusion-based synthetic augmentation over traditional data augmentation or even GAN-based approaches, which often suffer from mode collapse and unrealistic textures.

Another important contribution is the rigorous feature extraction procedure. Features such as oocyte deformation, needle entry angle, and speed were quantitatively derived from video sequences using optical flow and Haar-based tracking. Unlike prior works that emphasized static features or patient characteristics (e. g., age, hormonal profile) [4, 1], this study captures dynamic aspects of the surgical technique, offering a more useful insight for clinical feedback and training.

Moreover, the use of ensemble-based gradient boosting classifiers resulted in an F1-score of 0.833, which is notably higher than results typically reported in clinical metadata-based models, often in the range of 0.75–0.80. Feature importance analysis also provides a deeper understanding of procedural nuances—most notably, the degree of oocyte deformation and needle speed were shown to be the most predictive.

Despite these strengths, several limitations must be acknowledged. First, although 1600 manually labeled frames represent a solid dataset for segmentation, the classification dataset may still suffer from class imbalance and limited generalization to new clinical settings. While synthetic augmentation improved segmentation, it was not used to enrich the classification dataset, which could have limited classifier robustness.

Second, some of the generated synthetic images exhibited unrealistic structures (e. g., multiple pipettes or needles), requiring manual post-filtering. Although these artifacts were removed, it raises concerns about full automation scalability without human intervention.

Third, the segmentation of the microinjection needle remains the weakest among all object classes, with performance still affected by blur and motion artifacts. In real-time applications, this could impact the reliability of downstream feature extraction and classification. Although we intentionally excluded synthetic data from test sets to ensure fairness, further validation on external datasets from different clinics is essential for assessing model generalizability.

From a technical standpoint, this work aligns with the trend of moving away from supervised-only models toward hybrid generative-discriminative frameworks. The successful application of DDPMs here offers a blueprint for other domains with limited annotated data and complex image characteristics.

To address current limitations, future work should focus on two directions: (1) augmenting the classification dataset with synthetic sequences using temporal diffusion models, and (2) integrating domain adaptation techniques to validate cross-clinic generalizability.

5 Conclusion

This chapter provides a novel, end-to-end system for assessing procedural quality in ICSI, bridging computer vision and reproductive medicine.

It is demonstrated that fully automated classification of intracytoplasmic sperm injection outcomes from surgical video is not only feasible but highly accurate.

The work highlights the critical role of motion-based features, such as oocyte deformation and needle dynamics, in determining ICSI success, underscoring procedural skill over patient-based predictors.

The approach sets a new benchmark for video-based artificial reproductive technology (ART) outcome prediction.

Bibliography

[1] Z. Abbas. *Predicting the outcome of intracytoplasmic sperm injection in women using machine learning combining Doppler and clinical data: SPIRL*. Theses, Le Mans Université, December 2023. https://theses.hal.science/tel-04412683.

[2] S. S. Ali, J. H. A. Ameri, and T. Abbas. Face detection using Haar cascade algorithm. In *2022 Fifth College of Science International Conference of Recent Trends in Information Technology (CSCTIT)*, pages 198–201, AL-Mustansiriyah University, Baghdad-Iraq, August 2022.

[3] C. E. Daniel, C. Hickman, T. Wilkinson, O. Oliana, D. Gwinnett, G. Trew, and S. Lavery. Maximising success rates by improving ICSI technique: Which factors affect outcome? *Fertility and Sterility*, 104(3), 2015.

[4] S. Dehghan et al. Machine learning in predicting infertility treatment success: A systematic literature review of techniques. *Journal of Education and Health Promotion*, 14:103, 2025.

[5] T. Ebner, M. Moser, M. Sommergruber, K. Jesacher, and G. Tews. Complete oocyte activation failure after ICSI can be overcome by a modified injection technique. *Human Reproduction*, 19(8):1837–1841, 2004. 08.

[6] P. He et al. Semantic video segmentation for intracytoplasmic sperm injection procedures, 2021. https://arxiv.org/abs/2101.01207.

[7] J. Ho, A. Jain, and P. Abbeel. Denoising diffusion probabilistic models, 2020. https://arxiv.org/abs/2006.11239.

[8] Y. Kovalev and A. Shishkin. Automatic segmentation of intracytoplasmic sperm injection images. *Computer Optics*, 46:628–633, 2022.

[9] T. -Y. Lin, P. Dollár, R. Girshick, K. He, B. Hariharan, and S. Belongie. Feature pyramid networks for object detection. In *2017 IEEE Conference on Computer Vision and Pattern Recognition (CVPR)*, pages 936–944, Honolulu, HI, USA, July 2017. IEEE Xplore.

[10] Z. Lu, X. Zhang, C. Leung, N. Esfandiari, R. F. Casper, and Y. Sun. Robotic ICSI (Intracytoplasmic Sperm Injection). *IEEE Transactions on Biomedical Engineering*, 58(7):2102–2108, 2011.

[11] B. D. Lucas and T. Kanade. An iterative image registration technique with an application to stereo vision. In *Proceedings of Imaging Understanding Workshop*, pages 121–130, Washington, D.C., USA, April 1981.

[12] S. A. Mirroshandel and F. Ghasemian. Automated morphology detection from human sperm images. In G. D. Palermo and E. S. Sills, editors, *Intracytoplasmic Sperm Injection – Indications, Techniques and Applications*. Springer International Publishing AG, 2018.

[13] A. Mohammad, H. Awad, and H. I. Ashqar. Comparing machine learning approaches for predicting the success of ICSI treatment: A study on clinical applications. *Intelligence-Based Medicine*, 11:100204, 2025.

[14] J. Peng, X. Geng, Y. Zhao, et al. Machine learning algorithms in constructing prediction models for assisted reproductive technology (ART) related live birth outcomes. *Scientific Reports*, 14:32083, 2024.

[15] O. Ronneberger, P. Fischer, and T. Brox. U-Net: Convolutional networks for biomedical image segmentation. In *Medical Image Computing and Computer-Assisted Intervention–MICCAI 2015*, pages 234–241. Munich, Germany, October 2015. Springer.

[16] P. Rubino, P. Viganò, A. Luddi, and P. Piomboni. The ICSI procedure from past to future: A systematic review of the more controversial aspects. *Human Reproduction Update*, 22(2):194–227, 11 2015.

[17] M. Saadat, A. M. Hajiyavand, and A. P. Singh Bedi. Oocyte positional recognition for automatic manipulation in ICSI. *Micromachines*, 9(9):429, 2018.

[18] S. Shen, A. Khabani, N. Klein, and D. Battaglia. Statistical analysis of factors affecting fertilization rates and clinical outcome associated with intracytoplasmic sperm injection. *Fertility and Sterility*, 79(2):355–360, 2003.

[19] A. Targosz, P. Przystałka, R. Wiaderkiewicz, et al. Semantic segmentation of human oocyte images using deep neural networks. *BioMedical Engineering Online*, 20:40, 2021.

[20] A. W. Tiegs and R. T. Scott. Evaluation of fertilization, usable blastocyst development and sustained implantation rates according to intracytoplasmic sperm injection operator experience. *Reproductive Biomedicine Online*, 41(1):19–27, 2020.

[21] M. Zhao, H. Li, R. Li, Y. Li, X. Luo, T. C. Li, T. L. Lee, W. J. Wang, and D. Y. L. Chan. Automated and precise recognition of human zygote cytoplasm: A robust image-segmentation system based on a convolutional neural network. *Biomedical Signal Processing and Control*, 67:102551, 2021.

Part III-C: Data-analytic sciences

Rationale: 9

Part III-C: **Data-analytic sciences**

Conceptually, design and analysis of empirical investigations.

Specifically, this category includes:
- identification and estimation of systems parameters
- design of empirical investigations of systems parameters
- statistical estimation, hypothesis testing, and regression analyses, in general

In this volume:
- Bioequivalence testing based on the Kullback–Leibler divergence

In former volume BIOKYBERNETIKA:
- A new test to assess bioequivalence of two different drugs in the presence of missing values
- Estimation of adjusted relative risks in log-binomial regression using the Bekhit–Schöpe–Wagenpfeil algorithm.

https://doi.org/10.1515/9783112216620-027

Pavel V. Panov and Tatyana V. Zakharova

Bioequivalence testing based on the Kullback–Leibler divergence

Abstract: The paper proposes a reliable nonparametric procedure for testing pharmacokinetic bioequivalence (BE) based on the difference between the distributions of key characteristics for tested and reference drugs according to the Kullback–Leibler (KL) criterion. The KL method is based on a point estimate. We use a weighted kernel density estimation (KDE) to estimate the entire distribution, thereby taking into account even minor deviations from the norm, as well as missing values and random anomalies. In simulation studies simulating standard intersection patterns, the proposed weighted KDE approach reduced the number of false BE reports and increased the accuracy of decision making compared to classical parametric point estimates.

Keywords: Bioequivalence testing, pharmacokinetics, Kullback–Leibler divergence, kernel density estimation, robustness, crossover designs

MSC 2020: 62P10, 62G07

1 Introduction

Bioequivalence (BE) evaluation is a critical component in the development and approval of generic pharmaceutical formulations. It assesses whether a test formulation (T), such as a generic drug, produces pharmacokinetic (PK) profiles that are therapeutically equivalent to those of a reference formulation (R), typically the original branded product. This comparison ensures that patients receive the same therapeutic effect when switching between formulations.

Traditionally, BE has been evaluated using statistical criteria based on key PK metrics, such as the area under the concentration-time curve (AUC), which reflects total drug exposure, and the maximum observed concentration ($C_{\max}$), which indicates peak drug levels. These metrics are often log-transformed due to their skewed distribution in real-world data, and the resulting log AUC and log $C_{\max}$ are assumed to follow approximately normal distributionsThe current regulatory standard for average bioequivalence (ABE). The current regulatory standard for average bioequivalence (ABE) requires that confidence interval (CI) for the geometric mean ratio of T to R lies within the equivalence range $[0.80, 1.25]$ for both AUC and $C_{\max}$ [5]. This criterion is implemented via the two one-sided tests (TOST) procedure on the log scale.

Pavel V. Panov, Tatyana V. Zakharova, Faculty of Computational Mathematics and Cybernetics, Lomonosov Moscow State University, 119991, Leninskie Gory 1-52, MSU, 119991, Moscow, Russia, e-mails: mazeinmind@yandex.ru, tvzaharova@mail.ru

https://doi.org/10.1515/9783112216620-028

While this approach is widely accepted, it focuses primarily on point estimates of location (mean) and assumes known or estimated variance components, potentially overlooking broader distributional differences between the test and reference products. To address this limitation, Dragalin and colleagues [3] proposed using the Kullback–Leibler (KL) divergence as a unified measure of similarity between the full probability distributions of PK characteristics under T and R. However, their method relies on closed-form expressions that depend on point estimates of means and variances, inheriting sensitivity to outliers. As demonstrated in [7] for the TOST procedure, such reliance on point estimates can lead to inflated type I error when data are incomplete — a concern that extends to KL-based methods as well.

In this chapter, we present a robust framework for bioequivalence assessment based on nonparametric estimation of the underlying distributions using kernel density estimation (KDE). Instead of relying on parametric assumptions, we estimate the densities of log AUC for both formulations using weighted KDE, incorporating robustness against anomalies and handling missing data through inverse probability weighting. Our approach provides a more comprehensive evaluation of bioequivalence, also offering improved resilience to data irregularities.

The motivation of this work is to move beyond traditional point-based comparisons and toward a distributional perspective that better reflects the complexity of pharmacokinetic data. We demonstrate that this KDE-based KL divergence method improves decision accuracy, particularly in small samples with missing values or outliers, compared to classical parametric methods. This chapter shows how a unified, flexible, and robust statistical framework can enhance the reliability and interpretability of bioequivalence conclusions.

2 Methodology

2.1 Classical bioequivalence testing

Let R denote the reference product and T denote the test product.

Let X_T and X_R denote pharmacokinetic characteristics (e. g., log AUC or $\log C_{\max}$) for T and R. A common assumption is

$$X_T \sim \mathcal{N}(\mu_T, \sigma_T^2), \quad X_R \sim \mathcal{N}(\mu_R, \sigma_R^2),$$

arising from log-normality of the original-scale measures. In replicate crossover designs, mixed-effects models further decompose variability into within-subject, between-subject, and (for IBE) subject-by-formulation interaction components [5].

2.1.1 Two One-Sided Tests (TOST) procedure

In current practice (the 80/125 % rule), bioequivalence is assessed for key pharmacokinetic characteristics using two one-sided tests [6] on the log scale.

Statistical setting

Let μ_T and μ_R be the means (on the log scale) of the pharmacokinetic characteristic under T and R, respectively. Typical characteristics are log AUC and log $C_{\max}$. It is assumed that, on the log scale, the measurements are approximately normal; consequently, the original-scale values are log-normal.

As an illustration (for $C_{\max}$), assume

$$\ln(C_{\max,T}) \sim \mathcal{N}(a_1, \sigma_1^2), \quad \ln(C_{\max,R}) \sim \mathcal{N}(a_2, \sigma_2^2).$$

Then on the original scale

$$C_{\max,T} \sim \log\mathcal{N}(e^{a_1+\frac{1}{2}\sigma_1^2},\ (e^{\sigma_1^2}-1)\,e^{2a_1+\sigma_1^2}),$$
$$C_{\max,R} \sim \log\mathcal{N}(e^{a_2+\frac{1}{2}\sigma_2^2},\ (e^{\sigma_2^2}-1)\,e^{2a_2+\sigma_2^2}).$$

Let θ denote the ratio of expected values on the original scale,

$$\theta = \frac{\mathbb{E}[C_{\max,T}]}{\mathbb{E}[C_{\max,R}]} = \exp\Big((a_1-a_2)+\frac{1}{2}(\sigma_1^2-\sigma_2^2)\Big),$$

and let $\Delta = a_1 - a_2$ be the difference of the log means. Introduce logarithmic margins $\delta_1 = \ln\theta_1$ and $\delta_2 = \ln\theta_2$. Under the 80/125 % standard,

$$\delta_1 = \ln(0.80) \approx -0.223, \quad \delta_2 = \ln(1.25) \approx +0.223.$$

Hypotheses

The composite null hypothesis of nonequivalence is split into two one-sided null *hypotheses*:

$$H_{01}: \theta \le \theta_1 \iff H_{01}: \Delta \le \delta_1,$$
$$H_{02}: \theta \ge \theta_2 \iff H_{02}: \Delta \ge \delta_2,$$

with the alternative in each case asserting that Δ lies inside the equivalence region. Equivalently, the joint alternative is $|\Delta| < \delta_0$ with $\delta_0 = \ln(1.25)$.

Test statistics

Let $\bar{Y}_{\mathrm{T}}$ and $\bar{Y}_{\mathrm{R}}$ be the sample means of the log-transformed characteristic under T and R, and let n_1 and n_2 denote the numbers of subjects in the two sequences of a two-period crossover. Under standard assumptions (equal within-subject variances across formulations), the variance of the within-subject difference is $\sigma_d^2 = \sigma_w^2/2$, and its unbiased estimator is $\widehat{\sigma}_d^2$ (pooled over both sequences). The TOST statistics are

$$T_1 = \frac{(\bar{Y}_T - \bar{Y}_R) - \delta_1}{\widehat{\sigma}_d \sqrt{\frac{1}{n_1} + \frac{1}{n_2}}} \quad \text{and} \quad T_2 = \frac{(\bar{Y}_T - \bar{Y}_R) - \delta_2}{\widehat{\sigma}_d \sqrt{\frac{1}{n_1} + \frac{1}{n_2}}}.$$

Bioequivalence at level α (typically $\alpha = 0.05$) is concluded if

$$T_1 > t_{1-\alpha,\,\nu} \quad \text{and} \quad T_2 < -t_{1-\alpha,\,\nu},$$

where $t_{1-\alpha,\,\nu}$ is the $(1-\alpha)$ quantile of the t-distribution with ν degrees of freedom (e. g., $\nu = n_1+n_2-2$ in the simple pooled-t setting, or model-based ν in a mixed-effects analysis).

Confidence-interval equivalence

The TOST decision is equivalent to requiring that the $(1-2\alpha)100\,\%$ confidence interval for Δ lies entirely within (δ_1, δ_2):

$$\left((\bar{Y}_T - \bar{Y}_R) - t_{1-\alpha,\,\nu}\,\widehat{\sigma}_d \sqrt{\frac{1}{n_1} + \frac{1}{n_2}},\ (\bar{Y}_T - \bar{Y}_R) + t_{1-\alpha,\,\nu}\,\widehat{\sigma}_d \sqrt{\frac{1}{n_1} + \frac{1}{n_2}} \right) \subset (\delta_1, \delta_2).$$

In practice, the procedure is applied separately to log AUC and log $C_{\max}$; in replicate designs, the estimates and $\widehat{\mathrm{SE}}$ are obtained from an appropriate linear mixed-effects model, with $\widehat{\sigma}_d$ reflecting the relevant within-subject variance component.

2.2 Kullback–Leibler divergence

We will use Kullback–Leibler divergence for bioequivalence testing.

2.3 Definition and symmetrization

Let P and Q be probability distributions on $(\Omega, \mathcal{F})$ with densities p and q. The KL divergence of P from Q is

$$D_{\mathrm{KL}}(P\|Q) = \int p(x)\,\log\frac{p(x)}{q(x)}\,dx,$$

assuming absolute continuity and finiteness. Although not a metric (it is asymmetric and does not satisfy the triangle inequality), D_{KL} is nonnegative and equals zero if and only if $P = Q$ almost surely.

A commonly used symmetrization is the *Jeffreys divergence*

$$\mathrm{KLD}_\Sigma(P,Q) \;=\; D_{\mathrm{KL}}(P\|Q) + D_{\mathrm{KL}}(Q\|P).$$

This quantity is always nonnegative and equals zero only when $P = Q$.

2.4 Normal model: Closed-form expression

For univariate normals, the KL divergence admits a closed form. If $P = \mathcal{N}(\mu_T, \sigma_T^2)$ and $Q = \mathcal{N}(\mu_R, \sigma_R^2)$, then

$$D_{\mathrm{KL}}(P\|Q) = \log\frac{\sigma_R}{\sigma_T} + \frac{\sigma_T^2 + (\mu_T - \mu_R)^2}{2\sigma_R^2} - \frac{1}{2}.$$

Summing forward and reverse terms yields the symmetric divergence

$$\mathrm{KLD}_\Sigma(P,Q) = \frac{1}{2}\left(\frac{\sigma_T^2}{\sigma_R^2} + \frac{\sigma_R^2}{\sigma_T^2} - 2\right) + \frac{(\mu_T - \mu_R)^2}{2}\left(\frac{1}{\sigma_T^2} + \frac{1}{\sigma_R^2}\right).$$

A particularly transparent special case is homoscedasticity, $\sigma_T^2 = \sigma_R^2 = \sigma^2$, where

$$\mathrm{KLD}_\Sigma(P,Q) = \frac{(\mu_T - \mu_R)^2}{\sigma^2}.$$

Since PK characteristics are analyzed on the log scale, $\mu_T - \mu_R = \Delta$ and σ^2 can be interpreted (in simple models) as a relevant variance component (e. g., within-subject variance for ABE in a standard 2×2 design).

2.5 Important Kullback–Leibler divergence properties

The KL framework quantifies the information lost when Q is used to approximate P. In BE, P and Q represent the full distributions of PK characteristics for T and R (or predictive distributions at the subject level for IBE). The divergence therefore reflects discrepancies in both location and dispersion (and, when modeled hierarchically, variability components relevant to PBE and IBE). Key features for BE include:

1. *Nonnegativity and identifiability:* $\mathrm{KLD}_\Sigma \geq 0$ with equality if the distributions coincide.
2. *Additivity over independent observations:* for i. i. d. data, $D_{\mathrm{KL}}(P^{\otimes n}\|Q^{\otimes n}) = n\,D_{\mathrm{KL}}(P\|Q)$, aligning with information accumulation as sample size grows.
3. *Unified treatment of ABE/PBE/IBE:* under normal modeling, KLD_Σ decomposes naturally into terms involving means and variances (and can embed subject-level random effects), placing ABE, PBE, and IBE on a single quantitative scale.

4. *Invariance to monotone reparameterizations:* working on log-transformed characteristics retains interpretability and does not require retuning when switching between log AUC and log $C_{\max}$.

3 Results

We use the KL divergence for bioequivalence testing.

3.1 Hypothesis and decision threshold

The unified BE hypothesis is formulated as

$$H_0 : \mathrm{KLD}_\Sigma \geq c \quad \text{vs.} \quad H_1 : \mathrm{KLD}_\Sigma < c,$$

where $c > 0$ is a prespecified tolerance for distributional discrepancy. A principled way to calibrate c is to match the classical ABE boundary. Under homoscedasticity with unit-variance scaling (or, more generally, after standardizing by a relevant variance component), the 80/125 % margin $\Delta = \pm\delta_0$ with $\delta_0 = \log(1.25) \approx 0.223$ implies

$$\mathrm{KLD}_\Sigma = \frac{\Delta^2}{\sigma^2}.$$

3.2 Drawbacks of using the closed-form expression for KLD

3.2.1 Testing scenario used in Dragalin et al. (2003)

In their paper on Kullback–Leibler divergence (KLD) for bioequivalence, Dragalin, Fedorov, Patterson, and Jones (2003) introduce a unified criterion and evaluate its operating characteristics via simulation and retrospective analyses. The simulation scenarios are the standard ones for bioequivalence: balanced crossover designs, normality on the log scale, and complete data (i. e., no explicit mechanism for generating missing values is implemented). The retrospective comparisons likewise use complete-case data sets as published at the time. Consequently, missing samples or irregular profiles are not part of that study design and can be explored as an extension in our work.

3.2.2 Sensitivity of point estimators and distributional dependence

The KLD-based decision in practice relies on point estimators of means and variance components for the pharmacokinetic characteristics, and then on a confidence bound for KLD. Point estimators are inherently sensitive to anomalies in the data. For skewed, heavy-tailed, or mixture-like distributions, the sample mean and variance may be influenced even by a small number of anomalies. These considerations motivate exploring other ways to model the distribution of pharmacokinetic characteristics and then mapping them into KLD.

3.2.3 Nonparametric KLD via weighted KDE with missing data and anomalies

We estimate the distributions of the pharmacokinetic characteristics on the log scale for the test and reference products using a weighted Gaussian kernel density estimator (KDE) that explicitly attenuates the impact of missing characteristics and occasional anomalous observations.

Let $\{X_i^{(T)}\}_{i=1}^{n_T}$ and $\{X_j^{(R)}\}_{j=1}^{n_R}$ be the observed subject-level characteristics on the log scale. Each observation is assigned a nonnegative weight

$$w_i = a_i \times \frac{1}{\widehat{\pi}_{\text{a}}}, \quad a_i = \text{Huber}\left(\frac{X_i - \text{median}(X)}{1.4826\,\text{MAD}(X)};\, c_H\right),$$

where $a_i \in (0, 1]$ are Huber robustness weights (we use $c_H = 1.5$) and $\widehat{\pi}_{\text{a}}$ is a stabilized inverse-probability factor equal to the observed fraction within the treatment formulation (T or R).

The weighted KDE for T is

$$\widehat{f}_T(x; h) = \frac{\sum_{i=1}^{n_T} w_i^{(T)} K(\frac{x - X_i^{(T)}}{h})}{h \sum_{i=1}^{n_T} w_i^{(T)}}, \quad K(u) = \frac{1}{\sqrt{2\pi}} e^{-u^2/2},$$

and analogously $\widehat{f}_R(x; h)$ for R. To stabilize the ratio $\widehat{f}_T / \widehat{f}_R$, we use a single common bandwidth h for both formulations, selected by log-likelihood cross-validation on the pooled sample $\{X^{(T)}\} \cup \{X^{(R)}\}$:

$$\widehat{h} \in \arg\min_{h \in \mathcal{H}} \left\{ -\sum_{k=1}^{n_T + n_R} \log \widehat{f}_{-k}(X_k; h) \right\}$$

$$\widehat{f}_{-k}(X_k; h) = \frac{\sum_{\ell \neq k} K((X_k - X_\ell)/h)}{h(n_T + n_R - 1)}.$$

Given $\widehat{f}_T$ and $\widehat{f}_R$ evaluated on a common numerical grid, we compute the symmetric Kullback–Leibler divergence

$$\widehat{\text{KLD}}_\Sigma = \int \widehat{f}_T(x) \log \frac{\widehat{f}_T(x)}{\widehat{f}_R(x)}\, dx + \int \widehat{f}_R(x) \log \frac{\widehat{f}_R(x)}{\widehat{f}_T(x)}\, dx,$$

by trapezoidal quadrature. For decision making, we compare $\widehat{\mathrm{KLD}}_\Sigma$ with a prespecified tolerance $c = \log(1.25)^2$ that aligns with the 80/125 % margin under the usual ABE mapping; we declare bioequivalence if $\widehat{\mathrm{KLD}}_\Sigma < c$.

3.3 Model study results

3.3.1 Design

We simulated balanced samples with $n \in \{15, 20, 25, 30\}$ subjects per formulation on the log scale. Two ground-truth scenarios were considered: *True BE* (KLD slightly below c) and *True non-BE* (KLD above c). In each replication, we injected 1–2 missing values across $T+R$ and at most one outlier (probability 0.5). We compared (i) a parametric point estimation of KLD and (ii) the proposed weighted KDE. Performance metrics were the mean absolute error of $\widehat{\mathrm{KLD}}_\Sigma$ and decision accuracy against the threshold c.

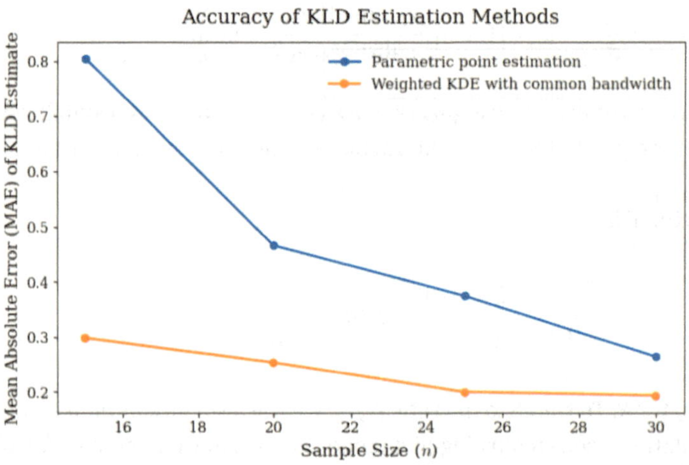

Figure 19.1: Mean absolute error (MAE) of symmetric KLD estimation by sample size (n). Two methods compared: parametric point estimation and weighted KDE with Huber×IPW and common bandwidth. Results averaged across two scenarios: true bioequivalence (BE) and true non-BE. Lower MAE indicates better accuracy.

3.3.2 Classification under true non-BE

Table 19.1 summarizes decision accuracy and the number of false BE declarations (i. e., misclassifying a truly non-BE case as BE). The proposed weighted KDE consistently reduced false BE counts and increased accuracy relative to the parametric point estimation across all sample sizes.

Table 19.1: Decision accuracy under the *true non-BE* scenario for 2 methods of calculating bioequivalence statistics $n_{\text{repetitions}} = 120$.

Sample size	Estimation method	Accuracy	False BE decisions	Repetitions
15	Parametric point estimation	0.958	5	120
15	Weighted KDE	0.983	2	120
20	Parametric point estimation	0.900	12	120
20	Weighted KDE	0.992	1	120
25	Parametric point estimation	0.942	7	120
25	Weighted KDE	0.992	1	120
30	Parametric point estimation	0.908	11	120
30	Weighted KDE	1.000	0	120

Interpretation

Across the simulated settings with small samples ($n \in \{15, 20, 25, 30\}$), occasional missing values and rare anomalies, the proposed weighted KDE procedure showed clear decision advantages under the *true non-BE* scenario. In particular, weighted KDE achieved higher decision accuracy and substantially fewer false BE outcomes than the parametric point estimator. This indicates that mild robustness combined with a common smoothing parameter stabilizes the density ratio $\widehat{f}_T/\widehat{f}_R$ and discourages spurious BE calls when the ground truth is non-BE.

4 Discussion

The results of this study provide new insights into the statistical properties of equivalence testing under log-normal models with missing data. We have proposed a kernel density estimation (KDE) based approach to evaluate the Kullback–Leibler divergence (KLD), which allows for a more robust assessment of bioequivalence (BE) compared to traditional procedures.

One of the main advantages of the proposed method is its ability to account for missing observations without introducing significant bias. Classical procedures, such as the two one-sided tests (TOST), assume complete data and, therefore, tend to inflate the type I error rate when data are missing. Similar concerns were later discussed by Donner et al. [1], who showed that missing data can have a significant impact on bioequivalence conclusions. In contrast, our approach preserves the nominal significance level more accurately by relying on nonparametric density estimates that incorporate all available information.

Another important feature is the robustness of the KDE-based estimator to heavy tails and outliers. Standard parametric models, including the log-normal and gamma families [4], are typically limited to unimodal distributions and may underestimate extreme values. By combining log-normal and hyperbolic components, the proposed mixture model is capable of reproducing multimodal profiles, which are often observed in pharmacokinetic data due to the presence of distinct absorption and elimination phases.

Comparison with other approaches

When compared with existing approaches, our method demonstrates several improvements. Dragalin et al. [2] proposed the use of KL-based statistics computed from point estimates, which makes the procedure sensitive to sampling variability. The KDE approach reduces this sensitivity and yields more stable estimates. Moreover, unlike closed-form likelihood-based procedures, the nonparametric nature of the estimator provides additional flexibility in modeling heterogeneous concentration profiles.

Limitations

At the same time, some limitations should be noted. The accuracy of the proposed procedure depends on the choice of bandwidth in the KDE, which may affect the finite-sample behavior of the test. Computational complexity is also higher than in classical TOST or parametric likelihood-based methods, which may limit its routine application in large-scale simulation studies. Furthermore, the current evaluation was restricted to simulated data; validation on real clinical datasets remains an important future step.

Directions for future work

There are several mathematical directions in which this study can be further developed. First, the problem of bandwidth selection in KDE should be addressed more rigorously, for example by deriving data-driven rules with provable optimality properties or by studying the asymptotic risk of different selectors. Second, it is of interest to extend the methodology to more complex experimental designs, such as replicate and higher-order crossover trials, and to investigate how the weighting scheme interacts with design-specific variance components. Third, a detailed theoretical analysis of the proposed test is required: in particular, its asymptotic distribution, power properties under local alternatives, and robustness to model misspecification should be established. Finally, an important step will be to validate the procedure on real pharmacokinetic data, comparing its performance against standard parametric approaches and assessing its stability in practice.

5 Conclusions

The proposed bioequivalence testing procedure is based on weighted kernel density estimation of the Kullback–Leibler divergence. *In silico* studies demonstrated reduced false-positive bioequivalence decisions and higher decision accuracy, particularly in small-sample scenarios, for an improved robustness to missing data, outliers, and heavy-tailed distributions when compared to classical approaches.

Notwithstanding some qualifications in its practical application, the approach provides a promising framework for unifying average bioequivalence assessment within a distributional perspective.

Bibliography

[1] A. Donner and N. Klar. The impact of drop-outs on bioequivalence studies. *Statistics in Medicine*, 24(1):1–10, 2005. https://doi.org/10.1002/sim.1934.

[2] V. Dragalin, V. Fedorov, et al. Recent developments in bioequivalence assessment using Kullback-Leibler divergence. *Statistics in Biopharmaceutical Research*, 11(3):287–299, 2019. https://doi.org/10.1080/19466315.2018.1553740.

[3] V. Dragalin, V. Fedorov, S. Patterson, and B. Jones. Kullback–Leibler divergence for evaluating bioequivalence. *Statistics in Medicine*, 22(6):913–930, 2003. https://doi.org/10.1002/sim.1451.

[4] C. M. Metzler. Statistical methods in pharmacokinetics. *Drug Metabolism Reviews*, 14(4):1017–1052, 1983. https://doi.org/10.3109/03602538309029971.

[5] J. pei Liu and S.-C. Chow. *Design and Analysis of Bioavailability and Bioequivalence Studies*. Chapman & Hall/CRC, Boca Raton, 3 edition, 2014.

[6] D. J. Schuirmann. A comparison of the two one-sided tests procedure and the power approach for assessing the equivalence of average bioavailability. *Journal of Pharmacokinetics and Biopharmaceutics*, 15(6):657–680, 1987. https://doi.org/10.1007/BF01068419.

[7] T. V. Zakharova, A. A. Tarkhov. Evaluation of the significance level in Schuirmann's test for checking the bioequivalence hypothesis in missing data conditions. *Inform. Primen.*, 13(3):58–62, 2019. https://doi.org/10.14357/19922264190309.

Part IV: Clinical medicine

Rationale: 10

Part IV: **Clinical medicine**

For infusion of mathematics into the health sciences, opportunities for introduction of mathematical models into clinical patient management shall be explored, covering screening, diagnostics, therapeutic interventions and follow-up for sequelae, remission, and relapse.

Four broad classes of disorders are of main interest:
- diseases of the cellular system
- disorders of cerebral control
- their interactions with other subsystems of the human body
- disorders for holistic interventions in Chinese medicine

In this volume:
- Optimization of adaptive treatment protocols for melanoma patients

In former volume BIOKYBERNETIKA:
- Finding optimal two-stage combined treatment protocols
- Mathematical perspectives on traditional Chinese medicine meridians

https://doi.org/10.1515/9783112216620-029

Evgenii N. Khailov, Nikolay L. Grigorenko, and Ellina V. Grigorieva

Optimization of adaptive treatment protocols for melanoma patients

Abstract: Standard treatment for melanoma, which involves continuous infusion of maximally tolerated doses of chemotherapeutic drugs, rapidly depletes the population of drug-sensitive cancer cells in favor of drug-resistant ones, reducing the effectiveness of further therapy. Adaptive therapy is a novel approach that preserves the proportion of drug-sensitive cells by using treatment protocols that combine intervals between drug administration and drug withdrawal. Thus, drug-sensitive cells, in turn, compete with drug-resistant cells and suppress their growth. The success of adaptive therapy depends on the precise timing of drug administration and withdrawal, which is ideally determined based on real-time patient data. To model and optimize these synchronization strategies, we describe the dynamics of competition between two cancer cell populations using the Lotka–Volterra equations. A time-dependent control function reflects the switching between drug administration and drug withdrawal periods. The optimal switching times and treatment protocol for melanoma are determined by minimizing the total cancer burden both during the treatment period and only at the end of the treatment period, using analytical methods such as Pontryagin's maximum principle and nonoscillation theory. Modeling results using real clinical data confirm that adaptive therapy, given an optimal protocol, can prolong cancer progression and reduce the total drug dose compared with other methods.

Keywords: Melanoma, adaptive therapy, two-dimensional controlled Lotka–Volterra competition model, optimal solution, Pontryagin's maximum principle, switching function, nonoscillation criterion, Riccati equation, stability analysis

MSC 2020: 34H05, 37N25, 37N40, 92D25

1 Introduction

Melanoma is a malignant neoplasm that develops from pigment cells of the skin (melanocytes) and mucous membranes. Usually, the tumor forms on the surface of the skin, but melanoma can form in the mouth, intestines, and retina. The histology of the disease is an abnormal degeneration and active division of these cells. The occurrence of pathology may be due to external influences and hereditary factors. Melanoma is considered the most aggressive type of skin cancer: the tumor grows rapidly and

Evgenii N. Khailov, Nikolay L. Grigorenko, Lomonosov Moscow State University, Moscow, 119991, Russia, e-mails: khailov@cs.msu.ru, grigor@cs.msu.ru
Ellina V. Grigorieva, Texas Woman's University, Denton, TX 76204, USA, e-mail: egrigorieva@twu.edu

https://doi.org/10.1515/9783112216620-030

metastasizes quickly into organs. When such a disease is first diagnosed, the most common treatment for melanoma is surgery. With repeated relapses, systemic drug therapy is also indicated.

Standard cancer treatments rely on using the maximum tolerated dose (MTD) of the drug to kill as many cancer cells as possible. Although this method has produced some good results, many patients develop resistance to the drugs used and relapse. This can be explained by the growth advantage of those cancer cell populations that evade or negate the actions of the drugs through specific resistance mechanisms such as [6, 24, 31],

- active efflux of drugs: drug-resistant cells can pump chemotherapy drugs out of the cell using specialized proteins, reducing the drug's effectiveness;
- enhanced drug detoxification: some cancer cells can chemically modify and neutralize the drug, rendering it harmless;
- alterations in drug targets: changes in cellular molecules targeted by the drug can make the therapy less effective;
- improved DNA repair: surviving cancer cells can repair the damage caused by chemotherapeutics more efficiently;
- genetic and epigenetic changes: mutations and gene expression changes can allow continued survival and growth even when drugs are present.

In an attempt to combat this phenomenon, a new type of chemotherapy called metronomic chemotherapy (MCT) [2] has emerged, in which anticancer drugs are administered either at a dose below the MTD or at a dose at the MTD repeatedly during the course of cancer treatment. Some researchers believe that the use of the lower dosage of cytotoxic drugs during MNC can add to activating of the immune system and minimizing the induction of therapeutic resistance [16]. However, ultimately, neither of the two methods is completely effective.

Adaptive therapy is a new treatment strategy based on real-time tracking of the dynamics of cancer [4, 9, 13, 18, 25, 30, 32, 33]. This strategy focuses on maintaining control of cancer rather than attempting to eradicate all cancer cells. The fundamental principle of adaptive therapy is to maintain the viability of drug-sensitive cancer cells, which inhibit the growth of drug-resistant cancer cells by competing for the same limited resources. This can be done either by reducing and modulating the dose of the active drug, or by introducing alternating "on" (treatment) and "off" (no treatment) intervals. An important feature of adaptive therapy is the active feedback of the patient's body to the treatment process. For the doctor, it is she who acts as the driving force in making a decision to temporarily stop or, conversely, resume the process of taking medications. Adaptive therapy is used, e. g., in the treatment of metastatic melanoma.

There are a number of important issues along the way:

- tracking the dynamics of cancer over time;
- correct interpretation of the received data for making a decision on treatment in real time;

– optimization of the patient's treatment regimen by finding the optimal alternation of the time of taking and a break in taking medications;
– the correctness of the prognosis in the selection of patients for whom adaptive therapy will bring the maximum benefit; the effectiveness of adaptive therapy will vary in different patients.

To overcome these problems, various mathematical models are widely used. In particular, models defined by systems of ordinary differential equations (ODE models) of the Lotka–Volterra type, used to describe various modes of competition between populations of healthy and cancer cells or within different classes of cancer cells, can be found in [7, 10, 11]. Some ODE models of the dynamics of melanoma and its treatment are given in [1, 12, 15, 17, 19, 20, 28, 29].

Studying mathematical models and finding model parameters that are appropriate for specific cancer patients is very important to understand the possible dynamics of cancer and healthy cells. However, such models alone do not provide an answer to the question of finding the right and effective treatment for a patient. Mathematically, the problem of finding the optimal treatment protocol for a patient with melanoma or other cancers is an optimal control problem in which either the concentration of the chemotherapeutic agent or the process of switching between taking and not taking the drug is controlled.

In [10, 11] we used Lotka–Volterra-type models to describe the competition between healthy and cancerous blood cells. By adding a control function (the concentration of the chemotherapeutic drug), we simulated the effect of the chemotherapy dosage on the dynamics of healthy and cancerous cells and found the optimal protocol for drug administration. In [17], we created a melanoma controlled model that describes the codynamics of drug-susceptible and drug-resistant cancer cells by combining two systems of differential equations (with and without drug administration) into a bilinear control system. In that work, we consider the competition not between healthy and cancerous cells, but between two types of cancer cells—susceptible and resistant to the chemotherapeutic drug. We also control not the drug dose, but the alternation of periods of drug administration and breaks in drug administration ("drug holidays"), maintaining the dose of the chemotherapeutic drug at the same level. Using optimal control theory and Pontryagin's maximum principle, we found optimal treatment protocols depending on various model parameters or the optimal schedule of drug administration and withdrawal, minimizing the total tumor burden. However, in [17] we ignored the competition between the two types of cancer cells. The novelty of this research is that we consider the competition between drug-susceptible and drug-resistant cancer cells and their interactions during adaptive therapy using a Lotka–Volterra controlled model, and then apply optimal control theory and Pontryagin's maximum principle to analytically predict possible optimal control types depending on the model parameters and its initial conditions. By writing computer programs in MAPLE and BOCOP and given available clinical data, we find optimal treatment protocols for some melanoma patients.

2 Methods

2.1 Statement of the minimization problem

To mathematically describe the interaction of drug-sensitive and drug-resistant cancer cell populations in adaptive therapy of metastatic melanoma, we use the ODEs Lotka–Volterra modell proposed in [18]:

$$\left.\begin{aligned} S'(\tau) &= r_S S(\tau)\left(1 - \frac{S(\tau)+R(\tau)}{K}\right) - \delta S(\tau), \\ R'(\tau) &= r_R R(\tau)\left(1 - \frac{\kappa S(\tau)+R(\tau)}{K}\right), \end{aligned}\right\} \tag{20.1}$$

in which
- $S(\tau)$ is the population size of drug-sensitive cancer cells;
- $R(\tau)$ is the population size of drug-resistant cancer cells;
- r_S is the intrinsic growth rate of drug-sensitive cancer cells;
- r_R is the intrinsic growth rate of drug-resistant cancer cells;
- K is the maximum tumor size due to nutrient and space constraints (these populations have the same carrying capacity);
- δ is the death rate of drug-sensitive cancer cells due to therapy;
- κ is the degree to which drug-sensitive cancer cells inhibit the growth rate of drug-resistant cancer cells.

We assume that
- parameters δ and K are positive;
- parameters r_S, r_R are positive and $r_S \neq r_R$;
- parameter κ is positive and $\kappa \neq 1$. If $\kappa > 1$, then drug-sensitive cancer cells have a greater competitive effect on drug-resistant cancer cells than drug-resistant cancer cells have on themselves. If $\kappa < 1$, then drug-sensitive cancer cells have a lesser competitive effect on drug-resistant cancer cells than drug-resistant cancer cells have on themselves.

Within the framework of the model (20.1), adaptive therapy looks like an alternation of two stages:
- **Stage 1.** No medication (anticancer drug) is taken. Then $\delta = 0$ and the system (20.1) has the form:

$$\left.\begin{aligned} S'(\tau) &= r_S S(\tau)\left(1 - \frac{S(\tau)+R(\tau)}{K}\right), \\ R'(\tau) &= r_R R(\tau)\left(1 - \frac{\kappa S(\tau)+R(\tau)}{K}\right); \end{aligned}\right\} \tag{20.2}$$

- **Stage 2.** The medication is taken. Then $r_S = 0$ and the system (20.1) has the form:

$$\left.\begin{aligned} S'(\tau) &= -\delta S(\tau), \\ R'(\tau) &= r_R R(\tau)\left(1 - \frac{\kappa S(\tau) + R(\tau)}{K}\right). \end{aligned}\right\} \tag{20.3}$$

  This stage simulates the use of drugs—BRAF/MEK inhibitors, which have been experimentally shown to suppress growth and induce apoptosis of cancer cells.

In order to understand in what sequence these two stages should alternate and how long each of them should last, we introduce the control function of time $v(\tau) \in \{0; 1\}$ that will unite the systems (20.2) and (20.3). When $v(\tau) = 0$, there is no medication and we use the system (20.2), and when $v(\tau) = 1$, on the contrary, medication is taken and we use the system (20.3).

By combining the models of the two stages, we obtain a control system:

$$\left.\begin{aligned} S'(\tau) &= r_S(1 - v(\tau))S(\tau)\left(1 - \frac{S(\tau) + R(\tau)}{K}\right) - \delta v(\tau)S(\tau), \\ R'(\tau) &= r_R R(\tau)\left(1 - \frac{\kappa S(\tau) + R(\tau)}{K}\right), \end{aligned}\right\} \tag{20.4}$$

to which initial conditions are added:

$$S(0) = S_0 > 0, \quad R(0) = R_0 > 0, \quad S_0 + R_0 < K. \tag{20.5}$$

We will also consider the system (20.4) and (20.5) for a given period of time $[0, \Theta]$, which is the total period of melanoma treatment.

Let us perform scaling of variables, time, as well as parameters and the initial conditions in this system:

$$x = \frac{S}{K}, \quad y = \frac{R}{K}, \quad x_0 = \frac{S_0}{K}, \quad y_0 = \frac{R_0}{K}, \quad t = r_R \tau, \quad T = r_R \Theta, \quad \rho = \frac{r_S}{r_R}, \quad \sigma = \frac{\delta}{r_R}.$$

Here, it is obvious that $\rho \neq 1$. In addition, the values x and y become concentrations of drug-sensitive and drug-resistant cancer cells.

Finally, we have the control system:

$$\left.\begin{aligned} & x'(t) = \rho(1 - v(t))x(t)(1 - x(t) - y(t)) - \sigma v(t)x(t), \ t \in [0, T], \\ & y'(t) = y(t)(1 - \kappa x(t) - y(t)), \\ & x(0) = x_0, \ y(0) = y_0; \ x_0, y_0 > 0, \ x_0 + y_0 < 1, \end{aligned}\right\} \tag{20.6}$$

which we will consider further.

By the set of admissible controls $\Omega(T)$, we mean all possible Lebesgue measurable functions $v(t)$, which for almost all $t \in [0, T]$, satisfy the inclusion $v(t) \in V = \{0; 1\}$. Here, V is a control set.

Let us introduce the set:

$$\Lambda = \{(x, y) : 0 < x < 1,\ 0 < y < 1,\ x + y < \kappa^{-1}\},$$

which is an invariant set of the system (20.6).

Then, for an arbitrary admissible control $v(t)$ from $\Omega(T)$, the corresponding absolutely continuous solution $(x(t), y(t))$ of the system (20.6) is defined on the entire interval $[0, T]$ and satisfies the inclusion:

$$(x(t), y(t)) \in \Lambda, \quad t \in [0, T]. \tag{20.7}$$

For system (20.6) on the set of admissible controls $\Omega(T)$, we introduce the objective function:

$$J(v) = \alpha(x(T) + y(T)) + \beta \int_0^T (x(t) + y(t))dt, \quad \alpha \geq 0,\ \beta \geq 0,\ \alpha + \beta = 1, \tag{20.8}$$

which is a weighted sum of the concentrations of drug-sensitive and drug-resistant cancer cells both at the final moment T of the total treatment period $[0, T]$ and throughout its entire duration. The medical meaning of this objective function is the cancer load in the patient's body both at the final moment T and over the entire interval $[0, T]$.

Then we consider the problem of minimizing the objective function (20.8):

$$J(v) \to \min_{v(\cdot) \in \Omega(T)}. \tag{20.9}$$

The control set V is non-convex, which can lead to the absence of an optimal solution to the minimization problem (20.9) in the classes of admissible regimes traditional for applications. To avoid this problem, the convex hull of the control set V is taken. As a result, the two-point set V turns into the segment $U = [0, 1]$. Then, instead of the set of admissible controls $\Omega(T)$, we consider the set of admissible controls $\Psi(T)$. It consists of all possible Lebesgue measurable functions $v(t)$ that satisfy the inclusion $v(t) \in U$ for almost all $t \in [0, T]$. Here, U is a new control set.

As a consequence of this, the relaxed minimization problem arises, which consists in minimizing the objective function (20.8) for the control system (20.6) on the set of admissible controls $\Psi(T)$:

$$J(v) \to \min_{v(\cdot) \in \Psi(T)}, \tag{20.10}$$

which we will now consider.

Since the inclusion (20.7) remains valid for an arbitrary admissible control $v(t)$ from $\Psi(T)$, then together with the convexity of the control set U and linearity of the system (20.6) in the control $v(t)$, they guarantee the existence of an optimal solution in the relaxed minimization problem (20.10), which consists of the optimal control $v_*(t)$ and the optimal solution $(x_*(t), y_*(t))$ of the system (20.6) corresponding to it [22].

It is important to note that if the inclusions $v_*(t) \in V$ and $v_*(t) \in U$ are simultaneously satisfied for the optimal control $v_*(t)$ (this will be the case for us), then the optimal solution in the relaxed minimization problem (20.10) is simultaneously the optimal solution in the original minimization problem (20.9).

2.2 Pontryagin's maximum principle

To analyze the optimal control $v_*(t)$ and the corresponding optimal solution $(x_*(t), y_*(t))$ of the system (20.6) in the relaxed minimization problem (20.10), we apply Pontryagin's maximum principle [27]. First, we write the corresponding Hamiltonian:

$$H(x, y, v, \psi_1, \psi_2) = (\rho(1-v)(1-x-y) - \sigma v)x\psi_1 + (1 - \kappa x - y)y\psi_2,$$

where ψ_1 and ψ_2 are the adjoint variables. Second, we use it to calculate the required partial derivatives:

$$\begin{aligned}
H'_x(x, y, v, \psi_1, \psi_2) &= (\rho(1-v)(1-x-y) - \sigma v)\psi_1 - \rho(1-v)x\psi_1 - \kappa y\psi_2 - \beta, \\
H'_y(x, y, v, \psi_1, \psi_2) &= (1 - \kappa x - y)\psi_2 - \rho(1-v)x\psi_1 - y\psi_2 - \beta, \\
H'_v(x, y, v, \psi_1, \psi_2) &= -(\rho(1-x-y) + \sigma)x\psi_1.
\end{aligned}$$

Then, in accordance with Pontryagin's maximum principle, for the optimal control $v_*(t)$ and the optimal solution $(x_*(t), y_*(t))$ corresponding to it, there exits such a vector function $\psi(t) = (\psi_1(t), \psi_2(t))$ that:

- $\psi(t)$ is a nontrivial solution of the adjoint system:

$$\left.\begin{aligned}
\psi'_1(t) &= -(\rho(1-v_*(t))(1-x_*(t)-y_*(t)) - \sigma v_*(t))\psi_1(t) \\
&\quad + \rho(1-v_*(t))x_*(t)\psi_1(t) + \kappa y_*(t)\psi_2(t) + \beta, \\
\psi'_2(t) &= -(1-\kappa x_*(t) - y_*(t))\psi_2(t) \\
&\quad + \rho(1-v_*(t))x_*(t)\psi_1(t) + y_*(t)\psi_2(t) + \beta, \\
\psi_1(T) &= -\alpha, \ \psi_2(T) = -\alpha;
\end{aligned}\right\}$$

- the control $v_*(t)$ maximizes the Hamiltonian $H(x_*(t), y_*(t), v, \psi_1(t), \psi_2(t))$ with respect to the variable $v \in [0, 1]$ for almost all $t \in [0, T]$ and, therefore, it satisfies the relationship:

$$v_*(t) = \begin{cases} 1, & \text{if } -(\rho(1 - x_*(t) - y_*(t)) + \sigma)x_*(t)\psi_1(t) > 0, \\ \text{any } v \in [0,1], & \text{if } -(\rho(1 - x_*(t) - y_*(t)) + \sigma)x_*(t)\psi_1(t) = 0, \\ 0, & \text{if } -(\rho(1 - x_*(t) - y_*(t)) + \sigma)x_*(t)\psi_1(t) < 0. \end{cases}$$

It is convenient to introduce new adjoint variables:

$$\phi_1(t) = -x_*(t)\psi_1(t), \quad \phi_2(t) = -y_*(t)\psi_2(t).$$

Then there exists the vector function $\phi(t) = (\phi_1(t), \phi_2(t))$ such that

- $\phi(t)$ is the nontrivial solution of the transformed adjoint system:

$$\left.\begin{aligned} &\phi_1'(t) = \rho(1 - v_*(t))x_*(t)\phi_1(t) + \kappa x_*(t)\phi_2(t) - \beta x_*(t), \\ &\phi_2'(t) = \rho(1 - v_*(t))y_*(t)\phi_1(t) + y_*(t)\phi_2(t) - \beta y_*(t), \\ &\phi_1(T) = \alpha x_*(T), \ \phi_2(T) = \alpha y_*(T); \end{aligned}\right\} \tag{20.11}$$

- the control $v_*(t)$ satisfies the relationship:

$$v_*(t) = \begin{cases} 1, & \text{if } L_v(t) > 0, \\ \text{any } v \in [0,1], & \text{if } L_v(t) = 0, \\ 0, & \text{if } L_v(t) < 0, \end{cases} \tag{20.12}$$

  in which the function

$$L_v(t) = (\rho(1 - x_*(t) - y_*(t)) + \sigma)\phi_1(t), \tag{20.13}$$

  is the switching function. It describes the behavior of the control $v_*(t)$ using the formula (20.12).

It can be seen that for the function $g(t)$ given by the formula:

$$g(t) = \rho(1 - x_*(t) - y_*(t)) + \sigma,$$

the following two cases take place.

Case 1. The function $g(t)$ is positive everywhere on the interval $[0, T]$ when

- $\kappa > 1$,
- $0.5 \le \kappa < 1$ and $\kappa \ge \rho(\sigma + \rho)^{-1}$,
- $0 < \kappa < 0.5$ and $\sigma \ge \rho$.

Case 2. The function $g(t)$ can vanish on the interval $[0, T]$ when

- $0.5 \le \kappa < 1$ and $\kappa < \rho(\sigma + \rho)^{-1}$,
- $0 < \kappa < 0.5$ and $\sigma < \rho$.

These two cases in the coordinate system σ/ρ and $1/\kappa$ are shown in Figure 20.1.

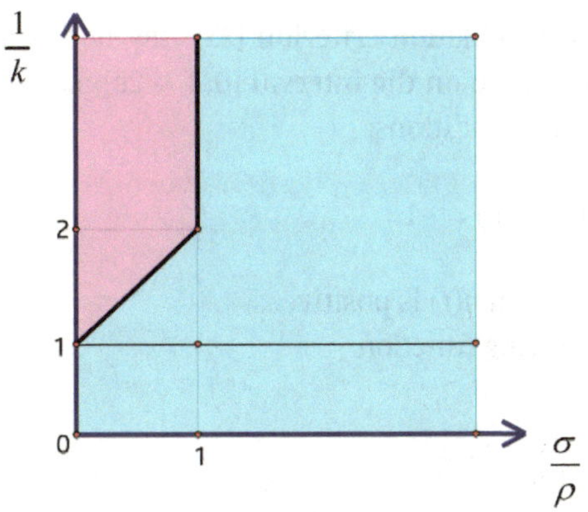

Figure 20.1: In the coordinate system σ/ρ and $1/\kappa$, region of Case 1 is shown in blue, and region of Case 2 is shown in pink.

3 Results

3.1 The case of positive function $g(t)$

3.1.1 Behavior of the switching function in the case of positive function $g(t)$

First, let us consider Case 1 in which the behavior of the switching function $L_v(t)$, given by the formula (20.13), is determined only the adjoint variable $\phi_1(t)$. An analysis of the equations of the system (20.11) shows that the function $\phi_1(t)$ does not vanish identically on any subinterval of the interval $[0, T]$. It can become equal to zero only at some points of this interval.

Now we will show how the number of zeros of the adjoint variable $\phi_1(t)$ can be estimated. To do this, consider on the interval $[0, T]$ the linear homogeneous differential equation of the second order:

$$\begin{aligned} z''(t) + \left(-\frac{x'_*(t)}{x_*(t)} - y_*(t) + \rho(1 - v_*(t))x_*(t) \right) z'(t) \\ - \kappa\rho(1 - v_*(t))x_*(t)y_*(t)z(t) = 0. \end{aligned} \tag{20.14}$$

We extend by continuity the optimal control $v_*(t)$ to the interval $[0, T + 2\epsilon]$, where ϵ is some given small positive number. Then the optimal solution $(x_*(t), y_*(t))$ is also extendable to this time interval. Thus, equation (20.14) is defined on the interval $[0, T + 2\epsilon]$.

Let us consider the function $w(t) = 1$ for all $t \in [0, T + 2\epsilon]$. It is positive and after its substitution into the left side of equation (20.14) we have the inequality:

$$-\kappa\rho(1 - v_*(t))x_*(t)y_*(t) \leq 0, \quad t \in [0, T + 2\epsilon].$$

Then, due to the de La Vallée–Poussin nonoscillation criterion [23], any nonzero solution of the equation (20.14) has at most one zero on the interval $[0, T + 2\epsilon]$. Let us take the solution $\eta(t)$ of this equation with initial conditions:

$$\eta(T + \epsilon) = 0, \quad \eta'(T + \epsilon) = -1.$$

Then everywhere on the interval $[0, T]$ the function $\eta(t)$ is positive.

Let us define on the interval $[0, T]$ the following function

$$h(t) = \frac{1}{\kappa x_*(t)} \frac{\eta'(t)}{\eta(t)}.$$

It is easy to see that this function satisfies the Riccati equation [14] on the entire interval $[0, T]$:

$$q'(t) = -\kappa x_*(t)q^2(t) + (y_*(t) - \rho(1 - v_*(t))x_*(t))q(t) + \rho(1 - v_*(t))y_*(t). \tag{20.15}$$

The transformed adjoint system (20.11) is the linear inhomogeneous system of differential equations. Let us make the following change of variables in it:

$$\chi_1(t) = \phi_1(t), \quad \chi_2(t) = \phi_2(t) - h(t)\phi_1(t) - \beta\kappa^{-1}.$$

Using the equations of system (20.11), as well as equation (20.15), for the new variables $\chi_1(t)$ and $\chi_2(t)$ we find the following system:

$$\left.\begin{aligned} &\chi_1'(t) = (\kappa h(t) + \rho(1 - v_*(t)))x_*(t)\chi_1(t) + \kappa x_*(t)\chi_2(t),\\ &\chi_2'(t) = (y_*(t) - \kappa x_*(t)h(t))\chi_2(t) + \beta(\kappa^{-1} - 1)y_*(t),\\ &\chi_1(T) = \alpha x_*(T), \end{aligned}\right\}$$

defined over the entire interval $[0, T]$. Involving the generalized Rolle's theorem [8] in it, one can show that the function $\phi_1(t) = \chi_1(t)$ has at most one zero on the interval $[0, T]$ for $\beta = 0$, and at most two different zeros on this interval for $\beta > 0$.

Based on this, we can draw the following final conclusions:

- for $\alpha = 1$ and $\beta = 0$, the function $\phi_1(t)$ has at most one zero on the interval $[0, T]$. At the same time, we conclude that either $\phi_1(t) > 0$ for all $t \in (0, T]$, or

$$\phi_1(t) \begin{cases} < 0, & \text{if } 0 \le t < \theta,\\ = 0, & \text{if } t = \theta,\\ > 0, & \text{if } \theta < t \le T; \end{cases}$$

- for $\alpha = 0$ and $\beta = 1$, the function $\phi_1(t)$ has at most two different zeros on the interval $[0, T]$. At the same time, we conclude that either $\phi_1(t) > 0$ for all $t \in (0, T)$, or

$$\phi_1(t) \begin{cases} < 0, & \text{if } 0 \le t < \theta, \\ = 0, & \text{if } t = \theta, \\ > 0, & \text{if } \theta < t < T, \\ = 0, & \text{if } t = T; \end{cases}$$

- for $\alpha > 0, \beta > 0$ and $\alpha + \beta = 1$, the function $\phi_1(t)$ has at most two different zeros on the interval $[0, T]$. At the same time, we conclude that either $\phi_1(t) > 0$ for all $t \in (0, T]$, or

$$\phi_1(t) \begin{cases} < 0, & \text{if } 0 < t < \theta, \\ = 0, & \text{if } t = \theta, \\ > 0, & \text{if } \theta < t \le T, \end{cases}$$

or

$$\phi_1(t) \begin{cases} > 0, & \text{if } 0 \le t < \theta_1, \\ = 0, & \text{if } t = \theta_1, \\ < 0, & \text{if } \theta_1 < t < \theta_2, \\ = 0, & \text{if } t = \theta_2, \\ > 0, & \text{if } \theta_2 < t \le T. \end{cases}$$

The analysis of the formula (20.12) shows that the optimal control $v_*(t)$ is a piecewise constant function and switches between the values 0 and 1. This implies that the control $v_*(t)$ obtained from the analysis of the relaxed minimization problem (20.10) is also the optimal control in the original minimization problem (20.9). Moreover, depending on the values of the weighting coefficients α and β, this control has the following forms:

- for $\alpha = 1$ and $\beta = 0$ (objective function (20.8) has only terminal part),

$$v_*(t) = \begin{cases} 0, & \text{if } 0 \le t \le \theta_*, \\ 1, & \text{if } \theta_* < t \le T; \end{cases}$$

- for $\alpha = 0$ and $\beta = 1$ (objective function (20.8) has only integral part),

$$v_*(t) = \begin{cases} 0, & \text{if } 0 \le t \le \theta_*, \\ 1, & \text{if } \theta_* < t \le T, \end{cases}$$

  where $\theta_* \in [0, T)$ is the moment of switching;
- for $\alpha > 0, \beta > 0$ and $\alpha + \beta = 1$ (objective function (20.8) has terminal and integral parts),

$$v_*(t) = \begin{cases} 1, & \text{if } 0 \le t \le \theta_1^*, \\ 0, & \text{if } \theta_1^* < t \le \theta_2^*, \\ 1, & \text{if } \theta_2^* < t \le T, \end{cases} \tag{20.16}$$

where θ_1^* and θ_2^*, $0 \le \theta_1^* \le \theta_2^* < T$ are the moments of switching.

3.1.2 Numerical results in the case of positive function $g(t)$

For our numerical calculations, we used the values of parameters, initial conditions, and treatment interval lengths for eight patients published in the paper [18] and described by the model (20.6). We solved the optimal control problem by two different methods: (a) by writing a finite-dimensional optimization program in MAPLE and (b) using the BOCOP environment [5]. The numerical calculations obtained by the two methods are identical. The graphs obtained by BOCOP are presented below. In these numerical calculations, we considered the general case for the objective function, so we solved the problem of minimizing the total cancer load over the entire time interval and at the end of the treatment. The values of the weighting coefficients of the objective function (20.8) are taken to be $\alpha = 0.5$ and $\beta = 0.5$. Theoretically, we have a case with at most two switchings, θ_1 and θ_2 in $[0, T]$, given by formula (20.16). Therefore, depending on the initial conditions and parameters of the model (a specific case of a cancer patient), there can be one of the following optimal treatment scenarios:

- drug intake – “drug holidays” – drug intake;
- “drug holidays” – drug intake;
- drug intake is “on” throughout the treatment period.

Note that for all eight patients, the value of $\kappa > 1$. This means that we are within the already considered Case 1. Suppose that the treatment time for three hypothetical patients is 18 units. Now consider the following three examples.

Example 1. For this example, the values of the parameters and initial conditions of the system (20.6), as well as the final moment of time, are as follows:

$$\rho = 1.4, \quad \sigma = 1.0, \quad \kappa = 21.0, \quad x_0 = 0.026, \quad y_0 = 0.003, \quad T = 18. \tag{20.17}$$

Graphs of the corresponding optimal solutions $x_*(t)$, $y_*(t)$ and control $v_*(t)$ are presented in Figure 20.2. In all these graphs, the horizontal axis is the axis of the variable t, and the vertical axes are the axes of the variables x, y, z, respectively. Here, z is the auxiliary variable defined through the variables x and y by the formula:

$$z(t) = \int_0^t (x(s) + y(s))ds, \quad t \in [0, T]. \tag{20.18}$$

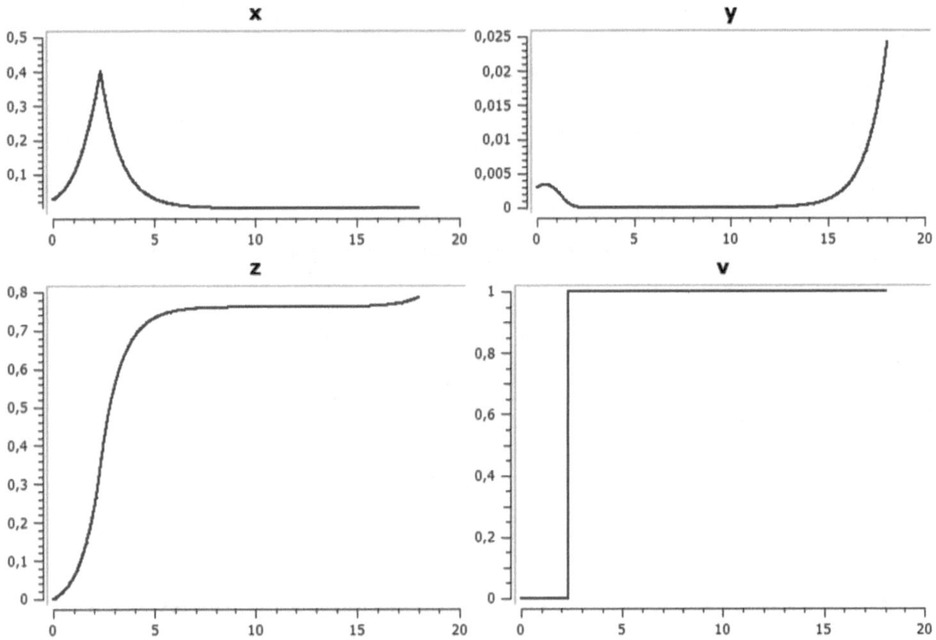

Figure 20.2: Optimal solutions $x_*(t), y_*(t), z_*(t)$, and control $v_*(t)$ in the minimization problem (20.10) for the values (20.17): upper row: $x_*(t), y_*(t)$; lower row: $z_*(t), v_*(t)$. Here, $x_*(t)$ and $y_*(t)$ are the optimal concentrations of drug-sensitive and drug-resistant cancer cells, respectively, $z_*(t)$ is the auxiliary function corresponding to them, and $v_*(t)$ is the optimal regimen of alternating between drug administration and drug withdrawal periods.

As can be seen from the graphs of upper row, the drug-sensitive cancer cells grow faster than they die under the therapy and $x_0/y_0 = 8.7$. The minimum value of the objective function (20.8) is $J_* = 0.404696$.

In this case, the optimal treatment protocol involves no drug for the first 2.5 units and using the maximum dose of the drug for the rest of the treatment period. We can see clearly that the drug-sensitive cancer cells continue to grow for $0 \leq t < 2.5$ units and then begin to die rapidly as soon as the drug is given. Interestingly, the drug-resistant cells first decrease slightly during $0 \leq t < 2.5$ and remain at this low concentration at the same level until unit 15, and then begin to grow rapidly thereafter.

Example 2. For this example, the values of the parameters and initial conditions of system (20.6), as well as the final moment of time, are as follows:

$$\rho = 1.4, \quad \sigma = 1.587, \quad \kappa = 4.5, \quad x_0 = 0.13, \quad y_0 = 0.0075, \quad T = 18. \tag{20.19}$$

Graphs of the corresponding optimal solutions $x_*(t), y_*(t)$ and control $v_*(t)$ are shown in Figure 20.3. In all these graphs, the horizontal axis is the axis of the variable t, and the vertical axes are the axes of the variables x, y, z, respectively. Here, z is the auxiliary

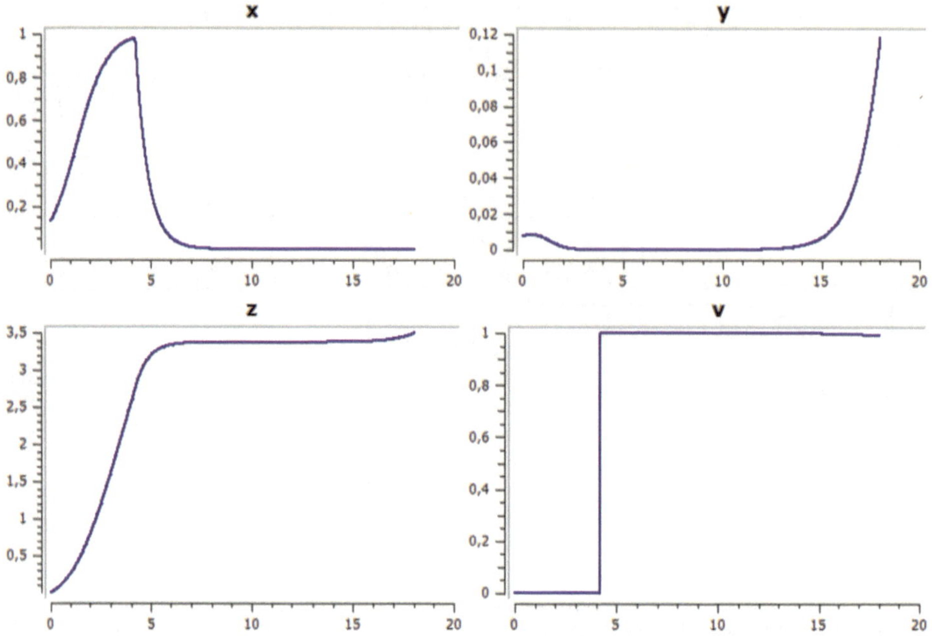

Figure 20.3: Optimal solutions $x_*(t)$, $y_*(t)$, $z_*(t)$, and control $v_*(t)$ in the minimization problem (20.10) for the values (20.19): upper row: $x_*(t)$, $y_*(t)$; lower row: $z_*(t)$, $v_*(t)$. Here, $x_*(t)$ and $y_*(t)$ are the optimal concentrations of drug-sensitive and drug-resistant cancer cells, respectively, $z_*(t)$ is the auxiliary function corresponding to them, and $v_*(t)$ is the optimal regimen of alternating between drug administration and drug withdrawal periods.

variable defined through the variables x and y by the formula (20.18). It is easy to see from the graphs of upper row that the death rate of drug-sensitive cancer cells due to drug therapy is higher than its growth rate and $x_0/y_0 = 17.33$. The minimum value of the objective function (20.8) is $J_* = 1.80658$.

In this case, the optimal treatment protocol involves no drug for the first 4 units and using the maximum dose of the drug for the remainder of the treatment period, up to 18 units. We see that, as in Example 1, the drug-sensitive cancer cells continue to grow for $0 \leq t < 4$ units and then begin to die rapidly as soon as the drug is administered. Interestingly, the drug-resistant cells first decrease slightly for $0 \leq t < 4$ and remain at this low concentration at the same level again until unit 15, and then begin to grow rapidly thereafter.

Example 3. For this example, the values of the parameters and initial conditions of system (20.6), as well as the final moment of time, are as follows:

$$\rho = 1.4, \quad \sigma = 0.714, \quad \kappa = 1.1, \quad x_0 = 0.316, \quad y_0 = 0.078, \quad T = 18. \tag{20.20}$$

Graphs of the corresponding optimal solutions $x_*(t)$, $y_*(t)$ and control $v_*(t)$ are given in Figure 20.4. In all these graphs, the horizontal axis is the axis of the variable t, and

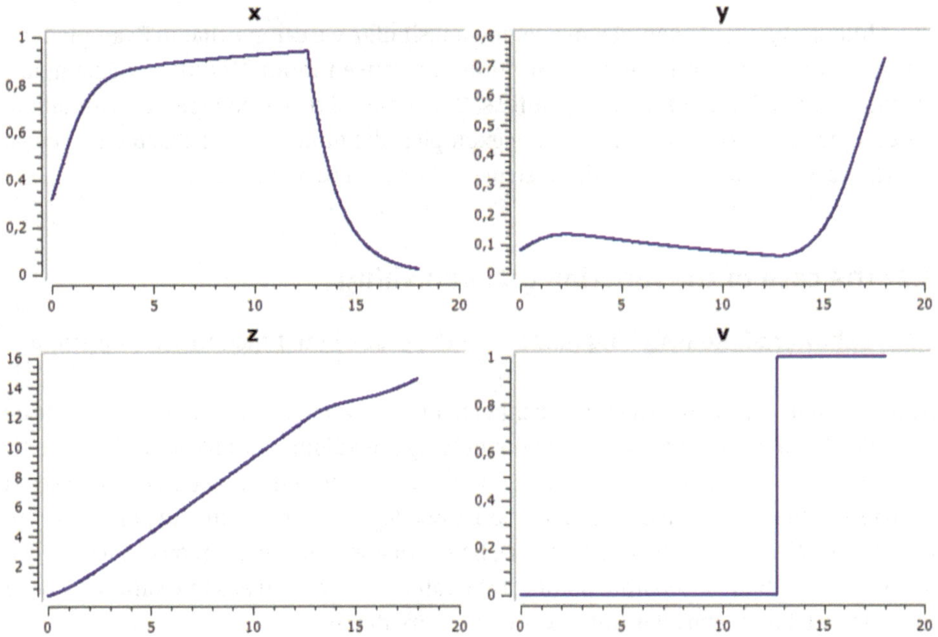

Figure 20.4: Optimal solutions $x_*(t), y_*(t), z_*(t)$, and control $v_*(t)$ in the minimization problem (20.10) for the values (20.20): upper row: $x_*(t), y_*(t)$; lower row: $z_*(t), v_*(t)$. Here, $x_*(t)$ and $y_*(t)$ are the optimal concentrations of drug-sensitive and drug-resistant cancer cells, respectively, $z_*(t)$ is the auxiliary function corresponding to them, and $v_*(t)$ is the optimal regimen of alternating between drug administration and drug withdrawal periods.

the vertical axes are the axes of the variables x, y, z, respectively. Here, z is the auxiliary variable defined through the variables x and y by the formula (20.18). We can see from the graphs of upper row that the drug-sensitive cancer cells grow faster than they die under the therapy and $x_0/y_0 = 4.05$. The minimum value of the objective function (20.8) is $J_* = 7.6632$.

In this case, we see that, as in Examples 1 and 2, the drug-sensitive cancer cells continue to grow when the drug is not administered (for about the first 13 units), and then begin to die quickly once the drug is used. On the other hand, the drug-resistant cells first decrease slightly during units $0 \leq t < 13$ and remain at almost the same low level again until unit 15, and then begin to grow.

Graphs are presented in order of the time of the first use of the drug in the optimal treatment protocols, as well as in descending order of the value of κ, the rate at which the drug-sensitive cells inhibit the growth rate of drug-resistant cancer cells.

For those values of the parameters and initial conditions of system (20.6), as well as the final moment of time, which are given in [18], it was not possible to find a situation with two switchings. All situations correspond to optimal control $v_*(t)$ with one switching. In all graphs $\rho = 1.4 > 1$, which means that $r_S > r_R$.

Although all three examples are somewhat similar, we can see that in Example 1 the drug-sensitive cancer cells reach their minimum concentration by unit 5, in Example 2 by unit 7, and in Example 3 only by unit 18. Of course, all three examples correspond to different model parameters. In all three examples, the numerical results did not reveal a single case with no switching (MTD schedule) or with two switchings.

3.2 The case of the function $g(t)$ vanishing

3.2.1 Behavior of the switching function in the case of the function $g(t)$ vanishing

Now consider Case 2, in which the behavior of the switching function $L_v(t)$ given by formula (20.13) is already determined not only by the adjoint variable $\phi_1(t)$, but also by the function $g(t)$. It is important to note that the behavior of the function $\phi_1(t)$ does not change. It is the same as in Case 1, since it does not depend on the values of parameters of the system (20.6). While the function $g(t)$, being an absolutely continuous function, can already vanish both at separate points of the interval $[0, T]$ and at some subintervals of this interval. Let us consider this question in more detail.

Let us introduce the auxiliary functions $p(t)$ and $r(t)$ corresponding to the function $g(t)$:

$$p(t) = \sigma y_*(t) + \sigma\rho x_*(t) - \rho(1-\kappa)x_*(t)y_*(t),$$
$$r(t) = \sigma\rho^{-2}(1-\kappa)^{-1}(\sigma+\rho) - x_*^2(t).$$

Then, using equations of the system (20.6), we obtain the system of differential equations for the function $g(t)$ and the corresponding auxiliary functions:

$$\left.\begin{aligned} g'(t) &= -(\rho(1-v_*(t))x_*(t) + y_*(t))g(t) + p(t), \\ p'(t) &= (\rho(1-v_*(t))x_*(t)(\sigma-(1-\kappa)y_*(t)) \\ &\quad + y_*(t)(\sigma\rho^{-1} - (1-\kappa)x_*(t)) - \sigma^2\rho^{-1})g(t) \\ &\quad - (\sigma + (\sigma\rho^{-1} - (1-\kappa)x_*(t)))p(t) + \sigma\rho(1-\kappa)r(t), \\ r'(t) &= -2(1-v_*(t))x_*^2(t)g(t) + 2\sigma x_*^2(t). \end{aligned}\right\} \tag{20.21}$$

An analysis of system (20.21) shows that the function $g(t)$ does not vanish identically on any subinterval of the interval $[0, T]$. It can vanish only at a finite number of points in this interval. Then, from formula (20.12), it follows that the optimal control $v_*(t)$ is a piecewise constant function and switches between the values 0 and 1. This means that the control $v_*(t)$ obtained from the analysis of the relaxed minimization problem (20.10) is also the optimal control in the original minimization problem (20.9). The question of the analysis of the number of zeros of the function $g(t)$ on the interval $[0, T]$ is still open.

Note that Case 2 describes the situation when drug-sensitive cells do not die faster than they grow under the influence of treatment, and also when they do not suppress (weaken) drug-resistant cells well enough ($\kappa < 1$). It is clear that in this case the concentration of cancer cells will eventually increase, which allows us to speak about the ineffectiveness of the treatment. In our numerical calculations, despite the change in parameters, the optimal control took the value of only 1 throughout the treatment. Clearly, this is better than no treatment, as can be seen numerically for the control function taking the value 0 throughout the interval. This may be the case when the patient requires continuous MTD therapy.

3.3 Stability analysis of equilibrium positions

To explain the results obtained in the previous sections, it is necessary to find the equilibrium positions of the system (20.6) for the cases $v(t) = 1$ and $v(t) = 0$ and find out whether they are stable or not.

- First, we determine the equilibrium positions of the system (20.6) when the control $v(t) = 1$:

$$\left.\begin{aligned} x'(t) &= -\sigma x(t), \\ y'(t) &= y(t)(1 - \kappa x(t) - y(t)). \end{aligned}\right\} \tag{20.22}$$

It is easy to see that this system on the set $\overline{\Lambda}$ has two equilibrium positions $(0, 0)$ and $(0, 1)$. Here, $\overline{\Lambda}$ is the closure of the set Λ. The corresponding Jacobian matrix is

$$J_1(x, y) = \begin{pmatrix} -\sigma & 0 \\ -\kappa y & (1 - \kappa x - 2y) \end{pmatrix}.$$

Then for the equilibrium position $(0, 0)$ this matrix looks as follows:

$$J_1(0, 0) = \begin{pmatrix} -\sigma & 0 \\ 0 & 1 \end{pmatrix}$$

and, therefore, such an equilibrium position is unstable.

For the equilibrium position $(0, 1)$, the Jacobian matrix $J_1(x, y)$ takes the form:

$$J_1(0, 1) = \begin{pmatrix} -\sigma & 0 \\ -\kappa & -1 \end{pmatrix}$$

and, therefore, such an equilibrium position is locally asymptotically stable.

To justify the global asymptotic stability of the equilibrium position $(0, 1)$, we introduce the Lyapunov function:

$$V_1(x, y) = 0.5\kappa^2 x + \sigma((y - 1) - \ln y), \quad (x, y) \in \Lambda.$$

Calculating the derivative of this function by virtue of the system (20.22), we find the relationship:

$$\left.\frac{dV_1}{dt}\right|_{(20.22)} = -\sigma\big((y(t) - 1 + 0.5\kappa x(t))^2 + 0.25\kappa^2 x(t)(2 - x(t))\big) \le 0.$$

This derivative vanishes only at the point $(0,1)$. Therefore, using the Lyapunov–LaSalle's invariance principle [21], we conclude that the equilibrium position $(0,1)$ is globally asymptotically stable for $\kappa \neq 1$.

– Second, we determine the equilibrium positions of the system (20.6) when the control $v(t) = 0$:

$$\left.\begin{aligned} x'(t) &= \rho x(t)(1 - x(t) - y(t)), \\ y'(t) &= y(t)(1 - \kappa x(t) - y(t)). \end{aligned}\right\} \tag{20.23}$$

It can be seen that this system on the set $\overline{\Lambda}$ has three equilibrium positions $(0,0)$, $(1,0)$ and $(0,1)$. The corresponding Jacobian matrix is

$$J_0(x,y) = \begin{pmatrix} \rho(1 - 2x - y) & -\rho x \\ -\kappa y & (1 - \kappa x - 2y) \end{pmatrix}.$$

Then for the equilibrium position $(0,0)$ this matrix looks as follows:

$$J_0(0,0) = \begin{pmatrix} \rho & 0 \\ 0 & 1 \end{pmatrix}$$

and, therefore, such an equilibrium position is unstable.

For the equilibrium position $(0,1)$, the Jacobian matrix $J_0(x,y)$ takes the form:

$$J_0(0,1) = \begin{pmatrix} 0 & 0 \\ -\kappa & -1 \end{pmatrix}$$

and, therefore, nothing can be said about the stability of this equilibrium position.

For the equilibrium position $(1,0)$, this matrix is

$$J_0(1,0) = \begin{pmatrix} -\rho & -\rho \\ 0 & (1 - \kappa) \end{pmatrix}$$

and, therefore, such an equilibrium position is unstable at $\kappa < 1$, and locally asymptotically stable at $\kappa > 1$.

To justify the global asymptotic stability of the equilibrium position $(1,0)$ at $\kappa > 1$, we consider the Lyapunov function:

$$V_0(x,y) = \kappa\big((x - 1) - \ln x\big) + \rho y, \quad (x,y) \in \Lambda.$$

Calculating the derivative of this function by virtue of the system (20.23), we obtain the relationship:

$$\left.\frac{dV_0}{dt}\right|_{(20.23)} = -\rho((\kappa-1)y(t)(1-y(t)) + \kappa(y(t)-1+x(t))^2) \le 0.$$

This derivative vanishes only at the points $(0,1)$ and $(1,0)$. Therefore, using again the Lyapunov–LaSalle's invariance principle, we conclude that the equilibrium position $(1,0)$ is globally asymptotically stable for $\kappa > 1$.

Now let us justify the global asymptotic stability of the equilibrium position $(0,1)$ at $\kappa < 1$. To do this, we introduce the Lyapunov function:

$$V_0(x,y) = \kappa x + \rho((y-1) - \ln y), \quad (x,y) \in \Lambda.$$

Calculating the derivative of this function by virtue of system (20.23), we have the relationship:

$$\left.\frac{dV_0}{dt}\right|_{(20.23)} = -\rho(\kappa(y(t)-1+x(t))^2 + (1-\kappa)(1-y(t))^2) \le 0.$$

This derivative vanishes only at the point $(0,1)$. Therefore, using again the Lyapunov–LaSalle's invariance principle, we conclude that the equilibrium position $(0,1)$ is globally asymptotically stable for $\kappa < 1$.

The equilibrium position $(0,1)$ at $\kappa > 1$ has a more complex structure. Therefore, we use a different approach ([3]) to analyze it. Namely, first, using a change of variables $\tilde{x} = x$ and $\tilde{y} = 1-y$, we transfer the point $(0,1)$ to the origin. As a result, the system (20.23) is rewritten as

$$\left.\begin{aligned} \tilde{x}'(t) &= -\rho\tilde{x}^2(t) + \rho\tilde{x}(t)\tilde{y}(t), \\ \tilde{y}'(t) &= \kappa\tilde{x}(t) - \tilde{y}(t) - \kappa\tilde{x}(t)\tilde{y}(t) + \tilde{y}^2(t). \end{aligned}\right\}$$

Then we simplify the resulting system by replacing variables $\hat{x} = \tilde{x}$ and $\hat{y} = \tilde{y} - \kappa\tilde{x}$ in it. After the necessary transformations, we have the system:

$$\left.\begin{aligned} \hat{x}'(t) &= (\kappa-1)\rho\hat{x}^2(t) + \rho\hat{x}(t)\hat{y}(t), \\ \hat{y}'(t) &= -\hat{y}(t) - \kappa(\kappa-1)\rho\hat{x}^2(t) - \kappa(\rho-1)\hat{x}(t)\hat{y}(t) + \hat{y}^2(t). \end{aligned}\right\} \tag{20.24}$$

Following the arguments from [3], we equate to zero the right-hand side of the second equation of the system (20.24) and from here find the dependence $\hat{y}(\hat{x})$. After that, we substitute it into the right-hand side of the first equation of this system. By expanding the obtained expression in powers of the small value $\hat{x}$, we finally find the relationship:

$$(\kappa-1)\rho\hat{x}^2 - 0.25\kappa\rho(\kappa(\rho-1)^2 + 4(\kappa-1)\rho)\hat{x}^3 + \cdots.$$

Due to the features of the first term of this relationship and the minus sign of the value $\hat{y}(t)$ in the second equation of the system (20.24), as well as the results from [3], we conclude that the equilibrium position $(0, 1)$ at $\kappa > 1$ is a saddle node. It has one stable nodal sector and two saddle sectors and, therefore, such an equilibrium position is unstable.

The obtained results can be combined into the Table 20.1.

Table 20.1: Stability analysis of nonzero equilibrium positions.

	$v(t) = 0$ **position** $(1, 0)$	$v(t) = 0$ **position** $(0, 1)$	$v(t) = 1$ **position** $(0, 1)$
$\kappa < 1$	unstable	globally asymptotically stable	globally asymptotically stable
$\kappa > 1$	globally asymptotically stable	unstable	globally asymptotically stable

An analysis of this table leads us to the following conclusions:

- Case $\kappa < 1$ has no medical meaning, since the equilibrium position $(0, 1)$ is globally asymptotically stable both for the control $v(t) = 0$, in which there is no drug, and for the control $v(t) = 1$, in which, on the contrary, the drug is taken. This means that whatever treatment schedule the doctor chooses, everything will lead to the extinction of the population of drug-sensitive cancer cells and the achievement of the maximum possible population of drug-resistant cancer cells. The drug used will no longer have the desired therapeutic effect.
- Case $\kappa > 1$ is preferable, since depending on which value the control $v(t)$ takes: 0 or 1, either the equilibrium position $(1, 0)$ or the equilibrium position $(0, 1)$ is globally asymptotically stable.

4 Discussion

4.1 Mathematical aspects

For a given time-interval, representing the total period of melanoma treatment, a mathematical Lotka–Volterra ODE controlled model was considered. The model described the competition between drug-sensitive and drug-resistant cancer cells during adaptive therapy, in which the time-dependent control function was responsible for the transition from the stage of active implementation of adaptive therapy to the stage of its absence and vice versa. To find the optimal moments of switching between these stages, it was necessary to minimize the overall load of cancer cells both during the entire period of melanoma treatment and at the time of its completion.

An analytical study of the optimal control minimization problem was carried out using Pontryagin's maximum principle and nonoscillation theory for ODEs. The analytical results were confirmed by numerical calculations performed using BOCOP and

MAPLE computer programs for the values of the parameters and the initial conditions of the Lotka–Volterra competition model taken from real clinical practice data.

The types of optimal control function that solve the stated minimization problem for specific clinical data were determined analytically and numerically. Different optimal melanoma treatment protocols were discussed.

4.2 Clinical aspects

The ineffectiveness and sometimes failure of melanoma treatment is partly due to the development of resistance of cancer cells to chemotherapeutic drugs. The goal of this study was to investigate the codynamics of drug-susceptible and drug-resistant melanoma cells within the framework of the ODE Lotka–Volterra competition model. By introducing the control function responsible for alternating drug intake and "drug holidays" intervals, we solved the optimal control problem of cancer burden minimization and found the corresponding optimal control analytically and numerically. The types of optimal control functions reflecting the most effective treatment protocols for specific (available) clinical data were determined.

This optimal adaptive treatment protocol differs from standard practice methods such as the maximum tolerated dose (MTD) protocol or metronomic chemotherapy (MNC). In the Appendix, we consider MTD and three MNC scenarios for the patient parameters in Example 1 and show that the optimal solution obtained in this paper indeed minimizes the total cancer burden much better than either of the other methods.

Let us discuss the effectiveness of adaptive therapy and the possibility of finding optimal treatment protocols for real cancer patients. How can we interpret our theoretical and numerical results shown in Examples 1, 2, and 3? Assume that three different cancer patients come to the doctor for diagnostics at the same time. Their tests are ready and the corresponding system parameters for each patient have been obtained. For simplicity, we will call the units of time "days." So, on the 1st day, all the characteristics are known, and by running the optimization program, we obtain the following most effective drug therapy protocols for each case. Patient 1 starts taking the treatment on the 2nd day and continues it for the remaining 15–16 days. Patient 2 starts taking the same drug on the 4th day and takes it for the next 14 days. Patient 3 does not start taking the drug until day 13 and continues until day 18 (up to 6 days).

Thus, the optimal protocols for administering drugs to patients in Examples 1, 2, and 3 can be understood as "treatment delay" protocols ([26]), which are also known in oncology practice. We know that chemotherapy primarily targets rapidly dividing cancer cells. In some aggressive forms of cancer, early treatment does not provide any survival benefit, and anticancer therapy can be delayed until there is clear evidence of cancer progression.

Our numerical results are derived from published clinical data. It is theoretically possible that for another melanoma patient, treatment would begin immediately and continue for the full 18 days, but for another patient, the drug would be administered for the first week, then interrupted for a few days, and then resumed. Will it be possible in the near future, e. g., to take a patient's blood and other tests, fit a mathematical model, find the optimal treatment protocol and treat the patient in real time?

Call for clinical implementation
We hope that by having access to real-world medical data and collaborating with physicians in clinical practice, we can help improve the melanoma treatment protocol tailored to the individual patient.

5 Conclusions

We solved the optimal control problem of cancer burden minimization by switching between drug administration and pausing both analytically and numerically. Our optimal adaptive treatment protocol differs from standard practice methods such as the maximum tolerated dose (MTD) protocol or metronomic chemotherapy (MNC). We could demonstrate that our optimal adaptive protocol minimizes a patient's total cancer burden much better than either of the other protocols used in clinical practice.

Bibliography

[1] M. Albrecht, P. Lucarelli, D. Kulms, and T. Sauter. Computational models of melanoma. *Theoretical Biology and Medical Modelling*, 17:8, 2020.

[2] D. Barbolosi, J. Ciccolini, C. Meille, X. Elharrar, C. Faivre, B. Lacarelle, N. André, and F. Barlesi. Metronomics chemotherapy: Time for computational decision support. *Cancer Chemotherapy and Pharmacology*, 74(3):647–652, 2014.

[3] N. N. Bautin and E. A. Leontovich. *Methods and Ways of the Qualitative Analysis of Dynamical Systems in a Plane*. Nauka, Moscow, 1990.

[4] S. Belkhir, F. Thomas, and B. Roche. Darwinian approaches for cancer treatment: Benefits of mathematical modeling. *Cancers*, 13(4448):1–19, 2021.

[5] F. Bonnans, P. Martinon, D. Giorgi, V. Grélard, S. Maindrault, O. Tissot, and J. Liu. *BOCOP 2.0.5 — user guide [e-resource]*. INRIA, France, 8 February 2017. http://bocop.org.

[6] K. Bukowski, M. Kciuk, and R. Kontek. Mechanisms of multidrug resistance in cancer chemotherapy. *International Journal of Molecular Sciences*, 21(3233):1–24, 2020.

[7] H. Cho, A. L. Lewis, K. M. Storey, and H. M. Byrne. Designing experimental conditions to use the Lotka-Volterra model to infer tumor cell line interaction types. *Journal of Theoretical Biology*, 559(111377):1–24, 2023.

[8] A. V. Dmitruk. A generalized estimate on the number of zeros for solutions of a class of linear differential equations. *SIAM Journal on Control and Optimization*, 30:1087–1091, 1992.

[9] R. A. Gatenby, A. S. Silva, R. J. Gillies, and B. R. Frieden. Adaptive therapy. *Cancer Research*, 69(11):4894–4903, 2009.

[10] N. L. Grigorenko, E. N. Khailov, E. V. Grigorieva, and A. D. Klimenkova. Optimal strategies in the treatment of cancers in the Lotka-Volterra mathematical model of competition. *Proceedings of the Steklov Institute of Mathematics*, 313(Suppl. 1):S100–S116, 2021.

[11] N. L. Grigorenko, E. N. Khailov, E. V. Grigorieva, and A. D. Klimenkova. Lotka-Volterra competition model with nonmonotone therapy function for finding optimal strategies in the treatment of blood cancers. *Proceedings of the Steklov Institute of Mathematics*, 317(Suppl. 1):S71–S89, 2022.

[12] G. Haddad, A. Kebir, N. Raissi, A. Bouhali, and S. B. Miled. Optimal control model of tumor treatment in the context of cancer stem cell. *Mathematical Biosciences and Engineering*, 19(5):4627–4642, 2022.

[13] E. Hansen and A. F. Read. Modifying adaptive therapy to enhance competitive suppression. *Cancers*, 12(3556):1–13, 2020.

[14] P. Hartman. *Ordinary Differential Equations*. John Wiley & Sons, New York, 1964.

[15] A. Hodgkinson, D. Trucu, M. Lacroix, L. Le Cam, and O. Radulescu. Computational model of heterogeneity in melanoma: Designing therapies and predicting outcomes. *Frontiers in Oncology*, 12(857572):1–10, 2022.

[16] I. Kareva, D. J. Waxman, and G. L. Klement. Metronomic chemotherapy: An attractive alternative to maximum tolerated dose therapy that can activate anti-tumor immunity and minimize therapeutic resistance. *Cancer Letters*, 358:100–106, 2015.

[17] E. Khailov and E. Grigorieva. Optimal melanoma treatment protocols for a bilinear control model. *Mathematics*, 11(3289):1–29, 2023.

[18] E. Kim, J. S. Brown, Z. Eroglu, and A. R. A. Anderson. Adaptive therapy for metastatic melanoma: Predictions from patient calibrated mathematical models. *Cancers*, 13(823):1–15, 2021.

[19] E. Kim, V. W. Rebecca, I. V. Fedorenko, J. L. Messina, R. Mathew, S. S. Maria-Engler, D. Basanta, K. S. M. Smalley, and A. R. A. Anderson. Senescent fibroblasts in melanoma initiation and progression: An integrated theoretical, experimental, and clinical approach. *Cancer Research*, 73(23):6874–6885, 2013.

[20] E. Kim, V. W. Rebecca, K. S. M. Smalley, and A. R. A. Anderson. Phase *i* trials in melanoma: A framework to translate preclinical findings to the clinic. *European Journal of Cancer*, 67:213–222, 2016.

[21] J. P. LaSalle. *The Stability of Dynamical Systems*. SIAM, Philadelphia, PA, USA, 1976.

[22] E. B. Lee and L. Marcus. *Foundations of Optimal Control Theory*. John Wiley & Sons, New York, 1967.

[23] A. Yu. Levin. Non-oscillation of solutions of the equation $x^{(n)} + p_1(t)x^{(n-1)} + \cdots + p_n(t)x = 0$. *Russian Mathematical Surveys*, 24:43–99, 1969.

[24] B. Mansoori, A. Mohammadi, S. Davudian, S. Shirjang, and B. Baradaran. The different mechanisms of cancer drug resistance: A brief review. *Advanced Pharmaceutical Bulletin*, 7:339–348, 2017.

[25] C. McGehee and Y. Mori. A mathematical framework for comparison of intermittent versus continuous adaptive chemotherapy dosing in cancer. *Systems Biology and Applications*, 10(140):1–11, 2024.

[26] H. C. Monro and E. A. Gaffney. Modelling chemotherapy resistance in palliation and failed cure. *Journal of Theoretical Biology*, 257:292–302, 2009.

[27] L. S. Pontryagin, V. G. Boltyanskii, R. V. Gamkrelidze, and E. F. Mishchenko. *Mathematical Theory of Optimal Processes*. John Wiley & Sons, New York, 1962.

[28] G. Rodrigues, J. G. Silva, D. S. Santurio, and P. F. A. Mancera. A mathematical model to describe the melanoma dynamics under effects of macrophage inhibition and car t-cell therapy. *Trends in Computational and Applied Mathematics*, 25:e01734:1–21, 2024.

[29] I. Smalley, E. Kim, J. Li, P. Spence, C. J. Wyatt, Z. Eroglu, V. K. Sondak, J. L. Messina, N. A. Babacan, S. S. Maria-Engler, L. De Armas, S. L. Williams, R. A. Gatenby, Y. A. Chen, A. R. A. Anderson, and K. S. M. Smalley. Leveraging transcriptional dynamics to improve braf inhibitor responses in melanoma. *eBioMedicine*, 48:178–190, 2019.

[30] M. A. R. Strobl, J. West, Y. Viossat, M. Damaghi, M. Robertson-Tessi, J. S. Brown, R. A. Gatenby, P. K. Maini, and A. R. A. Anderson. Turnover modulates the need for a cost of resistance in adaptive therapy. *Cancer Research*, 81(4):1135–1147, 2021.

[31] W. H. Talib, A. R. Alsayed, M. Barakat, M. I. Abu-Taha, and A. I. Mahmod. Targeting drug chemo-resistance in cancer using natural products. *Biomedicines*, 9(1353):1–90, 2021.
[32] J. Wang, Y. Zhang, X. Liu, and H. Liu. Optimizing adaptive therapy based on the reachability to tumor resistant subpopulation. *Cancers*, 13(5262):1–19, 2021.
[33] J. West, F. Adler, J. Gallaher, M. Strobl, R. Brady-Nicholls, J. Brown, M. Roberson-Tessi, E. Kim, R. Noble, Y. Viossat, D. Basanta, and A. R. A. Anderson. A survey of open questions in adaptive therapy: Bridging mathematics and clinical translation. *eLife*, 12:e84263:1–25, 2023.

Appendix A. Supplement

In this work, we wanted to understand how the mathematical modeling and optimal control theory can optimize the administration of anticancer drugs in adaptive therapy. Importantly, we control only the rotation and duration of drug administration and the intervals of drug interruptions, analytically finding the type of optimal control and the number of possible switchings. Then, the optimal control for a certain set of model parameters or the optimal treatment protocol is found numerically. In our controlled model, we do not vary the drug dose. We assume that the chemotherapeutic drug is supplied and administered in ampoules with the same dose, and the content of each ampoule should not be divided and should be used at once.

When we talk about the same dose of the drug, it is easier to use similarity with tablets: for example, you can take 10 mg tablet every day, 7 per week. Or, in a different case for a different illness, as prescribed by another doctor, one can take 10 mg tablets for 3 days, then take a break for 1 day, then take 10 mg tablets again, one per day, for 3 days. And thus, in both cases we will have 7 day schedules of taking the drug. We do not change the dose of the tablet. Well, it is clear that the total amount of the drug intake during the week will be different. In the second case, the total amount of the drug is less (6/7 of the continuous case). Or you can just take one pill for 2 days. Then take a 3-day break and take another 2 pills, thus also participating in the weekly pill plan. But here it is already clear that the total number of pills taken in 7 days will be only 4, so you would take 4/7 if compared to, say, daily pill intake for 7 days.

It is known that if for some diseases, such as infectious diseases, the total amount of drug is the most important in treatment, then in the case of cancer chemotherapy it was noted that MTD can lead to the growth of drug-resistant cancer cells. It was found that simply reducing the dose of the drug with continuous administration of this drug also does not lead to significant success, while the metronomic schedule sometimes shows better results.

Consider the maximum tolerated dose therapy (MTD) and metronomic therapy (MNC) applied to the patient from Example 1 and compare them with the optimal adaptive therapy schedule obtained in this study using optimal control theory and Pontryagin's maximum principle. If we apply the optimal adaptive therapy schedule, the minimum value of the objective function (20.8) is $J_* = 0.4047$ with the concentration of drug-sensitive and drug-resistant cancer cells given by the inequalities $0 < x(t) < 0.43$ and $0 < y(t) < 0.025$, respectively.

If we use MTD ($v(t) = 1$) over the entire period of 18 units, the objective function value is $J_{\text{MTD}} = 6.34$, which is much larger than the value 0.4047 obtained by solving the optimal control problem. This schedule is not optimal for the values (20.17) of Example 1. Moreover, drug-sensitive cancer cells will vanish, and drug-resistant cancer cells will increase to the maximum level of 1 very quickly.

A metronomic application of a chemotherapy (MNC) schedule with two switchings at $\tau_1 = 6$ and $\tau_2 = 12$ will lead to the objective function value $J_{\text{MNC}} = 6.32282$, which is a little less than under MTD therapy. Having two switchings means that overall a lower dose of chemotherapy is used. With this schedule, the patient receives only $2/3 \approx 0.67$ of the chemotherapy dose that she would receive under MTD therapy. The concentration of drug-sensitive cancer cells decreases rapidly $x(t) \ll x_0 = 0.026$. However, the concentration of drug-resistant cancer cells ($y(t)$) increases from 0.003 to its maximum of 1 at time of 10 units (see Figure A.1).

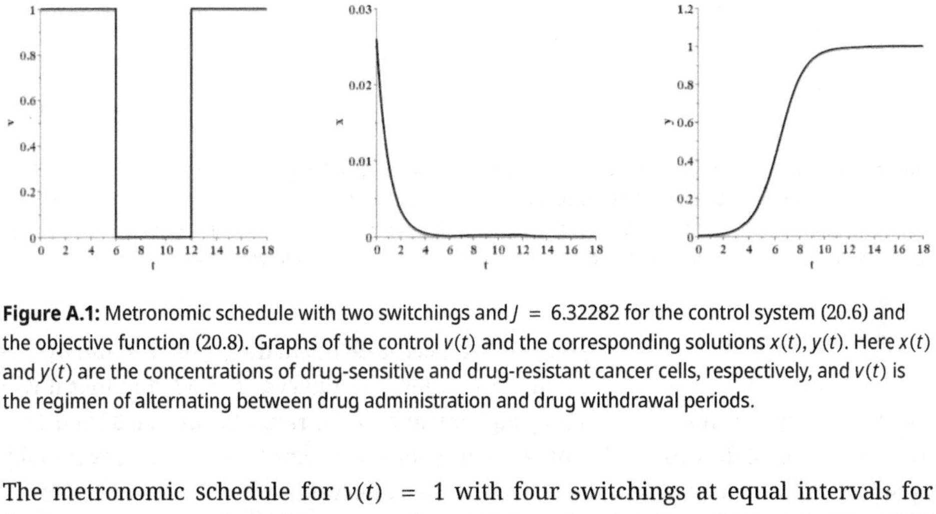

Figure A.1: Metronomic schedule with two switchings and $J = 6.32282$ for the control system (20.6) and the objective function (20.8). Graphs of the control $v(t)$ and the corresponding solutions $x(t)$, $y(t)$. Here $x(t)$ and $y(t)$ are the concentrations of drug-sensitive and drug-resistant cancer cells, respectively, and $v(t)$ is the regimen of alternating between drug administration and drug withdrawal periods.

The metronomic schedule for $v(t) = 1$ with four switchings at equal intervals for the same treatment period decreases the total drug dose intake to $3/5 = 0.6$ of the MTD and so as the objective function value ($J = 5.7061253$) compared to the schedule with two switchings ($J = 6.32282$). It is still 14 times higher than the minimum value of the objective function $J_* = 0.404696$ obtained by solving the optimal control problem (see Figure A.2).

Increasing the number of metronomic switchings from 4 to 8 at $v(t) = 1$ reflects the case when only $5/9 \approx 0.56$ of the MTD is taken over the treatment period of 18 units. This does improve the situation, and the total cancer load given by the objective function ($J = 3.46438054$) is lower than in the previous case with four switchings ($J = 5.7061$) (see Figure A.3).

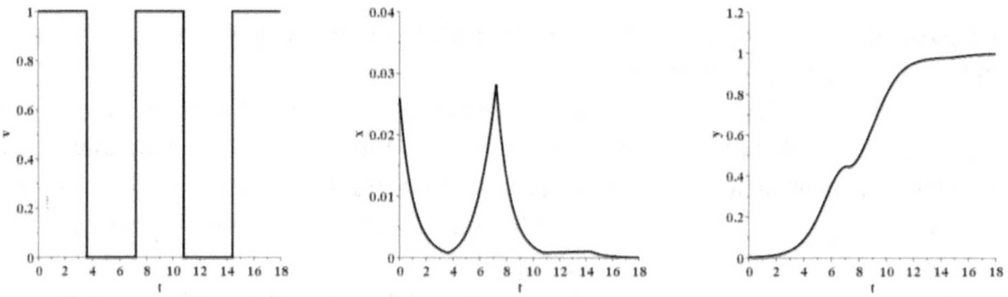

Figure A.2: Metronomic schedule with four switchings and $J = 5.7061$ for the control system (20.6) and the objective function (20.8). Graphs of the control $v(t)$ and the corresponding solutions $x(t)$, $y(t)$. Here $x(t)$ and $y(t)$ are the concentrations of drug-sensitive and drug-resistant cancer cells, respectively, and $v(t)$ is the regimen of alternating between drug administration and drug withdrawal periods.

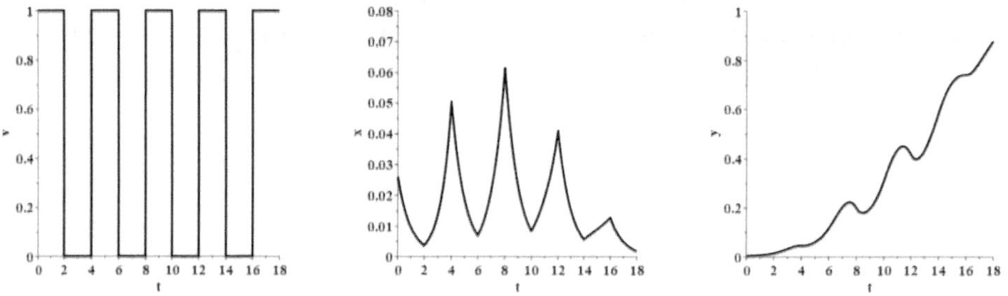

Figure A.3: Metronomic schedule with eight switchings and $J = 3.46438054$ for the control system (20.6) and the objective function (20.8). Graphs of the control $v(t)$ and the corresponding solutions $x(t)$, $y(t)$. Here $x(t)$ and $y(t)$ are the concentrations of drug-sensitive and drug-resistant cancer cells, respectively, and $v(t)$ is the regimen of alternating between drug administration and drug withdrawal periods.

We can observe that the drug-sensitive cancer cell population will eventually decrease from 0.026 to 0.003 at the end of the treatment interval $T = 18$, and that it will fluctuate during the interval, reaching higher values in the range between 0.04 and 0.06. The population of the drug-resistant cells will go through some local maxima, eventually approaching its value of 0.9 at the end of the time interval.

As we found in this study, the optimal treatment protocol can be one of the three options given by formula (20.16). Though MTD treatment ($v(t) = 1$ for all $t \in [0, T]$) is not the optimal solution for Example 1, but regarding (20.16) it can be the optimal solution for some other parameters of the system (20.6) or other patient.

Theoretically, metronomic treatment protocol with more than two switchings is never the optimal solution for the system (20.6). Although increasing the number of switchings in the metronomic schedule from 2 to 4 and from 4 to 8 only slightly reduces the total chemotherapy dose from 0.67 to 0.6 and from 0.6 to 0.56, respectively, it results in a decrease in the total cancer load and delays the moment when drug-resistant cancer cells reach their maximum capacity. Therefore, if following the optimal treatment schedule is not possible, MNC, presumably due to alternating drug “on” and “off” intervals, may be convenient to reduce the number of drug-resistant cancer cells and improve the efficacy of chemotherapy.

In any case, the optimal solution obtained using Pontryagin’s maximum principle provides the lowest value of the objective function (total cancer load) and should be considered as the best scenario.

Index

adaptive therapy 438
assembly theory
– *see* evolution 7
assisted reproductive technology (ART)
– intracytoplasmic sperm injection (ICSI) 403

Bateman function
– *see* epidemic models, pharmakcokinetic (PK) models 252
bioequivalence 421–431
– assessment 421
  – area under the concentration–time curve (AUC) 421
  – assessment rationale 421
  – average bioequivalence (ABE) 421
  – crossover design 422, 424, 426
  – equivalence range [0.80, 1.25] 421
  – improved resilience to data irregularities 422
  – kernel density estimation (KDE) 422
  – Kullback–Leibler (KL) divergence 422
  – maximum observed concentration (C_{max}) 421
  – missing observations 422
  – multimodal profiles 430
  – outliers 422, 429
– classical test procedure 422–424
  – two one-sided tests (TOST) procedure 421, 423, 424, 429
– Kullback–Leibler divergence 424
  – definition and symmetrization 424
  – drawbacks of using the closed-form expression for KLD 426
  – hypothesis and decision threshold 426
  – important properties 425
  – kernel density estimation (KDE) 427, 429
  – model study, decision accuracy under the “true non-bioequivalence” scenario 429
  – nonparametric KLD via weighted KDE 427, 428
  – normal model: closed-form expression 425
  – sensitivity of point estimators and distributional dependence 426
  – weighted KDE estimator, choice of bandwidth 430
  – weighted KDE estimator, improvements 430
  – weighted KDE estimator, limitations 430
  – weighted KDE estimator, missing observations 429
  – weighted KDE estimator, robustness 429
  – weighted KDE model study 428, 429
  – weighted KDE model study, classification under true non-bioequivalence 428
  – weighted KDE model study, design 428
biomedical imaging
– *see* assisted reproductive technology (ART) 403
– *see* magnetoencephalography (MEG) 387
biophysics, structural
– protein structure 95
  – secondary structures *see* protein structures 112
  – structure-function relationship 93
  – structure-stability relationship 112
  – supersecondary structures *see* protein structures 95
body energy
– budget
  – accounting 39
  – coordination 32, 35, 36, 38, 39
  – energy-by-entropy plot 37, 39
  – functional activity 38, 39
  – integration 29, 39
  – maintenance 29, 35, 36, 38, 39
  – reserve 29, 34, 36, 39
  – systemic readiness 30
– cellular energy
  – expenditure 39
  – supply grid 38
– demand
  – functional activity 30
  – human body 30
  – maintenance 30
– expenditure
  – *see* physiology, metabolism, BMR 31
  – scaling 35–38
  – systemic 31, 39
body function
– enabled by energy 30
– functional aggregate 30, 32, 33, 37, 38
– functional axis 32–34
– functional dynamics
  – axiomatic biodynamics 30
– functional level 32–34
– functional structure 30
  – aggregation 29, 35–38
  – decomposition 29, 30, 32, 33, 38, 39
  – hierarchy 30, 35
  – System Functional Architecture (SFA) 30, 35

https://doi.org/10.1515/9783112216620-031

- functional unit 30–33, 36–38
body structure
- entropy
  - environmental entropy 30
  - internal 35
  - maximum entropy 30
- free energy 30
- information 30
body system functional design concept
- canonical functions
  - cognitive task-solving 29, 30, 32, 33
  - physical task-solving 30, 32, 33
  - reproductive task-solving 29, 32, 33
  - vital task-solving 30, 32, 33
- central coordination 30, 32–34

cancer
- *see* melanoma 437
- *see* tumor growth 75
cardiovascular circulation
- Cardiovascular Simulating System (CVSS)
  - *see* substance transfer modeling 139
- hemodynamics
  - *see* substance transfer modeling 139
- quasi-one-dimensional model
  - *see* substance transfer modeling 139
chirality *see* protein structures 99
civilization
- *see* ecological economics 313
- *see* urbanization and industrialization 291
classification
- ICSI outcome prediction 413
coiled coil
- heptad 95
- hydrophobic core 96
- knobs-into-holes packing 96
- structures database (CCdb) 99, 103, 106
cortisol (CORT)
- adrenal cortex
  - secretion into circulation 154
- modeling of cortisol dynamics 153–175
- pharmacokinetics
  - clearance 154
  - clearance coefficient 166
  - half-life 166
- properties
  - free cortisol 153, 156
  - steroid hormone 154
  - total cortisol 157
- protein binding 156–159
  - albumin 156
  - binding affinity 154, 157, 158
  - binding capacity 154, 157
  - binding dissociation function 159
  - corticosteroid-binding globulin (CBG) 153, 156
  - corticosteroid-binding globulin (CBG) deficiency 170
  - equilibrium dissociation constant 158
  - free hormone hypothesis 157
  - reservoir hypothesis 154, 157
  - SERPINA6 mutations 170
- rhythms
  - amplitude-modulating function 160
  - circadian modulation 156
  - external circadian drive 154
  - Michaelis–Menten function 163
  - shaping function 162
  - ultradian pulses 156
- secretion 153
  - adrenal cortex 154
  - dense-core secretory vesicle (DCV) release 154
  - exocytosis 154–156, 161
COVID-19 Pandemic
- *see* epidemiological modeling 205
- *see* evolution, ecosystem 13
- *see* Shanghai 2022 omicron outbreak 243

Darwin
- theory of evolution 20
delay differential equations (DDE)
- *see* epidemiological modeling, delay models 207
- traveling wave solutions
  - *see* virus infection model 191

ecological dynamical systems 43
- bifurcation
  - Arnold tongue 61
  - chaos 43
  - circle map 59–62
  - Feigenbaum constant 62
  - Hopf bifurcation 51, 54, 55
  - period-doubling 62
  - period-doubling bifurcation 43
- cyclic competition
  - Rock–Paper–Scissors (RPS) 47
  - Rock–Paper–Scissors–Lizard–Spock 47
- dissipative stage
  - *see* nonlinear dynamics, attractors 58

- multi-species competition 45–47
  - Lotka–Volterra system 44–46, 50–54, 58, 67
  - May–Leonard system 44, 67
  - symmetric May–Leonard system 46–49
- multi-year cycles 43
- nonlinear dynamics
  - attractors 45–47, 53, 54, 56
  - attractors, conservative stage 63–65
  - attractors, dissipative model 50–54
  - attractors, dissipative stage 58–62
  - cyclic dynamics 45–47
  - heteroclinic cycle 45, 47, 50–55, 64, 65, 67
  - limit cycle 45, 54
  - phase space 54, 58, 63
  - stability analysis 47
- quasi-periodic behavior 43
- seasonal forcing
  - model switching 45
  - piecewise-constant forcing 44
  - time-dependent parameters 44
- seasonality 43
- systems with forcing
  - seasonal forcing 43
  - seasonal forcing, model switching 56–65
  - stroboscopic map 56–58, 60
  - time-dependent parameters 44

ecological economics
- ocean, wetland, farmland, forest economies
  - *see* fisheries 345
  - *see* invasive plants control 313

ecosystem
- thermodynamic cycles
  - Maximum Power Principle 20

elderly people
- community-based study
  - *see* longevity township 367
- longevity
  - *see* longevity township 367
- social support 368

entropy
- entropy production
  - *see* evolution, thermodynamics 9
- informational entropy
  - *see* evolution, thermodynamics 7, 8

enzymes
- cooperative regulation 128
- kinetics
  - Hill model 125
  - Lineweaver–Burk plot 124, 127, 129, 131
  - Michaelis–Menten model 126, 129
  - new universal model 123
  - non-Michaelis kinetics 124, 128–132
- lysozyme 124, 130–132
- multiprotein complexes 124
- nicotinamide adenine dinucleotide (NAD^+)-kinase 124, 128–130
- supercooperativity 124
  - stepwise supercooperativity 124
- transport protein 124, 126
  - ATPase 124

epidemic data models
- chemical kinetics
  - mass action 244
- mass action models
  - SEIR model 282, 283
  - SIR model 281, 282
- multi-state models
  - four-state MARKOV jump process 267, 283–286
  - state diagram 246, 247, 264, 275, 277, 280, 283–285
  - transition intensity 246, 247, 267, 284–286
- pharmacokinetic (PK) models 244
  - area-under-the-curve (AUC) 244, 253
  - Bateman effectiveness 253
  - Bateman function 252
  - Bateman function fits 251–253
  - power-series approximation 252
- phenomenological growth 265, 266

epidemic models
- mechanistic models
  - *see* epidemiological modeling 205
- phenomenological models
  - *see* epidemic data models 243

epidemic outbreaks
- *see* Shanghai 2022 omicron outbreak 244, 267
- phenomenological growth
  - sugar-loaf shape 252, 266, 267
  - volcano eruptions 266

epidemiological modeling
- compartmental models
  - classical susceptible-infected-recovered (SIR) 209
  - susceptible-infectious-immune-suceptible (SIRS) 208
- delay models 205–239
  - delay differential equations (DDE) 207, 208
  - delay-infection model 208

- delayed Susceptible-Infected-Recovered (SIR) 206
- distributed (time-dependent rates) model 206, 210–224
  - basic reproduction number $\mathcal{R}_0$ 217
  - comparison of integro-differential equation model, delay model, and SIR model 222
  - COVID-19 epidemic data 213
  - delay model 215–217
  - epidemic characteristics 217–222
  - estimation, gamma-distribution 213–215
  - final size of epidemic 218
  - integro-differential equation model 210–213
  - maximum number of infected individuals 220
  - reduction to the SIR model 212, 213
- infectious disease dynamics
  - immunity waning 208
- multi-compartment models 206
- spatio-temporal models 206
- state-occupancy times models 224–239
  - comparison with the Influenza A and B data 238
  - single strain delay model 224–229
  - two-strain delay model with cross-immunity 230–239

evolution
- adaptive radiation 18
- adiabatic compression 13
- assembly theory 7
- biological complexity 7
  - Second Law of Intellect 19
- Darwin 5, 6, 16, 20
- dinosaurs 14
- ecosystem 6, 8, 9, 11–15, 18–20
  - apex predators 16
  - competitive exclusion 16
  - COVID-19 Pandemic 13
  - dinosaurs 17
  - ecological vacuum 13
  - ecospace expansion 13
  - energy budget 15
  - entropy 7
  - equilibrium 8, 10, 18, 20
  - equilibrium, high-entropy 8, 15
  - equilibrium, punctuated 6
  - erosion of resilience 17
  - extinction 9, 11–13, 16, 18, 20
  - extinction, marine 10
  - extinction, mass 6, 8, 16, 19, 20
  - extinction, Permian 17
  - fitness 15
  - fragility 16
  - frailty 21
  - genetic complexity building 7
  - genetic diversity 16
  - high entropy 20
  - high productivity 15
  - high-entropy energy exportation 7
  - higher complexity 12
  - Kullback–Leibler divergence 15
  - low-entropy energy consumption 7
  - mass extinction 12
  - Minimum Entropy Production Principle (PEPP) 20
  - overpopulation, collapse 12
  - Panarchy adaptive cycle 19
  - planetary sustainability 21
  - population expansion 11, 12
  - population growth 12
  - reorganizaton 12
  - replicator equation 15
  - stability by simplicity 16
- entropy production
  - *see* evolution, thermodynamics 8
- Erwin Schrödinger 6
- evolutionary cycle 8, 11–13
- exaptation 13, 14
  - antifreeze proteins 14
- gene duplication 7, 13
- gene pool 9
  - Horizontal Gene Transfer (HGT) 11, 16
- genes
  - jumping genes 16
- genetic material 5, 11, 20
  - mobile genetic elements 13
- genome 7, 11, 12, 16
  - entropy 7, 8
  - entropy, bacterial 7
  - entropy, eukaryotic 7
  - temporal stability 12
- Helmholtz 8
- Landauer's principle 8
- Maximum Power Principle 6, 10
- Minimum Entropy Production Principle 9
- Minimum Entropy Production Principle (PEPP) 6
- multicellularity 14, 17
- Panarchy adaptive cycle 6
- thermodynamic cycles 17, 18, 21

- Cambrian explosion 14, 17
- collapse and mass extinction 16
- conservation and rigidity 15, 17
- consolidation and expansion 14, 17
- Cretaceous–Paleogene extinction 17
- Ediacaran explosion 14
- extinction and collapse 18
- Galápagos Finches 18
- reorganization and exploration 13, 17
- social temperature 7–9, 11–15, 19
- thermodynamic reset
  - extinction, mass 17
- thermodynamics
  - energy and information 7–9
  - entropy 7–9
  - entropy production 8
  - free energy 8, 13
  - informational entropy 7, 8
  - Maximum Entropy Production 14
  - Prigogine's Theorem 6, 10
  - reversed Carnot cycle 7, 11, 17
  - Second Law 6
- two arrows of time 19

fisheries
- multi-site single-species and predator-prey fisheries
  - *see* multi-site fishery management 345
- Brouwer's fixed point theorem
  - *see* invasive plants control 331
- Browder's fixed point theorem
  - *see* invasive plants control 336

frailty
- *see* geriatric syndrome 367
- adverse effects 367
- frailty index (FI) 369
- psychosocial factors 368
- risk factors 368
- WHO definition 369

game theory
- noncooperative differential game
  - *see* invasive plants control 313

games
- Rock–Paper–Scissors
  - *see* ecological dynamical systems 47
- Rock–Paper–Scissors–Lizard–Spock
  - *see* ecological dynamical systems 47

geriatric syndrome
- *see* elderly people 367

HPA axis
- anatomy
  - adrenal cortex *see* cortisol (CORT) 153
  - adrenal gland 153
  - hypothalamus 153
  - pituitary gland 153
- biological rhythms 153–175
  - Frequency Modulated Möbius (FMM) model 163
  - integrate-and-fire (IAF) model 160
- hormones
  - adrenocorticotropic hormone (ACTH) 153, 154
  - corticotropin-releasing hormone (CRH) 153, 154
  - negative feed-back 153

human body system
- *see* body energy 29
- *see* body function 29
- *see* body structure 29
- *see* body system functional design concept 29

imaging technology
- automated video analysis
  - *see* intracytoplasmic sperm injection (ICSI) 403
- brain mapping
  - *see* magnetoencephalography (MEG) 387

intracytoplasmic sperm injection (ICSI)
- automated video analysis 403–415
  - computer vision methods 404
  - data pre-processing 405, 406
  - overall model accuracy 412
  - semantic segmentation 412
  - synthetic augmentation of the training set 405, 412
  - video preprocessing and noise removal 404
- classification model 405, 410–412
  - gradient boosting 411, 412
- clinical success rate 404
- critical features 412
  - degree of oocyte deformation 405, 410, 413
  - entry angle 405, 410
  - Haar Cascades algorithm 410, 413
  - needle extraction 404, 405
  - needle extraction speed 410
  - needle insertion angle 403, 410, 413
  - needle insertion speed 403, 410, 413
  - needle speed 413

- number of puncture attempts 405, 410
- oocyte size 405, 410
- optical flow measurement 411, 413
- velocity of needle penetration 405
- procedure outcome prediction 413
- attribute selection methods 404
- Intersection over Union (IoU) score 413
- segmentation classes 405
- background 405
- microinjection needle 405, 407, 413
- micropipette 405
- oocyte 405
- semantic segmentation 404, 406, 407
- Feature Pyramid Network (FPN) 406, 407, 413
- Intersection over Union (IoU) metric 406, 409
- segmentation accuracy 407
- U-Net 406–408
- synthetic augmentation of the training set 407–410
- Denoising Diffusion Probabilistic Model (DDPM) 407
- denoising diffusion probabilistic models (DDPM) 408, 409
- synthetic ICSI frames 407, 409

invasive plants control 313–343
- comparison of control strategies 336–340
- centralized control, benefits 336
- centralized control, coordination costs 337
- decentralized control, fiscal transfers 339
- control strategies 325–340
- centralized control 331–336
- decentralized control 325–331
- ecological threat 313
- economical burden 314
- extinction and invasion dynamics 318–323
- modeling
- Brouwer's fixed point theorem 331
- Browder fixed point theorem 336
- characterizing equilibria as fixed-points 330
- cost function 326
- coupled population dynamics 332
- discrete spatial network 315
- equilibrium controls 333
- existence of centralized control solution 336
- fixed-point operator representation 334
- Nash equilibrium 327
- optimality conditions 327
- Pontryagin's Maximum Principle 327
- recursive solution 335
- strategic coupling 327
- policy frameworks 314
- spatially dynamic model 315–318

Kullback–Leibler divergence 424, 429
- *see* bioequivalence 421
- *see* evolution, ecosystem 15

liver
- cirrhosis
- Child–Pugh classification *see* substance transfer modeling 139
- detoxification
- indocyanine green (ICG) *see* substance transfer modeling 139

longevity township 367–382
- community-based study
- association of frailty and social support 373–378
- Chinese characteristics 380
- confounding factors 370
- data collection 369
- demographics 372, 373
- frailty measurement 369
- social support 370
- statistical methods 370
- study population 369

lysozyme
- *see* enzymes 124
- *see* proteins 112

machine learning
- gradient boosting 412

magnetoencephalography(MEG) 387–401
- computational methods
- continuous source of activity 392
- dipole clusters 390, 391
- dipole localization 389
- independent component analysis (ICA) 389, 399
- noise 390
- number of voxels 392
- robustness 390
- source decomposition 389
- neuroimaging methods
- dipole model 389
- inverse problem 389
- magnetic resonance imaging (MRI) 387
- magnetoencephalography (MEG) 387

- MEG direct problem 388
- MEG inverse problem with dipolar sources 388
- MEG/EEG inverse problem 388, 400
- source localization methods
- dipole localization *see* - computational methods 389
- hybrid ICA–LORETA algorithm 397, 400
- independent component analysis (ICA) 389
- low-resolution electromagnetic tomography (LORETA) 388, 400
- Multiple Signal Classification (MUSIC) 388, 400
mathematical model
- Frequency Modulated Möbius (FMM) *see* HPA axis, biological rhythms 163
- integrate-and-fire (IAF) *see* HPA axis, biological rhythms 160
- Lotka–Volterra (LV) model *see* ecological dynamical systems 43, *see* melanoma 439
- predator-prey LV model *see* multi-site fishery management 345
- May–Leonard model
- symmetric ML system *see* ecological dynamical systems 46
- of New Moscow development 303
- ordinary differential equations (ODE)
- *see* ecological dynamical systems 45, 439
- reaction diffusion models
- *see* territorial development of cities 291
mathematical modeling
- autowave self-organization in active media 291
- ecological model
- *see* invasive plants control 315
- epidemic models
- compartmental models *see* epidemiological modeling 206
- models based on sojourn times *see* epidemiological modeling, state-occupancy time models 207
- one-dimensional spatiotemporal model *see* territorial development of cities 298
- optimal treatment protocols *see* melanoma 437, 463
- spatio-temporal autowave model
- urban development *see* territorial development of cities 291
- spatio-temporal modeling
- *see* reaction diffusion equations 179
- virus infection progression
- *see* virus infection model 178, 179
Maximum Power Principle (MPP)
- *see* evolution 6, 10
melanoma 463
- adaptive therapy 438–463
- active body feedback 438
- alternating treatment “on” / “off” intervals 439, 440, 456, 457
- inhibition of drug-resistant cancer cell 438
- optimal control function 447
- optimal treatment protocols 439
- selective patients 439
- tracking dynamics of cancer cell population 438
- viability of drug-sensitive cancer cells 438
- BRAF/MEK inhibitors 441
- eradication therapy
- maximum tolerated dose (MTD) 438, 457, 460, 461
- Lotka–Volterra ODE controlled model 456
- death rate of drug-sensitive cancer cells 440
- drug-susceptible versus drug-resistant cancer cells 440
- inhibition of drug-resistant cells 440
- intrinsic growth rate drug-resistant 440
- intrinsic growth rate drug-sensitive 440
- maximum tumor size, external constraints 440
- population size drug-resistant 440
- population size drug-sensitive 440
- metronomic chemotherapy (MCT) 438, 457, 460, 461
- optimal control problem
- control function 439
- control function of time 441
- de La Vallée-Poussin non-oscillation criterion 446
- equilibrium positions 453–456
- minimization problem 440–443, 456
- minimizing the objective function 442
- Pontryagin’s maximum principle 439, 443, 444
- Riccati equation 446
- switching function 444, 445, 452
- total period of treatment 441, 456
- optimal treatment protocol
- BOCOP software 439
- MAPLE software 439
- numerical results 448–452
Minimum Entropy Production Principle (PEPP)
- *see* evolution 9

multi-site fishery management 345–362
- network of sites
  - connectivity 346, 354, 359, 360
  - fast migration 347, 354, 358, 359
  - maximal sustainable yield (MSY) 346, 349–351, 353, 354, 354, 359, 360
  - migration between sites 346
  - total equilibrium population 346, 359
- population decrease by fishing
  - catchability 346
  - fishing effort 346
  - Schaefer model 346
- population growth by reproduction
  - logistic model 346
  - resource carrying capacity 346
- predator-prey Lotka–Volterra model 351–359
  - Holling type I functional response 351–357
  - Holling type II functional response 357–359
- single species models 347–351
  - fast migration equilibrium 347
  - with harvesting 349–351
  - without harvesting 347, 348

myoglobin
- *see* proteins 112

Nash equilibrium
- *see* invasive plants control 313

necrosis
- necrotic core 77

nonlinear wave dynamics
- *see* tumor growth, spheroid model 78

organizational sciences
- administrative functional design concept 33
  - functional unit 33
  - levels of responsibility 33
- organizational complexity 36

pharmacokinetics (PK)
- area-under-the-curve (AUC)
  - *see* Shanghai 2022 omicron outbreak 244
- comparative bioavailability
  - *see* bioequivalence 421

physiology
- functional
  - axiomatic biodynamics 29–39
- metabolism
  - basal metabolic rate (BMR) 31

Pontryagin's maximum principle
- *see* invasive plants control 313

Prigogine's Theorem
- *see* evolution
  - thermodynamics 6, 10

protein structures
- chirality
  - enantiomer 94
  - general concepts 94, 95
  - pseudochirality 94
  - sign 94
  - value 94
- Protein Data Bank (PDB) 99
- secondary structures 112
  - 3_{10} helix 97, 112
  - α-helix 97, 112
  - β-strand 96
  - error 113
  - lysozyme 114
  - method of calculation 112
  - myoglobin 117
  - polyproline helix 96
  - protein stability 116
- supersecondary structures
  - β-hairpin 95, 96, 102
  - β-sheet 95–97
  - center of points 100
  - chirality 97
  - coiled coil superhelix 95–97, 100
  - collagen superhelix 95, 96, 102
  - heptad *see* coiled coil, heptad 95
  - hydrophobic core *see* coiled coil 96
  - method of calculations 99

proteins
- lysozyme 112
  - function 112
  - main types 112
  - structure 112, 116
- multiprotein complexes *see* enzymes 123
- myoglobin 112
  - function 112
  - structure 112
  - thermal stability 112
- transport protein *see* enzymes 123

public health
- infectious disease transmissions
  - COVID-19 (SARS-CoV-2) (2020-2023) 206
  - EVD (Ebola) (2014) 206
  - H1N1 (swine flu) (2009) 206

- H5N1 (avian flu) (2005) 206
- HIV (1980s-) 206
- SARS (SARS-CoV-1) (2002–2003) 206
- SARS-CoV-2 omicron (China 2022) 243
- seasonal influenza 238

reaction diffusion equations
- delay differential equations (DDE)
  - *see* virus infection model 187
- FitzHugh–Nagumo equation
  - spatially heterogeneous systems *see* territorial development of cities 291, 297
- parabolic
  - *see* tumor growth, spheroid model 78
- traveling wave solution 191
  - *see* virus infection model 188, 192

regulation of bodily functions
- neuroendocrine system
  - Hypothalamic–Pituitary–Adrenal (HPA) axis *see* HPA axis 153

reproductive medicine
- assisted reproductive technology (ART)
  - *see* intracytoplasmic sperm injection (ICSI) 403
- in vitro fertilization (IVF) 403
- video analysis 405

SARS-CoV-2
- China 2022
  - *see* Shanghai 2022 omicron outbreak 243

Second Law of Intellect
- *see* evolution 19

Second Law of Thermodynamics
- *see* evolution, thermodynamics 6

self-organization
- *see* territorial development of cities 292, 297
- autowave dissipative structures 296

Shanghai 2022 omicron outbreak 243–267, 286
- anti-epidemic response design
  - 2-week quarantine 244
  - dynamic zero-COVID policy 244
  - four-state model 246, 247
  - mass screening 244
- Bateman function 244
- Bateman function fits 260–263
  - Bateman effectiveness 261, 263
- daily numbers of new cases
  - asymptomatic infected cases (AIC) 246, 248, 249
  - clinically symptomatic COVID-19 by conversion from prior AIC 246, 249, 250
  - clinically symptomatic COVID-19 without prior AIC 246, 248, 250
  - collected from China National Health Commission 246, 272
- daily update analysis
  - graphical regression 250, 251
  - logistic difference 247, 249, 250
  - time series 247–250, 251, 251

social support
- social support scales 370
  - Shuiyuan Xiao's Social Support Rating Scale 370
  - Social Support Questionnaire 370

social temperature
- *see* evolution, thermodynamic cycles 7–9, 11–15

software
- BOCOP
  - *see* melanoma 439
- Cardiovascular Simulating System (CVSS)
  - *see* substance transfer modeling 139
  - anatomically detailed submodel–cerebral vasculature 145
  - anatomically detailed submodel–coronary circulation 145
  - anatomically detailed submodel–hepatic circulation 145
  - anatomically detailed submodel–Renal vascular networks 145
  - capable of patient-specific customization 145
  - compatible with clinical data integration 145
  - parameters adjustable across physiological ranges 145
- MAPLE
  - *see* melanoma 439

substance transfer modeling
- cardiovascular circulation
  - Cardiovascular Simulating System (CVSS) software package 145
  - convection-diffusion equation 139
  - quasi-one-dimensional model 139–141
- liver detoxification
  - Child–Pugh class versus portal pressure 148
  - Child–Pugh classification 147, 148, 151, 152
  - cirrhosis 139, 140, 147–150
  - hepatic clearance 150
  - *in-silico* experiments 149–151
  - indocyanine green (ICG) 148–151
  - portal hypertension 139, 149, 151

- transport model 141–145
  - approximation of transport equations: conservation properties 145
  - approximation of transport equations: coupling conditions atiInternal vertices 144
  - approximation of transport equations: discretization scheme 143
  - approximation of transport equations: finite-difference approximation 143
  - approximation of transport equations: heart balance equation 144
  - boundary conditions: coupling at internal graph vertices 142
  - liver modelling 146
  - model closure: heart balance equation 143
  - substance extraction by organs and tissues 146
  - transport equations on a vascular graph 141

synthetic augmentation
- see also diffusion models, segmentation 407

territorial development of cities 291
- activator-inhibitor system equations 297, 298
  - autowave 298
  - autowave excitable/bistable region 299, 300, 309
  - autowave front 298, 300, 302, 303
  - contrast structure 298, 300
  - cross-interactions 298
  - diffusion coefficients 298
- anthropogenic subsystem 293–296
  - activator diffusion coefficient 298, 304
  - activator intensity 298
  - activator intensity function 300
  - activator spatial distribution 302, 303
  - activators 300, 307, 308
  - chemical pollution 294
  - electromagnetic fields (EMF) of industrial frequencies 295
  - microorganisms 296
  - municipal solid waste and wastewater 294
  - static loads 293
  - thermal loads 293
  - water basin 294
- contrast structures 297
- coupled natural and anthropogenic subsystem 297
- FitzHugh–Nagumo equations 297, 298
- geobiocenosis 304–306
- natural subsystem
  - inhibitor diffusion coefficient 298, 304
  - inhibitor intensity 298
  - inhibitor intensity function 300
  - inhibitor spatial distribution 302, 303
  - inhibitors 307, 308
- urboecosystem (UES) 291, 292, 296, 300
  - fragmentation 296
  - megacity formation 294, 303
  - Moscow region 302, 303
  - New Moscow development 305
  - self-organization 292, 296
  - Shanghai development 307–309

thermodynamics
- concept of structure 31–37
- entropy 31, 34, 35, 38
  - informational 31
  - internal 31, 34, 35, 38
  - reverse entropy 34, 39

tumor growth
- mathematical model
  - parabolic equation system 80–82
- microenvironment
  - angiogenesis 76, 86
  - glycolytic metabolism 79
  - metastasis 76
  - necrotic core 79
  - oxygen availability 78
  - pH lowering 76
  - tumor cell proliferation 78
- spatio-temporal 3D dynamics 75, 88
  - cell density dynamics 81
  - nonlinear diffusion coefficient 82
  - oxygen concentration dynamics 81
  - pH dynamics 81
  - traveling wave regime 82, 85
- spheroid model 76, 78–82, 88
  - biological rationale 78, 79
  - cell density 81
  - necrotic core 77, 80, 81
  - necrotic core formation 79, 85, 86
  - oxygen availability 77, 86
  - oxygen gradients 77, 85
  - proliferative zone 86
  - proliferative zone volume 80
  - quiescent cell zones 83, 85
  - spheroid growth 86
  - spheroid volume 80, 86
  - tumor cell density 77, 78
  - tumor cell proliferation 79

- spheroid model parameters 79
  - cell diffusion coefficient 79, 86
  - cell division rate 79, 83
  - glycolysis rate 79
  - hydrogen ion concentration 78, 79
  - necrotic core volume 80
  - oxygen concentration 78, 79, 81, 82, 86
  - oxygen consumption rate 88
  - oxygen diffusion 86
  - oxygen uptake rate 79
  - pH concentration 81
  - time delay to glycolysis 79
- spheroid model specifications 79–82
  - calculation outcomes 83–88
  - calibration 82, 83
  - cell concentration threshold 86
  - cell density threshold 81
  - hypoxic threshold 80
  - necrosis threshold 80, 82
  - oxygen threshold 81, 82
  - pH threshold 82
  - proliferation inhibition threshold 80
  - proliferative zone 80
- tumor metabolism
  - aerobic glycolysis (Warburg effect) 76
  - anaerobic glycolysis 81, 85

tumor spheroid 78

urbanization and industrialization
- *see* territorial development of cities 291

urboecosystem (UES) 297, 298

virus infection
- progression
  - virus replication number 184

virus infection model
- delay differential equations (DDE)
  - exact solution 187
  - linear stability 184
  - Lyapunov function 185
  - traveling wave solutions 191, 192
- immune response
  - antibodies 179, 182
  - cytotoxic T lymphocytes (CTL) 179, 182
  - interferon 179, 180, 182
- infectivity 178
- progression 178–194
  - infection wave speed 188, 191
  - virus replication number 178, 184
- reaction diffusion equations
  - basic model 183–194
  - basic model with diffusion 188–194
  - basic model without diffusion 183–188
  - general model 178
- total viral load 178, 188, 190

virus infections
- in population
  - *see* epidemiological modeling 205
  - *see* public health 205

virus infections
- in tissue *see* virus infection model 177

De Gruyter Series in Mathematics and Life Sciences

Volume 13
Jules J. Berman
Fundamentals of Reality. Emergent Systems Development from Vacuum and Waves
ISBN 978-3-11-149814-0, e-ISBN (PDF) 978-3-11-149869-0, e-ISBN (EPUB) 978-3-11-149941-3

Volume 12
Olga Vasieva
Origins of Biocomplexity. The Fundamental Principles of Life in Terms of Physics
ISBN 978-3-11-133472-1, ISBN (PDF) 978-3-11-133479-0, ISBN (EPUB) 978-3-11-133483-7

Volume 11
Apul Dev, Sayantan Guha (Eds.)
Applied Engineering Mathematics. Fluid and Solid Mechanics in Life Science
ISBN 978-3-11-163853-9, ISBN (PDF) 978-3-11-163878-2, ISBN (EPUB) 978-3-11-163868-3

Volume 10
Fateh Mebarek-Oudina (Eds.)
CFD Simulation. Thermo-Fluids and Nanofluids in Engineering and Biomedicine, 2025
ISBN 978-3-11-140454-7, ISBN (PDF) 978-3-11-140509-4, ISBN (EPUB) 978-3-11-140523-0

Volume 9
Jochen Mau, Sergey Mukhin, Guanyu Wang, Shuhua Xu (Eds.)
BIOKYBERNETIKA. Mathematics for Theory and Control in the Human and in Society, 2024
ISBN 978-3-11-134197-2, ISBN (PDF) 978-3-11-134199-6, ISBN (EPUB) 978-3-11-134204-7

Volume 8/2
Igor M. Rouzine
Mathematical Modeling of Evolution. Volume 2: Fitness Landscape, Red Queen, Evolutionary Enigmas, and Applications to Virology, 2023
ISBN 978-3-11-069731-5, ISBN (PDF) 978-3-11-069738-4, ISBN (EPUB) 978-3-11-069746-9

Volume 8/1
Igor M. Rouzine
Mathematical Modeling of Evolution. Volume 1: One-Locus and Multi-Locus Theory and Recombination, 2020
ISBN 978-3-11-060789-5, ISBN (PDF) 978-3-11-061545-6, ISBN (EPUB) 978-3-11-060819-9

Volume 7
George Dassios, Athanassios S. Fokas
Electroencephalography and Magnetoencephalography, 2020
ISBN 978-3-11-054583-8, ISBN (PDF) 978-3-11-054753-5, ISBN (EPUB) 978-3-11-054578-4

De Gruyter Series in Mathematics and Life Sciences